Biological Control
of Plant Diseases

S.B. Chincholkar
K.G. Mukerji
Editors

CRC Press
Taylor & Francis Group
Boca Raton London New York

CRC Press is an imprint of the
Taylor & Francis Group, an informa business

Reprinted 2010 by CRC Press

CRC Press
6000 Broken Sound Parkway, NW
Suite 300, Boca Raton, FL 33487

270 Madison Avenue
New York, NY 10016

2 Park Square, Milton Park
Abingdon, Oxon OX14 4RN, UK

For more information on this book or to order, visit
http://www.haworthpress.com/store/product.asp?sku=5682

or call 1-800-HAWORTH (800-429-6784) in the United States and Canada
or (607) 722-5857 outside the United States and Canada

or contact orders@HaworthPress.com

Published by

Haworth Food & Agricultural Products Press™, an imprint of The Haworth Press, Inc.. 10 Alice Street, Binghamton, NY 13904-1580.

PUBLISHER'S NOTE
The development. preparation. and publication of this work has been undertaken with great care. However. the publisher, employees, editors, and agents of The Haworth Press are not responsible for any errors contained herein or for consequences that may ensue from use of materials or information contained in this work. The Haworth Press is committed to the dissemination of ideas and information according to the highest standards of intellectual freedom and the free exchange of ideas. Statements made and opinions expressed in this publication do not necessarily reflect the views of the Publisher. Directors, management, or staff. of The Haworth Press, Inc., or an endorsement by them.

Cover design by Jennifer M. Gaska.

Library of Congress Cataloging-in-Publication Data

Biological control of plant diseases / S.B. Chincholkar. K.G. Mukerji, editors.
    p. cm.
Includes bibliographical references and index.
ISBN-13: 978-1-56022-327-6 (hard : alk. paper)
ISBN-10: 1-56022-327-8 (hard . alk. paper)
ISBN-13: 978-1-56022-328-3 (soft · alk. paper)
ISBN-10: 1-56022-328-6 (soft . alk. paper)
1. Phytopathogenic microorganisms—Biological control. I. Chincholkar, S. B. II. Mukerji, K. G.

SB732.6.B5814 2007
632'.96—dc22

2006025374

CONTENTS

ABOUT THE EDITORS

S. B. Chincholkar, PhD, is Professor and Director of the School of Life Sciences at North Maharashtra University, Jalgaon in India, where he teaches advanced courses in microbial biotechnology. He has authored more than 60 research papers for international journals and has delivered numerous lectures on steroid and antibiotic biotransformations, microbial siderophores, and biological control. Dr. Chincholkar is General Secretary of the Biotech Research Society of India and Vice-Chair of Crop Management for the International Society of Food, Agriculture and Environment. He is an editorial board member of several journals, including the *Journal of Food Science and Technology*.

K. G. Mukerji, PhD, is retired Senior Professor at the University of Delhi, India. A distinguished mycologist and microbial ecologist, he has published more than 600 research papers, is an editorial board member of several national and international journals, and is an honorary member of the research board of advisors for the American Biological Institute. Dr. Mukerji is co-author of *Taxonomy of Indian Myxomycetes, Index to Plant Diseases of India,* and *The Haustorium,* and editor of numerous books, including *Biocontrol of Plant Diseases, Fruit and Vegetable Diseases,* and *Recent Developments in Biocontrol of Plant Diseases.*

Biological Control of Plant Diseases
© 2007 by The Haworth Press, Inc. All rights reserved.
doi:10.1300/5682_a

xi

CONTRIBUTORS

A. Akkopru, University of Yuzuncu Yil, Faculty of Agriculture, Department of Plant Pathology, 65080, Van/Turkey. E-mail: emdem@hotmail.com; emrademir@yyu.edu.tr

Tanya Azarova, All-Russia Research Institute for Agricultural Microbiology, Podbelsky Shossee 3, Pushkin 8, Saint Petersburg, 189620, Russian Federation

Gabriele Berg, Universität Rostock, FB Biowissenschaften, Mikrobiologie, Albert-Einstein-Str. 3, 18051, Rostock. E-mail: gabriele.berg@biologie.uni-rostock.de

Guido V. Bloemberg, All-Russia Research Institute for Agricultural Microbiology, Podbelsky Shossee 3, Pushkin 8, Saint Petersburg, 189620, Russian Federation

Sandhya, Chandran, Biotechnology Division, Regional Research Laboratory, CSIR. Trivandrum 695019, India. E-mail: pandey@csrrltrd.ren.nic.in

B.L. Chaudhari, School of Life Sciences, North Maharashtra University, PO Box 80, Jalgaon 425001, Maharashtra, India

C. Cortes-Penagos, Escuela de Quimico-Farmacobiologia, Universidad Michoacana de San Nicolas de Hidalgo, Tzintzuntzan 173, Morelia, Mich. 58324, Mexico. E-mail: neocces@yahoo.com

Sandra de Weert, Institute Biology, Leiden University, Clusius Laboratory, Wassenaarseweg 64, 2333, AL Leiden, The Netherlands. E-mail: weert@rulbim.leidenuniv.nl

S. Demir, University of Yuzuncu Yil, Faculty of Agriculture, Department of Plant Pathology, 65080, Van/Turkey. E-mail: emdem@hotmail.com; emrademir@yyu.edu.tr

Kevin Eijkemans, Hogeschool Leiden, Zernikedreef 11, 2333 CK, Leiden, the Netherlands

Deena Errampalli, Agriculture and Agri-Food Canada, Southern Crop Protection and Food Research Centre, 4902 Victoria Avenue North, Vineland Station Ontario, Canada L0R 2E0. E-mail: errampallid@agr.gc.ca

Atimanav Gaur, Senior Scientist, Amity Institute of Herbal & Microbial Studies, U.P., Sector-125, New Super Express Highway, NOIDA (UP). E-mail: atimanav@aihmr.amity.edu

A. Herrera-Estrella, Department of Plant Genetic Engineering, Centro de Investigación y Estudios Avanzados, Apartado Postal 629, Irapuato, Gto. 36500, MEXICO. E-mail: aherrera@ira.cinvestav.mx

Mitsuro Hyakumachi, Laboratory of Plant Disease Science, Faculty of Agriculture, Gifu University, Yanagido 1-1, Gifu, 501-1193 Japan. E-mail: hyakumac@cc.gifu-u.ac.jp

Faina D. Kamilova, All-Russia Research Institute for Agricultural Microbiology, Podbelsky Shossee 3, Pushkin 8, Saint Petersburg, 189620, Russian Federation

Gopal Kapooria Ram, Department of Biological Sciences, University of Zambia, Box 32379, Lusaka, Zambia. E-mail: rkapooria@natsci.unza.zm

Lev Kravchenko, All-Russia Research Institute for Agricultural Microbiology, Podbelsky Shossee 3, Pushkin 8, Saint Petersburg, 189620, Russian Federation

Irene Kuiper, Nederlands Forensisch Instituut. Volmerlaan 17, 2288 GD, Rijswijk, the Netherlands

Woong Lee Min, Department of Biology, Dongguk University, Seoul 100-715, Korea

Su Lee Youn, Division of Applied Plant Sciences, College of Agriculture and Life Sciences, Kangwon National University, Chuncheon 200-701, Korea

M. Lorito, Dipartimento di Arboricoltura, Botanica e Patologia vegetale and CNR-IPP, Facoltà di Agraria, Università degli Studi di Napoli "Federico II," Via Università 100, 80055 Portici, Italy. E-mail: lorito@unina.it

J.J. Lugtenberg, Service XS B.V., Wassenaarseweg 72, 2333 AL Leiden, the Netherlands

K.L. McLean, National Centre for Advanced Bio-Protection Technologies, PO Box 84, Lincoln University, Canterbury, New Zealand. E-mail: Mcleankl@lincoln.ac.nz

Ine H.M. Mulders, All-Russia Research Institute for Agricultural Microbiology, Podbelsky Shossee 3, Pushkin 8, Saint Petersburg, 189620, Russian Federation

V. Olmedo-Monfil, Department of Plant Genetic Engineering, Centro de Investigación y Estudios Avanzados, Apartado Postal 629, Irapuato, Gto. 36500, Mexico. E-mail: volmedo@ira.cinvestav.mx

Ashok Pandey, Biotechnology Division, Regional Research Laboratory, CSIR, Trivandrum 695019, India. E-mail: pandey@csrrltrd.ren.nic.in

Binod Parmeswaran, Biotechnology Division, Regional Research Laboratory, CSIR, Trivandrum 695019, India. E-mail: pandey@csrrltrd.ren .nic.in

Gail M. Preston, Department of Plant Sciences, University of Oxford South Parks Road, Oxford OX1 3RB, UK

Paul Rainey, Department of Plant Sciences, University of Oxford, South Park Road, Oxford OX1 3RB, UK and School of Biological Sciences, University of Auckland, Private Bag 92019, Auckland, New Zealand

A. Raio, Dipartimento di Arboricoltura, Botanica e Patologia vegetale and CNR-IPP, Facoltà di Agraria, Università degli Studi di Napoli "Federico II," Via Università 100, 80055 Portici, Italy. E-mail: lorito@unina.it

M.R. Rane, School of Life Sciences, North Maharashtra University, PO Box 80, Jalgaon 425001, MS, India

S. Rosas, Lab Fisiologia Vegetal., Fac Cs Exactas Fco Qcas y Naturales., Campus Universitario 5800, Río Cuarto (Cba), Argentina. E-mail: srosas@ exa.unrc.edu.ar

P.D. Sarode, School of Life Sciences, North Maharashtra University. PO Box 80, Jalgaon 425001, Maharashtra, India

F. Scala, Dipartimento di Arboricoltura, Botanica e Patologia vegetale and CNR-IPP, Facoltà di Agraria, Università degli Studi di Napoli "Federico II," Via Università 100, 80055 Portici, Italy. E-mail: lorito@unina.it

Mahavee P. Sharma, Senior Scientist, Microbiology Section, National Research Centre for Soybean (IC AR), Khandwa Road, Indore-452017 (M.P.). E-mail: mahaveer620@yahoo.com

A. Stewart, Centre Director, Bio-Protection, PO Box 84, Lincoln University, Canterbury, New Zealand. E-mail: Stewarta@lincoln.ac.nz

Yoshihiro Taguchi, Gifu Research Institute for Agricultural Sciences, Matamaru 729, Gifu, 501-1152, Japan

Igor Tikhonovich, All-Russia Research Institute for Agricultural Microbiology, Podbelsky Shossee 3, Pushkin 8, Saint Petersburg, 189620, Russian Federation

André H.M. Wijfjes, Service XS B.V., Wassenaarseweg 72, 2333 AL Leiden, the Netherlands

Eva Wilhelm, Biotechnology, Environmental and Life Science Division, ARC Seibersdorf research GmbH, A-2444 Seibersdorf, Austria. E-mail: Eva.Wilhelm@arcs.ac.at

Hai Yu, Agriculture and Agri-Food Canada, Food Research Program, 93 Stone Road West, Guelph, Ontario, Canada N1G 5C9. E-mail: zhout@ agr.gc.ca

Ting Zhou, Agriculture and Agri-Food Canada, Food Research Program, 93 Stone Road West, Guelph, Ontario, Canada N1G 5C9. E-mail: zhout@ agr.gc.ca

A. Zoina, Dipartimento di Arboricoltura, Botanica e Patologia vegetale and CNR-IPP, Facoltà di Agraria, Università degli Studi di Napoli "Federico II," Via Università 100, 80055 Portici, Italy. E-mail: lorito@unina.it

Preface

Since humans started performing agricultural practices during prehistoric times, control of plant diseases has remained a challenge. In spite of large-scale application of agrochemicals and pesticides during recent decades, the number of plant diseases is still beyond control. Every year, agricultural damages due to pests and pathogens cause enormous economic losses, and exports are restricted owing to poor quality. Adverse short- and long-term effects of agrochemicals on the ecosystem have prompted exploration of natural alternatives for the biocontrol of plant diseases on a sustainable basis. Safety and environmental concerns have forced the withdrawal of a number of synthetic agrochemicals from the market. As biological control alone is the logical path forward for a sustainable ecosystem, there is no viable alternative except to give it a fair chance to prove its ability. The use of biological control strategies offers several advantages over chemical control, since it is economical, self-perpetuating, and usually free from residual side effects. Biological approaches for biocontrol envisages principles, concepts, mechanisms, and current practices using predators, antagonistic microbes, rhizosphere microflora, genetic engineering, and so forth.

The book consists of 15 chapters. The first chapter deals with antagonistic action of rhizosphere microbes in control of soilborne pathogens of strawberries. Chapters 2 and 3 discuss the role of AMF in control of soilborne pathogens and the mechanisms involved in the process. The fourth chapter emphasizes the role of rhizobacteria in the biocontrol of soilborne pathogens. Different mechanisms involved in control are discussed. The fifth chapter highlights the importance of competitive root tip colonization in the control of foot and root rot of tomato by bacteria-producing antifungal metabolites and discusses the utilization of organic acids as nutritional basis of competitive tomato root tip colonization by biocontrol strain *Pseudomonas fluorescens* WCS 365. Chapter 6 discusses strategies for the use of biocontrol agents in the control of white rot of onion. Chapter 7 highlights the biological control of fruit and vegetable diseases using *Trichoderma* and *Agrobacterium* as antagonists. Chapter 8 describes the use of phyllosphere, rhizosphere, and manure-based microbes as biocontrol agents for the control of fruit and vegetable diseases. Chapter

9 describes the biological control of gray mold disease of fruits and vegetables using *Bacillus subtilis* IK-1080. Chapter 10 highlights the strategies used for developing effective biocontrol systems against major fungal diseases of fruits in temperate regions. Chapter 11 describes the biological control of vegetable diseases in Korea. Chapter 12 discusses mycoparasitism as a criteria for biological control. Parasitism among fungi is generally called mycoparasitism. The phenomenon is common among all groups of fungi, from simple chytrids to higher basidiomycetes. Chapter 13 emphasizes the application of plant tissue culture in biocontrol. Tissue culture methods offer possibilities of precisely controlling physical and chemical environments, thus providing a good system for studying plant–microbe interactions. Plant tissue culture, combined with in vitro manipulation of cells, has opened entirely new approaches to disease control, crop production, propagation, and preservation. Chapters 14 and 15 describe biocontrol tools such as microbial chitinases and siderophores.

We are thankful to Robert M. Owen and Dr. Amarjit S. Basra at The Haworth Press for their encouragement, active support, cooperation, and dedicated assistance in editorial structuring. We are looking forward to working with them on future volumes and enhancing the literature on various aspects of plant disease management. We are also thankful to several of our colleagues and students who helped us in various ways during the preparation of this book.

We are grateful to our contributors who responded instantaneously to our request for writing comprehensive reviews as well as sticking to a time schedule. It was because of their active cooperation that we completed this uphill task in the stipulated time. The chapters are original, and some aspects have been included for the first time in any book on plant pathology. Since these chapters have been written by independent authors, there is possibility of slight overlap/repetition of certain facts, but this is unavoidable in a task like this. We are sure that this volume will act as a catalyst in ushering in newer ideas to provide meaningful solutions to intricate problems in plant disease biocontrol technology. We also hope that the information provided in this book will be of immense value to scholars, scientists, agriculturists, plant pathologists, administrators, and enlightened farmers for the biological control of plant diseases.

A word of appreciation is also due to Mr. Mohd Akhtar, Vision Graphics (India), for active cooperation in preparing the electronic copies of the text of the chapters.

Chapter 1

Biological Control of Fungal Soilborne Pathogens in Strawberries

Gabriele Berg

INTRODUCTION

Strawberry (*Fragaria* × *ananassa* Duch.) is an important, high-value culturable crop. Over the last ten years, strawberry production has increased by 35 percent, and in 2002, 3.25 million t were produced worldwide (FAO, Statistical Databases, available at http://www.apps.fao.org). However, strawberry roots are attacked by several fungi (Maas, 1998), which cause high yield losses: for example, *Verticillium dahliae* can affect up to 80 percent (Kurze et al., 2001), and root rot caused by *Pythium, Rhizoctonia,* and *Cylindrocarpon* spp. up to 85 percent (Martin and Bull, 2002). In the coming years, the loss of methyl bromide as a control measure for soilborne pathogens will have a strong negative impact on the occurrence of diseases they cause (Martin, 2003). Although differences in field resistance against soilborne pathogens exist, host resistance is currently not a viable option for disease control, as no resistant cultivars are available (Harris and Yang, 1996; Shaw et al., 1997). Hence, new efficacious methods of control are urgently needed for commercial strawberry production.

I would like to thank my co-workers, Hella Goschke, Stefan Kurze, Jana Lottmann, Henry Müller, Irina Richter, Nicolle Roskot, Nicolas Sauerbrunn, Katja Scherwinski, Maren Schröder, Arite Wolf, Christin Zachow (Rostock), Kornelia Smalla (Braunschweig), and Robert Dahl Hilmar von Mansberg (Strawberry farm, Rövershagen), who formerly worked or are currently working with strawberry- associated microorganisms, for their excellent cooperation.

An environment-friendly alternative to protect roots against fungal pathogens is based on using naturally occurring rhizosphere-associated bacteria and fungi with antagonistic properties as biological control agents (BCAs) (Whipps, 1997; Emmert and Handelsman, 1999). Such bacteria and fungi are able to build a protection layer around the roots which helps to prevent infection of the hyphae by soilborne pathogens (Weller, 1988).

SOILBORNE PATHOGENS OF STRAWBERRY

Understanding the ecology and pathological mechanisms of pathogens is the first step toward managing the diseases they cause. An excellent overview of the soilborne pathogens of strawberry can be found in the *Compendium of Strawberry Diseases* (Maas, 1998). The most important diseases responsible for dramatic yield losses in commercial strawberry production worldwide include red stele root rot and rhizome rot caused by *Phytophthora* species, Verticillium wilt caused by *V. dahliae* Kleb., black root rot originating from a pathogen complex, and *Phythium* and *Rhizoctonia* root rot. Other diseases result from a complex of pathogens, that is, fungi as well as nematodes and bacteria (McKinley and Talboys, 1979). In addition, abiotic parameters such as oxygen and drought stress can strongly influence disease progress.

MICROBIAL ECOLOGY
OF ANTAGONISTIC MICROORGANISMS
IN THE STRAWBERRY RHIZOSPHERE

Biological control of soilborne pathogens cannot be properly understood or optimized without increased knowledge of the biology and ecology of the naturally occurring organisms, pathogens, as well as beneficials. Microorganism–host interactions can be symbiotic, commensal, or pathogenic. Antagonistic microorganisms form an important functional group of plant-associated beneficial microorganisms which are involved in plant growth promotion and pathogen defense (Berg et al., 2003). Mechanisms of microbial antagonism toward plant pathogenic fungi include competition for nutrients and space, production of antibiotics, and/or the production of fungal cell wall–degrading enzymes (Fravel, 1988; Chet et al., 1990). The production of antifungal metabolites is subject to complex regulation, allowing the bacteria to sense their own population density and to respond to different environmental stimuli (Bloemberg and Lugtenberg, 2001; Lugtenberg et al., 2001). Indirect disease control is achieved by mechanisms modulating the plant immune response, including the induction of systemic acquired resistance (Kloepper et al., 1992; Van Wees et al., 1999). The rhizosphere, which is the microenvironment directly influenced by root exudates, is a "hot spot" for antagonistic microorganisms and therefore an important reservoir for BCAs (Sørensen, 1997; Berg et al., 2003). In the strawberry rhizosphere, diverse interactions between the microorganisms themselves and the roots take place. The relationship between the antagonistic microorganisms' interactions and mechanisms along with the resulting potential for uses as BCAs and/or in biotechnology are depicted in Figure 1.1.

The Strawberry Root System

Strawberry plants produce extensive fibrous root systems. The perennial structural roots originate from the crown and can penetrate the soil to a depth of as much as 2.5 m, although most of the root system is confined to the upper 10 to 30 cm of the soil. The perennial roots produce fascicles of transient, multibranched, and short-lived feeder rootlets that are very important for root health and overall fruit productivity. Feeder rootlets that die naturally provide an entrance for microorganisms into the heartwood of older structural roots (Maas, 1998). Therefore, the strawberry root forms a

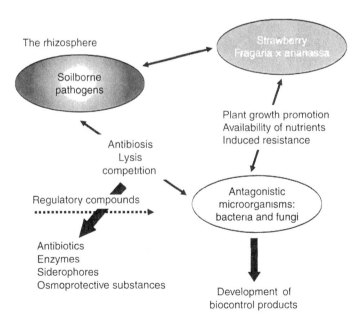

FIGURE 1.1. Interactions in the rhizosphere of strawberry: Potential of antagonistic metabolites and microorganisms in biotechnology.

typical microenvironment for bacteria and fungi. Little is known about root exudates, but they may have an important influence on microflora.

Bacterial Antagonists in the Strawberry Rhizosphere

The plant species or cultivar plays a key role in the composition of rhizobacterial populations colonizing the roots (Smalla et al., 2001). Antagonistic microorganisms found in the strawberry rhizosphere are typical for this crop and are completely different from those of potato and oilseed rape (Berg, Roskot, et al., 2002). The bacterial antagonistic community of strawberry was characterized by a high proportion (9.5 percent) of isolates with antifungal activity and a low diversity of species. Approximately 90 percent of antagonists isolated from the rhizosphere of field-grown strawberries belong to the *Pseudomonas* cluster and consist of *P. putida* A/B, *P. fluorescens, P. syringae,* and *P. corrugata.* Other antagonistic species found in the strawberry rhizosphere are *Acinetobacter* species, *Bacillus megaterium, Burkholderia cepacia, Comamonas acidovorans, Serratia grimesii,* and *Stenotrophomonas maltophilia.* A large proportion of the antagonists showed antifungal activity only against *V. dahliae* (81/125; 65 percent), and most of them were identified as *P. putida* B. About 30 percent of isolates could suppress *R. solani* (37/125), whereas activity against *Sclerotinia sclerotorium* and *P. cactorum* was found in only 7 percent and 6.4 percent, respectively.

Fluorescent pseudomonads were also revealed to be highly abundant in strawberry rhizospheres grown in California soils. Surprisingly, however, plate counts were enhanced in fumigated soils in comparison to nonfumigated soils (Martin and Bull, 2002). Myxobacteria that are able to produce fruiting bodies under starvation conditions are also known for their diverse antibiotic production; however, to date, they constitute part of the rhizosphere community which is still widely not understood, despite their obvious potential (Martin and Bull, 2002).

The proportion of *Verticillium* antagonists was higher for the Garden rather than the Greenish Strawberry species (Berg, Kurze, et al., 2000). The Greenish Strawberry species was obtained from a natural ecosystem (dry calcareous grassland) in a conserved area where they are native and form part of the natural vegetation. In contrast, the Garden Strawberry is a hybrid of *F. chiloensis* (L.) Duchesne and *F. virginiana* Duchesne (both of American origin) and was bred for cultivation in Europe in the eighteenth century. The Garden Strawberry was grown under field conditions with a high population density of *Verticillium* microsclerotia in the soil. That antagonistic bacteria accumulating in the pathogen-infested soil has been described previously (Berg, 1996). Besides many *Pseudomonas* species, a number of interesting *Streptomyces* species, including *S. albidoflavus, S. diastatochro-*

mogenes, and *S. rimosus,* were isolated from both the plant species. Some of the selected BCAs (*P. fluorescens* P6, P10, and *S. diastatochromogenes* S9) showed suppressive effects on Verticillium wilt in both greenhouse and field trials (Berg, Kurze, et al., 2000).

Fungal Antagonists in the Strawberry Rhizosphere

In contrast to the case for bacteria, the percentage of culturable fungal antagonists was generally higher and reached between 10 and 39 percent (Wolf et al., 2004). The dominant fungal genera with antagonistic properties in the strawberry rhizosphere currently known are *Trichoderma, Penicillium,* and *Paecilomyces* (Zachow, unpublished data). Moreover, the diversity of antagonistic strains was relatively high, with diversity indices (based on molecular fingerprints obtained by BOX-PCR and calculated according to Shannon and Weaver [1949]) being between 1.9 and 7.5; these indices were strongly influenced by soil parameters and the vegetation period (Zachow, unpublished data).

The characteristics of bacterial and fungal antagonists in the strawberry rhizosphere are presented in Figure 1.2. A strain collection of around 750 antagonistic strawberry-associated fungi and bacteria is available at Rostock University (Strain Collection of Antagonistic Microorganisms, SCAM).

SCREENING STRATEGIES FOR ANTAGONISTIC MICROORGANISMS AS BIOLOGICAL CONTROL AGENTS

One problem in developing an effective biological control system against soilborne pathogens lies in ascertaining the best screening method to use against a pathogen. As mentioned in the previous section, many antagonistic microorganisms occur in the rhizosphere of strawberry; however, the problem is to select the best one for product development. Indeed, several studies have highlighted discrepancies between the antagonistic effects under in vitro conditions and the corresponding in vivo efficacy (Weller and Cook, 1983; Reddy et al., 1993). Ideally, the candidate organisms should be screened on whole plants rather than, say, in cell culture, or as leaf disks in vitro (Weller, 1988). However, the drawbacks of large-scale screening trials on whole plants are they are time-consuming and expensive. To overcome these drawbacks, new in vitro screening methods that allow more efficient selection of potential antagonists for greenhouse and field experiments have to be devised. Other possibilities include using a combination of assays or developing a hierarchical screening scheme that involves selection after each step for the desired attributes. To generate valid data, all the tests

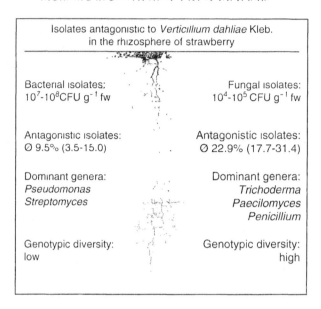

Isolates antagonistic to *Verticillium dahliae* Kleb. in the rhizosphere of strawberry	
Bacterial isolates: 10^7-10^8CFU g^{-1} fw	Fungal isolates: 10^4-10^5 CFU g^{-1} fw
Antagonistic isolates: Ø 9.5% (3.5-15.0)	Antagonistic isolates: Ø 22.9% (17.7-31.4)
Dominant genera: *Pseudomonas* *Streptomyces*	Dominant genera: *Trichoderma* *Paecilomyces* *Penicillium*
Genotypic diversity: low	Genotypic diversity: high

FIGURE 1.2. Characterization of bacterial and fungal antagonists from the strawberry cv. Elsanta established in a three-year field trail in Münster, Germany.

have to be adapted to the ecological behavior of pathogens and performed under realistic abiotic conditions (e.g., temperature, moisture).

The classical dual culture assay is a popular screening method. In this assay, the production of clear inhibition zones (halos) indicates the production of antibiotics, toxic metabolites, or siderophores as the mechanism for biological control (Swadling and Jeffries, 1996). However, as discrepancies are frequently found between the results of the dual culture assay and field trials, this assay should be combined with the evaluation of other parameters to make the data more reliable. Moreover, BCAs that can control more than one pathogen (for example, both fungi and nematodes) are of key interest in the field of pathogen control (McKinley and Talbot, 1979). High-throughout screening of large numbers of pathogens should, therefore, increase the chances of identifying potentially highly effective BCAs. In addition, as the mode of action is important for BCA selection as well as rhizosphere competence, these two should also be taken into account (Lugtenberg and Dekkers, 1999). In the future, identification based on 16S or 18S rDNA or fingerprinting would be helpful to avoid contaminations or problems with risk to human health (Anonymous, 1990; Rademaker and De

Bruijn, 1997) as well as to determine the level of resistance against antibiotics to exclude "risky" BCAs (Berg, 2000).

Many root-associated bacteria have a direct positive influence on plant growth and can indirectly stimulate plant health (Höflich et al., 1994), which is also an important criterion for a BCA. To test the plant-growth-promoting effect, a microplate assay with strawberry seedlings was developed (Berg et al., 2001). The newly developed plant growth promotion assay in microplates is an easier in planta test than a whole plant system in terms of time, plant material, and growth facilities. In addition, it has the advantages of allowing many repetitions and high-throughout screening of a large number of bacterial isolates. On the basis of in vitro tests, an assessment system can be developed to select for the most efficient BCAs for greenhouse trials performed under defined field abiotic conditions. Different cultivars should be assessed, as the genotype of plants also influences the biological control effect (Alström and Gerhardson, 1988). Ideally, greenhouse trials should be followed by field trials under different climatic conditions and in diverse soil qualities, as strawberries are grown worldwide, from Norway to New Zealand, under different conditions. Therefore, BCAs that can function under different climatic and soil conditions are of key scientific interest, since production systems for strawberries that are very specific for each region would probably require a specific solution, which is not a viable option. Further, strawberry production methods can be highly diverse, from high-yielding intensive production systems using methyl bromide soil fumigation and a complete plastic covering for soil in California to organically produced strawberry production systems in untreated soil in Finland (Kivijari et al., 2002; Martin and Bull, 2002). These highly different production systems require specific BCAs to suit their circumstances.

APPLICATION TECHNIQUES

In general, strawberries are planted as plants, sometimes as bare-rooted plants without leaves (transplants), or as runner plants. Prior to planting, many growers use a root-dipping bath to apply fungicides (e.g., Alliette effective against *Phytophthora*), which is an excellent approach to introduce BCAs, since the nearly microorganism-free roots are more easily colonizable. For other plants, it is sometimes difficult to introduce BCAs into stable microbial ecosystems such as the rhizosphere. However, because rhizosphere establishment is a key aspect of effective biological control, this makes strawberry an ideal host plant for BCAs. Another possibility for BCA application is via drip irrigation systems; these are also common in strawberry production systems worldwide.

BIOLOGICAL CONTROL OF STRAWBERRY DISEASES: EXAMPLES AND PRODUCTS

Biological Control Studies

Overall, the goal of a biologically based strategy is to promote rhizo-sphere populations so as to optimize plant growth and to stabilize plant health; this strategy will most probably include enhancing beneficial microbes and reducing pathogens. In contrast to research on the suppression of fruit diseases in strawberry, only a few studies have addressed the introduction of antagonistic microorganisms in the strawberry rhizosphere.

A biocontrol effect of antagonistic rhizobacteria *Erwinia herbicola* G-584, *Paenibacillus macerans* G-V 1, and *P. chlororaphis* I-112 on diseases caused by *Phytophthora* species was a reduction of up to 50 percent (Hessenmüller and Zeller, 1996; Gulati et al., 1998).

Significant reduction of Verticillium wilt by bacterial dipping-bath treatment was recorded in the greenhouse and in fields naturally infected by *V. dahliae*. The relative increase in yield ranged from 113 percent by *S. albidoflavus* S1 to 247 percent by *P. fluorescens* P6 and correlated with a suppression of pathogens in field trials integrated in commercial strawberry production (Berg, Kurze, et al., 2000).

The *Serratia plymuthica* strain HRO-C48 was evaluated for plant growth promotion of strawberries and biological control of the fungal pathogens, *V. dahliae* and *P. cactorum* (Kalbe et al., 1996; Berg, 2000). The antifungal activity of the strain was mainly based on an efficient chitinolytic system (Berg, Frankowski, and Bahl, 2000; Frankowski et al., 2001). The *S. plymuthica* strain HRO-C48 was shown to produce indole-3-acetic acid in vitro and to enhance plant growth in a strawberry seedling assay (Kurze et al., 2001). In three consecutive vegetation periods, field trials were carried out in soils naturally infested by soilborne pathogens. Dipping plants in a suspension of *S. plymuthica* prior to planting reduced Verticillium wilt compared with the nontreated control by 0 to 37.7 percent (mean 24.2 percent), whereas the increase in yield ranged from 156 to 394 percent (mean 296 percent). *Serratia* treatment also reduced *Phytophthora* root rot by 45 to 60 percent compared with the nontreated control (Berg et al., 1999). It is important to note that the results of field trials were strongly affected by abiotic and biotic conditions (Berg, Kurze, et al., 2002). In addition, the influence of the biocontrol agent on the bacterial communities (nontarget microorganisms) of the strawberry rhizosphere was monitored by the analysis of polymerase chain reaction (PCR)-amplified 16S rRNA genes of the whole bacterial community after separation by single-strand conformation polymorphism (SSCP) analysis (Figure 1.3). Using general universal and *Serratia* specific primers, no dif-

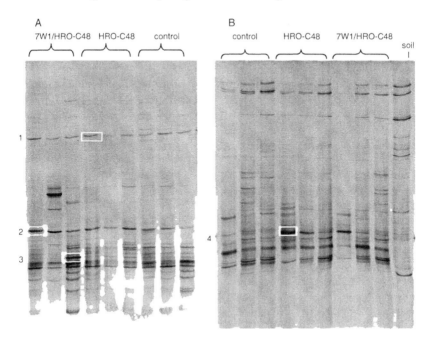

FIGURE 1.3. Comparison between the bacterial communities recovered from the rhizosphere of BCA-treated (*Serratia plymuthica* HRO-C48 and *Streptomyces rimosus* 7W1), nontreated strawberries and bulk soil using universal primers (A) and *Serratia*-specific primers (B), and separated by SSCP. Bands which are indicated were sequenced and identified as *Actinetobacter* sp. (1), *Pseudomonas chlororaphis* (2), *Serratia plymuthica* (3), and *Serratia plymuthica* (4) (Scherwinski and Berg, unpublished data).

ference in the bacterial community between BCA-treated and nontreated plants was observed. Under field conditions, the strain survived at approximately $\log_{10} 3$ to 7 CFU g^{-1} root in the strawberry rhizosphere at 14 months after root application.

Products

Some products on the market are suitable for controlling soilborne pathogens of strawberry. One product called Bio-Fungus (or Supresivit, Anti-Fungus) contains the BCA *Trichoderma* spp. and, on strawberry, is able to suppress pathogens such as *Sclerotinia, Phytophthora, Rhizoctonia solani,*

Pythium spp.. *Fusarium*, and *Verticillium* (BioPlant, Arhus, Denmark). Another product, Polyversum (formerly Polygandron), is based on the BCA *Pythium oligandrum*. This product, which is available as wettable powder, was developed as a broad-based antifungicid and is effective against pathogens such as *Pythium* spp., *Fusarium* spp., *Botrytis* spp., *Phytophthora* spp., *Alternaria* spp., *Tilletia,* and *R. solani* (Biopreparaty Ltd., Prague, Czech Republic). Another product based on fungal antagonists is Root-Shield (PlantShield, T-22 Planter box), and it uses the active strain *Trichoderma harzianum* Rifai strain KRL-AG2 (T-22) to avoid diseases caused by *Pythium* spp., *R. solani,* and *Fusarium* spp. in strawberries (Bioworks, Inc., Geneva, USA). Rhizo-Plus, which promotes plant growth, is a water-dispersible granule containing spores of *Bacillus subtilis* FZB24. This product can enhance fruit production in strawberries by up to 54 percent (KFZB Biotechnik GmbH, Berlin, Germany). RhizoStar is based on *Serratia plymuthica* HRO-C48, which builds a protection layer in the rhizosphere; it is currently in the registration procedure (e-nema, Raisdorf, Germany).

Mycorrhiza

Many reports have demonstrated the positive effects of mycorrhiza introduction on the growth and health of strawberries (Taylor and Harrier, 2001). After the treatment of strawberry plants with mycorrhiza fungi, benefits were observed in fruit production and/or runner plants development (Taube-Baab et al., 1996). Taken together, antagonistic bacteria that simultaneously work as mycorrhiza-helper bacteria (MHB) are therefore very promising candidates for future biocontrol strategies (Mamatha et al., 2002).

ALTERNATIVE OR COMBINED METHODS
TO CONTROL SOILBORNE PATHOGENS

The development and implementation of efficient biological control strategies will require an integrated system approach that takes into consideration aspects specific to the strawberry production system. Methods that transform resident microbial communities in a manner which induces natural soil suppressiveness also have a high potential as components of environmentally sustainable systems for soilborne plant pathogen management (Mazzola, 2002).

A cocktail of BCAs might be more effective in suppressing disease than a single one. In addition to the additive or synergistic effects of using more than one BCA at a time, it could be possible to use agents that would become active (when others become inactive) due to changes in environmental conditions or at various time intervals as host organs develop, and thus concomitantly to initiate different microniches (Baker, 1991; Guetsky et al., 2001). Moreover, a combination of *S. plymuthica* HRO-C48 and *P. fluorescens* E11 as well as *S. plymuthica* HRO-C48 and *S. rimosus* 7W1 stimulated enhanced effects in comparison to single-isolate applications (Sauerbrunn, 2000).

Crop rotation systems provide an effective means of controlling soilborne pathogens. One example is the use of broccoli to reduce *Verticilllium* inoculum in soilborne pathogens (Subbarao and Hubbard, 1996; Subbarao et al., 1999). However, conversely, other host plants of *Verticillium*, such as oilseed rape, enhanced disease in strawberry (Kurze et al., 2001).

Soil amendments offer another way to suppress soilborne pathogens, as shown for *Verticillium* (Conn and Lazarovits, 2000) and *Rhizoctonia* (Mazzola et al., 1997).

Solarization can suppress soilborne pathogens (Katan et al., 1976) and was also successful in strawberry production (Pinkerton et al., 1997). However, the use is restricted in some areas by the length and intensity of sunshine as well as temperature.

Organically produced strawberries are now a viable option for many growers worldwide. The introduction of BCAs in this system would be interesting not only for antipathogen effects but also for plant nutrition (Kivijari et al., 2002). Many BCAs also work as plant-growth-promoting agents as they are able to fix nitrogen or enhance phosphorus availability. Phytohormone production by bacteria may also encourage plant growth (Höflich et al., 1994).

CONCLUSION

Successful biological control of strawberry pathogens requires not only a better understanding of the complex regulation of antifungal metabolite production by antagonists in response to environmental factors but also a more comprehensive picture of the triggers of root colonization as well as the dynamics and composition of bacterial rhizosphere communities. Enhanced research activity in this area in recent years has greatly augmented our knowledge of the molecular ecology of plant-associated bacteria and fungi as well as of the microorganisms present in the soil. By continuing research efforts in this area, we can be optimistic that the progress in understanding the mechanisms of microbial ecology will translate into new and innovative concepts for biological control of strawberry pathogens.

REFERENCES

Alström. B. and B. Gerhardson. (1988). Differential reactions of wheat and pea genotypes to root inoculation with growth-affecting rhizobacteria. *Plant and Soil* 109: 263-269.

Anonymous. (1990). *Berufsgenossenschaft der chemischen Industrie: Sichere Biotechnologie.* Einstufung biologischer Agenzien: Bakterien: Merkblatt B 006. Jedermann-Verlag Heidelberg.

Baker, R. (1991). Diversity in biological control. *Crop Protection* 10: 85-94.

Berg, G. (1996). Rhizobacteria of oilseed rape antagonistic to *Verticillium dahliae* var. *longisporum* STARK. *Journal of Plant Diseases and Protection* 103: 20-30.

Berg, G. (2000). Diversity of antifungal and plant-associated *Serratia plymuthica* strains. *Journal of Applied Microbiology* 88: 952-960.

Berg, G., J. Frankowski, and H. Bahl. (2000). Interactions between *Serratia plymuthica* and the soilborne pathogen *Verticillium longisporum.* In *Advances in Verticillium Research and Disease Management,* eds., E.C. Tjamos, R.C. Rowe, J.B. Heale, and D.R. Fravel. APS Press. The American Phytopathological Society, St. Paul, MN, pp. 269-273.

Berg, G., J. Frankowski, A. Krechel, J. Lottmann, K. Ribbeck, A. Roder, and A. Wolf. (2003). Plant-associated bacteria: Diversity, antagonistic interactions and biotechnological applications. *Research Advances in Microbiology* 3: 23-33.

Berg, G., A. Fritze, N. Roskot, and K. Smalla. (2001). Evaluation of potential biocontrol rhizobacteria from different host plants of *Verticillium dahliae* Kleb. *Journal of Applied Microbiology* 91: 963-971.

Berg, G.. S. Kurze, A. Buchner, E.M. Wellington, and K. Smalla. (2000). Successful strategy for the selection of new strawberry-associated rhizobacteria antagonistic to Verticillium wilt. *Canadian Journal of Microbiology* 46: 1128-1137.

Berg, G.. S. Kurze, and R. Dahl. (1999). Rhizobacterial isolates for use against phytopathogenic soil fungi and process for applying these rhizobacterial isolates. European Patent Number 98124694.

Berg, G.. S. Kurze, J. Frankowsi, I. Richter. and R. Dahl. (2002). Biological control agent *Serratia plymuthica* strain HRO-C48: Performance in relation to environmental factors. *IOBC Bulletin* 25: 111-115.

Berg, G., N. Roskot. A. Steidle, L. Eberl, A. Zock, and K. Smalla. (2002). Plant-dependent genotypic and phenotypic diversity of antagonistic rhizobacteria isolated from different *Verticillium* host plants. *Applied and Environmental Microbiology* 68: 3328-3338.

Bloemberg, G.V. and B.J.J. Lugtenberg. (2001). Molecular basis of plant growth promotion and biocontrol by rhizobacteria. *Current Opinion in Plant Biology* 4: 343-350.

Chet, I.. A. Ordentlich. R. Shapira, and A. Oppenheim. (1990). Mechanisms of biocontrol of soilborne plant pathogens by rhizobacteria. *Plant and Soil* 129: 85-92.

Conn, K.L. and G. Lazarovits. (2000). Soil factors influencing the efficacy of liquid swine manure added to soil to kill *Verticillium dahliae. Canadian Journal of Plant Pathology* 21: 400-406.

Emmert, E.A.B. and J. Handelsman. (1999). Biocontrol of plant disease: A (Gram) positive perspective. *FEMS Microbiological Letters* 171: 1-9.

Frankowski, J., M. Lorito, R. Schmid, G. Berg, and H. Bahl. (2001). Purification and properties of two chitinolytic enzymes of *Serratia plymuthica* HRO-C48. *Archives of Microbiology* 176: 421-426.

Fravel, D.R. (1988). Role of antibiosis in the biocontrol of plant diseases. *Annual Review of Phytopathology* 26: 75-91.

Guetzky, R., D. Shtienberg, Y. Elad, E. Fischer, and A. Dinoor. (2001). Improving biological control by combining biological control agents each with several mechanisms of disease suppression. *Phytopathology* 92: 976-985.

Gulati, M.K., E. Koch, F. Huth, and W. Zeller. (1998). Biologische Bekämpfung der Roten Wurzelfäule *(Phytophthora fragariae* var. *fragariae)* und der Rhizomfäule *(Phytophthora cactorum)* der Erdbeere mit bakteriellen Antagonisten. *Mitteilungen der Biologischen Bundesanstalt* 357: 338.

Harris., D.C. and J.R. Yang. (1996). The relationship between the amount of *Verticillium dahliae* in soil and the incidence of strawberry wilt as a basis for disease risk prediction. *Plant Pathology* 45: 106-114.

Hessenmüller, A. and W. Zeller. (1996). Biologische Bekämpfung von bodenbürtigen *Phytophtora* Arten an der Erdbeere. *Journal of Plant Diseases and Protection* 103: 602-609.

Höflich, G., W. Wiehe, and G. Kühn. (1994). Plant growth stimulation by inoculation with symbiotic and associative rhizosphere microorganisms. *Experientia* 50: 897-905.

Kalbe, C., P. Marten, and G. Berg. (1996). Members of the genus *Serratia* as beneficial rhizobacteria of oilseed rape. *Microbiological Research* 151: 433-439.

Katan, J., A. Greenberger, H. Laon, and A. Grinstein. (1976). Solar heating by polyethylene mulching for the control of diseases caused by soilborne pathogens. *Phytopathology* 66: 683-688.

Kivijari, P., S. Prokkola, A. Aflatuni, P. Parikka, and T. Tuovinen. (2002). Cultivation techniques for organic strawberry production in Finnland. *ISHS Acta Horticulturae* 567: 531-534.

Kloepper, J.W., G. Wei, and S. Tzun. (1992). Rhizosphere population dynamics and internal colonization of cucumber by plant growth-promoting rhizobacteria, which induce systemic resistance to *Colletotrichum orbiculare*. In *Biological Control of Plant Diseases: Progress and Challenges for the Future*, eds., E.C. Tjamos, G.C. Papavizas, and R.J. Cook. NATO ASI Series, Series A, Vol. 230, Plenum Press, New York, pp. 185-192.

Kurze, S., R. Dahl, H. Bahl, and G. Berg. (2001). Biological control of fungal strawberry diseases by *Serratia plymuthica* HRO-C48. *Plant Diseases* 85: 529-534.

Lugtenberg, B.J.J., and L.C. Dekkers (1999). What makes *Pseudomonas* bacteria rhizosphere competent? *Environmental Microbiology* 1: 9-13.

Lugtenberg, B.J.J., L. Dekkers, and G.V. Bloemberg. (2001). Molecular determinants of rhizosphere colonization by *Pseudomonas*. *Annual Review of Phytopathology* 39: 461-490.

Maas, J.L. (1998). *Compendium of Strawberry Diseases*. APS Press, St. Paul, MN.

Mamatha, G., D.J. Bagyaraj, and S. Jaganath (2002). Inoculation of field-established mulberry and papaya with arbuscular mycorrhizzal fungi and a mycorrhiza helper bacterium. *Mycorrhiza* 12: 313-316.

Martin, F.N. (2003). Development of alternative strategies for management of soil-borne pathogens currently controlled with methyl bromide. *Annual Review of Phytopathology* 41: 325-350.

Martin, F.N. and C.T. Bull. (2002). Biological approaches for control root pathogens of strawberry. *Phytopathology* 92: 1356-1362.

Mazzola, M. (2002). Mechanisms of natural soil suppressiveness to soil-borne diseases. *Antonie van Leeuwenhoek* 81: 557-564.

Mazzola, M., T.E. Johnson, and R.J. Cook. (1997). Influence of field burning and soil treatments on growth of wheat after Kentucky bluegrass, and effect of *Rhizoctonia cerealis* on bluegrass emergence and growth. *Plant Pathology* 46: 708-716.

McKinley, R.T. and P.W. Talboys. (1979). Effects of *Pratylechus penetrans* on development of strawberry wilt caused by *Verticillium dahliae*. *Annales Applied Biology* 92: 347-357.

Pinkerton, J.N., K.L. Ivors, P.W. Reeser, P.R. Bristow, and G.E. Windom. (1997). The use of soil solarization for the management of soilborne plant pathogens in strawberry and red raspberry production. *Plant Diseases* 86: 645-651.

Rademaker, J.L.W. and F.J. De Bruijn. (1997). Characterization and classification of microbes by REP-PCR genomic fingerprinting and computer-assisted pattern analysis. In *DNA Markers: Protocols, Applications and Overviews,* eds., G. Caetano-Anollés and P.M. Gresshoff. J. Wiley & Sons, Inc., New York.

Reddy, M.S., R.K. Hynes, and G. Lazarovits. (1993). Relationship between in vitro growth inhibition of pathogens and suppression of preemergence damping-off and postemergence root rot of white bean seedlings in the greenhouse by bacteria. *Canadian Journal of Microbiology* 40: 113-119.

Sauerbrunn, N. (2000). Molekulare Analyse des Einflusses verschiedener Biological Control Agents auf die Rhizosphärenpopulationen der Erdbeere. Diplom Universität Rostock.

Shannon, C.E. and W. Weaver. (1949). *The Mathematical Theory of Communication.* University Illinois Press, Urbana.

Shaw, D.V., W.D. Gubler, and J. Hansen (1997). Field resistance of California strawberries to *Verticillium dahliae*. *Horticultural Science* 32: 711-713.

Smalla, K., G. Wieland, A. Buchner, A. Zock, J. Parzy, N. Roskot, H. Heuer, and G. Berg. (2001). Bulk and rhizosphere soil bacterial communities studied by denaturing gradient gel electrophoresis: Plant dependent enrichment and seasonal shifts. *Applied and Environmental Microbiology* 67: 4742-4751.

Sørensen, J. (1997). The rhizosphere as a habitat for soil microorganisms. In *Modern Soil Microbiology,* eds., J.D. Van Elsas, J.T. Trevors, and E.M.H. Wellington. Marcel Dekker, Inc., New York, pp. 21-45.

Subbarao, K.V. and J.C. Hubbard. (1996). Interactive effects of broccoli residue and temperature on *Verticillium dahliae* microsclerotia in soil and on wilt in cauliflower. *Phytopathology* 86: 1303-1309.

Subbarao, K.V., J.C. Hubbard, and S.T. Koike. (1999). Evaluation of broccoli residue incorporation into field soil for Verticillium wilt control in cauliflower. *Plant Disease* 83: 124-129.

Swadling, I.R. and P. Jeffries. (1996). Isolation of microbial antagonists for biocontrol of grey mold disease of strawberry. *Biocontrol Science Technology* 6: 125-136.

Taube-Baab, H., H. Ahrweiler, and H. Baltruschat. (1996). Ertragssteigerung von Erdbeerpflanzen nach Inokulation mit VA-Mykorrhiza-Pilzen. *Erwerbsobstanbau* 38: 144-147.

Taylor, J. and L.A. Harrier. (2001). A comparison of development and mineral nutrition of micropropagated *Fragaria* × *ananassa* cv. Elvira (strawberry) when colonised by nine species of arbuscular mycorrhizal fungi. *Applied Soil Ecology* 18: 205-215.

Van Wees, S.C., M. Luijendijk, I. Smoorenburg, L.C. van Loon, and C.M. Pieterse. (1999). Rhizobacteria-mediated induced systemic resistance (ISR) in *Arabidopsis* is not associated with a direct effect on expression of known defense-related genes but stimulates the expression of the jasmonate-inducible gene *Atvsp* upon challenge. *Plant Molecular Biology* 41: 537-549.

Weller, D.M. (1988). Biological control of soilborne plant pathogens in the rhizosphere with bacteria. *Annual Review of Phytopathology* 26: 379-407.

Weller, D.M. and R. Cook. (1983). Suppression of take-all of wheat by seed-treatment with fluorescent pseudomonads. *Phytopathology* 73: 463-469.

Whipps, J.M. (1997). Ecological considerations involved in commercial development of biological control agents for soil-borne diseases. In *Modern Soil Microbiology*, eds., J.D. Van Elsas, J.T. Trevors, and E.M.H. Wellington. Marcel Dekker, Inc. New York, pp. 525-545.

Wolf, A., S. Scherwinski, H. Müller, A. Golly, K. Smalla, and G. Berg. (2004). Impact of application of biocontrol agents to plant root on the natural occurring microbial community. *IOBC Bulletin*. 27: 297-300.

Chapter 2

Use of Arbuscular Mycorrhizal Fungi for Biocontrol of Soilborne Fungal Plant Pathogens

S. Demir
A. Akkopru

INTRODUCTION

A key and universally accepted concept is that natural microbial populations in soil or other living substrates are associated with growth around developing plant roots, giving rise to the so-called rhizosphere (Azcón-Aguilar and Barea, 1992). As a rhizosphere develops at the root–soil interface, microorganisms there interact with both plant roots and soil constituents. Communication among these components is mainly via chemical or biochemical signals, although physical interaction is also possible (Azcón-Aguilar and Barea, 1996).

The most important interactions developing in the rhizosphere can be classified into three main groups: (1) plant–plant interactions (competition for nutrients); (2) root–microorganism interactions (microorganisms can stimulate, inhibit, or be without effect on root growth and resistance, depending on the type of microorganism, plant species, and environment conditions); and (3) microorganism–microorganism interactions (which include both synergistic and antagonistic activities) (Lynch, 1990; Marschner, 1995; Azcón-Aguilar and Barea, 1996).

Soil microorganisms are paramount in the biogeochemical cycling of both inorganic and organic nutrients in the soil and in the maintenance of soil quality. Microbial activity in the rhizosphere, in particular, is a major factor that determines the availability of nutrients to plants and has a significant influence on plant health and productivity (Jeffries et al., 2003). From

Biological Control of Plant Diseases
© 2007 by The Haworth Press, Inc. All rights reserved.
doi:10.1300/5682_02

this point of view, it is important to use a broad definition of the rhizosphere to include the rhizosphere soil, the volume of soil adjacent to and influenced by the root, the root surface or rhizoplane, and the root itself, which includes the cells of the root cortex where invasion and colonization by endophytic microorganisms have occurred (Jeffries et al., 2003).

Soil–plant–microorganism interactions are complex and can influence the health, growth, and productivity of plants. The result may be harmful (root pathogens, subclinical pathogens, detrimental rhizobacteria, cyanide producers), neutral, or beneficial (rhizobium bacteria, mycorrhizae, antagonists [biocontrol] of detrimental microorganisms, hormone producers, plant growth-promoting rhizobacteria [PGPR]) to the plants (Marschner, 1995; Jeffries et al., 2003).

Mycorrhizal fungi are relevant members of the rhizosphere mutualistic microsymbiont populations known to carry out many critical ecosystem functions, such as improvement of plant establishment, enhancement of plant nutrient uptake, plant protection against cultural and environmental stresses, and improvement of soil structure (Smith and Read, 1997; Barea et al., 2002).

When arbuscular mycorrhizae (AM) are formed, there are some changes in the root morphology, but root physiology changes significantly, as does the physiology of the rest of the AM-colonized plant. For example, when the plants are colonized by the mycorrhizal fungus, there are changes in concentrations of growth-regulating compounds, such as auxins, cytokinins, and gibberellins; the photosynthetic rates increase; and the portioning of photosynthates to shoots and roots changes. The nutritional status of the host tissues changes in response to altered uptake of minerals from the soil, and this in turn can change structural and biochemical aspects of root cells that can alter membrane permeability and thus the quality and quantity of root exudation. Altered exudation induces changes in the composition of microorganisms in the rhizosphere soil, known as the "mycorrhizosphere" (Linderman, 1988, 1992). In the mycorrhizosphere, two different zones are discernible: the first is under the joint influence of the root and fungal components of the mycorrhiza, and the second is affected by the mycelium of the mycorrhizal fungus only. This second zone, called the "hyphosphere," is part of the overall mycorrhizosphere but may support biotic activities distinct from those in soil under joint root and fungal influence (Linderman, 1988; Marschner, 1995; Andrade et al., 1998).

When AM fungi (AMF) establish a symbiotic association with host plant roots, they live both within root tissues and external to those tissues in the soil. They could have direct interactions with other soil organisms, or they could influence those organisms indirectly by changing host plant physiology including root physiology, and thus the patterns of exudation into the mycorrhizosphere (Linderman, 1992).

Because arbuscular mycorrhizae are established in the roots of host plants, research on mycorrhizae by disease interactions has been concentrated on diseases caused by soilborne fungi pathogens. Although most of the experiments differ in approaches and methods, diseases caused by soilborne fungi can be influenced by the formation of arbuscular mycorrhizae in the root system. In general, mycorrhizal plants suffer from less damage, the incidence of disease is decreased, or pathogen development is inhibited (Dehne, 1982; Bethlenfalvay and Linderman, 1992).

The study of the possible role of AM symbiosis in the protection against plant pathogens began in the 1970s. Although a great deal of information has been published on this subject, we still know very little about the mechanism involved (Linderman, 1994; Azcón-Aguilar and Barea, 1996; Mukerji, 1999). For this reason, in the present review, we will analyze the possibilities of using AM fungi against soilborne fungal pathogens with mechanisms and demonstrate possible fruitful research approaches.

GENERAL CONCEPTS IN THE ARBUSCULAR MYCORRHIZAL STATUS

The term "mycorrhiza" refers to the association between fungi and roots. This association is generally considered a mutualistic symbiosis because of the highly interdependent relationship established between both partners. The function of all mycorrhizal systems depends on the ability of the fungal symbiont in the absorption of nutrients available in inorganic and/or organic forms in soil. In most mycorrhizal types, organic C, which is derived from photosynthesis, is also transferred from the plant to the fungus, followed by translocation to the growing margins of the extraradical mycelium and to the developing spores and fruit bodies (Smith and Read, 1997). The fungus, in fact, becomes an integral part of the root system. The symbiosis is considered the most metabolically active component of the absorbing organs of the autrophic host plant which, in addition to furnishing the heterotrophic fungal associate with organic nutrients, is an ecologically protected habitat for the fungus (Azcón-Aguilar and Barea, 1997).

Mycorrhizae can be found in almost any kind of soil. All but a few vascular plant species (those belonging mainly to the families Cruciferae, Chenopodiaceae, Cyperaceae, Caryophyllaceae, and Juncaceae) are able to form mycorrhizae (Marschner, 1995; Azcón-Aguilar and Barea, 1997). The physiology of the plant is highly affected by the presence of the fungal symbionts (Harley and Smith, 1983; Smith and Read, 1997).

Several types of mycorrhizae exist, defined by the plant root-fungus combination and the symbiotic structure (Strack et al., 2003). For example, ectomycorrhizal fungi (ECMF) form a particular extracellular morphological complex with the roots of many temperate forest trees. Many thousands of fungi can form these relationships, in contrast to the restricted range of plant species that are involved. Even more specific and unique mycorrhizal relationships are formed by the arbutoid, orchid, and ericoid plant families (Peterson and Farquhar, 1994; Mukerji et al., 2000, 2002; Jeffries et al., 2003).

The endotophic arbuscular mycorrhiza is the most common type, occurring in about 80 percent of plant species (Jeffries et al., 2003; Strack et al., 2003). AM fungi are described by more than 150 species of the Zygomycota included in the order of Glomales (Morton and Benny, 1990). Recent work on the taxonomy and phylogeny of AM fungi provided a basis for a new classification, removing these fungi from the polyphyletic Zygomycota and placing them into a new monophyletic phylum, the Glomeromycota (Schüßler et al., 2001, 2002; Strack et al., 2003).

Beneficial effects for AM fungi, apart from supplying plants with phosphate and other nutrients, have been described. The symbiosis has a positive effect on plant water potential, especially for plants under drought stress. AM-colonized plants show a significant degree of biocontrol against various pathogens. Positive effects of AM fungi on soil structure have been described (Tisdall, 1994). As a consequence, AM symbiosis is regarded as a key component of sustainable agriculture, whereas under conventional agricultural conditions, AM fungi seem to be of only minor importance when we compare conventional and organic-biological agricultural systems directly (Bethlenfalvay and Linderman, 1992; Strack et al., 2003).

Apart from agricultural systems, the application of AM fungi is experimented for the revegetation of desertified areas and in cultivation of microporopagated plantlets. In this context, major technological problems are the form of application of the AM inoculum and the combinations of AM inoculum with other microorganisms that are beneficial for plant growth (Strack et al., 2003).

Formation of AM Symbiosis

AM colonization begins with hyphae that arise from soilborne propagules (large resting spores of the AMF or mycorrhizal root fragments) or from an AM plant growing nearby.

Upon the arrival of the fungal hyphae at the root surface, an appressorium is usually formed on the epidermal cells. The colonizing hyphae originating from appressoria pass through the intercellular spaces and then enter root tissues, spreading between and through cells of the cortical root layers. Once the hyphae reach the inner cortex, they grow into the cells by means of repeated dichotomous branching and form treelike structures called "arbuscules," which are believed to be the site of phosphate exchange between the fungus and the plant. The life span of individual arbuscules is from 4 to 14 days. As internal colonization spreads, the extraradical hyphae ramify and grow along the root surface and then form more penetration points. They also grow outward into the surrounding soil, thus developing an extensive tridimensional network of mycelium interfacing with soil particles. The fungus in some genera may form "vesicles": oval-to-spherical structures with a storage (mainly lipids and carbon) function. Most AMF form large resting spores on the external mycelium (Azcón-Aguilar and Barea, 1997; Hodge, 2000; Strack et al., 2003). The term "arbuscular mycorrhiza" replaced the earlier term "vesicular–arbuscular mycorrhiza" (VAM) because not all endomycorrhizae of this type develop vesicles, but all form arbuscules (Strack et al., 2003).

The fungus in the intracellular colonization, as in the case of arbuscules, is always surrounded by the intact host-cell plasmalemma. Arbuscule formation, therefore, represents a large surface of cellular contact between host and fungus. In fact, the arbuscule is probably the main transfer site of mineral nutrients from the fungus to the plant and of C compounds to the fungus (Azcón-Aguilar and Barea, 1997; Smith and Read, 1997). The mycelial network of AM fungus can extend several centimeters outward from the root surface, bridging over the zone of nutrient depletion around roots to absorb low-mobility ions (mineral plant nutrients) from the bulk soil. The extraradical hyphae also interact with components of the rhizosphere microbiota, and together they contribute to the formation of water-stable aggregates that are critical for a good soil structure. As a result, AMF act as a major interface between plants and the biotic and abiotic component of the complex soil system (Azcón-Aguilar and Barea, 1997).

Some features of the host plant, depending primarily (but not exclusively) on the morphological and physiological characteristics of the root system, condition its ability to obtain nutrients and, consequently, the host derives benefit from its mycorrhizal status. Certain factors affecting mycorrhiza formation can also indirectly influence functioning of the symbiosis (Azcón-Aguilar and Barea, 1997).

In general, those plants which are able to absorb relatively immobile soil resources, such as phosphorus, are less dependent on mycorrhiza formation (Brundrett, 1991). It is known that the ability of the root system to absorb low-mobility nutrients from the soil solution is positively correlated with the root surface area, which is in turn a consequence of root system architecture (branching pattern, length and diameter of the roots, and length and number of root hairs). Thus, plants with low branching frequency, low number of lateral roots, and short root hairs appear to be more dependent on mycorrhizae for mineral nutrient uptake (Brundrett, 1991).

The Role of AM Fungi on the Plant-Soil Ecosystem

The key effects of AM symbiosis can be summarized as follows: (1) improved rooting and plant establishment; (2) enhanced uptake of low-mobility ions; (3) improved nutrient cycling; (4) enhanced plant tolerance to biotic and a biotic stress factors; (5) enhanced quality of soil structure; and (6) improved plant community diversity (Linderman, 1994; Azcón-Aguilar and Barea, 1997). For this reason, AM symbiosis influences several aspects of plant physiology, such as mineral nutrition uptake, plant growth, and plant resistance.

The primary effect of AM symbiosis is to increase the supply of mineral nutrients to the plant, particularly those whose ionic forms have a poor mobility rate, or those which are present in low concentration in the soil solution (Azcón-Aguilar and Barea, 1997). This chiefly concerns phosphorus, ammonium, zinc, and copper (Marschner, 1995).

AM-colonized plants show a significant degree of biocontrol against biotic stress, including the root fungal pathogens and nematodes (Linderman, 1994). AM fungi also enable plants to cope with a biotic stress by alleviating nutrient deficiencies, improving drought tolerance, overcoming the detrimental effects of salinity, enhancing tolerance to pollution, and helping adaptation of sterile micropropagated plants to nonsterile substrates and field conditions (Brundrett, 1991; Azcón-Aguilar and Barea, 1997). Furthermore, AM fungi have an indirect influence on plant growth due to their stabilizing effects on soil structure (for example, on soil aggregate formation and humic substances accumulation) (Tisdall, 1994).

As a consequence, AM symbiosis is regarded as a key component of sustainable agriculture, whereas under conventional agricultural conditions, AM fungi seem to be only of minor importance. Mäder et al. (2000) have compared conventional and organic-biological agricultural system directly. They found stronger mycorrhizal colonization for the organic-biological system and preliminary evidence for a partial compensation for the disadvantages of the organic-biological system by AM fungi. In this context, major technological problems are the form of application of the AM inoculum and the combinations of AM inoculum with other microorganisms that are beneficial for plant growth (Bethlenfalvay and Linderman, 1992; Linderman, 1994; Strack et al., 2003).

ARBUSCULAR MYCORRHIZAL FUNGI
IN HORTICULTURAL SYSTEMS

Most of the major plant families are able to form mycorrhize, the arbuscular mycorhizal association being the most common mycorrhizal type involved in agricultural systems. AM technology is feasible for crops using a transplant stage, as is the case with horticultural systems. Recent developments regarding and insights into the potential of AM symbiosis in horticultural practices are discussed. In general, fruit crops have received more attention than vegetable and ornamental crops (Azcón-Aguilar and Barea, 1997).

Clearly, the interest of horticulturists in AM technology is due to the ability of AMF to increase the uptake of phosphorus and other nutrients and the resistance to biotic and abiotic stresses (Azcón-Aguilar and Barea, 1997).

Principally, plant propagation in horticultural systems usually starts from seedlings, cuttings, or graftings produced or developed in soil or in substrates that have been treated to lower the level of pathogenic organisms. In the case of micropropagation, sterility is obviously a key component. Thus, the most common propagation techniques include treatments whose aim is to eliminate or reduce the microbial population in the rooting medium, although this also obviously affects the beneficial ones, such as AMF. However, these practices provide the most rewarding situation for the application of mycorrhizal biotechnology by reintroducing appropriately selected AMF at suitable stages of the plant production schedules (Azcón-Aguilar and Barea, 1997).

Many researchers have shown that application of AMF in horticulture can be very successful (Gianinazzi et al., 1990; Barea et al., 1993; Lovato et al., 1995; Azcón-Aguilar and Barea, 1997). In summary, the information reported and recorded in the above review papers refers to (1) vegetable crops and species, such as lettuce, onion, leek, celery, asparagus, tomato, pepper, cucumber, beans, strawberry, muskmelon, watermelon, and potato; (2) temperate fruit crops such as cherry, citrus, apple, almond, peach, olive, grapevine, blackberry, pear, kiwifruit, raspberry, and plum; (3) tropical plantation crops such as coffee, rubber, cacao, papaya, pineapple, oil palm, cocoa, avocado, and pineapple; and (4) floricultural crops such as rose, lilac, verbena, primula, cyclamen, geranium, begonia, gerbera, and marigold.

Consequently, it can be confirmed that a careful selection of a functionally compatible host/fungus/substrate combination is critical for success; after sowing, or at outplanting, an early formation of the AM status is a key factor to enhance plant performance in horticultural practices (Gianinazzi et al., 1990; Azcón-Aguilar and Barea, 1997).

ARBUSCULAR MYCORRHIZAE
AS BIOCONTROL AGENTS

As arbuscular mycorrhizae are a major component of rhizopheres of plants, it is logical that they could affect the incidence and severity of root diseases (Linderman, 1992). Comprehensive reviews summarizing and discussing results of AM fungi and possibilities of their biocontrol include those by Schönbeck (1979), Dehne (1982), Bagyaraj (1984), Smith (1988), Caron (1989), Paulitz and Linderman (1991), Linderman (1992, 1994), Azcón-Aguilar and Barea (1996), Smith and Read (1997), and Mukerji (1999). The main results that can be drawn are (1) AM associations can reduce damage caused by soilborne plant pathogens (fungal diseases and nematodes); (2) the abilities of AM symbiosis to enhance resistance or tolerance in roots are not equal for the different AM fungi so far tested; (3) protection is not effective for all pathogens; and (4) protection is modulated by soil and other environmental conditions (Azcón-Aguilar and Barea, 1996).

AM Fungi in Biocontrol of Soilborne Fungal Pathogens

Consistent reduction of disease symptoms has been described for soilborne fungal pathogens, such as *Phytophthora* (Ross, 1972; Davis and Menge, 1980; Pozo et al., 1998, 2002; Fusconi et al., 1999; Vigo et al., 2000), *Gaeumannomyces* (Graham and Menge, 1982), *Fusarium* (Zambolim and Schenck, 1983; Caron et al., 1986 a,b,c; Datnoff et al., 1995; St-Arnaud et al., 1995; Dar et al., 1997), *Thielaviopsis* (Baltruschat and Schönbeck, 1972, 1975; Schönbeck and Dehne, 1977), *Cylindrocladiun* (Barnard, 1977; Declerck et al., 2002), *Pythium* (Stewart and Pfleger, 1977; Afek and Menge, 1990; St-Arnaud et al., 1994), *Rhizoctonia* (Zambolin and Schenck, 1983; Abdalla and Abdel-Fattah, 2000; Guillon et al., 2002), *Verticillium* (Davis et al., 1979; Bååth and Hayman, 1983; Matsubara et al., 1995; Demir, 1998; Demir and Onogur, 1999; Karagiannidis et al., 2002) *Aphanomyces* (Bødker et al., 1998), *Macrophomina* (Zambolin and Schenck, 1983; Demir, 1998; Demir and Onogur, 1999). Some interactions between AM fungi and soilborne fungal pathogens, particularly in vegetable and fruit crops, are given in Table 2.1.

It does not mean that AM establishment will be effective against root pathogens under all circumstances. Actually, an AM-induced increase in resistance or decrease in susceptibility requires preformation of AM and extensive development of symbiosis before pathogen attack. The reactions of mycorrhizal and nonmycorrhizal plants to pathogens become distinct (Schönbeck, 1979; Azcón-Aguilar and Barea, 1996).

TABLE 2.1. Some interactions between AM fungi and soilborne plant pathogens in vegetable and fruit crops.

Pathogen	Host plant	AM fungus	Author(s)
Aphanomyces euteiches	Pea	*Glomus intraradices*	Kjøller and Rosendahl, 1996; Bødker et al., 1998
Cylindrocarpon destructans	Peach	*G. aggregatum*	Traquair, 1995
Cylindrocladium spathiphyllii	Banana	Mix of 4 *Glomus* spp.	Declerck et al., 2002
Fusarium oxysporum f.sp. *chrysanthemi*	Carrot	*G. intraradices*	St-Arnaud et al., 1995
F.o. f.sp. *cucumerinum*	Cucumber	*G. mosseae*	Dehne and Schönbeck, 1979
F. o. f.sp. *radicis-lycopercici*	Tomato	*G. intraradices*	Caron et al. 1986 a,b,c; Datnoff et al., 1995
F. solani	Common bean	*G. mosseae*	Dar et al., 1997
Macrophomina pheaseolina	Melon	*G. intraradices*	Demir and Onogur, 1999
Phytophthora cinnamomi	Pineapple	*Glomus* spp.	Guillemin et al., 1994
P. cinnamomi	Wild cherry	*G. mosseae*	Cordier, Trouvelot, et al., 1996
P. fragariae	Strawberry	*G. fasciculatum and g. etunicatum*	Norman et al., 1996
P. nicotianae	Tomato	*G. mosseae*	Cordier, Gianinazzi, and Gianinazzi-Pearson, 1996; Trotta et al., 1996
P. parastica	Tomato	*G. intradices*	Pozo et al., 1999
P. parastica	Tomato	*G. mosseae*	Fusconi et al., 1999; Vigo et al., 2000; Pozo et al.. 2002
P. parastica	Tomato	*Glomus mosseae, Gigaspora roseae*	Pozo et al., 1998
P. parasitica	Sweet orange	*G. constrictus*	Davis and Menge, 1981
P. parasitica	Sweet orange	*G. fasciculatus*	Davis and Menge, 1980
P. parasitica	Sweet orange	*G. intraradices*	Graham and Egel, 1988; Nemec et al., 1996
Pythium ultimum	Onion	*G. intraradices*	Afek and Menge, 1990
P. ultimum	Pepper	*G. intraradices*	Afek and Menge. 1990
Rhizoctonia solani	Bean	*G. intraradices*	Guillon et al., 2002
R. solani	Celery	*G. intraradices*	Nemec et al., 1996
R. solani	Potato	*G. etunicatum, G. intraradices*	
Verticillium albo-atrum	Tomato	*G. mosseae and G. caledonium*	Bååth and Hayman, 1983
V. dahliae	Eggplant	*G. etunicatium, Gigaspora margarita*	Matsubara et al., 1995
V. dahliae	Eggplant	*G. intraradices*	Demir and Onogur, 1999

Although only the root is infected by the AM fungus, the entire plant is influenced. Enhanced resistance of mycorrhizal plants to fungal soilborne diseases was observed in most cases. This was exposed mostly by less host damage, decreased infection (especially in wilt diseases), inhibited spore production, increased fresh weight, increased yield, and so forth (Schönbeck, 1979).

An exchange of nutrients and metabolic products between the symbiotic partners, AM fungi and plant roots, takes place. The AM fungus provides the plant with nutrients, particularly phosphorus, and receives it from organic substances (especially carbohydrates) (Schönbeck, 1979; Smith and Read, 1997). AM fungi are obligate biotrophic organisms that depend on their host. They are usually beneficial for the plant but can cause reduced plant growth and possibly become pathogenic under certain conditions. Beneficial effects on the host can become obvious only after the formation of the AM fungus in the root. In the early stages of colonization, the AM fungus may thus act as a parasite rather than a symbiont (Schönbeck, 1979).

Mechanisms of AM Association on Control of Soilborne Fungal Pathogens

Mechanisms that could account for the protective activity ascribed to AM fungi include enhanced plant nutrition, root damage compensation, competition for nutrients and infection/colonization site, competition for host photosynthates, anatomical or morphological changes in the root system, microbial changes in the mycorrhizosphere, and activation of plant defense mechanisms (Schönbeck, 1979; Linderman, 1994; Azcón-Aguilar and Barea, 1996).

Enhanced Plant Nutrient Status

The obvious AM contribution to reduction of root disease is to increase nutrient uptake, particularly P and other minerals, because AM symbiosis results in more vigorous plants; the plant itself may thus be more resistant or tolerant to pathogen attacks (Linderman, 1994). The evidence to support the improved nutrition idea comes from experiments where effects comparable to AM effects were observed when more P fertilizer was added. Davis (1980) found this type of response on *Thielaviopsis* root rot of citrus where AM plants were larger than nonmycorrhizal plants until the latter were fertilized with additional P. Declerck et al. (2002) suggested a similar effect where AM fungi or added P reduced root rot of bananas caused by *C. spathiphyylii*. In another study to clarify the confusion about the role of enhanced P nutrition in the interaction of AM and root disease incidence,

Graham and Egel (1988) showed that there was no difference between *Phytophthora* root rot levels on mycorrhizal and nonmycorrhizal citrus seedlings fertilized to be of equal size and the P content.

Although in many studies of enhanced nutrition as a mechanism for disease control improved P nutrition could account for the higher tolerance of mycorrhizal plants to fungal root pathogens, there are a number of contradictory reports (Davis et al., 1979; Linderman, 1994). Caron et al. (1986 a,b,c) compared responses between mycorrhizal and nonmycorrhizal tomato plants with a relatively low P threshold requirement to root and crown rot disease caused by *Fusarium oxysporum* f. sp. *radicis-lycopersici* Jarvis & Shoemaker. Added P did not reduce disease response and pathogen populations in the rhizosphere soil of nonmycorrhizal plants, but it did with AM plants, even if plant growth and tissue P content were not different in both treatments. This work suggests the involvement of some mechanism of disease suppression other than improved P uptake.

Whether or not enhanced P uptake by AM fungi was involved either directly or indirectly in disease expression remains controversial. The possibility that enhanced uptake of other mineral elements from soil could be involved has not been explored (Linderman, 1994).

Root Damage Compensation

It has been suggested that AM fungi increase host tolerance to pathogen attack by compensating for the loss of root biomass or function caused by soilborne fungal pathogens (Linderman, 1994). This represents an indirect contribution to biocontrol through the conservation of root system function both by fungal hyphae growing out into soil and increasing the absorbing surface of the roots, and by the maintenance of root-cell activity through arbuscule formation (Azcón-Aguilar and Barea, 1996; Cordier, Gianinazzi, and Gianinazzi-Pearson, 1996).

Competition for Host Photosynthates

Arbuscular mycorrhizal fungi develop intensively inside roots and within the soil by forming an extensive extraradical network, and this helps plants noticeably in exploiting mineral nutrients and water from the soil. Phosphorus is the key element obtained by plants through the symbiosis, and there is plenty of supportive evidence for that. In exchange, mycorrhizal plants provide the fungus with photosynthetic C which is recycled and delivered to the soil via fungal hyphae. The extraradical hyphae of AMF, therefore, act as a direct conduit for host C into the soil and contribute directly to its C pools, bypassing the decomposition process (Jeffries et al.,

2003). As a consequence of this, the amount and activity of other soil biota are stimulated; however, this seems to be a selective phenomenon, since it stimulates in particular the microorganisms having antagonistic activity against soilborne pathogens (Linderman, 2000). It has been proposed that the growth of both AM fungi and root pathogens depend on host photosynthesis; therefore, they compete for the carbon compounds reaching to the roots (Linderman, 1994). When AM fungi have primary access to photosynthates, the higher carbon demand may inhibit pathogen growth.

The reason for this phenomenon is unknown, but this observation clearly indicates that AMF could be useful biological tools for maintaining healthy soil systems (Barea et al., 2002; Jeffries et al., 2003).

Competition for Infection/Colonization Sites

Pioneering research illustrated how fungal root pathogens and AM fungi, though colonizing the same host tissues, usually develop in different root cortical cells, indicating some sort of competition for space (Dehne, 1982; Linderman,1994). When arbuscular mycorrhizae are reported to suppress fungal root disease, they generally must be established and functioning before invasion by pathogens. This has been demonstrated by Stewart and Pfleger (1977) on *Pythium* and *Rhizoctonia* root rot of poinsettia, by Bartschi et al. (1981) on *Phytophthora* root rot of Lawson cypress, and by Rosendahl (1985) on *Aphanomyces* root rot of pea. That this would be the case seems logical, considering both the faster infection rate of most fungal root pathogens compared with AM fungi and the time needed for AM effects on the host physiology to occur (Linderman, 1994).

In addition, other reports have shown that established root infection by various pathogens can reduce the colonization by AM fungi and, therefore, its potential for positive effects on disease incidence or severity (Bååth and Hayman, 1983; Rosendahl, 1985; Afek and Menge, 1990). Cordier, Gianinazzi, and Gianinazzi-Pearson (1996) indicated that *Phytophthora* development in the tomato plants was reduced in AM-colonized and adjacent uncolonized regions of AM root systems, and that in the former, the pathogens did not penetrate arbuscule-containing cells. This means that localized competition occurs, and that even systemic resistance was still induced at some distance from the AM-colonized tissue. Fusconi et al. (1999) also demonstrated that in the interaction of AM fungus *(G. mossea)* and a pathogenic fungus *(P. nicotianae* var. *parasitica* in tomato), AM fungi increased at the root apex site. They also provided protection against infection, as shown by the smaller percentage of necrosis when tomato plants were treated with both the fungi than when they were treated only with *P. parasitica*. The apices do not represent a preferential site of attack and are, therefore, protected by all the mechanisms that limit the spread of *P. parasitica* into the root (Gnavi et al., 1996). The increased diameter of AM roots resulting from larger apices could, in turn, be a protection factor, because a thicker cortex hinders invasion of whole root segments and their necrosis. In fact, it has been shown that *P. parasitica* hyphae rarely occur in zones colonized by AM fungi and are always excluded from arbuscule-containing cells (Cordier et al., 1998).

It is obvious that the effect of AMF on the number of infection loci is a mechanism through which biocontrol of fungal root pathogens (e.g., *P. parasitica*) can occur. This finding will have important implications for future mechanism-led research as well as research into biocontrol applications. For example, for biocontrol applications, these data suggest that the timing of inoculation with AMF and the dynamics of colonization are likely to be very important, with the protective effect dependent on reducing primary and secondary infection by pathogens at the root epidermis, rather than retarding development in planta (Vigo et al., 2000).

Anatomical and Morphological Changes
in the Root Systems

It has been shown that localized morphological effects have occurred in AM roots. For example, Dehne and Schönbeck (1979) indicated increased lignifications of tomato and cucumber root cells of the endodermis and reported that such response accounted for reduced Fusarium wilt (*Fusarium oxysporum* Schlet). Becker (1976) found similar effects studying pink rot of onion [*Pyrenocheta terrestris* (Hans.) Garenz, Walker & Larson]. Dar et al. (1997) also showed that enhanced plant growth, improved nutrient assimilation, and, possibly, a physical barrier probably conferred altered resistance in the plants, as the disease incidence was significantly reduced in plants inoculated with AM fungus.

It has been demonstrated that AM colonization induces remarkable changes in root system morphology as well as in the meristematic and nuclear activities of root cells. This might affect rhizosphere interactions and, particularly, pathogen-infection development. The most frequent consequence of AM colonization is an increase in branching, resulting in a relatively larger proportion of higher-order roots in the root system. However, the significance of this finding for plant protection has not yet been sufficiently considered. In most studies on AM fungi and biocontrol, the roots have not been for anatomical changes. More attention needs to be given, therefore, to root system morphology in the future, because it could modify the infection dynamics of pathogens as well as the pattern of resistance of AM roots to pathogen attack (Azcón-Aguilar and Barea, 1996).

Activation of Plant Defense Mechanism

In most pathogenic interactions, plant resistance is associated with a multiple defense response which might include (1) a hypersensitive response, which is characterized by rapid, localized, and chemical defenses, and death of plant cells surrounding the infection site; (2) structural defensive barriers such as lignin- and hydroxyproline-rich cell wall proteins; and (3) production of new enzymes, which later show antifungal activity (Kapulnik et al., 1996; Guenoune et al., 2001). Current research using molecular biology techniques and immunological and histochemical analyses will probably provide more information about these mechanisms. This methodology may detect substances and reactions elicited only to low levels by AM formation but still in some way involved in plant protection (Azcón-Aguilar and Barea, 1996).

It is likely that AM associations, as agents in biological control, will be acting in more than one mechanism. The activation of specific plant defense mechanisms as a response to AM colonization is an obvious basis for the protective capacity of AM fungi (Azcón-Aguilar and Barea, 1996). Modern analyses of plant microbe interactions suggest that root infection by pathogenic fungi and AM fungi differ at the molecular level. The elicitation by an AM symbiosis of specific plant defense reactions could predispose the plant to an early response to attack by a pathogenic root fungus. However the general conclusion is that only a weak local transient activation of plant defense mechanisms occurs during AM formation (Volpin et al., 1995; Gianinazzi-Pearson et al., 1996).

During their life cycle, plants evolve a number of defense responses elicited by various signals, including those associated with pathogen attack. Among the compounds involved in plant defense studied in relationship to AM formation are phytoalexins, enzymes of the phenylpropanoid pathway, chitinases, β-1,3-glucanases, peroxidases, pathogenesis-related (PR) proteins, callose, hydroxyproline-rich glycoprotein (HRGP), and phenolics (Morandi, 1996; Pozo et al., 1998, 1999, 2002; Guenoune et al., 2001; Garcia-Garrido and Ocampo, 2002). In the early stages of root colonization by AMF, the plant defense response is characterized by a weak and transient activation, but in the later stages (for example, during the arbuscule formation) this activation appears to become stronger, although only in those cells that contain fungal structures (Blee and Anderson, 2001; Garcia-Garrido and Ocampo, 2002). For instance, the use of specific probes in studies of in situ expression revealed that some mRNAs of genes associated with plant defense response specifically accumulated in plant cells containing arbuscules. Likewise, members of different classes of plant defense genes, including the genes encoding hydroxyproline-rich glycoproteins, phenylpropanoid metabolism enzymes, enzymes which are involved in the metabolism of reactive oxygen species and plant hydrolase, have been detected in plant cells containing arbuscules (Blee and Anderson, 2001; Garcia-Garrido and Ocampo, 2002).

Phytoalexins, low-molecular-weight toxic compounds usually accumulating with pathogen attack and released at the sites of infection, are not detected during the first stages of AM formation but can be detected in the later stages of symbiosis (Morandi et al., 1984; Morandi, 1996). The main phytoalexins of soybean glyceollin could not be detected during the first 30 days of AM inoculation, but there was an evident increase in the compound in roots infected by *R. solani* (Azcón-Aguilar and Barea, 1996). On the other hand, the level of the phytoalexins increased transiently during the early stages of AM colonization in plants, but decreased to very low levels during the early stages of symbiosis development (Harrison and Dixon, 1993; Guenoune et al., 2001; Garcia-Garrido and Ocampo, 2002). Results obtained on the relationship between phytoalexins and mycorrhizal infection lead to the conclusion that mycorrhizal fungi are able to cause an accumulation of phytoalexins in the roots of their host plants. The magnitude of plant reaction is always low, compared with pathogenic interactions. Although elicitation is transitory and suppressed in the mature stage of infection for some specific phytoalexins, other phenolics with antimicrobial properties are consistently higher in mycorrhizal roots. For this reason, mycorrhizal roots are possibly a less suitable host for pathogens than for mycorrhizal fungi. Conjugate isoflavanoids that are not toxic and that accumulate at higher levels in mycorrhizal roots could serve as the storage form of phytoalexins. These could be quickly hydrolyzed in the case of a pathogen attack (Morandi, 1996).

Increased levels of phytoalexins in mature mycorrhizal roots could be elicited by the release of fungal cell walls in senescent arbuscules. Several steps of mycorrhizal development may, therefore, be explained in relation to phenolic compounds. These steps are as follows: (1) the primary contact of the AM fungus with the roots that induce a specific defense mechanism; (2) a specific active suppression of phytoalexins biosynthesis; and (3) an elicitation of phytoalexins in aging arbuscules due to the disintegration of the arbuscule and the release of fungal cell wall components (Morandi, 1996).

Much as in the case of phytoalexins, there seems to be low activation of the phenylpropanoid-related enzymes. In particular, both phenylalanine ammonium-lyase (PAL), the first of the phenylpropanoid pathway, and chalcone isomerase, the second enzyme specific for flavonoid/isoflavonoid biosynthesis, increase in amount and activity during early colonization of plant roots by AM fungus *(G. intraradices)*, but then decrease sharply to levels equal to or below those in noninoculated controls (Lambais and Mehdy, 1993; Volpin et al., 1994; Azcón-Aguilar and Barea, 1996; Kapulnik et al., 1996). All available data suggest an activation of phenylpropanoid metabolism in mycorrhizal roots, characterized by the weak, localized (preferably in cells containing arbuscules), and uncoordinated induction of genes together with the accumulation of the phytoalexin product, some of which are found at high levels (Garcia-Garrido and Ocampo, 2002). Although no evidence exists supporting a specific role for flavonoids/isoflavonoids in AM symbiosis, there are several results suggesting that flavonoids can stimulate AM symbiosis in a way similar to the rhizobial symbiosis (Harrison, 1999).

Chitinases and β-1,3 glucanose are synergistically induced during an attack by fungal pathogens and by fungal elicitors. Their induction is generally considered to be part of a nonspecific defense response initiated in plants after pathogen attack, but is also a consequence of various physical, chemical, and environmental stresses (Pozo et al., 1998). However, the induction of new root-acidic chitinase isoforms during AM symbiosis appears to be a specific response, since differential induction of chitinase isoforms after symbiotic or pathogenic fungal interactions has been reported in various plants (Dumas-Gaudot, Furlan et al., 1992; Dumas-Gaudot, Grenier, et al., 1992; Dumas-Gaudot et al., 1996). This was later confirmed in tomato roots during symbiosis with *G. mosseae* after *P. parasitica* attack (Pozo et al., 1996). In another work a similar induction of specific acidic chitinase isoforms in tomato roots has been shown to be a consequence of interactions with a different AM fungal species *(G. intraradices)*, and even with a species belonging to a different genus *(G. rosea)* (Pozo et al., 1998). Pozo et al. (2002) compared the abilities of two AMF *(G. intraradices* and *G. mosseae)* to induce local or systemic resistance to *P. parasitica* in tomato roots, using a split root experimental system. *G. mosseae* was effective in reducing disease symptoms produced by *P. parasitica* infection, and evidence points to a combination of local and systemic mechanisms being responsible for this bioprotector effect. The biochemical analysis of different plant defense-related enzymes showed a local induction of mycorrhiza-related new isoforms of the hydrolytic enzymes, chitinase, chitosonase, and β-1,3 glucanase, as well as superoxide dismutase, an enzyme involved in cell protection against oxidative stress. According to the results of this study, systemic alterations of the activity of some of the constitutive isoforms were also observed in the nonmycorrhizal roots of mycorrhizal plants. Total chitinase activity is higher in AM roots than in a nonmycorrhizal control. A decrease in β-1,3-glucanese activity has also been reported at specific stages during mycorrhiza development (Lambais and Mehdy, 1993). The decreases were accompanied by differential reductions in the levels of mRNAs encoding for different endochitinase and endoglucanase isoforms (Azcón-Aguilar and Barea, 1996; Pozo et al., 1998; Garcia-Garrido and Ocampo, 2002). These observations suggest a systemic suppression of the defense reaction when the AM symbiotic interactions begin to take place. However, mRNAs encoding chitinases and β-1,3 glucanoses have been reported to accumulate in and around cells containing arbuscules (Garcia-Garrido and Ocampo, 2002). This suggests localized induction of specific defense-related genes that might be involved in the regulation of AM development by controlling intraradical fungal colonization (Lambais, 2001).

Wall-bound peroxidase activity that later decreases has been detected during the initial stage of AM colonization (Azcón-Aguilar and Barea, 1996; Morandi, 1996). However peroxidase activity associated with epidermal and hypodermal cells increased in mycorrhizal roots, a process that can contribute to higher resistance to certain root pathogens. Peroxidase activity was not detectable in cells containing intracellular arbuscular hyphae (Spanu and Bonfante-Fasolo, 1988). Thus, in contrast to pathogen infection, peroxidases do not appear to be associated with plant control of AM fungal development in the root cortex or linked to arbuscule senescence and death within host cells (Garcia-Garrido and Ocampo, 2002).

PR proteins and HRPG are only synthesized locally and in very low amounts (Azcón-Aguilar and Barea, 1996). In particular, the PR-b_1 proteins are only synthesized in cells containing living arbuscules (Gianinazzi-Pearson et al., 1992). In a study, Vierheilig et al. (1995) reported that the constitutive expression of different acidic isoforms of PR proteins of tobacco did not affect the time course or final level of colonization by the AM fungus, with the exception of transgenic plants expressing PR-2. Actually, the fact that none of the PRs tested present root colonization by the mycorrhizal fungi suggests that they were not effective in degrading the AM fungal cell wall or a compartmental separation between the fungus and its host (Kapulnik et al., 1996).

The strong accumulation of HRPG infections by pathogenic fungi contrasts with the weak reaction observed during AM colonization, although these molecules were regularly distributed around the arbuscular hyphae (Azcón-Aguilar and Barea, 1996).

One of the best ways to assess the implication of phenolics in mycorrhizally improved resistance or tolerance of plant to soilborne pathogens is to compare the accumulation of phenolic compounds in mycorrhizal and nonmycorrhizal roots when challenged by pathogens (Dehne, 1982; Morandi, 1996). Unfortunately, very few studies have been published on this subject, and the role of phenolics on biocontrol by AM is still a matter of speculation. Dehne and Schönbeck (1979) have studied the influence of the AM fungus *Glomus mosseae* and *F. oxysporum*, and they detected that the combination of them increased the PA and β glucosidase activities and total phenol contents of roots more than the mycorrhizal fungus or the pathogen alone. In addition to the increased lignifications of endodermis and stele by mycorrhizal infection, these findings present a good argument in favor of the hypothesis that enhanced resistance to *Fusarium* wilt due to mycorrhiza was mediated by induced phenolic metabolism in roots. In spite of the fact that phenolic compounds have not been detected in significant amounts, increased lignifications of root endodermal cells induced by AM colonization has been suggested (Dehne, 1982). This deserves further investigation, as the process would make penetration of pathogens into root tissue more difficult (Azcón-Aguilar and Barea, 1996).

Finally, only weak responses to AM infection have been observed in some activities, such as lignification, production of phytoalexins and peroxidases, and the expression of genes coding for PR proteins, indicating that AM fungi do not elicit typical defense responses. However, these compounds could sensitize the root to pathogens and improve mechanisms of defense to subsequent pathogen infection (Azcón-Aguilar and Barea, 1996). This hypothesis was supported by the result of Benhamou et al. (1994). These researchers compared the responses of AM and nonmycorrhizal transformed carrot roots to the infection of *F. oxysporum* f. sp. *chrysanthemi*. In mycorrhizal roots, the growth of pathogens was usually restricted to the epidermis and cortical tissues, whereas in mycorrhizal roots, pathogens reached a higher development and infected even the vascular stele. The *Fusarium* hyphae inside mycorrhizal roots exhibited a high level of structural disorganization, probably induced by a strong reaction of the host cells characterized by the massive accumulation of phenolic-like compounds and the production of hydrolytic enzymes such as chitinases. This strong reaction was not induced by nonmycorrhizal roots, suggesting that the activation of plant defense responses by AM formation provides a certain protection against the fungal pathogens.

The results, although they need to confirmed using entire plants, clearly indicate that AM infection makes the root more responsive to pathogen attack and promotes a quicker and stronger reaction against pathogens (Azcón-Aguilar and Barea, 1996; Morandi, 1996; Garcia-Garrido and Ocampo, 2002).

Microbial Changes in the Mycorrhizosphere

While any of the above mechanisms, or combinations thereof, could be involved in AM suppression of root diseases, the thing that should be more carefully considered is the AM alteration of rhizosphere populations of antagonists (Linderman, 1994). There is an evidence that microbial shifts occur in the mycorrhizosphere and that the resulting microbial equilibrium could influence the growth and health of plants. Although this effect has not been specifically evaluated as a mechanism for AM-associated biological control, there are indications that such a mechanism does operate (Azcón-Aguilar and Barea, 1992, 1996; Linderman, 1994).

The concept of mycorrhizosphere implies that mycorrhizae significantly influence the microflora of the rhizosphere by altering root physiology and exudation (Linderman, 1994; Hodge, 2000; Barea et al., 2002). In addition, extraradical hyphae of AM fungi provide a physical or nutritional substrate for bacteria.

Meyer and Linderman (1986) observed no alteration in the total number of bacteria or actinomycetes isolated from the rhizosphere of *Zea mays* and *Trifolium subterraneum* L. colonized by the AM fungus *G. fasciculatum*. However, there was a change in the functional groups of these organisms, including more facultative anaerobic bacteria in the rhizoplane of AM-colonized *T. subterraneum*, but fewer fluorescent pseudomonades and chitinase-producing actinomycetes in the rhizosphere of AM-colonized *Z. mays*. The total number of bacteria isolated from the rhizoplane of both *T. subterraneum* and *Z. mays* increased as a result of AM colonization although total numbers of actinomycetes were unaffected. In addition, leachates from *Z. mays* rhizosphere soil reduced production of zoospores and sporangia by *P. cinnamomi* when colonized by *G. fasciculatum* unlike nonmycorrhizal *Z. mays* rhizosphere leachates, indicating a potential mechanism by which AM colonization may aid pathogen resistance. Secilia and Bagyaraj (1987) isolated more pathogen-antagonistic actinomycetes from the rhizosphere of AM plants than from nonmycorrhizal controls, an effect that also depended on the AM fungus involved. Furthermore, Caron (1989) indicated a reduction in *Fusarium* populations in the mycorrhizosphere soil of tomatoes and a corresponding reduction in root rot in AM plants relative to nonmycorrhizal plants. This study also showed that reduced disease incidence was independent of the level of P nutrition but dependent on the nature of the growth substrate. A study by Bartschi et al. (1981) found protection of host roots against *P. cinnamomi* root rot when preinoculated with a mixture of AM pot culture inoculum. The authors concluded that a mixture of AM fungi was more effective than a single fungus, but effects could also have been due to building up of antagonists in the pot cultures, as indicated by Secilia and Bagyaraj (1987). These studies merit further attention in order to elucidate how microbiota-mediated changes are involved in biological control by AM associations.

The prophylactic ability of some AM fungi could be exploited in association with other microorganisms known to be antagonistic to fungal soilborne pathogens that are being used as biological control agents, such as *Rhizobium* spp., PGPR, and biocontrol agents (Kloepper et al., 1991; Linderman, 1994).

The attribute "rhizobacteria" refers to the ability of some soil bacteria to colonize "aggressively" the root-soil interface, where they establish and maintain for a time a considerable number of cells (Azcón-Aguilar and Barea, 1997). These bacteria, which can promote plant growth by means of several mechanisms, are now commonly designated by the acronym PGPR (Kloepper et al., 1991; Linderman, 1992). The activity of PGPR is related to effect on seedling emergence, root formation, nutrient cycling, and nodulation promotion (Kloepper et al., 1991). However, one of the most remarkable effects of PGPR on plant growth issues from their ability as biological control agents for root pathogens, particularly fungal pathogens (Kloepper, 1992; Linderman, 1994; Edwards et al., 1998). Their effects are mainly due to the capacity of PGPR to colonize the rhizosphere soil and rhizosphere and to diminish the population of deleterious organism (Azcón-Aguilar and Barea, 1996). They do this mainly by the production of antibiotics, hormones, and cyanic acid (HCN), and by competition for colonization cites and carbon components (Kloepper, 1992; Leeman et al., 1996). Leeman et al. reported that systemically induced resistance has been proposed to be a mechanism of disease suppression by PGPR. In addition to their effect on plant performance through indirect biological control, some PGPR can directly promote root shoot, shoot growth, nodule formation by rhizobium (nodulation promoting rhizobacteria, NPR), seedling emergence (EPR) (Kloepper et al., 1991), and, in some cases, mycorrhiza formation (MPR, or, as they are usually known, mycorrhiza-helper-bacteria, MHB) (Garbaye, 1994).

AM fungi and PGPR may cooperate in several ways, including improvement in plant rooting, enhancement of plant growth and nutrition, biological control of root pathogens, and enhancement of nodulation in the case of legumes (Linderman, 1992). In addition, both groups of microorganisms (the saprophyte PGPR and the symbiont AMF) are able to interact with each other and improve their establishment in the rhizosphere (Paulitz and Linderman, 1989; Linderman, 1992; Azcón-Aguilar and Barea, 1996; Requena et al., 1997; Barea et al., 1998, 2002; Edwards et al., 1998; Budi et al., 1999; Hodge, 2000; Söderberg et al., 2002; Sood, 2003).

The prophylactic ability displayed by some AM fungi could be exploited through their cooperation with other rhizosphere microorganisms that show antagonistic activity against pathogens and are being used as biological control agents (Linderman, 1994). Among the microorganisms catalogued as antagonists to fungal pathogens, there are fungi such as *Trichoderma* spp. (Datnoff et al., 1995) and *Gliocladium* spp. (Paulitz and Linderman, 1991), and bacteria, which are actually PGPR, such as *Pseudomonas* spp. and *Bacillus* spp. (Kloepper et al., 1991; Linderman, 1994). It has been shown that microbial antagonists toward fungal pathogens do not necessarily have any detrimental effect against AM fungi (Azcón-Aguilar and Barea, 1997).

Some examples are available concerning horticultural plants: *Glomus* + *Trichoderma* + wilt pathogenic fungi (Calvet et al., 1990); *Glomus* + *Trichoderma* + *Fusarium* crown and root rot in tomato (Datnoff et al., 1995); and *Glomus* + *Trichoderma* + *Pythium,* in marigòld (Calvet et al., 1993).

The information related to interactions between biological control agents and soil fungi, whether pathogenic or mycorrhizal, is summarized by Azcón-Aguilar and Barea (1997).

It is noteworthy that microbial antagonists of fungal pathogens can improve AM formation (Linderman, 1994). The management of these interactions that improve plant growth and health, in integrated approach, should be one of the main objectives of sustainable agriculture (Bethlenfalvay and Linderman, 1992).

Current interest in this topic has led to research on the manipulation of soil microorganisms, particularly with regard to improving production, formulation, and practical use of efficient microbial inoculants. Recent developments in molecular biological techniques and the application of novel biotechnical approaches are facilitating a more accurate exploration of the natural diversity of soil microorganisms for the isolation of new strains and the generation of genetically modified superior rhizobacteria strains. Enhanced microbial inoculants may become available as biofertilizer or for the biocontrol of plant disease. However, the existing barriers to growing AM fungi in pure culture are still holding back a parallel AM inoculum development and application (Azcón-Aguilar and Barea, 1996).

CONCLUSIONS

Arbuscular mycorrhizal symbiosis must be considered an essential factor for promoting plant health and productivity. With few exceptions, crop plants have AM fungi, but the degree of root colonization by AMF and the effects of symbiosis may vary, depending on the overall interaction among host, symbiont, and environment. In most cases, it is evident that AM significantly change the physiology and chemical constituents of the host, the pattern of root exudation, and the microbial composition of mycorrhizosphere soil. These changes can greatly influence the growth and health of plants, in part due to the biological suppression of plant diseases (Linderman, 1994).

As is known, AM can increase plant tolerance to a number of stress factors, including high salt levels, toxicities with minimized operations, heavy metals, and mineral imbalances. Disease suppression may be the result of reduction of these environmental stresses that may limit plant growth and predispose the plants to infection by opportunistic pathogens. However, more important are the specific morphological and physiological changes that directly or indirectly result in reduced incidence and severity of plant diseases in AM plants compared with nonmycorrhizal plants.

Although it is difficult to get practical results because of the complexity of the microorganism-soil-plant system and the decisive influence of prevailing environmental conditions it may, nevertheless, be possible to find the right combination of factors to exploit the prophylactic ability of AM fungi. So far, examples of successful practical application are scarce (Linderman, 1994). Although further research is needed, existing knowledge suggests that management recommendations for the biological control of target diseases in sustainable agrosystems, particularly with nursery and horticultural crops (Azcón-Aguilar and Barea, 1997), could be made in the future. Appropriate AM fungi must be used, preferably in association with other pathogens of antagonistic members of soil microbiota (Azcón-Aguilar and Barea, 1996). It is also important to improve inoculum production techniques for the proper application of AM biotechnology in commercial horticultural and nursery plant production systems. The possible role of AMF in the biological control of soilborne fungal pathogens must also be considered and exploited in plant breeding programs aiming to select pathogen-resistant cultivars.

REFERENCES

Abdalla, M.E. and G.M. Abdel-Fattah. (2000). Influence of the endomycorrhizal fungus *Glomus mosseae* on the development of peanut pod rot disease in Egypt. *Mycorrhiza* 10: 29-35.

Afek, U. and J.A. Menge. (1990). Effect of *Pythium ultimum* and metalaxyl treatments on root length and mycorrhizal colonization of cotton, onion, and pepper. *Plant Disease* 74:117-120.

Andrade, G., R.G. Linderman, and G.J. Bethlenfalvay. (1998). Bacterial associations with the mycorrhizosphere and hyphosphere of the arbuscular mycorrhizal fungus *Glomus mossea. Plant and Soil* 202:79-87.

Azcón-Aguilar, C. and J.M. Barea. (1992). Interactions between mycorrhizal fungi and other rhizosphere microorganisms. In *Mycorrhizal Functioning: An Integrative Plant-Fungal Process,* ed., M.J. Allen. Chapman and Hall, New York, pp. 163-198.

Azcón-Aguilar, C. and J.M. Barea. (1996). Arbscular mycorrhizas and biological control of soil-borne plant pathogens: An overview of the mechanisms involved. *Mycorrhiza* 6: 457-464.

Azcón-Aguilar, C. and J.M. Barea. (1997). Applying mycorrhiza biotechnology to horticulture: Significance and potentials. *Scientia Horticulturae* 68: 1-24.

Bååth, E. and D.S. Hayman. (1983). Plant growth responses to vesicular-arbuscular mycorrhiza: XIV. Interactions with *Verticillium* wilt on tomato plants. *New Phytologist* 95: 419-426.

Bagyaraj, D.J. (1984). Biological interactions between mycorrhizal fungi and other rhizosphere microorganisms. In *VA Mycorrhiza*, eds., C.L. Powell, and D.J. Bagyaraj. CRC, Boca Raton, FL, pp. 131-153.

Baltruschat, H. and F. Schönbeck. (1972). The influence of endotrophic mycorrhiza on the infestation of tobacco by *Thielaviopsis basicola*. *Phytopathologische Zeitschrift* 84:172-188.

Baltruschat, H. and F. Schonbeck. (1975). Studies on the influence of endotrophic mycorrhiza on the infection of tobacco by *Thielaviopsis basicola*. *Phytopathdgische Zeitschrift* 84:172-188.

Barea, J.M., G. Andrade, V. Bianciotto, D. Dowling, S. Lohrke, P. Bonfante, F. O'Gara, and C. Azcón-Aguilar. (1998). Impact on arbuscular mycorrhiza formation of *Pseudomonas* strains used as inoculants for biocontrol of soil-borne fungal plant pathogens. *Applied and Environmental Microbiology* 64: 2304-2307.

Barea, J.M., R. Azcón, and C. Azcón-Aguilar. (1993). Mycorrhiza and crops. In *Advances in Plant Pathology*, Vol. 9, *Mycorrhiza: A Synthesis*, ed., I. Tommerup. Academic Press, London, pp. 167-189.

Barea, J. M., R. Azcón, and C. Azcón-Aguilar. (2002). Mycorrhizosphere interactions to improve plant fitness and soil quality. *Antonie van Leeuwenhoek* 81: 343-351.

Barnard, E.L. (1977). The mycorrhizal biology of *Liriodenron tulipifera* L. and its relationship to *Cylindrocladium* root rot. PhD dissertation, Duke University, Durham, NC.

Bartschi, H., V. Gianinazzi-Pearson, and I. Vegh. (1981). Vesicular-arbuscular mycorrhiza formation and root rot disease *(Phytophthora cinnamoi)* development in *Chamaecyparis lawsoniana*. *Phytopathologische Zeitschrift* 102: 213-218.

Becker, W.N. (1976). Quantification of onion vesicular-arbuscular mycorrhizae and their resistance to *Pyrenochaeta terrestris*. PhD dissertation, University. of Illinois, Urbana (Diss. Abstr. 76-24041).

Benhamou, N., J.A. Fortin, C. Hamel, M. St-Arnaud, and A. Shatilla. (1994). Resistance responses of mycorrhizal Ri T-DNA-transformed carrot roots to infection by *Fusarium oxysporum* f. sp. *Chrysanthemi*. *Phytopathology* 84: 958-968.

Bethlenfalvay, G.J. and R.G. Linderman (eds.). (1992). *Mycorrhizae in Sustainable Agriculture*. ASA Special Publication No. 54, Madison, WI.

Blee, K.K and A.J. Anderson. (2001). Defense responses in plants to arbuscular mycorrhizal fungi. In *Current Advances in Mycorrhizae Research*, eds. G.P. Podila and D.D. Douds Jr. APS Press, St. Paul, MN, pp. 45-59.

Bødker, L., R. Kjøller, and S. Rosendahl. (1998). Effect of phosphate and the arbuscular mycorrhizal fungus *Glomus intraradices* on disease severity of root rot of peas *(Pisum sativum)* caused by *Aphanomyces euteiches*. *Mycorrhiza* 8: 169-174.

Brundrett, M. (1991). Mycorrhizas in natural ecosystem. *Advances in Ecological Research* 21: 171-313.

Budi, S.W., D. Van Tuinen, G. Martinotti, and S. Gianinazzi. (1999). Isolation from the *Sorghum bicolor* mycorrhizosphere of a bacterium compatible with arbuscular mycorrhiza development and antagonistic towards soil-borne fungal pathogens. *Applied and Environmental Mycrobiology* 65: 5148-5150.

Calvet, C., J. Pera, and J.M. Barea. (1990). Interactions of *Trichoderma* spp. with *Glomus mosseae* and two wilt pathogenic. *Agricultural Ecosystems and Environment* 29: 59-65.

Calvet, C., J. Pera, and J.M. Barea. (1993). Growth response of marigold (*Tagetes erecta* L.) to inoculation with *Glomus mosseae*, *Trichoderma aureoviride* and *Pythium ultimum* in a peat-perlite mixture. *Plant Soil* 148: 1-6.

Caron, M. 1989. Potential use of mycorrhizae in control of soil-borne diseases. *Canadian Journal of Plant Pathology* 11: 177-179.

Caron, M., J.A. Fortin, and C. Richard. (1986a). Effect of *Glomus intraradices* on infection by *Fusarium oxysporum* f. sp. *radicis-lycopersici* in tomatoes over a 12-week period. *Canadian Journal of Botany* 4: 552-556.

Caron, M., J.A. Fortin, and C. Richard. (1986b). Effect of phosphorus concentration and *Glomus intraradices* on Fusarium crown and root rot of tomatoes. *Phytopathology* 76: 942-946.

Caron, M., J.A. Fortin, and C. Richard. (1986c). Effect of pre-infestation of the soil by a vesicular-arbuscular mycorrhizal fungus, *Glomus intraradices*, on *Fusarium* crown and root rot of tomatoes. *Phytoprotection* 67: 15-19.

Cordier, C., S. Gianinazzi, and V. Gianinazzi-Pearson. (1996). Colonization patterns of root tissues by *Phytophthora nicotianae* var. *parasitica* related to reduced disease in mycorrhizal tomato. *Plant and Soil* 185: 223-232.

Cordier, C., M.J. Pozo, J.M. Barea, S. Gianinazzi, and V. Gianinazzi-Pearson. (1998). Cell defense responses associated with localized and systemic resistance to *Phytophthora parasitica* induced in tomato by an arbuscular mycorrhizal fungus. *Molecular Plant-Microbe Interactions* 11: 1017-1028.

Cordier, D.A., A. Trouvelot, S. Gianinazzi, and V. Gianinazzi-Pearson. (1996). Arbuscular mycorrhiza technology applied to micropropagated *Prunus avium* and to protection against *Phytophthora cinnamomi*. *Agronomie* 16: 679-688.

Dar, G. H., M.Y. Zargar, and G.M. Beigh. (1997). Biocontrol of *Fusarium* root rot in the common bean (*Phaseolus vugaris* L.) by using simbiotic *Glomus moseae* and *Rhizobium leguminosarum*. *Microbial Ecology* 34: 74-80.

Datnoff, L.E., S. Nemec, and K. Pernezy. (1995). Biological control of *Fusarium* crown and root rot of tomato in Florida using *Trichoderma harzianum* and *Glomus intraradices*. *Biological Control* 5: 427-431.

Davis, R.M. (1980). Influence of *Glomus fasciculatus* on *Thielaviopsis basicola* root rot of citrus. *Plant Disease* 64: 839-840.

Davis, R.M. and J.A. Menge. (1980). Influence of *Glomus fasciculatus* and soil phosphorus on *Phytophthora* root rot of citrus. *Phytopathology* 70: 447-452.

Davis, R.M. and J.A. Menge. (1981). *Phytophthora parasitica* inoculation and intensity of vesicular-arbuscular mycorrhizae in citrus. *New Phytologist* 87: 705-715.

Davis, R.M., J.A. Menge, and D.C. Erwin. (1979). Influence of *Glomus fasciculatus* and soil phosphorus on *Verticillium* wilt of cotton. *Phytopathology* 69: 453-456.

Declerck, S., M.J. Risede, G. Rufyikiri, and B. Delvaux. (2002). Effects of arbuscular mycorrhizal fungi on severity of root rot of bananas caused by *Cylindrocladium spathiphylli. Plant Pathology* 51: 109-115.

Dehne, H.W. (1982). Interaction between vesicular-arbuscular mycorrhizal fungi and plant pathogens. *Phytopathology* 72: 1115-1119.

Dehne, R.W. and F. Schönbeck. (1979). Untersuchungen zum Einfluss der endotrophen Mycorrhiza auf Pflanzenkrankheiten: II. Phenolstoffwechsel und Lignifizierung. (The influence of endotrophic mycorrhiza on plant diseases: II. Phenol-metabolism and lignification.) *Phytopathologische Zeitschrift* 95: 210-216.

Demir, S. (1998). Studies on the formation of vesicular-arbuscular mycorrhizae (VAM) in some culture plants and its role on growth and resistance against plant pathogens. PhD thesis, University of Yuzuncu Yil, Van, Turkey.

Demir, S. and E. Onogur. (1999). An investigation on the use of vesicular- arbuscular mycorrhizae (VAM) (*Glomus intraradices* Schenck & Smith) against some root diseases in the ecological agriculture strategy. In *Proceedings of the National Symposium on Ecological Agriculture*, Izmir, Turkey, pp. 123-130.

Dumas-Gaudot, E., V. Furlan, J. Grenier, and A. Asselin, (1992). New acidic chitinase isoforms induced in tobacco roots by vesicular-arbuscular mycorrhizal fungi. *Mycorrhiza* 1: 133-136.

Dumas-Gaudot, E., J. Grenier, V. Furlan, and A. Asselin, (1992). Chitinase, chitosanase and β-1,3-glucanase activities in *Allium* and *Pisum* roots colonized by *Glomus* species. *Plant Science* 84: 17-24.

Dumas-Gaudot, E., S. Slezack, B. Dassi, M.J. Pozo, V. Gianinazzi-Pearson, and S. Gianinazzi (1996). Plant hydrolytic enzymes (chitinases and β-1,3-glucanases) in root reactions to pathogenic and symbiotic microorganisms. *Plant and Soil* 185: 211-221.

Edwards, S.G., E. Young, J.P.W. Alastair, and A.H. Fitter. (1998). Interactions between *Pseudomonas fluorescens* biocontrol agents and *Glomus mosseae,* an arbuscular mycorrhizal fungus, within the rhizosphere. *FEMS Microbiology Letters* 166: 297-303.

Fusconi, A., E. Gnavi, A. Trotta, and G. Berta. (1999). Apical meristems of tomato roots and their modifications induced by arbuscular mycorrhizal and soil-borne pathogenic fungi. *New Phytologist* 142: 505-516.

Garbaye, J. (1994). Helper bacteria: A new dimension to the mycorrhizal symbiosis. *New Phytologist* 128: 197-210.

Garcia-Garrido, J. M. and J.A. Ocampo. (2002). Regulation of the plant defence response in arbuscular mycorizzal symbiosis. *Journal of Experimental Botany* 53: 1377-1386.

Gianinazzi, S., A. Trouvelot, and V. Gianinazzi-Pearson. (1990). Role and use of mycorrhizas in horticultural crop production. In *Proceedings of the 23rd International Horticulture Conference,* Florence, Italy, pp. 25-30.

Gianinazzi-Pearson, V., A. Gollotte, C. Cordier, and S. Gianinazzi. (1996). Root defense responses in relation to cell and tissue invasion by symbiotic microorganisms: Cytological investigations. In *Histology, Ultrastructure and Molecular Cy-*

tology of Plant-Microorganism Interactions, eds., M. Nicole and V. Gianinazzi-Person. Kluwer Academic, Dordrecht, the Netherlands, pp. 177-191.

Gianinazzi-Pearson, V., A. Tahiri-Alaoui, J.F. Antoniw, S. Gianinazzi, and E. Dumas-Gaudot. (1992). Weak expression of to be pathogenesis related PR-b$_1$ gene and localization of related protein during symbiotic endomycorrhizal interact roots. Endocytobiosis Cell Research 8: 177-185.

Gnavi, E., A. Fusconi, C. Cordier, V. Gianinazzi-Pearson, and G. Berta. (1996). Possible involvement of root apical meristems of mycorrhizal tomato plants in resistance to the pathogenic fungus Phytophthora nicotianae var. parasitica. Giornale Botanico Italiano 130: 934-937.

Graham, J.H. and D.S. Egel. (1988). Phytophthora root rot development on mycorrhizal and phosphorus-fertilized non-mycorrhizal sweet orange seedlings. Plant Disease 72: 611-614.

Graham, J.H. and J.A. Menge. (1982). Influence of vesicular-arbuscular mycorrhizae and soil phosphorus on take-all disease of wheat. Phytopathology 72: 95-98.

Guenoune, D., S. Galili, D.A., Phillips, H. Volpin, I. Chet, Y. Okon, and Y. Kapulnik. (2001). The defense response elicited by the pathogen Rhizoctonia solani is suppressed by colonization of the AM-fungus Glomus intraradices. Plant Science 160: 925-932.

Guillemin, J.-P., S. Gianinazzi, V. Gianinazzi-Pearson, and J. Marchal. (1994). Contribution of arbuscular mycorhizas to biological protection of micropropagated pineapple (Ananas comosus (L.) Merr) against Phytophthora cinnamomi Rands. Agricultural Science in Finland 3: 241-251.

Guillon, C., M. St-Arnaud, C. Hamel, and S.H. Jabaji-Hare. (2002). Differential and systemic alteration of defence-related gene transcript levels in mycorrhizal bean plants infected with Rhizoctonia solani. Canadian Journal of Botany 80: 305-315.

Harley, J.L. and S.E. Smith (eds.). (1983). Mycorrhizal Symbiosis. Academic Press, New York.

Harrison, M. (1999). Molecular an cellular aspects of the arbuscular mycorrhizal symbiosis. Annual Review of Plant Physiology and Plant Molecular Biology 50: 361-389.

Harrison, M.J. and R.A. Dixon. (1993). Isoflavonoid accumulation and expression of defense gene transcripts during the establishment of vesicular-arbuscular mycorrhizal associations in roots of Medicago truncatula. Molecular Plant Microbe Interactions 6: 643-654.

Hodge, A. (2000). Mycrobial ecology of the arbuscular mycorrhiza. FEMS Microbiology Ecology 32: 91-96.

Jeffries, P., S. Gianinazzi, S. Perotto, K. Turnau, and J.M. Barea. (2003). The contribution of arbuscular mycorrhizal fungi in sustainable maintenance of plant health and soil fertility. Biology and Fertility of Soils 37: 1-16.

Kapulnik, Y., H. Vopin, H. Itzhaki, D. Ganon, S. Galili, R. David, O. Shaul, Y. Elda, I. Chet, and Y. Okon. (1996). Suppression of defence responses in mycorrhizal alfalfa and tobacco roots. New Phytologist 113: 59-64.

Karagiannidis, N., F. Bletsos, and N. Stavropoulos. (2002). Effect of Verticillium wilt (Verticillium dahliae Kleb.) and mycorrhiza (Glomus mosseae) on root colonization, growth and nutrient uptake in tomato and eggplant seedlings. *Scientia Horticulturae* 94: 145-156.

Kjøller, R. and S. Rosendahl. (1996). The presence of the arbuscular mycorrhizal fungus *Glomus intraradices* influences enzymatic activities of the root pathogen *Aphanomyces euteiches* in pea roots. *Mycorrhiza* 6: 487-491.

Kloepper, J.W. (1992). Plant growth-promoting rhizobacteria as biological control agents. In *Soil Microbial Ecology Applications in Agriculture, Forestry and Environmental Management,* eds. F. Blaine and J. Metting. Dekker, New York, pp. 255-274.

Kloepper, J.W., R.M. Zablotowick, E.M. Tipping, and R. Lifshitz. (1991). Plant growth promotion mediated by bacterial rhizosphere colonizers. In *The Rhizosphere and Plant Growth,* eds. D.L. Keister and P.B. Cregan. Kluwer Academic, Dordrecht, the Netherlands, pp. 315-326.

Lambais, M.R. (2001). Regulation of plant defense-related genes in arbuscular mycorrhizae. In *Current Advances in Mycorrhizae Research,* eds. G.P. Podila and D.D. Douds Jr. PS Press, St. Paul, MN, pp. 27-44.

Lambais, M.R. and M.C. Mehdy. (1993). Suppression of endochitinase, β-1,3-endoglucanase and chalcone isomerase expression bean vesicular-arbuscular mycorrbizal roots under different soil phosphate conditions. *Molecular Plant Microbe Interactions* 6: 75-83.

Leeman, M., F.M. Denouden, J.A. Valpelt, C. Cornelissen, A. Matamalgarros, P. Bakker, and B. Schippers. (1996). Suppression of Fusarium-wilt of radish by co-inoculation of fluorescent *Pseudomonas* spp. and root colonizing fungi. *European Journal of Plant Pathology* 102: 21-31.

Linderman, R.G. (1988). Mycorrhizal interactions with the rhizosphere microflora: The mycorrhizosphere effect. *Phytopathology* 78: 366-371.

Linderman, R.G. (1992). Vesicular-arbuscuiar mycorrhizae and soil, microbial interactions. In: *Mycorrhizae in Sustainable Agriculture,* eds. G.J. Bethlenfalvay and R.G. Linderman. ASA Special Publication No. 54, Madison, WI, pp. 45-70.

Linderman, R.G. (1994). Role of V AM fungi in biocontrol. In *Mycorrhizae and Plant Health,* eds. F.L. Pfleger and R.G. Linderman. APS Press, St. Paul, MN, pp. 1-26.

Linderman, R.G. (2000). Effects of mycorrhizas on plant tolerance to diseases. In *Arbuscular Mycorrhizas: Physiology and Function,* eds. Y. Kapulnik and D.D. Douds Jr. Kluwer Academic, Dordrecht, the Netherlands, pp. 345-365.

Lovato, P.E., H. Schüepp, A. Trouvelot, and S. Gianinazzi. (1995). Application of arbuscular mycorrhizal fungi (AMF) in orchard and ornamental plants. In *Mycorrhiza Structure, Function, Molecular Biology and Biotechnology,* eds. A. Varma and B. Hock. Springer, Heidelberg, pp. 521-559.

Lynch, J.M. (ed.). (1990). *The Rhizosphere.* Wiley, New York.

Mäder, P., S. Edenhofer, T. Boller, A. Wiemken, and U. Niggli. (2000). Arbuscular mycorrhizae in a long-term field trial comparing low-input (organic, biological) and high-input (conventional) farming system in a crop rotation. *Biology and Fertility of Soils* 31: 150-156.

Marschner, H. (ed.). 1995. *Mineral Nutrition of Higher Plants,* Second Edition. Academic Press, London.

Matsubara, Y., H. Tamura, and T. Harada. (1995). Growth enhancement and *Verti-cillium* wilt control by vesicular-arbuscular mycorrhizal fungus inoculation in eggplant. *Journal of Japanese Society of Horticulture Science* 64: 555-561.

Meyer, J.R. and R.G. Linderman. (1986). Selective influence on populations of rhizosphere or rhizoplane bacteria and actinomycetes by mycorrhizas formed by *Glomus fasciculatum*. *Soil Biology and Biochemistry* 18: 191-196.

Morandi, D. (1996). Occurence of phytoalexins and phenolic compounds in en-domycorrhizal interactions and their potential role in biological control. *Plant and Soil* 185: 241-251.

Morandi, D., J.A. Bailey, and V. Gianinazzi-Pearson. (1984). Isoflavonoid accumu-lation in soybean roots infected with vesicular-arbuscular mycorrhizal fungi. *Physiology of Plant Pathology* 24: 357-364.

Morton, J.B. and G.L. Benny. (1990). Revised classification of arbuscular my-corrhizal fungi (Zygomycetes): A new order, Glomales, two new suborders, Glom-ineae and Gigasporineae, and two new families Aculosporaceae and Gigaspor-aceae with an emendation of Glomaceae. *Mycotaxon* 37: 471-492.

Mukerji, K.G. (1999). Mycorrhiza in control of plant pathogens: Molecular ap-proach. In *Biotechnological Approaches in Biocontrol of Plant Pathogens*, eds. K.G. Mukerji. B.P. Chamola, and R.K. Upadhyay. Kluwer Academic/Plenum Publishers, New York, pp. 157-176.

Mukerji, K.G., B.P. Chamola, and Jagjit Singh. (eds.) (2000). *Mycorrhizal Biology*. Kluwer Academic / Plenum Publishers, New York, Dordrecht, London.

Mukerji, K.G., C. Manoharachary, and B.P. Chamola. (eds.) (2002). *Techniques in Mycorrhizal Studies*. Kluwer Academic Publishers, Dordrecht, the Netherlands.

Nemec, S., L.E. Datnoff, and J. Strandberg. (1996). Efficacy of biocontrol agents in planting mixes to colonize plant roots and control root diseases of vegetables and citrus. *Crop Protection* 15: 735-742.

Norman, J.R., D. Atkinson, and J.E. Hooker. (1996). Arbuscular mycorrhizal fungal induced alteration root architecture in strawberry and induced resistance to the root pathogens *Phytophthora fragariae*. *Plant and Soil* 185: 191-198.

Paulitz, T.C. and R.G. Linderman. (1989). Interactions between fluorescent pseu-domonads and V A mycorrhizal fungi. *New Phytologist* 113: 37-45.

Paulitz, T.C. and R.G. Linderman. (1991). Lack of antagonism between the bio-control agent *Gliocladium virens* and vesicular arbuscular mycorrhizal fungi. *New Phytologist* 117:303-308.

Peterson, R.L. and M.L. Farquhar. (1994). Mycorrhizas: Integrated development between root and fungi. *Mycologia* 86 (3): 311-326.

Pozo, M.J., C. Azcón-Aguilar, E. Dumas-Gaudot, and J.M. Barea. (1998). Chito-sanase and chitinase activites in tomato roots during interaction with arbuscular mycorrhizal fungi or *Phytophythora parasitica*. *Journal of Experimental Botany* 49: 1729-1739.

Pozo, M.J., C. Azcón-Aguilar, E. Dumas-Gaudot, and J.M. Barea. (1999). β-1,3-Glucanase activities in tomato roots inoculated with arbuscular mycorrhizal fungi and/or *Phytophthora parasitica* and their possible involvement in biopro-tection. *Plant Science* 141: 149-157.

Pozo, M.J., C. Cordier, E. Dumas-Gaudot, S. Gianinazzi, J.M. Barea, and C. Azcón-Aguilar. (2002). Localized versus sistemic effect of arbuscular mycorrhizal fungi on defence responses to *Phytophythora* infection in tomato plant. *Journal of Experimental Botany* 53: 525-534.

Pozo, M.J., E. Dumas-Gaudot, S. Slezack, C. Cordier, A. Asselin, S. Gianinazzi, V. Gianinazzi-Pearson, C. Azcón-Aguilar, and J.M. Barea. (1996). Induction of new chitinase isoforms in tomato roots during interactions with *Glomus mosseae* and/or *Phytophthora nicotianae* var. *parasitica*. *Agronomie* 16: 689-697.

Requena, N., I. Jimenez, M. Toro, and J.M. Barea. (1997). Interactions between plant-growth-promoting rhizobacteria (PGPR), arbuscular mycorrhizal fungi and *Rhizobium* spp. in the rhizosphere of *Anthyllis cytisoides,* a model legume for revegetation in mediterranean semi-arid ecosystems. *New Phytologist* 136: 667-677.

Rosendahl, S. (1985). Interactions between the vesicular-arbuscular mycorrhizal fungus *Glomus fasciculatum* and *Aphanomyces euteiches* root rot of peas. *Phytopathologische Zeitschrift* 114: 31-40.

Ross, J.P. (1972). Influence of *Endogone* mycorrhiza on Phytophthora rot of soybean. *Phytopathologische Zeitschrift* 67: 1507-1511.

Schönbeck, F. (1979). Endomycorrhiza in relation to plant diseases. In *Soilborne Plant Pathogens,* eds. B. Schippers and W. Gams. Academic Press, New York, pp. 271-280.

Schönbeck, F. and H.W. Dehne. (1977). Damage to mycorrhizal cotton seedlings by *Thielaviopsis basicola*. *Plant Disease Reports* 61: 266-268.

Schüßler, A. (2002). Molecular phylogeny, taxonomy, and evolution of *Geosiphon pyriformis* and arbuscular mycorrhizal fungi. *Plant Soil* 244: 75-83.

Schüßler, A., D. Schwarzott, and C. Walker. (2001). A new fungal phylum, the Glomeromycota: Phylogeny and evolution. *Mycological Research* 105: 1413-1421.

Secilia, J. and D.J. Bagyaraj. (1987). Bacteria and actinomycetes associated with pot cultures of vesicular-arbuscular mycorrhizas. *Canadian Journal of Microbiology* 33: 1069-1073.

Smith, G.S. (1988). The role of phosphorus nutrition in interactions of vesicular-arbuscular mycorrhizal fungi with soil-borne nematodes and fungi. *Phytopathology* 78: 371-374.

Smith, S. and D.J. Read (eds.). (1997). *Mycorrhizal Symbiosis,* Second Edition. Academic Press Publishers, London.

Söderberg, K.H., P.A. Olsson, and E. Baath. (2002). Structure and activity of the bacterial community in the rhizosphere of different plant species and the effect of arbuscular mycorrhizal colonisation. *FEMS Microbiology Ecology* 40: 223-231.

Sood, S.G. (2003). Chemotactic response of plant-growth-promoting bacteria towards root of vesicular-arbuscular mycorrhizal tomato plants. *FEMS Microbiology Ecology* 45: 219-227.

Spanu, P. and P. Bonfante-Fasolo. (1988). Cell-wall-bound peroxidase activity in roots of mycorrbizal *Allium porrum*. *New Phytologist* 109: 11-19.

St-Arnaud, M., C. Hamel, M. Caron, and J.A. Fortin. (1994). Inhibition of *Pythium ultimum* in roots and growth substrate of mycorrhizal *Tagetes patula* colonized with *Glomus intraradices*. *Canadian Journal of Plant Pathology* 16: 187-194.

St-Arnaud, M., C. Hamel, B. Vimard, M. Caron, and J.A. Fortin. (1995). Altered growth of *Fusarium oxysporum* f. sp. *chrysanthemi* in an in vitro dual culture system with the vesicular arbuscular mycorrhizal fungus *Glomus intraradices* growing on *Daucus carota* transformed roots. *Mycorrhiza* 5: 431-438.

Stewart, E.L. and F.L. Pfleger. (1977). Development of poinsettia as influenced by endomycorrhizae, fertilizer and root rot pathogens *Pythium ultimum* and *Rhizoctonia solani*. *Florist's Review* 159: 79-80.

Strack, D., T. Fester, B. Hause, W. Scheliemann, and M. Walter. (2003). Arbuscular mycorrhiza: Biological, chemical, and molecular aspects. *Journal of Chemical Ecology* 29: 1955-1979.

Tisdall, J.M. (1994). Possible role of soil microorganisms in aggregation in soils. *Plant and Soil* 159: 115-123.

Traquair, J.A. (1995). Fungal bicontrol of root diseases: Endomycorrhizal suppression of cylindrocarpon root rot. *Canadian Journal of Botany* 73: 89-95.

Trotta, A., G.C. Varese, E. Gnavi, A. Fusconi, S. Sampò, and G. Berta. (1996). Interactions between the soil-borne root pathogen *Phytophthora nicotianae* var. *parasitica* and the arbuscular mycorrhizal fungus *Glomus mosseae* in tomato plants. *Plant and Soil* 185: 199-209.

Vierheilig, H., M, J. Lange, M. Gut-Rella, A. Wiemken, and T. Boller. (1995). Colonisation of transgenic tobacco constitutively expressing pathogenesis-related proteins by the vesicular-arbuscular mycorrhizal fungus *Glomus mosseae*. *Applied and Environmental Microbiology* 61: 3031-3034.

Vigo, C., J.R. Norman, and J.E. Hooker. (2000). Biocontrol of the pathogen *Phytophthora parasitica* by arbuscular mycorrhizal fungi is a consequence of effects on infection loci. *Plant Pathology* 49: 509-514.

Volpin, H., Y. Elkind, Y. Okon, and Y. Kapulnik. (1994). A vesicular arbuscular mycorrhizal fungi *(Glomus intraradix)* induces a defense response in alfalfa roots. *Plant Physiology* 104: 683-689.

Volpin, H., D.A. Phillips, Y. Okon, and Y. Kapulnik. (1995). Suppression of an isoflavonoid phytoalexin defense response in mycorrhizal alfalfa roots. *Plant Physiology* 108: 1449-1454.

Yao, M.K., R.J. Tweddell, and H. Désilets. (2002). Effect of two vesicular-arbuscular mycorrhizal fungi on the growth of micropropagated potato plantlets and on the extent of disease caused by *Rhizoctonia solani*. *Mycorrhiza* 12: 243-247.

Zambolim, L. and N.C. Schenck. (1983). Reduction of the effects of pathogenic root infecting fungi on soybean by the mycorrhizal fungus, *Glomus mosseae Phytopathology* 73: 1402-1405.

Chapter 3

Arbuscular-Mycorrhiza-Mediated Plant-Pathogen Interactions and the Mechanisms Involved

Mahaveer P. Sharma
Atimanav Gaur
K.G. Mukerji

INTRODUCTION

Excessive use of chemical inputs to control diseases poses an important problem for present-day plant production systems. Research priorities for alternative management practices compatible with sustainable agriculture and the environment include the use of beneficial microorganisms as biological control agents. Biological control of plant pathogens is currently accepted as a key practice in sustainable agriculture and can be defined as direct and accurate management of common components of ecosystems to protect plants against pathogens. Biological control preserves environmental quality by a reduction in chemical inputs and is characteristic of sustainable management practices (Mukerji and Garg, 1988a,b; Mukerji et al., 1992, 1999; Altieri, 1994; Barea and Jeffries, 1995; Upadhyay et al., 2000, 2001).

Among symbiotic associations between soil microorganisms and plants, the endosymbiosis formed by plant roots and arbuscular mycorrhizal fungi (AMF) are very important, because they are widespread and involve most agronomically important crops. AM fungi constitute an important component of the microbial soil community, and several studies have shown that the presence of these fungi has considerable impact on plant health (Dehne, 1982; Perrin, 1990; Hooker et al., 1994; Linderman, 1994; Azcón-Aguilar and Barea, 1996; Mukerji, Chamola, and Sharma, 1997). Mycorrhizal inoc-

Biological Control of Plant Diseases
doi:10.1300/5682_03

ulation can be effective even against soilborne pathogens, which are difficult to control by chemical and physical treatments (Azcon-Aguilar and Barea, 1996, 1997; Cordier et al., 1997). It is evident that an increased capacity for nutrient acquisition resulting from mycorrhizal association could help the resulting stronger plants to resist stress. However, AM symbioses may also improve plant health through a more specific increase in protection, improved resistance, and/or tolerance against biotic and abiotic stresses (Bethlenfalvay and Linderman, 1992; Barea and Jeffries, 1995).

Strictly speaking, plants under natural conditions do not have roots; they have mycorrhizas: the roots of most flowering plants form mutualistic symbioses with certain soil fungi (Harley and Smith, 1983). Mycorrhizal associations are found in nearly all ecological situations, with AM being the most common type in normal cropping systems and natural ecosystems (Harley and Smith, 1983; Bethlenfalvay and Schüepp, 1994). AM fungi, which belong to the order Glomales of the Zygomycetes (Rosendahl et al., 1994), biotrophically colonize the root cortex and develop an extramatrical mycelium which helps the plant acquire mineral nutrients and water from the soil. AM symbioses play a key role in nutrient cycling in ecosystems (Barea and Jeffries, 1995), and the external mycorrhizal mycelium, in association with other soil organisms, forms water-stable aggregates necessary for good soil quality (Bethlenfalvay and Schüepp, 1994). Thus, there is growing awareness of the role of mycorrhizal fungi as bioprotectors, biostimulators, or biofertilizers (Azcón-Aguilar and Barea, 1997). This review will highlight the mechanism and strategies to understand the role of mycorrhizae as biocontrol agents against soilborne pathogens.

INTERACTION OF AM FUNGI WITH FUNGAL PATHOGENS ASSOCIATED WITH HORTICULTURAL PLANTS

AM fungi are important components of the soil that associate with most species of terrestrial plants (Allen et al., 1995). Nursery and growth chamber experiments have repeatedly demonstrated that these root symbionts can benefit host nutrition primarily by facilitating phosphorus utilization. The mode of increased P availability through hyphal structures of AM fungi has been established (Hayman, 1982). Despite the fact of P mobilization through AM fungi confirmed and established from laboratory studies, field studies suggest that the role of AM fungi in P facilitation is not universal (Newsham et al., 1995a). Because the association between AM fungi and plants is more common than can be explained by P facilitation alone, further hypotheses have been advanced to make this association mutually advantageous (Newsham et al., 1995b). A leading hypothesis proposes that AM fungi alleviate the effects of plant pathogens (Dehne, 1982; Smith, 1988).

The effects of AM fungi on pathogens are most likely indirect and result from enhanced nutrition or altered physiology of the host (Sharma et al., 1992). Most of the studies have concentrated on the effect of AM fungi on improved nutrition and growth of plants (Yao et al., 2002). Most commonly, AM fungi appear to enhance host tolerance by improving root growth and function (Smith, 1988). The most important interactions that develop in the rhizosphere can be classified into three groups (Azcón-Aguilar and Barea, 1996): (1) plant–plant interactions caused by overlapping rhizospheres, which results in competition for nutrients; (2) root–microorganism interactions, determined by plant activities that stimulate microorganisms to grow around the roots (rhizosphere effect), and by microbial activities that affect plant development, either by benefiting the plant or by inducing disease; and (3) microbe–microbe interactions, which include both synergistic and antagonistic activities (Lynch, 1990). A number of reports are available on the interactions of AM fungi with plant pathogens, such as fungi, nematodes, bacteria, and viruses associated with vegetables and fruits, which are summarized in Table 3.1. From the reports of all experiments, it can be speculated that interactions between different AM fungi and plant pathogens will vary with the host plant and the cultural conditions.

Since arbuscular mycorrhizae are established in the roots of plants, research on interactions between mycorrhizae and disease incidence has been concentrated on diseases caused mainly by soilborne pathogens (Singh et al., 2000). Most of the work focuses on increased P assimilation, which has been proposed as a theory for higher tolerance of mycorrhizal plants to pathogens (Azcón-Aguilar and Barea, 1996), but there are a number of contradictory reports (Linderman, 1994; Hooker et al., 1994). Higher disease resistance by mycorrhiza *(Glomus mosseae)*-colonized tomato toward soilborne root pathogen *Phytophthora nicotianae* var. *parasitica* has also been shown by Trotta et al. (1996). They suggested that the effect of P nutrition was only apparent in the form of stimulated plant growth. Overall, P appears to have fewer effects on *Phytophthora* diseases than other chemical factors (Schmitthenner and Canaday, 1983).

Karagiannidis et al. (2002), while working on Verticillium wilt (*Verticillium dahliae* Kleb.) and mycorrhiza *(G. mosseae)* on root colonization, growth, and nutrient uptake in tomato and eggplant seedlings found that, apart from increased growth of mycorrhizal plants, the beneficial effect of AM inculcation supersedes the pathogenic effect of *V. dahliae* and enhances the P and N uptake more significantly than do control plants. Similarly, Abdel-Fattah and Shabana (2002), worked on the efficacy of the arbuscular mycorrhizal fungus *Glomus clarum* in protection of cowpea

TABLE 3.1. Interaction of arbuscular mycorrhizal fungi with fungal pathogens associated with horticultural plants.

Fungal pathogen	Host plant	Disease incidence	Reference
Rhizoctonia solani	Potato	−	Yao et al., 2002
		−	Bodker et al., 2002
Fusarium solani f. sp. phaseoli	Bean	−	Dar et al., 1997; Filion et al., 2003
Aphanomyces euteiches	Pea	−	Rosendahl, 1992
Sclerotium cepivorum	Onion	−	Torres-Barragan et al., 1996
F. oxysporum f. sp. cepa	Onion	−	Dehne, 1982
F. oxysporum f. sp. cumini	Cumin	+	
F. oxysporum f. sp. lycopersici	Tomato	−	Dehne and Schönbeck, 1979
F. oxysporum f. sp. lycopersici	Tomato	−	
F. oxysporum f. sp. radicis lycopersici	Tomato	−	Caron et al., 1986a
Phytophthora nicotianae sp. parasitica	Tomato	−	Trotta et al., 1996
Phytophthora parasitica	Tomato	−	Pozo et al., 2002
F. oxysporum f. sp asparagi	Asparagus	−	Matsubara et al., 2002
		−	Matsubara et al., 2001
Colletotrichum lindemuthianum	French bean	+	Schoenbeck, 1979
Pyrenochaeta terrestris	Tomato	−	Safir, 1968
Pyrenochaeta lycopersici	Tomato	−	
Pythium aphanidermatum	Tomato	−	
Sclerotium rolfsii	Chilli	−	
Sclerotium cepivorum	Garlic	. . .	
Rhizoctonia solani	Tomato	−	
F. oxysporum f. sp medicaginis	Alfalfa	−	
P. parasitica	Citrus	−	Davis and Menge, 1980
P. ultimum	Cucumber	−	Rosendahl, 1992

Source: Modified from Sharma et al., 2004.

plants against root rot pathogen *Rhizoctonia solani* and found that, when challenged with *R. solani,* the mycorrhizal plants notably contained higher concentrations of P, N, K, and Mg compared with the nonmycorrhizal plants. The results suggested that (1) the ability of mycorrhizal roots to absorb phosphorus and other nutrients, and (2) the imperative competition for infection sites and antagonistic reactions between the mycorrhizal fungus and the pathogen may be the reasons for increasing disease resistance in mycorrhizal plants. Thus, the AM fungus acts as a biological control agent against *R. solani,* the causal agent for root rot disease in cowpea plants. However, other authors observed no effect of P fertilization on pathogen population (Caron et al., 1986a).

The tolerance to fungal pathogens of mycorrhizal plants also varied with the type of fungal species. Matsubara et al. (2001) showed that ten weeks after AM fungal inoculation, the inoculated plants were taller; they produced more shoots and roots,' and accumulated more dry matter in the shoots and roots than the noninoculated ones. The infection levels in a root system differed with the fungal species. Six weeks after *Fusarium oxysporum* f. sp. *asparagi* (Foa) inoculation, 90 percent of the noninoculated plants exhibited symptoms of fusarium root rot, whereas only 20 to 50 percent of the inoculated plants did. The effect was more pronounced in GR, GM, and GF, in that order. As for the disease indices, they were lower in the inoculated plants than in the noninoculated ones. The indices differed among the AM fungal species; it was significantly low in GR. The number of Foa hyphae invading feeder roots decreased in the inoculated plants, compared with the noninoculated ones. The results revealed that the tolerance to fusarium root rot was conditioned by AM fungal infection in asparagus seedlings, although the effect differed with the AM fungal species. It seems that the tolerance to fusarium root rot was partially caused by AM fungal preinfection in short cells that suppressed invasion by Foa in feeder roots.

Further, AM fungi were evaluated with various soil amendments to check the incidence of fusarium root rot of asparagus seedlings. They suggested that the addition of coconut charcoal or manure of coffee residue to bed soil is effective for increasing the tolerance to fusarium root rot in AM fungus-infected asparagus plants, though the effect differed with the quantity of soil amendment (Matsubara et al., 2002).

Besides P nutrition, Mohr et al. (1998) showed that plant defense genes are induced in the pathogenic interaction between bean roots and *Fusarium solani* but not in the symbiotic interaction with the arbuscular mycorrhizal fungus *G. mosseae.* The time–course study revealed that the symbiotic interaction between bean roots and *G. mosseae* covering all stages of mycorrhizal development found little change in the expression of the defense-related genes, chitinase, beta-1,3-glucanase, and phenylalanine ammonia-lyase, compared with nonmycorrhizal control roots. The only difference observed was a transient increase in chalcone synthase transcripts in the later stages of mycorrhizal root colonization. In interactions with the pathogen, a marked induction of chitinase and phenylalanine ammonia-lyase expression was observed at the level of both the transcripts and enzyme activities. Class I beta-1,3-glucanase levels strongly increased at the transcript level, but there was little change in the overall beta-1,3-glucanase enzyme activity. In the nonhost interaction between common bean and *F. solani* f. sp. *pisi*, defense responses increased only slightly and transiently, if at all.

Very recently, Harrier and Watson (2004) highlighted that bioprotection within AM fungus-colonized plants is the outcome of complex interactions between plants, pathogens, and AM fungi and justified the need for molecular tools to study the multifaceted interactions, which may aid optimization of the bioprotective responses and their utility within sustainable farming systems.

It is apparent from many reports that AM fungi can lead to lower disease incidence on mycorrhizal plants (Perrin, 1990; St Arnaud et al., 1994). Mycorrhiza-mediated tomato plants by *Glomus intraradices* significantly reduced the population of *Fusarium* root rot caused by *Fusarium oxysporum* f. sp. *radices-lycopersici* (Caron et al., 1985). St Arnaud et al. (1994) observed that inoculations of substrate with *G. intraradices* reduced the populations of *Pythium ultimum* on *Tagetes patula*. They showed that the extent of colonization bearing arbuscules or vesicles of AM was unrelated to P nutrition and to the observed reduction of *P. ultimum* in the roots or in the substrate. Recently, Bodker et al. (2002) showed that mycorrhization to field-grown pea significantly affects the life cycle of *Aphanomyces euteiches* (root rot of pea). The rate of increase in pathogen infection potential showed negative correlation with the levels of indigenous AM population in the soil. Torres-Barragan et al. (1996) also showed that onion plants inoculated with AM reduced the white rot incidence and delayed the disease development by two weeks over control plants. Also, the onion roots having higher colonization and spore density in the rhizosphere showed lower disease and vice versa. There are reports that contradict the idea that a high level of mycorrhization is necessary to induce protection against pathogens (Smith et al., 1986a,b; Smith, 1988), and these reports agree with the results of Caron et al. (1986b) that tomato plants could protect against *F. oxysporum* even at very low colonization levels. Dugassa et al. (1996) reported that the influence of AMF symbiosis on plant health depends more on the host-plant and pathogen genotype than on AM colonization level. Mycorrhizal inoculation with either *G. intraradices* or *G. etunicatum* on plantlets of potato cultivar Goldrush enhanced growth and yield, and improved resistance/ tolerance to *R. solani* infection. In natural conditions as well as through inoculation, most species or isolates of AM fungi can establish symbiosis with a very wide host range, including potato. However, many studies have demonstrated that the AM fungi-plant cultivar combination influences plant responses (Yao et al., 2002). The influence exerted by AM fungi by stimulating plant defense reactions was investigated using an in vitro system in which RiT-DNA-transformed carrot roots were infected with

F. oxysporum f. sp. *chrysanthemi*. In the noninoculated roots, the pathogen multiplied abundantly through much of the tissue, including the vascular stele, whereas in mycorrhizal plants, its growth was restricted to the epidermis and the outer cortex (Benhamou et al., 1994).

The reduction in the infection by *F. oxysporum* due to inoculations with *G. mosseae* in tomato and *Capsicum* roots was also reported by Al-Momany and Al-Radded (1988). In the tomato plants inoculated with *G. mosseae*, the incidence of *Fusarium* wilt was only 11 percent, as compared to 45 percent in nonmycorrhizal plants (Ramraj et al., 1988). Rosendahl and Rosendahl (1990) showed that inoculation of cucumber seedling with *G. etunicatum* and *Glomus* sp. before or simultaneously with *Pythium ultimum* increased the survival of the seedlings and saved the plants from damping off.

INTERACTION OF AM WITH SOILBORNE NEMATODES ASSOCIATED WITH HORTICULTURAL PLANTS

Mycorrhizal fungi and plant-parasitic nematodes are commonly found inhabiting the rhizosphere and colonizing the roots of their host plants. These two groups of microorganisms exert opposite effects on plant growth (Hussey and Roncondori, 1982). A large number of reviews have appeared on the interaction of nematodes and mycorrhizal fungi (Paulitz and Linderman, 1991; Mukerji, Chamola, and Sharma, 1997; Sharma and Adholeya, 2000). As a result of interactions, the severity of nematode diseases in general is reduced in mycorrhizal plants (Table 3.2). However, in some cases, root colonization by mycorrhizal fungi had no effect on nematode reproduction, while in others it had a stimulating effect (Siddiqui and Mahmood, 1995).

Parasitism of nematode eggs with AM fungi has been demonstrated, but the level of parasitism was not considered sufficient to negatively affect nematode activities (Francl and Dropkin, 1985). In most cases, AM fungi colonize only stressed or weakened nematode eggs (Siddiqui and Mahmood, 1995). The nematode parasitism by AM fungi is opportunistic and depends on carbon nutrition from autotrophic symbionts, rather than being representative of a true host–parasitic relationship (Siddiqui and Mahmood, 1995). The negative effect of AM fungi on nematode development and reproduction is mainly due to histopathological changes occurring in the cells, such as smaller syncytia and fewer giant cells, which confer resistance against *Meloidogyne* sp. on the host plants (Fassuliotis, 1970). The alteration of host physiology and biochemical changes brought by mycorrhizal plants has been found to be one of the factors conferring resistance against nematodes (Dehne, 1982; Morandi, 1987; Umesh et al., 1988; Calvet, Pinochet,

TABLE 3.2. Adverse effect of arbuscular mycorrhizal fungi on nematode diseases of horticultural crops.

Host plant	Nematode	Mycorrhizal fungi	Reference
Tomato	M. incognita	G. mosseae	Talavera et al., 2001
		Glomus sp. and Pasteuria penetrans	Talavera et al., 2002
		G. deserticola and Verticilium chalamydosporium	Rao et al., 1997
	M. javanica	G. mosseae, Paecilomyces lilacinus	Ahmad and Al-Raddad, 1995
		G. fasciculatum, G. mosseae	Singh et al., 1990
	M. incognita	G. fasciculatum	Sharma et al., 1994
	M. javanica	G. fasciculatum	
In vitro culture conditions	Radopholus similis	G. intraradices	Elsen et al., 2001
Carrot	Pratylenchus penetrans	G. mosseae	Talavera et al., 2001
Banana	Radopholus similis and Pratylenchus coffeae	G. mosseae	Elsen et al., 2003
Pyrethrum	M. hapla	Glomus sp.	Waceke et al., 2002
White clover	M. incognita	G. aggregatum, G. intraradices, G. mosseae	Habte et al., 1999
Peach almond	M. javanica	G. intraradices, G. mosseae, G. etunicatum	Calvet, Pinochet, Camprubi, et al., 2001; Calvet, Pinochet, Hernandez-Dorrego, et al., 2001
Coffee	M. exigua	Entrophosphora colombiana, Gigaspora margarita	Rivas-Platero and Andrade, 1998
Potato	Globodera pallida	Vaminoc (AM fungal commercial inoculum)	
Pepper	M. incognita	G. Fasciculatum, G. mosseae	Sivaprasad et al., 1990
Black pepper	M. incognita	G. fasciculatum, G. mosseae, G. margarita, Acaulospora laevis	

Camprubi, et al., 2001; Calvet, Pinochet, Hernandez-Dorrego, et al., 2001). Apart from the efficacy of AM fungi alone in reducing nematode infestations, studies have also focused on the influence of fertilization on their efficacy (Thomson-Cason et al., 1983). Recently, Waceke et al. (2002) worked with inorganic phosphatic fertilizers on the efficacy of an AM fungus against a root-knot nematode on pyrethrum. The fertilizers were applied at 150 and 300 kg · ha^{-1} at the time of fungus inoculation. Two months later, plants were inoculated with nematodes. The fertilizers at both levels were found to improve plant growth in all treatments. In general, the fungus improved plant growth on its own, or in the presence of nematodes, but not in the presence of fertilizers. The two fertilizers at both levels were more effective in improving plant growth than the fungus was. The fungus showed sensitivity to inorganic P fertilizers, in that the fertilizers significantly reduced fungal root colonization and its pyrethrum growth-stimulative effects. The fungus suppressed nematode disease severity, unlike the fertilizers. The suppressive effects of the fungus on the nematodes were in most cases reduced by the fertilizers. The nematodes, unlike the fertilizers, did not have any significant effects on root colonization by the fungus or on its ability to improve pyrethrum growth. The presence of nematodes in fertilizer or fertilizer-fungus-treated plants, however, significantly reduced pyrethrum growth. A number of recent reports are available on fruit and vegetable crops, where AM fungi were found to reduce nematode infestations; however, the degree of reduction was found to be varied due to AM application combinations (Rao et al., 1997; Habte et al., 1999; Elsen et al., 2001, 2003; Calvet, Pinochet, Hernandez-Dorrego, et al., 2001; Talavera et al., 2001, 2002).

MECHANISMS INVOLVED IN CONTROL OF ROOT PATHOGENS

The possible mechanism that could account for the role of AM fungi in control of root pathogens includes changes in mycorrhizosphere microbial population, enhancement in plant nutrition, root damage compensation, competition of colonization sites and photosynthates, production of changes in root systems, and activation of plant defense mechanisms.

Mycorrhizospheric Effect

The concept of "mycorrhizosphere" implies that mycorrhizae significantly influence the microflora of the rhizosphere by altering root physiology and exudation. In addition, extraradical hyphae of AM fungi provide a physical or nutritional substrate for bacteria. AM symbiosis can cause qualitative and quantitative changes in rhizospheric microbial populations; the resulting microbial equilibria could influence the growth and health of plants. These changes may result from AM fungus-induced changes in root exudation patterns (Smith, 1987; Azcón-Aguilar and Bago, 1994; Smith et al., 1994; Bansal et al., 2000). These changes in microbial populations induced by AM formation may lead to stimulation of the microbiota which, in turn, may be antagonistic to root pathogens. It is known that AM establishment can change both the total population and specific functional groups of microorganisms in the rhizoplane or the rhizosphere soil (Meyer and Linderman, 1986; Linderman, 1994). Pathogen-antagonistic actinomycetes were more in the rhizosphere of AM plants than in nonmycorrhizal controls (Secilia and Bagyaraj, 1987). They showed that pot cultures of *G. fasciculatum* harbored more actinomycetes antagonistic to *F. solani* and *Pseudomonas solanacearum* than those of nonmycorrhized plants. AM fungi can cooperate in several ways with plant growth-promoting rhizobacteria (PGPR) for biological control of root pathogens, in addition to improvement in plant root and enhancement in plant growth and nutrition. Other studies have indicated that pathogen suppression by AM fungi involved changes in mycorrhizosphere microbial populations. Caron et al. (1985, 1986a,b,c) showed a reduction in *Fusarium* population in the mycorrhizosphere soil of tomatoes and a corresponding reduction in root rot in AM plants compared with non-AM plants, probably due to the increased antagonism in the AM mycorrhizosphere.

Enhanced Plant Nutrition

Mycorrhizal plants in general are more tolerant to root pathogens due to the increased nutrient uptake made possible by AM symbiosis. Many of these factors are related to P physiology and nutrition (Graham et al., 1981; Hayman, 1982; Pacovsky et al., 1986). AM fungi enhance root growth, expand absorptive capacity, and affect cellular processes in roots (Hussey and Roncondori, 1982; Smith and Gianinazzi-Pearson, 1988). Davis (1980) showed that the effect of AM fungi on *Thieloviopsis* root rot of citrus was larger in noninoculated plants, unless the latter were supplied with additional P. These mycorrhizal-induced compensatory processes may explain the increased tolerance of mycorrhiza- and P-fertilized plants, because many plants can compensate for loss of root mass or function caused by pathogens (Wallace, 1983). Greater tolerance is also attributable to increased root growth and phosphate status of the plant (Cameron, 1986). In addition to P, AM fungi can enhance the uptake of Ca, Cu, Mn, S, and Zn (Pacovsky et al. 1986; Smith and Gianinazzi-Pearson, 1988). Host susceptibility to pathogens and the tolerance of the disease process can be influenced by the nutritional status of the host and the fertility status of the soil (Cook and Baker, 1982). For example, nematode-damaged plants frequently show deficiencies of B, N, Fe, Mg, and Zn (Good, 1968). In the absence of AM, phosphate can combine with minor elements to create deficiencies, which would predispose plants to root-knot nematodes (Smith et al., 1986a,b). Thus, AM fungi may increase tolerance to pathogens by increasing the uptake of essential nutrients other than P that would be deficient in a non-mycorrhizal plant. These reports indicate improved nutrition as a mechanism for disease control and could account for the higher tolerance of mycorrhizal plants to pathogens. However, there are contradictory reports, where benefits achieved by mycorrhizal inoculation against pathogens could not be duplicated by adding P fertilizer (Cooper and Grandison, 1987). Another report (Zambolin, 1987) has shown that P-tolerant AM fungi reduced nematode effects even under high P conditions, indicating that non-P-mediated mechanisms are involved.

Root Damage Compensation

AM fungi are reported to compensate for the root damage by pathogens with fungal hyphae growing out into the soil and increasing the absorbing surface of the roots, and with the maintenance of root-cell activity through arbuscule formation (Cordier et al., 1996). This represents an indirect contribution to biocontrol of root pathogens by maintaining the root-system function. It is known that AM fungi increase host tolerance of pathogen attack by compensating for the loss of root biomass or function caused by pathogens (Linderman, 1994), including nematodes (Pinochet et al., 1996) and fungi (Cordier et al., 1996).

Competition for Host Photosynthates

Both pathogens and mycorrhizal fungi depend on host photosynthates for their growth, and they may compete for the carbon compounds reaching the root (Smith, 1987; Linderman, 1994). The higher carbon demand due to AM fungal colonization may thus inhibit pathogen growth. However, there is little or no evidence to indicate that competition for carbon compounds is a generalized mechanism for the pathogen biocontrol activity of AM symbiosis. Dehne (1982) indicated that fungal root pathogens could occupy root cortical cells adjacent to those colonized by AM fungi, indicating a lack of competition. It has been suggested that nematodes, on the other hand, require host nutrients for reproduction and development, and direct competition with AM fungi has been hypothesized as a mechanism of their inhibition (Dehne, 1982; Smith, 1988). Since AM fungi, soilborne fungal pathogens, and plant-parasitic nematodes occupy similar root tissues, direct competition for space has been postulated as a mechanism of pathogen inhibition by AM fungi (Davis and Menge, 1980; Hussey and Roncondori, 1982; Linderman, 1994).

Competition for Infection Sites

Both localized and nonlocalized mechanisms could be responsible for competition for colonization sites, mainly depending on the pathogen (fungus, nematode). Fungal pathogens of root and AM fungi, although colonizing the same host tissues, usually develop in different root cortical cells, thus indicating some sort of competition for space (Dehne, 1982). Jalali and Jalali (1991) have reported a localized effect, while Dehne (1982) and Smith (1987) suggested that the extent of protection cannot be explained by a localized mechanism alone. *Phytophthora* development was reduced in AM fungus-colonized and the adjacent noncolonized root cells, and that, in the former, the pathogen does not penetrate arbuscule-containing cells (Cordier et al., 1996).

AM-Induced Changes in Roots

AM colonization induces remarkable changes in root system morphology, as well as in the meristematic and nuclear activities of the root. Lignification prevents penetration of mycorrhizal plants by pathogens. Dehne et al. (1978) showed increased lignification of cells in the endodermis of mycorrhizal tomato and cucumber plants and speculated that such responses accounted for reduced incidence of *Fusarium* wilt *(F. oxysporum* f. sp. *lycopersici)*. Increased wound-barrier formation inhibited *Thielaviopsis* black root rot *(T. basicola)* of mycorrhizal holly *(Ilex crenata)* plants. A stronger vascular system in mycorrhizal plants increases flow of nutrients, imparts greater mechanical strength, and reduces the effect of vascular pathogens (Schoenbeck, 1979). AM fungi enhance root growth, expand absorptive capacity, and affect cellular processes in roots (Hussey and Roncondori, 1982; Smith and Gianinazzi-Pearson, 1988).The most frequent consequence of AM colonization is an increase in branching, resulting in a relatively larger proportion of higher-order roots in the root system (Hooker et al., 1994). However, the significance of this finding for plant protection has not yet been sufficiently considered. This represents an indirect contribution to biocontrol through the conservation of root-system function, both by fungal hyphae growing out into the soil and increasing the absorbing surface of the roots and by the maintenance of root-cell activity through arbuscule formation (Cordier et al., 1996) in cells (Atkinson et al., 1994). This might affect rhizosphere interactions and, particularly, the pathogeninfection development.

Changes in Chemical Constituents of Plant Tissues

Various evidence has indicated structural and biochemical changes in the cell walls of the plants colonized by AM fungi. Dehne and Schönbeck (1979) and Becker (1976) reported enhanced lignification of endosperm cell walls and vascular tissues. Also, Dehne et al. (1978) demonstrated increased concentration of antifungal chitinase in AM roots. They also suggested that increased arginine accumulations in AM roots suppresses sporulation in *Thielaviopsis,* a mechanism reported earlier (Baltruschat and Schöenbeck, 1975). Synthesis of β-1,3-glucans has been reported in mycorrhizal pea-root, and also been detected within the structural host-wall material around the point of penetration of AM hyphae into plant cell (Gianinazzi-Pearson et al., 1996).

ROLE OF AM ESTABLISHMENT
IN TRIGGERING DEFENSE RESPONSES

It is most likely that more than one mechanism is responsible for the role of AM fungi as biocontrol agents (Mukerji, 1999). Triggering of a specific plant defense response due to AM colonization can be a most obvious basis for the protective capacity of AM fungi. Among the compounds involved in plant defense (Bowles, 1990) in relation to AM formation are phytoalexins, enzymes of the phenylpropanoid pathway, chitinases, β-1,3-glucanases, peroxidases, pathogenesis-related (PR) proteins, callose, hydroxyproline-rich glycoproteins (HRGP), and phenolics (Gianinazzi-Pearson et al., 1994).

Phytoalexins

Phytoalexins are low-molecular-weight, toxic compounds usually accumulating with pathogen attack and released at the sites of infection. These are not detected during the initial stages of AM formation but can be found in the later stages of symbiosis (Morandi et al., 1984). There is evidence for increase in phytoalexins in mycorrhizal roots infected with *R. solani* (Wyss et al., 1991). Harrison and Dixon (1993) reported an increase in the level of the phytoalexin "medicarpin" transiently in the early stages of AM colonization in *Medicago truncatula,* which decreased to very low levels in the later stages. In another study (Yao et al., 2003), the effect of colonization with the vesicular arbuscular mycorrhizal fungus *Glomus etunicatum* on the content of "rishitin" and "solavetivone" was determined in potato plants challenged with *R. solani.* Mycorrhization significantly stimulated the accumulation of both the phytoalexins in the roots of plantlets challenged with *R. solani,* but did not influence phytoalexin levels in nonchallenged plantlet roots. Lambias (2000) showed that the elicitor derived from an extract of extraradical mycelium of *G. intraradices* was able to induce phytoalexin synthesis in soybean cotyledons. Further results have shown that *G. intraradices* induces the expression of chalcone synthase, the first enzyme in the metabolism of flavonoid compound, such as phytoalexin, in *Medicago truncatula* (Bonanomi et al., 2001). Increased levels of phytoalexins in mature mycorrhizal soybean roots could be elicited by the release of fungal cell wall molecules in senescent arbuscules. Several steps of mycorrhizal development can, therefore, be explained in relation to phenolic compounds (Morandi, 1996):

1. primary contact of mycorrhizal fungus with the roots that induces a specific defense mechanism;
2. a specific suppression of phytoalexin biosynthesis; and
3. in aging arbuscules, an elicitation of phytoalexins due to the disintegration of the arbuscule and the release of fungal cell wall components.

Pathogenesis-Related Proteins

PR proteins are induced in plants during resistance to pathogen infection. They are synthesized locally and in small amounts; for example, PR-b_1 proteins are synthesized only in cells containing living arbuscules (Gianinazzi-Pearson et al., 1992).

Hydroxyproline-Rich Glycoproteins

HRGP-encoding mRNA has been found to accumulate in mycorrhizal parsley roots. These molecules were found to be regularly distributed around the arbuscular hyphae (Balestrini et al., 1994). An increase in the quantity of mRNA encoding HRGP has also been reported in the root cells of maize and bean that contained arbuscules (Balestrini et al., 1997; Blee and Anderson, 2000).

Enzymes

Plants control the ingress of potential fungal pathogens with increased activity of enzymes and accumulation of cell wall proteins associated with defense. These enzymes include phenylalanine ammonia-lyase (PAL) (Edwards et al., 1985) and chalcone synthase (CHS) (Ryder et al., 1987), which are involved in enhanced phenolic metabolism. Other enzymes, chitinase (CHT) (Hedrick et al., 1988) and glucanase (GLU) (Edington et al., 1991), which can degrade fungal cell walls, may accumulate. Enhanced accumulations of the structural protein HYP (hydroxyproline-rich glycoprotein) may increase the resistance of plant cell walls to enzymatic degradation (Corbin et al., 1987). Molecular techniques, like polyacrylamide gel electrophoresis, followed by densitometry of specific fungal enzymes, have proven successful for quantifying the activity of both the fungus and the host in AM symbioses (Gianinazzi-Pearson and Gianinazzi, 1976; Hepper et al., 1988; Rosendahl et al., 1989; Rosendahl, 1992; Thingstrup and Rosendahl, 1994).

Alterations in the isoenzymatic patterns and biochemical properties of some defense-related enzymes, such as chitinases (Pozo et al., 1996), chitosanases (Pozo et al., 1998), and β-1,3-glucanases (Pozo et al., 1999), have been shown earlier during the mycorrhizal colonization of tomato roots with the induction of new isoforms. These hydrolytic enzymes are believed to have a role in the defense against the invading fungal pathogens because of their potential to hydrolyze fungal cell wall polysaccharides (Grenier and Asselin, 1990; Simmons, 1994). The elicitation of both the enzymes, chitinases and/or β-1,3 glucanases, has been reported in AM roots, and these are antifungal against soil/root pathogens (Blee and Anderson, 1996). Mycorrhiza-induced chitinase isoforms appear to be a general phenomenon in AM roots. These chitinases release oligosaccharide elicitors from chitinous AM fungal cell walls which, in turn, stimulate the general defense responses of plants (Cordier et al., 1996). However, Dumas-Gaudot et al. (1992a,b) found that chitinases are only transiently or slightly induced by AM colonization. These chitinase isoforms differ from those elicited by root fungal pathogens, indicating a different pattern of plant response to pathogenic and mutualistic fungi (Dumas-Gaudot et al., 1992b). Further, it was shown that increased levels of chitinase activity are only detected in AM roots at the beginning of colonization (Spanu and Bonfante-Fasolo, 1988; Bonfante-Fasolo, and Spanu, 1992). A decrease in β-1,3-endoglucanase activity has also been reported at specific stages during mycorrhiza development (Lambais and Mehdy, 1993). The decreases were accompanied by differential reductions in the levels of mRNAs encoding for different endochitinase and endoglucanase isoforms. These observations indicate a systemic suppression of the defense reaction in the presence of AM symbiotic interactions. In another study by Pozo et al. (2002), the ability of two AMF (*G. mosseae* and *G. intraradices*) to induce local or systemic resistance to *P. parasitica* in tomato roots was compared using a split root experimental system. *G. mosseae* was effective in reducing disease symptoms produced by *P. parasitica* infection, and evidence points to a combination of local and systemic mechanisms being responsible for this bioprotector effect. The biochemical analysis of different plant defense-related enzymes showed a local induction of mycorrhiza-related new isoforms of the hydrolytic enzymes, chitinase, chitosanase, and β-1,3-glucanase, as well as superoxide dismutase, an enzyme which is involved in cell protection against oxidative stress.

Peroxidase activity has been associated with epidermal and hypodermal cells increased in mycorrhizal roots (Gianinazzi and Gianinazzi-Pearson, 1992), a process that can contribute to higher resistance to certain root pathogens. Also, wall-bound peroxidase activity has been detected during the initial stages of AM colonization, which later decreases (Spanu and Bonfante-Fasolo, 1988). However, peroxidase activity was not detectable in cells containing intracellular arbuscular hyphae.

Enhancement of the phenylalanine ammonium-lyase, the first enzyme of the phenylpropanoid pathway, and chalcone isomerase, the second enzyme specific for flavonoid/isoflavonoid biosynthesis, in amount and activity during the early colonization of plant roots by *G. intraradices* have been reported (Lambais and Mehdy, 1993; Volpin et al., 1994, 1995). These results suggest that AM fungi have an important role in initiating host defense response. In contrast, levels of transcripts encoding PAL and chalcone synthase, also involved in the flavonoid/isoflavonoid biosynthesis, increased in *Medicago truncatula* roots during colonization with *Glomus versiforme* (Harrison and Dixon, 1993).

Fungal enzyme activities were quantified in an interaction study between the fungus *G. intraradices* and the pea pathogen *Aphanomyces euteiches* (Kjoller and Rosendahl, 1996). Fungal and host enzymes were separated by polyacrylamide gel electrophoresis, and the activity of *A. euteiches*-specific glucose-6-phosphate dehydrogenase (Gd), phosphoglucomutase and peptidase (PEP) enzymes were quantified by densitometry. The PEP and Gd activities of *A. euteiches* were suppressed in mycorrhizal plants compared with nonmycorrhizal plants. The patterns of *A. euteiches* enzyme activity suggest that the initial, parasitic phase of the pathogen was relatively short, with oospores formed as the fungus entered its resting state, and that this occurred later in mycorrhizal than in nonmycorrhizal roots. The plants preinoculated with *G. intraradices* showed no symptoms of severe root rot, even though the pathogen was present and active in these plants. Thus, plants preinoculated with *G. intraradices* were more tolerant of infection with *A. euteiches* than nonmycorrhizal plants.

Also, some elements of signal transduction pathways activated after pathogen recognition by a plant have been observed transiently during the early stages of AM formation. In mycorrhizal tobacco plants, transient increase in catalase and peroxidase activity was observed coinciding with appressoria formation and fungal penetration into the root (Blilou, Bueno, et al., 2000). Similar results have been seen in onion and bean roots inoculated with AMF (Spanu and Bonfante-Fasolo, 1988; Lambais, 2000). The transient increase in catalase and peroxidase activity observed in tobacco mycorrhizal roots also coincided with the accumulation of salicylic acid (SA) (Blilou, Bueno et al., 2000). SA is a signal molecule involved in the signal transduction pathway activated in plant–pathogen reactions (Malamy et al., 1990; Métraux et al., 1990). A transient accumulation of SA during the early stages of infection has also been observed in the interaction between rice and *G. mosseae* (Blilou, Ocampo, et al., 2000). In this case, the accumulation of SA was also correlated with an increase in the expression of genes encoding lipid transfer protein (LTP) and phenylalanine ammonia-lyase (Blilou, Ocampo et al., 2000). This provides evidence that induction of PAL and LTP is part of the defense pathway (Blilou, Ocampo et al., 2000).

REFERENCES

Abdel-Fattah, G.M. and Y.M. Shabana. (2002). Efficacy of the arbuscular mycorrhizal fungus *Glomus clarum* in protection of cowpea plants against root rot pathogen *Rhizoctonia solani*. *Zeitschrift für Pflanzenkrankheiten und Pflanzenschutz* 109: 207-215.

Ahmad, M. and A. Al-Raddad. (1995). Interaction of *Glomus mosseae* and *Paecilomyces lilacinus* on *Meloidogyne javanica* of tomato. *Mycorrhiza* 5: 233-236.

Allen, E.B., A.F. Allen, D.J. Helm, J.M. Trappe, R. Molina, and E. Rincon. (1995). Pattern and regulation of mycorrhizal plant and fungal diversity. *Plant and Soil* 170: 47-62.

Al-Momany, A. and A. Al-Radded. (1988). Effect of vesicular-arbuscular mycorrhizae on *Fusarium* wilt of tomato and pepper. *Alexandria Journal Agricultural Research* 33: 249-261.

Altieri, M.A. (1994). Sustainable agriculture. *Encyclopedia of Agricultural Science* 4: 239-247.

Atkinson, D., G. Berta, and J.E. Hooker. (1994). Impact of mycorrhizal colonisation on root architecture, root longevity and the formation of growth regulators. In *Impact of Arbuscular Mycorrhizas on Sustainable Agriculture and Natural Ecosystems*, eds. S. Gianinazzi and H. Schüepp. Birkhäuser, Basel, pp. 89-99.

Azcón-Aguilar, C. and B. Bago. (1994). Physiological characteristics of the host plant promoting an undisturbed functioning of the mycorrhizal symbiosis. In *Impact of Arbuscular Mycorrhizas on Sustainable Agriculture and Natural Ecosystems*, eds. S. Gianinazzi and H. Schüepp. Birkhäuser, Basel, pp. 47-60.

Azcón-Aguilar, C. and J.M. Barea. (1996). Arbuscular mycorrhizas and the biological control of soil-borne plant pathogens—An overview of the mechanisms involved. *Mycorrhiza* 6: 457-464.

Azcón-Aguilar, C. and J.M. Barea. (1997). Applying mycorrhiza biotechnology to horticulture: Significance and potentials. *Science Horticulture* 68: 1-24.

Balestrini, R., M. José-Estanyol, P. Puigdoménech, and P. Bonfante. (1997). Hydroxyproline-rich glycoprotein mRNA accumulation in maize root cells colonized by an arbuscular mycorrhizal fungus as revealed by in situ hybridization. *Protoplasma* 198: 36-42.

Balestrini, R., C. Romera, P. Puigdomenech, and P. Bonfante. (1994). Location of a cell-wall hydroxyproline-rich glycoprotein, cellulose and β-1,3-glucans in apical and differentiated regions of maize mycorrhizal roots. *Planta* 195: 201-209.

Baltruschat, H. and F. Schoenbeck. (1975). The influence of endotrophic mycorrhiza on the infestation of tobacco by *Thielaviopsis basicola*. *Phytopathologische Zetschrift* 84: 172-188.

Bansal, M., B.P. Chamola, N. Sarwar, and K.G. Mukerji. (2000). Mycorrhizosphere: Interactions between rhizosphere microflora and VAM fungi. In *Mycorrhizal Biology*, eds. K.G. Mukerji, B.P. Chamola, and J. Singh. Kluwer Academic/Plenum Publishers, New York, pp. 143-152.

Barea J.M. and P. Jeffries. (1995). Arbuscular mycorrhizas in sustainable soil plant systems. In *Mycorrhiza Structure, Function, Molecular Biology and Biotechnology*, eds. B. Hock and A. Varma. Springer, Heidelberg, pp. 521-559.

Becker, W.N. (1976). Quantification of onion vesicular-arbuscular mycorrhizae and their resistance to *Pytrnochaeta terrestrae*. PhD dissertation, University of Illinois. Urbana.

Benhamou, N., J.A. Fortin, C. Hamel, M. St-Arnaud, and A. Shatilla. (1994). Resistance responses of mycorrhizal RiT-DNA-transformed carrot roots to infection by *Fusarium oxysporum* f. sp. *chrysanthemi*. *Phytopathology* 84: 958-968.

Bethlenfalvay, G.J. and R.G. Linderman (eds.). (1992). *Mycorrhizae in Sustainable Agriculture*. ASA Special Publication No. 54, Madison, WI.

Bethlenfalvay, G.J. and H. Schüepp. (1994). Arbuscular mycorrhizas and agrosystem stability. In *Impact of Arbuscular Mycorrhizas on Sustainable Agriculture and Natural Ecosystems*, eds. S. Gianinazzi and H. Schüepp. Birkhäuser, Basel, pp. 117-131.

Blee K.A. and A.J. Anderson. (1996). Defense-related transcript accumulation in *Phaseolus vulgaris* L. colonized by the arbuscular mycorrhizal fungus *Glomus intraradices* Schenck & Smith. *Plant Physiology* 110: 675-688.

Blee K.A. and A.J. Anderson. (2000). Defense responses in plants to arbuscular mycorrhizal fungi. *Current Advances in Mycorrhizae Research*, eds. G.K. Podila and D.D. Douds. The American Phytopathological Society, St. Paul, MN, pp. 27-44.

Blilou, I., P. Bueno, J.A. Ocampo, and J.M. García-Garrido. (2000). Induction of catalase and ascorbate peroxidase activities in tobaccco roots inoculated with the arbuscular mycorrhizal fungus *Glomus mosseae*. *Mycological Research* 104: 722-725.

Blilou, I., J.A. Ocampo, and J.M. García-Garrido. (1999). Resistance of pea roots to endomycorrhizal fungus or Rhizobium correlates with enhanced levels of endogenous salicylic acid. *Journal of Experimental Botany* 50: 1663-1668.

Blilou, I., J.A. Ocampo, and J.M. García-Garrido. (2000). Induction of Ltp (Lipid transfer protein) and Pal (phenylalanine ammonia-lyase) gene expression in rice roots colonized by the arbuscular mycorrhizal fungus *Glomus mosseae*. *Journal of Experimental Botany* 51: 1969-1977.

Bodker, L., R. Kjoller, K. Kristensen, and S. Rosendahl. (2002). Interactions between indigenous arbuscular mycorrhizal fungi and *Aphanomyces euteiches* in field-grown pea. *Mycorrhiza* 12: 7-12.

Bonanomi, A., J.H. Oetiker, R. Guggenheim, T. Boller, A. Wiemken, and R. Vögeli-Lange. (2001). Arbuscular mycorrhizas in mini-mycorrhizotrons: First contact of *Medicago truncatula* roots with *Glomus intraradices* induces chalcone synthase. *New Phytologist* 150: 573-582.

Bonfante-Fasolo, P. and P. Spanu. (1992). Pathogenic and endomycorrhizal associations. In *Methods in Microbiology,* Vol 24: *Techniques for the Study of Mycorrhiza,* eds. J.R. Norris, D.J. Read, and A.K. Varma. Academic Press, London, pp. 142-167.

Bowles, D.J. (1990). Defense related proteins in higher plants. *Annual Review Biochemistry* 59: 873-907.

Calvet, C., J. Pinochet, A. Camprubi, V. Estaun, and R. Rodriguez-Kabana. (2001). Evaluation of natural chemical compounds against root-lesion and root-knot nematodes and side-effects on the infectivity of arbuscular mycorrhizal fungi. *European Journal Plant Pathology* 107: 601-605.

Calvet, C., J. Pinochet, A. Hernandez-Dorrego, V. Estaun, and A. Camprubi. (2001). Field microplot performance of the peach-almond hybrid GF-677 after inoculation with arbuscular mycorrhizal fungi in a replant soil infested with root-knot nematodes. *Mycorrhiza* 10: 295-300.

Cameron, G.C. (1986). Interactions between two vesicular-arbuscular mycorrhizal fungi, the soybean cyst nematode, and phosphorus fertility on two soybean cultivars. MS thesis, University of Georgia, Athens.

Caron, M., J.A. Fortin, and C. Richard. (1985). Influence of substrate on the interaction of *Glomus intraradices and Fusarium oxysporum* f. sp. *radicis-lycopersici* on tomatoes. *Plant Soil* 87: 233-239.

Caron, M., J.A. Fortin, and C. Richard. (1986a). Effect of *Glomus intraradices* on infection by *Fusarium oxysporum* f. sp. *radicis-lycopersici* in tomatoes over a 12-week period. *Canadian Journal of Botany* 64: 552-556.

Caron, M., J.A. Fortin, and C. Richard. (1986b). Effect of phosphorus concentration and *Glomus intraradices* on Fusarium crown and root rot of tomatoes. *Phytopathology* 76: 942-946.

Caron, M., J.A. Fortin, and C. Richard. (1986c). Effect of preinfestion of the soil by a vesicular-arbuscular mycorrhizal fungus. *Glomus intraradices,* on Fusarium crown and root rot of tomatoes. *Phytoprotection* 67: 15-19.

Cook, R.J. and K.F. Baker. (1982). *The Nature and Practice of Biological Control of Plant Pathogens.* APS Press, St. Paul, MN.

Cooper, K.M. and G.S. Grandison. (1987). Effects of vesicular-arbuscular mycorrhizal fungi on infection of tamarillo *(Cyphomandra betacea)* by *Meloidogyne incognita* in fumigated soil. *Plant Disease* 71: 1101-1106.

Corbin, D.R., N. Sauer, and C.J. Lamb. (1987). Differential regulation of a hydroxyproline-rich glycoprotein gene family in wounded and infected plants. *Molecular Cell Biology* 7: 4337-4344.

Cordier, C., S. Gianinazzi, and V. Gianinazzi-Pearson. (1996). Colonisation patterns of root tissues by *Phytophthora nicotianae* var. *parasitica* related to reduced disease in mycorrhizal tomato. *Plant Soil* 185: 223-232.

Cordier, C., A. Trouvelot, S. Gianinazzi, and V. Gianinazzi-Pearson. (1997). Arbuscular mycorrhiza technology applied to micropropagated *Prunus avium* and to protection against *Phytophthora cinnamomi. Agronomie* 17: 256-265.

Dar, G.H., M.Y. Zargar, and G.M. Beigh. (1997). Biocontrol of Fusarium root rot in the common bean (*Phaseolus vulgaris* L) by using symbiotic *Glomus mosseae* and *Rhizobium leguminosarum. Microbial Ecology* 34: 74-80.

Davis, R.M. (1980). Influence of *Glomus fasciculatus* on *Thielaviopsis basicola* root rot of citrus. *Plant Disease* 64: 839-840.

Davis, R.M. and J.A. Menge. (1980). Influence of *Glomus fasciculatus* and soil phosphorus on *phytophthora* root rot of citrus. *Phytopathology* 70: 447-452.

Dehne, H.W. (1982). Interaction between vesicular-arbuscular mycorrhizal fungi and plant pathogens. *Phytopathology* 72: 1115-1119.

Dehne, H.W. and F. Schönbeck. (1979). Untersuchungen zum Einfluss der endotrophen Mykorrhiza auf Pflanzenkrankheiten: II. Phenolstoffwechsel und Lignifizierung. (The influence of endotrophic mycorrhiza on plant diseases: II. Phenolmetabolism and lignification). *Phytopathologisch Zeitscrift* 95: 210-216.

Dehne, H.W. F. Schonbeck, and H. Baltruschat. (1978). Untersuchungen zum einfluss der endotrophen Mycorrhiza auf Pflanzenkheiten: 3. Chitinase-aktivitat und Ornithinzyklus. (The influence of endotrophic mycorrhiza on plant diseases: 3. Chitniase-activity and ornithinecycle.) *Zeitschrift fur Pflanzenkrankheiten* 85: 666-678.

Dugassa G.D., H. von Alten, and F. Schonbeck. (1996). Effects of arbuscular mycorrhiza (AM) on health of *Linum usitatissimum* L. infected by fungal pathogens. *Plant and Soil* 185: 173-182.

Dumas-Gaudot, E., V. Furlan, J. Grenier, and A. Asselin. (1992a). Chitinase, chitosanase and β-1, 3-ghcanase activities in *Allium* and *Pisum* roots colonized by *Glomus* species. *Plant Science* 84: 17-24.

Dumas-Gaudot, E., V. Furlan, J. Grenier, and A. Asselin. (1992b). New acidic chitinase isoforms induced in tobacco roots by vesicular-arbuscular mycorrhizal fungi. *Mycorrhiza* 1: 133-136.

Edington, B.V., C.J. Lamb, and R.A. Dixon (1991). cDNA cloning and characterization of a putative 1, J-P-D-glucanase transcript induced by fungal elicitor in bean cell suspension cultures. *Plant Molecular Biology* 16: 81-94.

Edwards, K., C.L. Cramer, G.P. Bolwell, R.A. Dixon, W. Schuch, and C.J. Lamb. (1985). Rapid transient induction of phenylalanine ammonialyase mRNA in elicitor-treated bean cells. *Proceedings National Academy Sciences USA* 82: 6731-6735.

Elsen, A, H. Baimey, R. Sweenen, and D. De Waele. (2003). Relative mycorrhizal dependency and mycorrhiza-nematode interaction in banana cultivars (*Musa* spp.) differing in nematode susceptibility. *Plant and Soil* 256: 303-313.

Elsen, A., S. Declerck, and D. De Waele. (2001). Effects of *Glomus intraradices* on the reproduction of the burrowing nematode *(Radopholus similis)* in dixenic culture. *Mycorrhiza* 11: 49-51.

Fassuliotis, G. (1970). Resistance in *Cucumis* spp. to root-knot nematode *Meloidogyne incognita* acrita. *Journal of Nematology* 2: 174.

Filion, M., M. St-Arnaud, and S.H. Jabaji-Hare. (2003). Quantification of *Fusarium solani* f. sp *phaseoli* in mycorrhizal bean plants and surrounding mycorrhizosphere soil using real-time polymerase chain reaction and direct isolations on selective media. *Phytopathology* 93: 229-235.

Francl I.J. and V.H. Dropkin. (1985). *Glomus fasciculatum* a weak pathogen of *Heterodera glycines. Journal of Nematology* 17: 470-475.

Gianinazzi-Pearson, V., E. Dumas-Gaudot, A. Gollotte, A. Tahiri-Alaoui, and S. Gianinazzi. (1996). Cellular and molecular defencerelated root responses to invasion by arbuscular mycorrhizal fungi. *New Phytologist* 133: 45-57.

Gianinazzi-Pearson, V. and S. Gianinazzi. (1976). Enzymatic studies on the metabolism of vesicular-arbuscular mycorrhizal: I. Effect of mycorrhiza formation and phosphorus nutrition on soluble phosphatase activities in onion roots. *Physiologie Vegetable* 14: 833-841.

Gianinazzi-Pearson, V., A. Gollotte, E. Dumas-Gaudot, P. Franken, and S. Gianinazzi. (1994). Gene expression and molecular modifications associated with plant responses to infection by arbuscular mycorrhizal fungi. In *Advances in Molecular Genetics of Plant-Microbe Interactions,* eds. M. Daniels, J.A. Downic, and A.E. Osbourn. Kluwer, Dordrecht, the Netherlands, pp. 179-186.

Gianinazzi-Pearson, V., A. Tahiri-Alaoui, J.F. Antoniw, S. Gianinazzi, and E. Dumas-Gaudot. (1992). Weak expression of the pathogenesis related PR-b1 gene and localization of related protein during symbiotic endomycorrhizal interactions in tobacco roots. *Endocytobiosis Cell Research* 8: 177-185.

Good, J.M. (1968). Relation of plant parasitic nematodes to soil management practices. In *Tropical Nematology,* eds. G.S. Smart and V.G. Perry. University of Florida, Gainsville, pp. 113-138.

Graham, J.H., R.T. Leonard, and J.A. Menge. (1981). Membrane mediated decreases in root exudation responsible for phosphorus inhibition of vesicular-arbuscular mycorrhiza formation. *Plant Physiology* 68: 548-552.

Grenier, J. and A. Asselin. (1990). Some pathogenesis-related proteins are chitosanases with lytic activity against fungal spores. *Molecular Plant-Microbe Interactions* 3: 401-407.

Habte, M., Y.C. Zhang, and D.P. Schmitt. (1999). Effectiveness of *Glomus* species in protecting white clover against nematode damage. *Candian Journal of Botany* 7: 135-139.

Harley, J.L. and S.E. Smith. (1983). *Mycorrhizal Symbiosis.* Academic Press, New York.

Harrier, L.A. and C.A. Watson. (2004). The potential role of arbuscular mycorrhizal (AM) fungi in the bioprotection of plants against soil-borne pathogens in organic and/or other sustainable farming systems. *Pest Management Science* 60: 149-157.

Harrison, M.J. and R.A. Dixon. (1993). Isoflavonoid accumulation and expression of defense gene transcripts during the establishment of vesicular-arbuscular mycorrhizal associations in roots of *Medicago truncatula*. *Molecular Plant-Microbe Interactions* 6: 643-654.

Hayman, D.S. (1982). The physiology of vesicular-arbuscular endo-mycorrhizal symbiosis. *Canadian Journal of Botany* 6: 944-963.

Hedrick, S.A., J.N. Bell, T. Boller. and C.J. Lamb. (1988). Chitinase cDNA cloning and mRNA induction by fungal elicitor, wounding and infection. *Plant Physiology* 86: 182-186.

Hepper, C.M., C. Azcón-Aguilar, S. Rosendahl, and R. Sen. (1988). Competition between three species of *Glomus* used as spatially separated, introduced and indigenous mycorrhizal inocula for leek (*Allium porrum* L.). *New Phytology* 93: 401-413.

Hooker, J.E., M. Jaizme-Vega. and D. Atkinson. (1994). Management of positive interactions of arbuscular mycorrhizal fungi with essential groups of soil micro-organisms. In *Impact of Arbuscular Mycorrhizas on Sustainable Agriculture and Natural Ecosystems*, eds. S. Gianinazzi and H. Schüepp. Birkhäuser, Basel, pp. 191-200.

Hussey, R.S. and R.W. Roncondori. (1982). Vesicular-arbuscular mycorrhizae may limit nematode activity and improve plant growth. *Plant Disease* 66: 9-14.

Jalali, B. L. and Jalali, I. (1991). Mycorrhiza in plant disease control. In *Handbook of Applied Mycology. Soil and Plants*, Vol.1, eds. D.K. Arora, B. Rai, K.G. Mukerji, and G.R. Knudsen. Marcel Dekker, New York, pp. 131-154.

Karagiannidis, N., F. Bletsos. and N. Stavropoulos (2002). Effect of Verticillium wilt (*Verticillium dahliae* Kleb.) and mycorrhiza (*Glomus mosseae*) on root colonization, growth and nutrient uptake in tomato and eggplant seedlings. *Scientia Horrticulture* 94: 145-156.

Kjoller, R. and Rosendahl, S. (1996). The presence of the arbuscular mycorrhizal fungus *Glomus intraradices* influences enzymatic activities of the root pathogen *Aphanomyces euteiches* in pea roots. *Mycorrhiza* 6: 487-491.

Lambais, M.R. (1985). Microbial interactions in the mycorrhizosphere. *Proceedings 6th North American Conference on Mycorrhizae*, ed. R. Molina, Oregon, pp. 117-120.

Lambais, M.R. (2000). Regulation of plant defence-related genes in arbuscular mycorrhizae. In *Current Advances in Mycorrhizae Research*, eds. G.K. Podila and D.D. Douds. The American Phytopathological Society, St. Paul, MN, pp. 45-59.

Lambais, M.R. and M.C. Mehdy. (1993). Suppression of endochitinase, β-1,3-endoglucanase and chalcone isomerase expression in bean vesicular-arbuscular mycorrhizal roots under different soil phosphate conditions. *Molecular Plant-Microbe Interactions* 6: 75-83.

Linderman, R.G. (1994). Role of AM fungi in biocontrol. In *Mycorrhizae and Plant Health*, eds. F.L. Pfleger and R.G. Linderman. American Phytopathological Society, St. Paul, MN, pp. 1-25.

Lynch, J.M. (1990). *The Rhizosphere*. Wiley, New York.

Malamy, J., J.P. Carr, D.F. Klessig, and I. Raskin (1990). Salicylic acid: A likely endogenous signal in the resistance response of tobacco to viral infection. *Science* 250: 1002-1004.

Matsubara, Y., N. Hasegawa, and H. Fukui. (2002). Incidence of Fusarium root rot in asparagus seedlings infected with arbuscular mycorrhizal fungus as affected by several soil amendments. *Journal Japanese Society of Horticultural Sciences* 71: 370-374.

Matsubara, Y., N. Ohba, and H. Fukui. (2001). Effect of arbuscular mycorrhizal fungus infection on the incidence of fusarium root rot in asparagus seedlings. *Journal Japanese Society Horticultural Sciences* 70: 202-206.

Métraux, J.P., H. Signer, J. Ryals, E. Ward, M. Wyss-Benz, J. Gaudin, K. Raschdorf, E. Schmid, W. Blum, and B. Inverardi (1990). Increase in salicylic acid at the onset of systemic acquired resistance in cucumber. *Science* 250: 1004-1006.

Meyer, J.R. and R.G. Linderman. (1986). Selective influences on populations of rhizosphere or rhizoplane bacteria and actinomycetes by mycorrhizas formed by *Glomus fasciculatum*. *Soil Biology Biochem* 18: 191-196.

Mohr, U., J. Lange, T. Boller, A. Wiemken, and R. VogeliLange. (1998). Plant defence genes are induced in the pathogenic interaction between bean roots and *Fusarium solani,* but not in the symbiotic interaction with the arbuscular mycorrhizal fungus *Glomus mosseae. New Phytologist* 138: 589-598.

Morandi, D. (1987). VA mycorrhizae, nematodes, phosphorus and phytoalexins on soybean. In *Mycorrhizae in the Next Decade, Practical Application and Research Priorities,* eds. D.M. Sylvia, L.L. Hung, and J.H. Graham. *Proceedings of the 7th North American Conference on Mycorrhiza,* Institute of Food and Agricultural Sciences, University of Florida, Gaineville.

Morandi, D. (1996). Occurrence of phytoalexins and phenolic compounds in endomycorrhizal interactions, and their potential role in biological control. *Plant and Soil* 185: 241-251.

Morandi, D., J.A. Bailey, and V. Gianinazzi-Pearson. (1984). Isoflavonoid accumulation in soybean roots infected with vesicular-arbuscular mycorrhizal fungi. *Physiology Plant Pathology* 24: 357-364.

Mukerji, K.G. (ed.). (1999). Mycorrhiza in control of plant pathogens: Molecular approaches. In *Bio-technological Approaches in Biocontrol of Plant Pathogens,* eds. K.G. Mukerji, B.P. Chamola, and R.K. Upadhyay. Kluwer Academic/ Plenum Publishers, New York, pp. 135-155.

Mukerji, K.G., B.P. Chamola, and M. Sharma. (1997). Mycorrhiza in control of plant pathogens. In *Management of Threatening Plant Diseases of National Importance,* eds. V.P. Agnihotri, A.K. Sarbhoy, and D.V. Singh. Mehrotra Publishing House, New Delhi, pp. 297-314.

Mukerji, K.G., B.P. Chamola, and R.K. Upadhyay (eds.). (1999). *Biotechnological Approaches in Biocontrol of Plant Pathogens.* Kluwer Academic/Plenum Publishers, New York.

Mukerji, K.G. and K.L. Garg (eds.). (1988a). *Biocontrol of Plant Diseases,* Vol. I. CRC Press, Boca Raton, FL.

Mukerji, K.G. and K.L. Garg (eds). (1988b). *Biocontrol of Plant Diseases,* Vol. II. CRC Press, Boca Raton, FL.

Mukerji, K.G., J.P. Tewari, D.K. Arora, and G. Saxena (eds.). (1992). *Recent Developments in Biocontrol of Plant Diseases.* Aditya Books Ltd., New Delhi, India.

Mukerji, K.G., R.K. Upadhyay, and A. Kaushik. (1997). Mycorrhiza and integrated disease management. In *IPM System in Agriculture*, Vol. 2: *Biocontrol in Emerging Biotechnology*, eds. R.K. Upadhyay, K.G. Mukerji, and R.L. Rajak. Aditya Books Ltd., New Delhi, India, pp. 423-452.

Newsham, K.K., A.H. Fitter, and A.R. Watkinson. (1995a). Arbuscular mycorrhiza protect an annual grass from root pathogenic fungi in the field. *Journal of Ecology* 83: 991-1000.

Newsham, K.K., A.H. Fitter, and A.R. Watkinson. (1995b). Multi-functionality and biodiversity in arbuscular mycorrhizas. *Trends in Ecology and Evolution* 10: 407-411.

Pacovsky, R.S., G.J. Bethelenfalvay, and E.A. Paul. (1986). Comparisons between P-fertilized and mycorrhizal plants. *Crop Science* 26: 151-156.

Paulitz, T.C. and R.G. Linderman. (1991). Lack of antagonism between the biocontrol agent *Gliocaldium vivens* and vesicular arbuscular-mycorrhizal fungi. *New Phytologist* 117: 303-308.

Perrin, R. (1990). Interactions between mycorrhizae and diseases caused by soilborne fungi. *Soil Use and Management* 6: 189-195.

Pinochet, J., C. Calvet, A. Camprubi, and C. Fernandez. (1996). Interaction between migratory endoparasitic nematodes and arbuscular mycorrhizal fungi in perennial crops. *Plant Soil* 185: 183-190.

Pozo, M.J., C. Azcón-Aguilar, E. Dumas-Gaudot, and J.M. Barea. (1999). β-1,3-glucanase activities in tomato roots inoculated with arbuscular mycorrhizal fungi and/or *Phytophthora parasitica* and their possible involvement in bioprotection. *Plant Science* 141: 149-157.

Pozo, M.J., C. Cordier, and E. Dumas-Gaudot. (2002). Localized versus systemic effect of arbuscular mycorrhizal fungi on defence responses to Phytophthora infection in tomato plants. *Journal Experimental Botany* 53: 525-534.

Pozo, M.J., E. Dumas-Gaudot, C. Azcón-Aguilar, and J.M. Barea. (1998). Chitosanase and chitinase activities in tomato roots during interactions with arbuscular mycorrhizal fungi or *Phytophthora parasitica*. *Journal of Experimental Botany* 49: 1729-1739.

Pozo, M.J., E. Dumas-Gaudot, S. Slezack, C. Cordier, A. Asselin, S. Gianinazzi, V. Gianinazzi-Pearson, C. Azcón-Aguilar, and J.M. Barea (1996). Detection of new chitinase isoforms in arbuscular mycorrhizal tomato roots: Possible implications in protection against *Phytophthora nicotianae* var. *parasitica*. *Agronomie* 16: 689-697.

Ramraj, B., N. Sahnmugam, and A. Dwarkanath Reddy. (1988). Biocontrol of *Macrophomina* root rot of cowpea and *Fusarium* wilt of tomato by using VAM fungi. *Mycorrhizae for Green Asia: Proceedings of the first Asian Conference on Mycorrhizae*, Madras University, pp. 250-251.

Rao, M.S., B.R. Kerry, S.R. Gowen, J.M. Bourne, and P.P. Reddy. (1997). Management of *Meloidogyne incognita* in tomato nurseries by integration of *Glomus deserticola* with *Verticillium chlamydosporium*. *Zeitschrift fur Pflanzenkrankheiten und Pflanzenschutz* 104: 419-422.

Rivas-Platero, G.G. and J.C. Andrade. (1998). Interaction of mycorrhizal fungi and *Meloidogyne exigua* in coffee. *Manejo Integrado de Plagas* (Costa Rica) 49: 68-72.

Rosendahl, C.N. and S. Rosendahl. (1990). The role of vesicular-arbuscular mycorrhiza in controlling damping-off and growth reduction in cucumber caused by *Pythium ultimum*. *Symbiosis* 9: 363-366.

Rosendahl, S. (1992). Influence of three vesicular-arbuscular mycorrhizal fungi (Glomaceae) on the activity of specific enzymes in the root system of *Cucumis sativus* L. *Plant Soil* 144: 219-226.

Rosendahl, S., J.C. Dodd, and C. Walker. (1994). Taxonomy and phylogeny of the Glomales. In *Impact of Arbuscular Mycorrhizas on Sustainable Agriculture and Natural Ecosystems*, eds. S. Gianinazzi and H. Schüepp. Birkhäuser, Basel, pp. 1-12.

Rosendahl, S., R. Sen, C.M. Hepper, and C. Azcón-Aguilar. (1989). Quantification of three vesicular-arbuscular mycorrhizal fungi (*Glomus* spp.) in roots of leek *(Allium porrum)* on the basis of activity of diagnostic enzymes after polyacrylamide gel electrophoresis. *Soil Biology Biochemistry* 21: 519-522.

Ryder, T.B., S.A. Hedrick, J.N. Bell, X. Liang, S.D. Clouse, and C.J. Lamb. (1987). Organization and differential activation of a gene family encoding the plant defense enzyme chalcone synthase in *Paseolus vulgaris*. *Molecular and General Genetics* 210: 219-233.

Safir, G. (1968). The influence of vesicular-arbuscular mycorrhiza on the disease of onion in *Pyrenocharia terrestris*. MS thesis, University of Illinois, Urbana.

Schmitthenner, A.F. and C.H. Canaday. (1983). Role of chemical factors in development of *Phytophthora* diseases. In *Phytoplithora: Its Biology, Taxonomy, Ecology, and Pathology,* eds. D.C. Erwin, S. Bartnicki-Garcia, and P.H. Tsao. American Phytopathological Society, St. Paul, MN, pp. 175-187.

Schoenbeck, F. (1979). Endomycorrhiza in relation to plant diseases. In *Soil-borne Plant Pathogens,* eds. B. Schipper and W. Gams. Academic Press, New York, pp. 271-280.

Secilia, J. and D.J. Bagyaraj. (1987). Bacteria and actinomycetes associated with pot cultures of vesicular-arbuscular mycorrhizas. *Candian Journal of Microbiology* 33: 1069-1073.

Sharma, A.K., B.N. Johri, and S. Gianinazzi. (1992). Vesicular-arbuscular mycorrhizae in relation to plant diseases. *World Journal of Microbiological Biotechnology* 8: 559-563.

Sharma, M.P. and A. Adholeya. (2000). Sustainable management of arbuscular mycorrhizal fungi in the biocontrol of soil-borne plant diseases. In *Biocontrol Potential and Its Exploitation in Sustainable Agriculture,* Vol. I: *Crop Diseases,* eds. R.K. Upadhaya, K.G. Mukerji, and B.P. Chamola. Kluwer Academic/Plenum Publishers, New York, pp. 117-138.

Sharma, M.P., S. Bhargava, M.K. Verma, and A. Adholeya. (1994). Interaction between the endomycorrhizal fungus *Glomus fasciculatum* and the root-knot nematode, *Meloidogyne incognita* on tomato. *Indian Journal of Nematology* 24: 34-39.

Sharma, M.P., A. Gaur, Tanu, and O.P. Sharma. (2004). Prospects of arbuscular mycorrhiza in sustainable management of root and soil borne diseases of vegetable crops. In *Disease Management of Fruits and Vegetables,* Vol. 1: *Fruit and*

Vegetable Diseases, ed. K.G Mukerji. Kluwer Academic Publishers, Dordrecht, the Netherlands, pp. 501-539.

Siddiqui, Z.A. and I. Mahmood. (1995). Some observations on the management of the wilt disease complex of pigeonpea by treatment with a vesicular arbuscular fungus and biocontrol agents for nematodes. *Bioresource Technology* 54: 227-230.

Sivaprasad, P., A. Jacob, S.K. Nair, and B. George. (1990). Influence of VA mycorrhizal colonization on root-knot nematode infestation in *Piper nigrum* L. In *Current Trends in Mycorrhizal Research,* eds. B.L. Jalali and H. Chand. *Proceedings of the National Conference on Mycorrhiza,* Haryana Agricultural University, Hisar, India, pp. 100-101.

Simmons, C.R. (1994). The physiology and molecular biology of plant 1,3-β-D-glucanases and 1,3;1,4-β-D-glucanases. *Critical Reviews in Plant Sciences* 13: 325-387.

Singh, R., A. Adholya, and K.G. Mukerji. (2000). Mycorrhiza in control of soil borne pathogens, In *Mycorrhizal Biology,* eds. K.G. Mukerji, B.P. Chamola, and J. Singh. Kluwer Academic/Plenum Publishers, New York, pp. 173-196.

Singh, Y.P., R.S. Singh, and K. Sitaramaiah. (1990). Mechanism of resistance of mycorrhizal tomato against root-knot nematode. In *Current Trends in Mycorrhizal Research,* eds. B.L. Jalali and H. Chand. *Proceedings of the National Conference on Mycorrhiza,* Haryana Agricultural University. Hisar, India, pp. 96-97.

Smith, G.S. (1987). Interactions of nematodes with mycorrhizal fungi. In *Vistas on Nematology,* eds. J.A. Veech and D.W. Dickon. Society of Nematology, Hyattsville, MD, pp. 292-300.

Smith, G.S. (1988). The role of phosphorus nutrition in interactions of vesicular-arbuscular mycorrhizal fungi with soilborne nematodes and fungi. *Phytopathology* 78: 371-374.

Smith, G.S., R.S. Hussey, and R.W. Roncadori. (1986a). Interaction of endomycorrhizal fungi, superphosphate and *Meloidgyne incognita* on cotton in microplot and field studies. *Journal of Nematology* 18: 208-216.

Smith, G.S., R.S. Hussey, and R.W. Roncadori. (1986b). Penetration and post-infection development of *Meloidogyne incognita* as affected by *Glomus intraradices* and phosphorus. *Journal of Nematology* 18: 429-435.

Smith, G.S. and D.T. Kaplan. (1988). Influence of mycorrhizal fungus, phosphorus and burrowing nematode interactions on growth of rough lemon citurs seedlings. *Journal of Nematology* 20: 539-544.

Smith, S.E. and V. Gianinazzi-Pearson. (1988). Physiological interactions between symbionts in vesicular-arbuscular mycorrhizal plants. *Annual Review of Plant Physiology Molecular Biology* 39: 221-244.

Smith, S.E., V. Gianinazzi-Pearson, R. Koide, and J.W.G. Cainey. (1994). Nutrient transports in mycorrhizas: Structure, physiology and consequences for efficiency of the symbiosis. In *Management of Mycorrhiza in Agriculture, Horticulture and Forestry,* eds. A.D. Robson, L.K. Abbott, and N. Malajczuk. Kluwer, Dordecht, the Netherlands, pp. 103-113.

Spanu, P. and P. Bonfante-Fasolo. (1988). Cell wall-bound peroxidase activity in roots of mycorrhizal *Allium porrum.* New *Phytologist* 109: 119-124.

St Arnaud, M., C. Hamel, M. Caron, and J.A. Fortin. (1994). Inhibition of *Pythium ultimum* in roots and growth substrate of mycorrhizal *Tagetes patula* colonized with *Glomus intraradices. Canadian Journal of Plant Pathology* 16: 187-194.

Talavera, M., K. Itou, and T. Mizukubo. (2001). Reduction of nematode damage by root colonization with arbuscular mycorrhiza (*Glomus* spp.) in tomato-*Meloidogyne incognita* (Tylenchida: Meloidognidae) and carrot-*Pratylenchus penetrans* (Tylenchida: Pratylenchidae) pathosystems. *Applied Entomology and Zoology* 36: 387-392.

Talavera, M., K. Itou, and T. Mizukubo. (2002). Combined application of *Glomus* sp. and *Pasteuria penetrans* for reducing *Meloidogyne incognita* (Tylenchida: Meloidogynidae) populations and improving tomato growth. *Applied Entomology and Zoology* 37: 61-67.

Thingstrup, I. and S. Rosendahl. (1994). Quantification of fungal activity in arbuscular mycorrhizal symbiosis by polyacrylamide electrophoresis and densitometry of malate dehydrogenase. *Soil Biology Biochemistry* 26: 1483-1489.

Thomson-Cason, K.M., R.B. Hussey, and R.W. Roncadori. (1983). Interaction of vesicular-arbuscular mycorrhizal fungi and phosphorus with *Meloidogyne incognita* on tomato. *Journal of Nematology* 15: 410-417.

Torres-Barragan, A., E. Zavaleta-Mejia, C. Gonzalez-Chavez, and R. Ferrera-Cerrato. (1996). The use of arbuscular mycorrhizae to control onion white rot (*Sclerotium cepivorum* Berk.) under field conditions. *Mycorrhiza* 6: 253-258.

Trotta, A., G.C. Varese, E. Gnavi, A. Fusconi, S. Sampo, and G. Berta. (1996). Interactions between the soilborne root pathogen *Phytophthora nicotianae* var. *parasitica* and the AM fungi *Glomus mosseae* in tomato plants. *Plant and Soil* 185: 199-209.

Umesh, K.C., K. Krishnappa, and D.J. Bagyaraj. (1988). Interaction of burrowing nematode, *Radopholus similis* (Cobb) Thorne and VA mycorrhiza, *Glomus fasciculatum* (Thaxt.) Gerd. and Trappe in banana (*Musa acuminata* Coll.). *Indian Journal of Nematology* 18: 6-11.

Upadhyay, R.K., K.G. Mukerji, and B.P. Chamola (eds). (2000). *Biocontrol Potential and Its Exploitation in Sustainable Agriculture,* Vol I: *Crop Diseases, Weeds and Nematodes.* Kluwer Academic/Plenum Publishers, New York.

Upadhyay, R.K., K.G. Mukerji, and B.P. Chamola (eds). (2001). *Biocontrol Potential and Its Exploitation in Sustainable Agriculture,* Vol II: *Insect Pests,* Kluwer Academic/Plenum Publishers, New York.

Volpin, H., Y. Elkind, Y. Okon, and Y. Kapulnik. (1994). A vesicular arbuscular mycorrhizal fungus *(Glomus intraradix)* induces a defense response in alfalfa roots. *Plant Physiology* 104: 683-689.

Volpin, H., D.A. Phillips, Y. Okon, and Y. Kapulnik. (1995). Suppression of an isoflavonoid phytoalexin defense response in mycorrhizal alfalfa roots. *Plant Physiology* 108: 1449-1454.

Waceke, J.W., S.W. Waudo, and R. Sikora. (2002). Effect of inorganic phosphatic fertilizers on the efficacy of an arbuscular mycorrhiza fungus against a root-knot nematode on pyrethrum. *International Journal of Pest Management* 48: 307-313.

Wallace, H.R. (1983). Interactions between nematodes and other factors on plants. *Journal of Nematology* 15: 221-227.

Wyss, P., T. Boller, and A. Wiemken. (1991). Phytoalexin response is elicited by a pathogen *(Rhizoctonia solani)* but not by a mycorrhizal fungus *(Glomus mosseae)* in soybean roots. *Experientia* 47: 395-399.

Yao, M.K., R.J. Tweddell, and H. Desilets. (2002). Effect of two vesicular-arbuscular mycorrhizal fungi on the growth of micropropagated potato plantlets and on the extent of disease caused by *Rhizoctonia solani. Mycorrhiza* 12: 235-242.

Zambolin, L. (1987). Tolerancia de plantas micorrizadas a fitonematóides. II. *Reuniao Brasileira sobre Micorrizas,* Sao Paulo, pp. 103-125.

Chapter 4

Role of Rhizobacteria in Biological Control of Plant Diseases

S. Rosas

INTRODUCTION

Pests such as microbial pathogens, insects, and weeds invade agricultural crops, causing significant losses in plant yield and quality. Disease supression can be achieved through management practices, such as use of crop rotation, soil amendments, and soil solarization. To combat these pests, producers basically rely on chemicals. In 1998, about $1.4 billion was spent on herbicides, insecticides, and fungicides throughout the world for control of diseases.

However, overuse of chemical pesticides to prevent plant diseases has caused soil pollution and environmental contamination, with ill effects on human health (Johnson and Ware, 1992; Andrés et al., 1999). It is desirable to replace, gradually, the use of agrochemicals with new methods to supplement existing disease control strategies to provide effective control and minimize the negative consequences (Reuveni, 1995).

The use of microorganisms that produce metabolites able to suppress plant diseases or products produced biologically constitutes a powerful alternative to synthetic chemicals. This method is known as biological control or biocontrol (Mukerji and Garg, 1988a,b).

At present, more emphasis is on the use of biocontrol agents for control of plant diseases originated from seeds or the rhizosphere (Kloepper and Schroth, 1978; Lemanceau, 1992; Budge et al., 1995; Glick, 1995; Backman et al., 1997; van Loon, 1997; Whipps, 1997a,b,c; Mathre et al., 1999). Most investigations about biocontrol of plant diseases have used a single biocontrol agent as the antagonist to a single pathogen. This fact constitutes a limitation because, generally, single biocontrol agents are not likely to be active in all soil environments or against all pathogens of plant diseases. Several strategies could be employed, including mixtures of antagonists that control different pathogens through different mechanisms.

It is likely that most cases of naturally occurring biological control result from mixtures of antagonististic microorganisms, rather than from a single antagonist. That is the case with disease-suppressive soils (Schippers, 1992). Previous studies on combinations of biological control agents have been reported, such as mixtures of fungi (Budge et al., 1995; Datnoff et al., 1995; De Boer et al., 1997), mixtures of fungi and bacteria (Janisiewicz, 1988; Lemanceau and Alabouvette, 1991; Duffy and Weller, 1995; Duffy et al., 1996; Leeman, de Ouden, van Pelt, Cornellissen, et al., 1996; Hassan et al., 1997; Leibinger et al., 1997), and mixtures of bacteria (Pierson and Weller, 1994; Janisiewicz and Bors, 1995; Mazzola et al., 1995; Raaijmakers, Leeman, et al., 1995; Raaijmakers, van der Sluis, et al., 1995; De Boer et al., 1997; Raupach and Kloepper, 1998).

It is important that the microorganisms utilized in this control system produce stable metabolites in the environment maintaining their effectiveness. An important prerequisite for successful development of strain mixtures is the compatibility of the inoculated microorganisms (Baker, 1990; De Boer et al., 1997). Compared with classical biological control, microbial control needs more human involvement and effort, and, in the wider developmental context, offers conditions for public–private partnerships.

If executed properly, biological control increases the ecological and economic sustainability of farming systems by reducing both the risk of crop losses and the risk to human health from insecticide use.

The diversity and complexity of organismal interactions, the participation in the numerous mechanisms of disease control of microorganisms, and their adaptedness to the environment in which they are used contribute to predict that biocontrol could be more perdurable and effective than synthetic chemicals (Cook, 1993; Benbrook et al., 1996).

As most soilborne plant pathogens are fungi, biocontrol in this aspect has been attempted intensively, utilizing bacteria such as *Pseudomonas, Bacillus,* and *Streptomyces* as effective biocontrol agents (Kloepper and Schroth, 1978; Lemanceau, 1992; Glick, 1995). Fluorescent *Pseudomonas* spp. are among the most effective PGPR (plant growth-promoting rhizobacteria) and have been shown to be responsible for the reduction of soilborne diseases. The biological control activity of selected *Pseudomonas* spp. strains are effective under field conditions (Tuzun and Kloepper, 1995; Wei et al., 1996) and in commercial greenhouse (Leeman et al., 1995). This efficiency can be the result of competition for nutrients, among other reasons (Baker, 1991).

Different mechanisms are involved in the action of the microorganisms against plant pathogens, such as inhibition of pathogens by antimicrobial compounds (antibiosis); competition for iron through production of side-rophores; competition for colonization sites and nutrients; parasitism and production of extracellular cell wall–degrading enzymes, such as chitinase and β-1,3 glucanase (Keel and Défago, 1997; Whipps, 1997a); induction of plant defense mechanism, such as induced systemic resistance (ISR) (Kloepper et al., 1992; Pieterse et al., 1996; Chen et al., 1999; de Meyer et al., 1999; Zehnder et al., 2000; Pieterse et al., 2003), and inactivation of patho-gens by germination factors present in seed or root exudates.

Of the mechanisms mentioned above, none is necessarily exclusive, and frequently, more than one is exhibited by a biocontrol agent. Combination of various mechanisms may be involved in the control of different plant dis-eases by biocontrol microorganisms.

In some soils, disease incidence is low despite the presence of aggressive populations of pathogens, susceptibility of host-plant, and climatic condi-tions. Most soils are naturally suppressive to some degree to soilborne plant pathogens. This widespread characteristic has been referred to as "general suppression." A soil that is suppressive to one pathogen is not necessarily suppressive to another; in each case, a specificity of the soil-plant-microbe interaction for disease suppression exists.

At present, the greatest interest lies in the development and application of specific biocontrol agents for the control of plant diseases. In the follow-ing pages of this chapter, different mechanisms or combinations of mecha-nisms that are involved in the control of soilborne pathogens will be discussed.

BACTERIAL-FUNGAL PATHOGEN INTERACTIONS: ANTIBIOSIS MECHANISMS

Antibiotic Production

Understanding of the molecular basis of antibiosis in disease suppres-sion has come from the application of genetic and biochemical research techniques that have led to the development of improved assay systems with high specificity and sensitivity not only in vitro, but also in situ within natu-ral ecosystems. Mutants in the production of biochemically or pheno-typically distinguishable metabolites were used for the molecular analysis of antibiosis (Bonsall et al., 1997; Chin-A-Woeng, 1998). Antibiotic pro-duction is most often associated with fungal suppression by fluorescent pseudomonads (Howell and Stipanovic, 1979; Voisard et al., 1989; Thoma-show et al., 1990; Shanahan et al., 1992; Lovic et al., 1993). Fluorescent pseudomonads are common inhabitants of the rhizosphere and are the most investigated group within the genus *Pseudomonas*. Fluorescent *Pseudomonads* comprise *P. aeruginosa, P. aureofasciens, P. aurantiaca* (Rovera et al., 2000; Rosas et al., 2001, 2002), and *P. fluorescens*, among others. Antibiotic production by them constitute the major factor in suppression of root pathogens. There are numerous reports about the production of antifungal metabolites produced by bacteria in vitro that may also have activity in vivo on fungal growth. These metabolites include ammonia, butyrolactones, 2,4-diacetylphloroglucinol (Phl or DAPG), HCN, kanosamine, oligomycin A, oomycin A, phenazine-1-carboxylic acid (PCA), pyoluterin (Plt), pyrrolnitrin (Pln), viscosinamide, xanthobaccin, and zwittermycin A, as well as several other uncharacterized compounds (Milner et al., 1996; Keel and Défago, 1997; Whipps, 1997a; Kang et al., 1998; Nielsen et al., 1998; Nakayama et al., 1999; Kim et al., 1999; Thrane et al., 1999). The natural decrease in "take-all" disease (TAD) of wheat root produced by *Gaeummanomyces graminis tritici* (Ggt) during extended monoculture of wheat is an interesting example of the natural biological control phenomenon. Fluorescent pseudomonads are thought to be responsible for the reported biocontrol. Furthermore, a minor class of antifungal compounds that does not have nitrogen, 2,4-diacetylphloroglucinol, has been extensively used in the biological control of TAD of wheat. Genes involved in the biosynthesis of these antibiotic molecules have been cloned and characterized (Walsh et al., 2001; Whistler and Pierson, 2003).

The antibiotics pyoluteorin (Plt), pyrrolnitrin (Prn), phenazine-1-carboxylic acid (PCA), and 2,4-diacetylphloroglucinol (Phl) are currently the major focus of research in biological control (Table 4.1).

Phl

Phl, a broad-spectrum antibiotic, is a phenolic molecule produced by many fluorescent pseudomonads, and it exhibits antifungal, antibacterial, antihelmenthic, and phytotoxic activities (Reddi et al., 1969; Vincent et al., 1991; Gaur, 2002; Levy et al., 1992; Harrison et al., 1993; Nowak-Thompson et al., 1994; Bangera and Thomashow, 1996; Raaijmakers and Weller, 1998; Abbas et al., 2002). Morever, it shows herbicidal activity, like 2,4 dichlorophenoxyacetic acid (2,4-D). Phl is a polyketide synthesized by the condensation of three molecules of acetyl CoA with one molecule of malonyl CoA to produce the precursor monoacetylphloroglucinol (MAPG), which is subsequently transacetylated to generate DAPG.

The root-associated fluorescent *Pseudomonas* spp. with the capacity to synthesize Phl are responsible for the biological control of Ggt. (Raaijmakers and Weller, 1998). Phl is also a major determinant in the biological control

TABLE 4.1. Chemical structure of major antibiotics of pseudomonads.

Products	Size (kb)	Antibiotic
Pl	6.5	
Plt	6	
Prn	5.8	
Phz	6.8	

activity of plant growth-promoting rhizobacteria. Biotic and abiotic factors associated with field location and cropping time affect the performance of fluorescent pseudomonads (Thomashow and Weller, 1995; Duffy and Défago, 1997; Notz et al., 2002). Biotic factors, such as plant species, plant age, host cultivar, and infection with the plant pathogen, can significantly alter the expression of the gene *phlA* (Notz et al., 2001). Among abiotic factors, carbon sources and various minerals influence production of Phl.

Fe^{3+} and sucrose have been reported to increase the levels of DAPG and MAPG in *P. fluorescens* F113, whereas in *P. fluorescens* Pf-5 and CHA0, Phl was stimulated by glucose (Nowak-Thompson et al., 1994; Duffy and Défago, 1999). In *P. fluorescens* strain S272, the highest DAPG yield was obtained with ethanol as the sole source of carbon (Yuan et al., 1999). Microelements, such as Zn^{2+}, Cu^{2+}, and Mo^{2+} stimulate Phl production in *P. fluorescens* CHA0 (Notz et al., 2001) The exact mechanism of DAPG action is still unclear, although it is known that disease suppression by this antifungal molecule is a result of interaction of specific root-associated microorganisms and the pathogen. Phl also appears to cause induced systemic resistance (ISR) in plants. Thus, Phl-producing bacteria used in biocontrol can serve as specific elicitors of phytoalexins; other similar molecules are involved in the transport of Phl out of the cell.

Prn

Pyrrolnitrin (3-chloro-4-[2'-nitro-3'-chloro-phenyl] pyrrole) is a broad-spectrum antifungal metabolite produced by many fluorescent and non-fluorescent strains of the genus *Pseudomonas* (Elander et al., 1968; Howell and Stipanovic, 1979). It was first described by Arima et al. (1964). This highly active metabolite has been primarily used as a clinical antifungal agent for the treatment of skin mycoses against dermatophytic fungi, particularly members of the genus *Trichophyton.*

A phenyl pyrrol derivative of Prn has been developed as an agricultural fungicide. Others like isopyrrrolnitrin, oxypyrrolnitrin, and mono dechloropyrrolnitrin have lower antifungal activities than Prn (Elander et al., 1968). Pyrrolnitrin persists actively in the soil for at least 30 days. It does not readily diffuse and is released only after the lysis of host bacterial cell. This property of slow release facilitates protection against *Rhizoctonia solani* as the cell dies (Schnider-Keel et al., 2000). The biological control agent, *P. fluorescens* BL915, contains four gene clusters involved in the biosynthesis of antifungal molecule Prn from the precursor tryptophan (Hamill et al., 1967, 1970; Lively et al., 1966).

Phenazines

Phenazines (Phz) are reported as N-containing heterocyclic pigments synthesized by *Brevibacterium, Burkholderia, Pseudomonas,* and *Streptomyces* (Leisinger and Margraff, 1979; Budzikiewicz, 1993; Stevans et al., 1994). At present, over 50 naturally occurring Phz compounds have been described, and mixtures of as many as ten different Phz derivatives can occur simultaneously in one organism (Wienberg, 1970; Smirnov and Kiprianova, 1990; Mazzola et al., 1992). Growth conditions determine the number and type of Phz synthesized by an individual bacterial strain. For example, *P. fluorescens* 2-79 produces mainly PCA (phenazine 1-carboxylic acid), whereas *P. aureofaciens* 30-84 produces not only PCA but also lesser amounts of 2-OH-phenazines. The major Phz synthesized by *P. aeruginosa* is pyocyanin (1-OH-5-methyl Phz) (Wienberg, 1970). Almost all Phz exhibit broad-spectrum activity against bacteria and fungi (Smirnov and Kiprianova, 1990). In addition to inhibiting fungal pathogenesis, Phz play an important role in microbial competition in the rhizosphere, including in survival and competence (Mazzola et al., 1992).

Plt

Pyoluteorin (Plt) is an aromatic polyketide antibiotic consisting of a resorcinol ring, which is derived through polyketide biosynthesis. This, in turn, is linked to a bichlorinated pyrrole moiety, whose biosynthesis remains unknown (Cuppels et al., 1986; Kitten et al., 1998; Nowak-Thompson et al., 1997, 1999). Several *Pseudomonas* spp. produce Plt, including strains that suppress plant diseases caused by phytopathogenic fungi (Maurhofer et al., 1994; Kraus and Loper, 1995). Plt mainly inhibits the oomycetous fungi, including *Pythium ultimum* against which it is strongly active, decreasing the severity of *Pythium* damping-off (Nowak-Thompson et al., 1999). Biosynthesis of Plt is initiated from proline or a related molecule, which condenses serially with three acetate equivalents, coupled to chlorination and oxidation at yet-unidentified stages. The formation and cyclization of the C-skeleton has been reported to proceed by the action of a single multienzyme complex (Cuppels et al., 1986; Nowak-Thompson et al., 1999).

To demonstrate a role for antibiotics in biocontrol, mutants lacking production of antibiotics or overproducing mutants have been used (Bonsall et al., 1997; Chin-A-Woeng et al., 1998; Nowak-Thompson et al., 1999). As an alternative, the use of reporter genes or probes to demonstrate production of antibiotics in the rhizosphere is becoming more common (Kraus and Loper, 1995; Raaijmakers et al., 1997; Chin-A-Woeng et al., 1998). Isolation and characterization of genes or gene clusters responsible for antibiotic production has now been achieved (Kraus and Loper, 1995; Bangera and Thomashow, 1996; Hammer et al., 1997; Kang et al., 1998; Nowak-Thompson et al., 1999). Significantly, both Phl and PCA have been isolated from the rhizosphere of wheat following introduction of biocontrol strains of *Pseudomonas* (Thomashow et al., 1990; Bonsall et al., 1997; Raaijmakers et al., 1999), finally confirming that such antibiotics are produced in vivo.

PARASITISM AND PRODUCTION OF EXTRACELLULAR ENZYMES

Chitin is the second most abundant polysaccharide in nature, surpassed only by cellulose. In addition to its ubiquitous occurrence in all true fungi (Chytridiomycetes, Zygomycetes, Deuteromycetes, Ascomycetes, and Basidiomycetes) (Bartnicki-Garcia and Lippman, 1982), chitin is also present in other groups of organisms such as diatoms, protozoa, insects, crustaceans, and nematodes (Gooday, 1990a). The annual production of chitin has been estimated to be in the order of 10^{10} to 10^{11} tons (Gooday, 1990c). However, there are no massive accumulations in nature, and it seems as if most chitin is rapidly recycled (Gooday, 1990a).

Chitin can be degraded via either of two main pathways. In one, degradation is initiated by the chitinase-induced hydrolysis of the β-1,4-glycosidic linkage. Alternatively, the polymer is first deacetylated and thereafter hydrolyzed by chitosanases. The latter pathway is primarily of importance in estaurine environments, where chitosan is a major organic constituent (Gooday et al., 1991).

The degradation of chitin is primarily carried out by microbes, although chitinolytic enzymes have been found in plants, invertebrates, and vertebrates. Chitinases in plants are inducible enzymes produced as a defense response to microbial infection or to elicitors, like chitin oligosaccharides from the fungal cell wall (Graham and Sticklen, 1994). Fungal chitinases are thought to have autolytic, nutritional, and morphogenetic roles, while bacterial chitinases are mainly secreted to obtain carbon, nitrogen, and energy. Among the prokaryotes, pseudomonads, enteric bacteria, gliding bacteria, actinomycetes, *Bacillus* spp., *Vibrio* spp., and *Clostridium* spp. are frequently isolated chitin degraders. Among the eukaryotic microorganisms, chitinase production has been found in Myxomycetes, Chytridiomycetes, Zygomycetes, Deuteromycetes, Ascomycetes, and Basidiomycetes (Goodday, 1990a).

Mitchell and Alexander (1961) published various papers about mycolytic activities of chitinolytic soil bacilli and pseudomonads. They postulated that fungal cell walls could be degraded by the action of bacterial chitinases. Several workers demonstrated that chitinases were involved in mycolysis. In addition to chitinases, the bacteria require other factors to lyse fungal hyphae. (Sundhein et al., 1988; Chet et al., 1990; Lim et al., 1991; Chernin et al., 1997).

Fungal growth in stored crops can lead to the production of mycotoxins and allergenic spores as well as to reduction in seed viability and nutritional value, among other problems. In recent years, there has been an increase in the frequency with which food- and feed-associated fungal species have been found to be involved in human diseases. Chitin-degrading bacteria have been suggested as biocontrol agents owing to their ability to degrade fungal cell wall components.

Bacteria produce several endo-chitinases and exo-chitinases, so the composition of the chitinase pool may also help to determine antifungal effects. More emphasis has been given to rhizospheric bacteria, as they are best adapted to the natural environment where plant-pathogenic fungi are established (Ordentlich et al., 1988; Inbar and Chet, 1991; Kobayashi et al., 1995). However, the application of mycolytic strains under field conditions has been far less successful (Maloy, 1993). Additional information is needed about the ecological function of the chitinase-producing bacteria and the impact of those mycolytic activities under natural environmental conditions.

The most obvious advantage for chitinolytic bacteria is the use of chitin as sources of carbon, nitrogen, and energy (Cohen-Kupiec and Chet, 1998). This role for chitinases is well established for aquatic bacteria (Gooday, 1990a). Bacteria and actinomycetes are able to parasitize and degrade spores of fungal plant pathogens (El-Tarabily et al., 1997). Assuming that nutrients pass from the plant pathogen to bacteria, and that fungal growth is inhibited, the spectrum of parasitism could range from simple attachment of cells to hyphae, as with the *Enterobacter cloacae–Pythium ultimum* interaction (Nelson et al., 1986), to complete lysis and degradation of hyphae, as found with the *Arthrobacter–Pythium debaryanum* interaction. If fungal cells are lysed and cell walls are degraded, then it is generally assumed that cell wall–degrading enzymes produced by the bacteria are responsible, even though antibiotics may be produced at the same time. Considerable effort has gone into identifying cell wall–degrading enzymes produced by biocontrol strains of bacteria even though relatively little direct evidence for their presence and activity in the rhizosphere has been obtained.

Iron Competition

Siderophores (Greek: "iron carriers") are defined as relatively low-molecular-weight, ferric ion-specific chelating agents produced by bacteria and fungi growing under low iron stress. A common structure of these compounds is shown in Figure 4.1.

FIGURE 4.1. General structure of the citrate-hidroxamate siderophores.

The role of these compounds is to scavenge iron from the environment and to make the mineral, which is almost always essential, available to the microbial cell. Siderophores have been related to virulence mechanisms in microorganisms pathogenic to both animals and plants. In addition, they have clinical applications and actually are important in agriculture (Winkelmann et al., 1987; Winkelmann, 1991).

The concentration of free ferric ion at neutral pH is determined by the solubility product constant of ferric hydroxide. Depending on the value selected for this constant, the maximum amount of uncomplexed ferric ion in solution at biological pH is probably not greater than 218 μM (Neilands et al., 1987). Microorganisms growing under aerobic conditions need iron for a variety of metabolic functions, including reduction of oxygen for synthesis of ATP, reduction of ribotide precursors of DNA, formation of heme, and for other essential purposes. For this reason, microorganisms form specific molecules that can compete effectively with hydroxyl ion for the ferric state of iron, a nutrient which is abundant but essentially unavailable.

The vast majority of microorganisms produce siderophores, a diverse class of high-affinity iron chelating compounds that are produced in response to iron limitation and exported from the cell where they chelate ferric iron. Ferric-siderophore complexes are recognized and bound by specific outer-membrane receptor proteins, and iron is transported into the cell where it becomes available for metabolic functions mentioned above. Among the siderophores produced by rhizosphere bacteria, only the pyoverdines (also called pseudobactins) produced by the fluorescent pseudomonads have been implicated in ISR (Métraux et al., 1990; Leeman, den Ouden, van Pelt, Dirkx, et al., 1996).

Although competition between bacteria and fungal plant pathogens for space or nutrients has been known to exist as a biocontrol mechanism for many years (Whipps, 1997a), the actual interest in siderophores lies in competition for iron. Under iron-limiting conditions, bacteria produce a range of iron chelating compounds or siderophores which have a very high affinity for ferric iron. These bacterial iron chelators are thought to sequester the limited supply of iron available in the rhizosphere, making it unavailable to pathogenic fungi, thereby restricting their growth (O'Sullivan and O'Gara, 1992; Loper and Henkels, 1999). The iron nutrition of the plant influences the rhizosphere microbial community structure (Yang and Crowley, 2000). Iron competition in pseudomonads has been intensively studied, and the role of the pyoverdine siderophore produced by many *Pseudomonas* species has been clearly demonstrated in the biocontrol of *Pythium* and *Fusarium* species, either by comparing the effects of purified pyoverdine with synthetic iron chelators or through the use of pyoverdine minus mutants (Loper and Buyer, 1991; Duijff et al., 1993). Pseudomonads also produce two

other siderophores, pyochelin and its precursor, salicylic acid. Some siderophores can be used only by the bacteria that produce them (Ongena et al.,
1999), whereas others can be used by many different bacteria (Loper and
Henkels, 1999). Different biotic and abiotic environmental factors can also
influence the quantity of siderophores produced (Duffy and Défago, 1999).
It has been reported that pyoverdine and salicylate may act as elicitors for
inducing systemic resistance against pathogens in some plants (Métraux
et al., 1990; Leeman, den Ouden, van Pelt, Dirkx, et al., 1996).

Induced Systemic Resistance

Plants have evolved complex and varied defense mechanisms for protection against disease. These mechanisms may be constitutive or induced following attack by pathogens. Recent studies have suggested that inducible
defenses in plants may have selective advantages over constitutive defenses
(Agrawal, 1998). While inducible defenses are often localized at the site of
attack, plant defense mechanisms may be activated systemically throughout the plant following a localized infection or attack (Kessman et al.,
1994).

One of the first published reports of systemic resistance in plants was by
Chester (1933), who used the term "acquired physiological immunity." Later,
Ross (1961) reported that tobacco plants exhibited "systemic acquired resistance" following local infection with tobacco mosaic virus. Other terms that
have been used to describe systemic resistance in plants include "translocated
resistance" (Hubert and Helton, 1967), plant immunization (Kúc, 1987), and
"induced systemic resistance" (ISR) (Hammerschmidt et al., 1982). The term
"induced systemic resistance" is used to denote induced systemic resistance
by nonpathogenic biotic agents (Kloepper et al., 1992) and may differ in the
mode of action from the resistance induced by other elicitors (van Loon et al.,
1998).

To protect themselves from pathogen attack, plants have developed
defense mechanisms in which several signal molecules, such as salicylic
acid, jasmonic acid, and ethylene, play crucial roles.

Fluorescent *Pseudomonas* spp. constitute the most important group of
bacteria with PGPR properties. Besides having a direct antagonistic effect
on soilborne plant pathogens, fluorescent *Pseudomonas* spp. strains are capable of triggering a plant-mediated resistance in response to pathogen
presence.

Induced resistance is a state of enhanced defensive capacity developed
by a plant when appropriately stimulated (van Loon et al., 1998). In 1991,
two research groups independently described ISR as the mode of action of
disease suppression by nonpathogenic rhizosphere bacteria (van Peer et al.,

1991; Wei et al., 1991). Since then, the involvement of ISR in disease suppression has been studied for a wide range of biological control microorganisms and, in many cases, ISR was found to be involved. This type of resistance has been demonstrated in various plant species, e.g., cucumber, tomato, tobacco radish, and the model plant *Arabidopsis thaliana* (van Loon et al., 1998).

Phenotypically, ISR is similar to systemic acquired resistance (SAR) that is triggered by necrotizing pathogens. Although the terms SAR and ISR are synonymous (Hammerschmidt et al., 2001), it is necessary to distinguish between pathogen- and rhizobacteria-induced resistances by using SAR for the pathogen-induced type and ISR for the rhizobacteria-induced type (Bakker et al., 2003).

SAR has been studied extensively and excellent reviews have been written (Ryals et al., 1996; Delaney, 1997; Sticher et al., 1997; Hammerschmidt, 1999; Métraux, 2001). SAR is characterized by an accumulation of salicylic acid (SA) and pathogenesis-related proteins (PRs) (Ward et al., 1991; Uknes et al., 1992; Kessman et al., 1994). Accumulation of SA occurs both locally and, at lower levels, systemically, concomitant with the development of SAR. Exogenous application of SA also induces SAR in several plant species (Gaffney et al., 1993; Ryals et al., 1996). Both pathogen- and SA-induced resistance are associated with the production of several families of PRs. Induction of PRs is invariably linked to necrotizing infections giving rise to SAR and has been taken as a marker of the induced state (Uknes et al., 1992; Kessman et al., 1994). Some of these PRs are β-1,3-glucanases and chitinases, and they are capable of hydrolyzing fungal cell walls. Other PRs have poorly characterized antimicrobial activities or unknown functions.

The association of PRs with SAR suggests an important contribution of these proteins to the increased defensive capacity of induced tissues. SAR has been documented in multiple plant species, whereas studies on ISR by rhizosphere bacteria are restricted to a few species. Notably, no ISR has yet been reported in monocotyledons. To study rhizobacteria-mediated ISR an *Arabidopsis*-based model system was assayed; in this model, the *Pseudomonas fluorescens* WCS417r was used as inducing agent (Pieterse et al. 1996), and the strain has been shown to trigger ISR in carnation (van Peer et al., 1991), radish (Leeman et al., 1995), tomato (Duijff et al., 1998), and bean (Bigirimana and Höfte, 2002).

Whereas SAR requires accumulation of salicylic acid (SA) in the plant (Sticher et al., 1997), ISR does not and, instead, is dependent on intact responses to ethylene and jasmonic acid (JA) (Pieterse et al., 1998, 2003). Although ISR in *Arabidopsis* was found to be dependent on JA and ethylene perception, bacterization of plants roots did not lead to increase in JA or ethylene levels in either roots or shoots. Treatments of plants with either methyl-JA (MeJA) or the precursor of ethylene, 1-aminocyclopropane-1-carboxylic (ACC), induced an identical resistance from rhizobacteria-mediated ISR. The similarity between rhizobacteria-mediated ISR and pathogen-induced SAR is that both types of resistance are effective against a wide spectrum of plant pathogens.

When these different signal transduction pathways are triggered simultaneously in *A. thaliana,* disease suppression is enhanced (van Wees et al., 2000). This suggests that combining bacterial traits that trigger either the SA-, or the ethylene- or JA-dependent response can improve biological control. Elucidating the nature of bacterial triggers of the different pathways will enable us to utilize ISR in a sensible way. Distinguishing between ISR and direct effect *Arabidopsis* mutants that are insensitive to ethylene (*etr*1) or JA (*jar*1) and the *Arabidopsis* mutant expressing nonpathogenesis-related proteins for resistance to *Pseudomonas* (*npr*1) do not express ISR after treatment with *Pseudomonas fluorescens* Migula WCS417r (Pieterse et al., 1998).

Role of Salicylic Acid

A rhizobacterial metabolite that was suggested to trigger the SA-dependent signal transduction pathway is SA itself (Maurhofer et al., 1994; Leeman, den Ouden, van Pelt, Dirkx, et al., 1996; De Meyer and Höfte, 1997). Salicylic acid production has been observed for several bacterial strains, and exogenously applied SA can induce resistance in many plant species. For example, the *P. fluorescens* strain WCS374 produces relatively large quantities of SA under conditions of iron limitation. Moreover, WCS374 is more effective in radish, in ISR against *Fusarium* wilt under iron-limited conditions compared with conditions of sufficient iron availability (Leeman et al., 1995).

Pseudobactin siderophores are low-molecular-weight molecules which are also produced by WCS374 under iron-limited conditions. However, a pseudobactin mutant of WCS374 was as effective as the parental strain in disease suppression. Therefore, it was concluded that SA was probably responsible for the increased effectiveness of WCS374 under conditions of iron limitation. However, no indications for activation of SA-dependent signaling in radish after WCS374 treatment were observed (Hoffland et al., 1995). Moreover, treatment with strain WCS374 cannot induce systemic resistance in *A. thaliana,* whereas the application of SA does (van Wees et al., 1999). Possibly, the iron-regulated ISR by WCS374 in radish is mediated by the SA-containing siderophore pseudomonine (Mercado-Blanco et al., 2001). Pseudomonine is a nonfluorescent siderophore, produced by WCS374, that was discovered when analyzing nonfluorescent mutants of this strain. Whether pseudomonine is active in ISR remains to be elucidated. The plant-root-colo-nizing *Pseudomonas aeruginosa* (Schroeter) Migula strain 7NSK2 induces resistance in tobacco, bean, and tomato. Its ability for ISR has been linked to the production of SA (De Meyer and Höfte, 1997). However, recent evidence strongly suggests that SA is not the inducing compound from 7NSK2, but that the compounds pyochelin and pyocyanin produced by this strain are a prerequisite for the induction of resistance (Audenaert et al., 2002). It was postulated that 7NSK2 does not produce SA in the rhizosphere, but instead, the SA is channeled into the siderophore pyochelin (Audenaert et al., 2002). Thus, for several bacterial strains, a role of produced SA in induced resistance has been suggested, but upon careful examination, SA seems not to be in-volved directly (van Loon et al., 1998; Audenaert et al., 2002).

Siderophores, which pathogens use as a mechanism for biocontrol by competition for iron, can induce systemic resistance (Loper and Buyer, 1991; Bakker et al., 1993; Duijff et al., 1999). Production of antibiotics has been de-scribed as a powerful mode of action in disease suppression by which devel-opment and activity of the pathogen is thought to be directly inhibited (Handelsman and Stabb, 1996). The involvement of antibiotic production in ISR has not been investigated in detail. Isolates of *G. graminis* var. *tritici,* which varied in their sensitivity to the antibiotics phenazine-1-carboxylic acid and 2,4-diacetylphloroglucinol, were also differentially sensitive to bio-logical control by *Pseudomonas* spp. strains, producing these same antibiot-ics (Mazzola et al., 1995). These results demonstrated that the effect of these antibiotics on the pathogen was direct and not mediated by the plant. However, antibiotics do have direct effects on plants and, therefore, might in-duce systemic resistance mediated by rhizobacteria The antibiotics pyoluteorin and 2,4-diacetylphloroglucinol reduced the growth of sweet corn, cress, and cucumber (Maurhofer et al., 1992) and, in turn, the stress caused by these antibiotics may also trigger resistance. To further study the possible role of antibiotics in ISR, strains of biocontrol bacteria that produce specific antibi-otics should be applied, with spatial separation from the pathogen on the plant surface, to prevent direct interactions between their populations (van Loon et al., 1998). If these strains induce systemic resistance, their mutants lacking antibiotic production should be tested. Conversely, the purified com-pound can be tested for resistance induction.

Lipopolysaccharides (LPS) have been implicated in ISR triggered by fluorescent pseudomonads in carnation (van Peer and Schippers, 1992), radish (Leeman et al., 1995), and *A. thaliana* (van Wees et al., 1999). In the case of ISR against *Globodera pallida* (Stone) Behrens on potato by *Rhizobium etli,* the LPS also plays a major role (Reitz et al., 2002). In those studies, either purified LPS was used to induce systemic resistance, or mutants that lack part of the LPS were compared with the parental strain regarding their abilities to induce resistance.

In a dose-response study of ISR mediated by *P. fluorescens* WCS374 in radish, it was revealed that population densities of 10^5 colony forming units per gram of root are required for significant suppression of disease (Raaijmakers, Leeman, et al., 1995). In situations where no bacteria are introduced, population densities of one single bacterial genotype probably never reach such high densities making siderophore produced by WCS374 (Mercado-Blanco et al., 2001).

Another bacterial determinant that was suggested to be involved in ISR is flagellin, the protein subunit of the flagella. It appears that in most inducing bacteria, more than one determinant is operative in triggering systemic resistance in the plant. The use of indicator plants that contain a reporter gene, expressed when ISR occurs, would be very helpful in studies for identifying the inducing bacterial determinants. A major problem that has hampered the development of such reporter lines is the lack of known, defense-related genes responding to the bacterial treatments (van Wees et al., 2000). The development of gene chip technology is a powerful tool to study gene expression patterns in response to ISR.

CONCLUSIONS

Fewer pesticides must be used in the future, and greater reliance will be placed on biological technologies, including the use of microorganisms as antagonists.

Interactions between microorganisms and plants have undoubtedly had major effects on the development of civilization from the time humans began to rely extensively on cultivated crops for food; so, it is necessary to have deeper knowledge of plant–microbe interactions to provide the necessary information to permit rational agronomic responses. Developing biocontrol strategies by manipulating plant–microbe interactions requires a comprehensive understanding of different mechanisms.

Microorganisms as biocontrol agents have a relatively narrow spectrum of activity compared with synthetic pesticides (Baker, 1991; Janisiewicz, 1996), and often exhibit inconsistent performance in practical agriculture. Application of a mixture of introduced biocontrol agents would more closely mimic the natural situation and might broaden the spectrum of biocontrol activity. A good colonization capacity and compatibility of the inoculated microorganisms constitute an important prerequisite for succesful development of biological control (Baker, 1990; De Boer et al., 1997).

This chapter was focused on recent research concerning interactions between biocontrol agents and pathogens in the rhizosphere and a large number of differing types of interaction operating through a variety of modes of action. The greatest interest lies in the antibiotic production and the phenomenon of induced resistance.

Integration of biological and chemical control systems may also be an approach that receives more attention in the future. If pesticide-tolerant isolates of biocontrol agents could be used to reduce the application of pesticides, then an environmental benefit would ensue. Thus, combined treatment of rosemary (*Rosemarinus officinalis* L.) with the biocontrol agent *Laetisaria arvalis* Burdsall and an experimental fungicide CGA 173506 at one-half the recommended rate reduced *Rhizoctonia* disease more than treatment with either fungus or fungicide alone (Conway, 1997). Several other combinations of biocontrol agents and pesticides have been tested, and others are under development (Harris and Nelson, 1999; Budge and Whipps, 2001).

Finally, many factors affecting the success of biological control need to be taken into account when working directly with producers. Hence, it is necessary to deal separately with each social and economic group. The key approach to technology transfer activities should be through demonstration of economic cost-benefit and the beneficial effects of biological control on the environment and human health.

REFERENCES

Abbas, A., J.P. Morrisey, P.C. Marquez, M.M. Sheehan, I.R. Delany, and F. O'Gara. (2002). Characterization of interaction between the transcriptional repressor PhlF and its binding site at the phlA promoter in *Pseudomonas fluorescens* F113. *Journal of Bacteriology* 184: 3008-3016.

Agrawal, A.A. (1998). Induced responses to herbivory and increased plant performance. *Science* 279: 1201-1202.

Andrés, J.A., N.S. Correa, and S.B. Rosas. (1999). El potencial mutagénico del fungicida thiran. *Revista Argentina de Microbiología* 31: 82-86.

Arima, K., H. Imanaka, M. Kausaka, A. Fukuda, and C. Tameera. (1964). Pyrrolnitrin, a new antibiotic substance, produced by Pseudomonas. *Agriculture and Biology Chemistry* 28: 575-576.

Audenaert, K., T. Pattery, P. Cornelis, and M. Höfte. (2002). Induction of systemic resistance to *Botrytis cinerea* in tomato by *Pseudomonas aeruginosa* 7NSK2: Role of salicylic acid, pyochelin, and pyocyanin. *Molecular Plant-Microbe Interaction* 15: 1147-1156.

Backman, P.A., M. Wilson, and J.F. Murphy. (1997). Bacteria for biological control of plant diseases. In *Environmentally Safe Approaches to Crop Disease Control,* eds. N.A. Rechcigl and J.E. Rechcigl. CRC Lewis Publishers, Boca Raton, FL, pp. 95-109.

Baker, R. (1990). An overview of current and future strategies and models for biological control. In *Biological Control of Soilborne Plant Pathogens,* ed. D. Hornby. CAB International, Wallingford, UK, pp. 375-388.

Baker, R. (1991). Diversity in biological control. *Crop Protection* 10: 85-94.

Bakker, P.A.H.M., J.M. Raaijmakers, and B. Schippers. (1993). Role of iron in the suppression of bacterial plant pathogens by fluorescent pseudomonads. In *Iron Chelation in Plants and Soil Microorganisms,* eds. L.L. Barton and B.C. Hemming. Academic Press, San Diego, CA, pp. 269-282.

Bakker, P.A.H.M., L.X. Ran, C.M.J. Pieterse, and L.C. van Loon. (2003). Understanding the involvement of rhizobacteria-mediated induction of systemic resistance in biocontrol of plant diseases. *Canadian Journal of Plant Pathology* 25: 5-9.

Bangera, M.G. and L.S. Thomashow. (1996). Characterization of a genomic locus required for synthesis of the antibiotic 2,4-diacetylphloroglucinol by the biological control agent *Pseudomonas fluorescens* Q2-87. *Molecular Plant-Microbe Interactions* 9: 83-90.

Bartnicki-Garcia, S. and E. Lippman. (1982). Fungal cell wall composition. In *CRC Handbook of Microbiology,* Vol. 4, eds. A.I. Laskin and H.A. Lechevalier. CRC Press, Boca Raton, FL, pp. 229-252.

Benbrook, C.M., E. Groth, J.M. Halloran, M.K. Hansen, and S. Marquardt. (1996). *Pest Management at the Cross-roads.* Consumers Union, Yonkers, NY.

Bigirimana, J. and M. Höfte. (2002). Induction of systemic resistance to *Colletotrichum lindemuthianum* in bean by a benzothiadiazole derivative abd rhizobacteria. *Phytoparasitica* 30: 159-168.

Bonsall, R.F., D.M. Weller, and L.S. Thomashow. (1997). Quantification of 2,4-diacetylphloroglucinol produced by fluorescent *Pseudomonas* spp. in vitro and in the rhizosphere of wheat. *Applied and Environmental Microbiology* 63: 951-955.

Budge, S.P., M.P. McQuilken, J.S. Fenlon, and J.M. Whipps. (1995). Use of *Coniothyrium minitans* and *Gliocladium virens* for biological control of *Sclerotinia sclerotiorum* in glasshouse lettuce. *Biological Control* 5: 513-522.

Budge, S.P. and J.M. Whipps. (2001). Potential for integrated control of *Sclerotinia sclerotiorum* in glasshouse lettuce using *Coniothirium minitans* and reduced fungicide application. *Phytopathology* 91: 221-227.

Budzikiewicz, H. (1993). Secondary metabolites: Fluorescent pseudomonads. *FEMS Microbiology Review* 104: 209-228.

Chen, C., R.R. Bélanger, N. Benhamou, and T.C. Paulitz. (1999). Role of salicylic acid in systemic resistance induced by *Pseudomonas* spp. against *Pythium aphanidermatum* in cucumber roots. *European Journal of Plant Pathology* 105: 477-486.

Chernin, L.S., L. De La Fuente, V. Sobolev, S. Haran, C.E. Vorgias, A.B. Oppenheim, and I. Chet. (1997). Molecular cloning, structural analysis, and expression in *Escherichia coli* of a chitinase gene from *Enterobacter agglomerans*. *Applied and Environmental Microbiology* 63: 834-839.

Chester, K. (1933). The problem of acquired physiological immunity in plants. *Quart. Review of Biology* 8: 129-151.

Chet, I., A. Ordentlich, R. Shapira, and A. Oppenheim. (1990). Mechanisms of biocontrol of soil-borne plant pathogens by rhizobacteria. *Plant and Soil* 129: 85-92.

Chin-a-Woeng, T.F.C., G.V. Bloemberg, A.J. van der Bij, K.M.G.M. van der Drift, J. Schripsema, B. Kroon, R.J. Scheffer, C. Keel. P.A.H.M. Bakker, J.V. Tichy, F.J. de Bruijn, J.E. Thomas-Oates, and B.J.J. Lugtenberg. (1998). Biocontrol by phenazine-1-carboxamide-producing *Pseudomonas chlororaphis* PCL 1391 of tomato root rot caused by *Fusarium oxysporum* f. sp. *radicis-lycopersici. Molecular Plant-Microbe Interactions* 11: 1069-1077.

Cohen-Kupiec, R. and I. Chet. (1998). The molecular biology of chitin digestion. *Current Opinions in Biotechnology* 9: 270-277.

Conway, K.E. (1997). Integration of biological and chemical controls for Rhizoctonia aerial blight and root rot of rosemary. *Plant Disease* 81: 795-798.

Cook, R.J. (1993). Making greater use of introduced microorganisms for biocontrol of plant pathogens. *Annual Review of Phytopathology* 31: 53-80.

Cuppels, D.A., C.R. Howell, R.D. Stipanovic, A. Stossel, and J.B. Stothers. (1986). Biosynthesis of pyoluteorin: A mixed polyketide tricarboxylic acid cycle origin demonstrated by [1,2-13C2] acetate incorporation. *Z Naturforsch* 41: 532-536.

Datnoff, L.E., S. Nemec, and K. Pernezny. (1995). Biological control of Fusarium crown and root rot of tomato in Florida using *Trichoderma harzianum* and *Glomus intraradices. Biological Control* 5: 427-431.

De Boer, M., I. van der Sluis, L.C. van Loon, and P.A.H.M. Bakker. (1997). In vitro compatibility between fluorescent *Pseudomonas* spp. strains can increase effectivity of Fusarium wilt control by combinations of these strains. In *Plant Growth Promoting Rhizobacteria-Present Status and Future Prospects.* Proceduring Int. Workshop on Plant Growth-Promoting Rhizobacteria, 4th. eds. A. Ogoshi, K. Kobayashi, Y. Homma, F. Kodama, N. Kondo, and S. Akino. Nakanishi Printing, Sapporo, Japan. pp. 380-382.

Delaney, T.P. (1997). Genetic dissection of acquired resistance to disease. *Plant Physiology* 113: 5-12.

de Meyer, G., K. Capieau, K. Audenaert, A. Buchala, J.P. Métraux, and M. Höfte. (1999). Nanogram amounts of salicylic acid produced by the rhizobacterium *Pseudomonas aeruginosa* 7NSK2 activate the systemic acquired resistance pathway in bean. *Molecular Plant-Microbe Interactions* 12: 450-458.

de Meyer, G. and M. Höfte. (1997). Salicyclic acid produced by the rhizobacterium *Pseudomonas aeruginosa* 7NSK2 induces resistance to leaf infection by *Botrytis cinerea* on bean. *Phytopathology* 87: 588-593.

Duffy, B.K. and G. Défago. (1997). Zinc improves biocontrol of Fusarium crown and root rot of tomato by *Pseudomonas fluorescens* and represses the production of pathogen metabolites inhibitory to bacterial antibiotic biosynthesis. *Phytopathology* 87: 1250-1257.

Duffy, B.K. and G. Défago. (1999). Environmental factors modulating antibiotic and siderophore biosynthesis by *Pseudomonas fluorescens* biocontrol strains. *Applied Environmental Microbiology* 65: 2429-2438.

Duffy, B.K., A. Simon, and D.M. Weller. (1996). Combination of *Trichoderma koningii* with fluorescent pseudomonads for control of take-all on wheat. *Phytopathology* 86: 188-194.

Duffy, B.K. and D.M. Weller. (1995). Use of *Gaeumannomyces graminis* var. *graminis* alone and in combination with fluorescent *Pseudomonas* spp. to suppress take-all of wheat. *Plant Disease* 79: 907-911.

Duijff, B.J., J.W. Meijer, P.A.H.M. Bakker, and B. Schippers. (1993). Siderophore-mediated competition for iron and induced resistance in the suppression of Fusarium wilt of carnation by fluorescent *Pseudomonas* spp. *Netherlands Journal of Plant Pathology* 99: 277-289.

Duijff, B.J., D. Pouhair, C. Olivain, C. Alabouvette, and P. Lemanceau. (1998). Implication of systemic induced resistance in the suppression of Fusarium wilt of tomato by *Pseudomonas fluorescens* WCS417r and by non-pathogenic *Fusarium oxysporum* Fo47. *European Journal of Plant Pathology* 104: 903-910.

Duijff, B.J., G. Recorbet, P.A.H.M. Bakker, J.E. Loper, and P. Lemanceau. (1999). Microbial antagonism at the root level is involved in the suppression of Fusarium wilt by the combination of nonpathogenic *Fusarium oxysporum* Fo47 and *Pseudomonas putida* WCS358. *Phytopathology* 89: 1073-1079.

Elander, R.P., J.A. Mabe, R.H. Hamill, and M. Gorman. (1968). Metabolism of tryptophans by *Pseudomonas aureofaceins:* VI. Production of pyrrolnitrin by selected *Pseudomonas* spp. *Applied Environmental Microbiology* 16: 753-758.

El-Tarabily, K.A., G.E.St.J. Hardy, K. Sivasithamparam, A.M. Hussein, and D.I. Kurtböke. (1997). The potential for the biological control of cavity-spot disease of carrots, caused by *Pythium coloratum,* by streptomycete and non-streptomycete actinomycetes. *New Phytologist* 137: 495-507.

Gaffney, T., L. Friedrich, B. Vernooij, D. Negrotto, G. Nye, S. Uknes, E. Ward, H. Kessmann, and J. Ryals. (1993). Requirement of salicylic acid for the induction of systemic acquired resistance. *Science* 261: 754-756.

Gaur, R. (2002). Diversity of 2,4-diacetylphloroglucinol and 1-amino-cyclopropane 1-carboxylate deaminase producing rhizobacteria from wheat rhizosphere. PhD thesis. G.B. Pant University of Agriculture and Technology, Pantnagar.

Glick, B.R. (1995). The enhancement of plant growth by free-living bacteria. *Canadian Journal of Microbiology* 41: 109-117.

Gooday, G.W. (1990a). The ecology of chitin degradation. In *Advances in Microbial Ecology,* vol. 14. K.C. Marshall (ed.). Plenum Press, New York, pp. 387-430.

Gooday, G.W. (1990b). Inhibition of chitin metabolism. In *Biochemistry of Cell Walls and Membranes in Fungi,* eds. P.J. Kuhn, A.P.J. Trinci, M.J. Jung, M.W. Goosey, and L.G. Copping. Springer-Verlag, Berlin, pp. 61-79.

Gooday, G.W. (1990c). Physiology of microbial degradation of chitin and chitosan. *Biodegration* 1: 177-190.

Gooday, G.W., J.I. Prosser, K. Hillman, and M.G. Cross. (1991). Mineralization of chitin in an estaurine sediment: The importance of the chitosan pathway. *Biochemical Systematic Ecology* 19: 395-400.

Graham, L.S. and M.B. Sticklen (1994). Plant chitinases. *Canadian Journal of Botany* 72: 1057-1083.

Hamill, R., R. Elander, J. Mabe, and M. Gorman. (1967). Metabolism of tryptophans by *Pseudomonas aureofaceins:* III. Production of substituted pyrrolnitrins from tryptophan analogues. *Antimicrobial Agents Chemotherapy* 2: 388-396.

Hamill, R.L., R.P. Elander, J.A. Mabe, and M. Goreman. (1970). Metabolism of tryptophans by *Pseudomonas aureofaceins:* V. Conversion of tryptophan to pyrrolnitrin. *Applied and Environmental Microbiology* 19: 721-725.

Hammer, P.E., D.S. Hill. S.T. Lam, K.H. van Pée. and J.M. Ligon. (1997). Four genes from *Pseudomonas fluorescens* that encode the biosynthesis of pyrrolnitrin. *Applied and Environmental Microbiology* 63: 2147-2154.

Hammerschmidt, R. (1999). Induced disease resistance: How do induced plants stop pathogens? *Physiologcal and Molecular Plant Pathology* 55: 77-84.

Hammerschmidt, R., J.P. Métraux, and L.C. van Loon. (2001). Inducing resistance: A summary of papers presented at the First International Symposium on Induced Resistance to Plant Diseases, Corfu, May 2000. *European Journal of Plant Pathology* 107: 1-6.

Hammerschmidt, R., E.M. Nuckles, and J. Kuc. (1982). Association of enhanced peroxidase activity with induced systemic resistance of cucumber to *Colletotrichum lagenarium. Physiological Plant Pathology* 20: 73-82.

Handelsman, J. and E.V. Stabb. (1996). Biocontrol of soilborne plant pathogens. *Plant Cell* 8: 1855-1869.

Harris, A.R. and S. Nelson (1999). Progress towards integrated control of damping-off disease. *Microbiological Research* 154: 123-130.

Harrison, L.A., L. Letrendre, P. Kovacevich, E.A. Pierson, and D.M. Weller. (1993). Purification of an antibiotic effective against *Gaumannomyces graminis* var *tritici* produced by a biocontrol agent, *Pseudomonas aureofaceins. Soil Biology and Biochemistry* 25: 215-221.

Hassan, D.G., M. Zargar, and G.M. Beigh. (1997). Biocontrol of Fusarium root rot in the common bean (*Phaseolus vulgaris* L.) by using symbiotic *Glomus mosseae* and *Rhizobium leguminosarum. Molecular Ecology* 34: 74-80.

Hoffland, E., C.M.J. Pieterse, L.M.J., L. Bik, and J.A. Van Pelt. (1995). Induced systemic resistance in radish is not associated with accumulation of pathogenesis-related proteins. *Physiological and Molecular Plant Pathology* 46: 309-320.

Howell, C.R. and R.D. Stipanovic. (1979). Control of *Rhizoctonia solani* on cotton seedlings with *Pseudomonas fluorescens* with an antibiotic produced by the bacterium. *Phytopathology* 69: 480-482.

Howell, C.R. and R.D. Stipanovic. (1995). Mechanisms in the biocontrol of *Rhizoctonia solani*-induced cotton seedling disease by *Gliocladium virens*: Antibiosis. *Phytopathology* 85: 469-471.

Hubert, J.J. and A.W. Helton. (1967). A translocatable-resistance phenomenon in *Prunus domestica* induced by initial infection with *Cytospora cineta. Phytopathology* 57: 1094-1098.

Inbar, J. and I. Chet. (1991). Evidence that chitinase produced by *Aeromonas caviae* is involved in the biological control of soil-borne plant pathogens by this bacterium. *Soil Biology and Biochemistry* 23: 973-978.

Janisiewicz, W.J. (1988). Biocontrol of postharvest diseases of apples with antagonist mixtures. *Phytopathology* 78: 194-198.

Janisiewicz, W.J. (1996). Ecological diversity, niche overlap, and coexistance of antagoinists used in developing mixtures for biocontrol of post-harvest disease of apples. *Phytopathology* 86: 473-479.

Janisiewicz, W.J. and B. Bors (1995). Development of a microbial community of bacterial and yeast antagonists to control wound-invading postharvest pathogens of fruits. *Applied and Environmental Microbiology* 61: 3261-3267.

Johnson, J.M. and G.W. Ware (eds). (1992). *Pesticides Litigation Manual*. Clark, Boardman, Callagan, USA.

Kang, Y., R. Carlson, W. Tharpe, and M.A. Schell. (1998). Characterization of genes involved in biosynthesis of a novel antibiotic from *Burkholderia cepacia* BC11 and their role in biological control of *Rhizoctonia solani*. *Applied and Environmental Microbiology* 64: 3939-3947.

Keel, C. and G. Défago. (1997). Interactions between beneficial soil bacteria and root pathogens: Mechanisms and ecological impact. In *Multitrophic Interactions in Terrestrial System*, eds. A.C. Gange and V.K. Brown. Blackwell Science, Oxford, UK, pp. 27-47.

Kessman, H., T. Staub, C. Hofmann, T. Maetzke, G. Herzog, E. Ward, S. Ukne, and J. Ryals. (1994). Induction of systemic acquired disease resistance in plants by chemicals. *Annual Review of Phytopathology* 32: 439-460.

Kim, B.S., S.S. Moon, and B.K. Hwang. (1999). Isolation, identification and antifungal activity of a macrolide antibiotic, oligomycin A, produced by *Streptomyces libani*. *Canadian Journal of Botany* 77: 850-858.

Kitten, T., T. Kinscherf, G. McEvoy, and D.K. Willis. (1998). A newly identified regulator is required for virulence and toxin production in *Pseudomonas syringae*. *Molecular Microbiology* 28: 917-929.

Kloepper, J.W. and M.N. Schroth. (1978). Plant growth-promoting rhizobacteria on radishes. In *Soil Microbial Ecology*, ed. F.B. Meting Jr., Marcal Dekker, Inc., New York, pp. 279-304.

Kloepper, J., S. Tuzun, and J. Ku. (1992). Proposed definitions related to induced disease resistance. *Biocontrol Science and Technology* 2: 347-349.

Kobayashi, D.Y., M. Guglielmoni, and B.B. Clarke. (1995). Isolation of the chitinolytic ssbacteria *Xanthomonas maltophilia* and *Serratia marcescens* as biological control agents for summer patch disease of turf grass. *Soil Biology and Biochemistry* 27: 1479-1487.

Kraus, J. and J.E. Loper. (1995). Characterization of a genomic locus required for production of the antibiotic pyoluteorin by the biological control agent *Pseudomonas fluorescens* Pf-5. *Applied and Environmental Microbiology* 61: 849-854.

Kúc, J. (1987). Plant immunization and its applicability for disease control. In *Innovative Approaches to Plant Disease Control*, ed. I. Chet. John Wiley, New York, pp. 225-274.

Leeman, M., F.M. den Ouden, J.A. van Pelt, C. Cornellissen, A. Matamala-Garros, P.A.H.M. Bakker, and B. Schippers. (1996). Suppression of Fusarium wilt of radish by co-inoculation of fluorescent *Pseudomonas* spp. and root-colonizing fungi. *European Journal of Plant Pathology* 102: 21-31.

Leeman, M., F.M. den Ouden, J.A. van Pelt, F.P.M. Dirkx, H. Steijl, P.A.H.M. Bakker, and B. Schippers. (1996). Iron availability affects induction of systemic resistance to Fusarium wilt radish by *Pseudomonas fluorescens*. *Phytopathology* 86: 149-155.

Leeman, M., J.A. van Pelt, F.M. den Ouden, M. Heinsbroek, P.A.H.M. Bakker, and B. Schippers. (1995). Induction of systemic resistance against Fusarium wilt of radish by lipopolysaccharides of *Pseudomonas fluorescens*. *Phytopathology* 85: 1021-1027.

Leibinger, W., B. Beuker, M. Hahn, and K. Mendgen. (1997). Control of postharvest pathogens and colonization of the apple surface by antagonistic microorganisms in the field. *Phytopathology* 87: 1103-1110.

Leisinger, T. and R. Margraff. (1979). Secondary metabolites of fluorescent pseudomonads. *Microbiological Review* 43: 422-442.

Lemanceau, P. (1992). Effects bénéfiques de rhizobacteéries sur les plantes exemple des *Pseudomonas* spp. *Fluorescents Agronomie* 12: 413-417.

Lemanceau, P. and C. Alabouvette. (1991). Biological control of Fusarium diseases by fluorescent *Pseudomonas* and nonpathogenic *Fusarium*. *Crop Protection* 10: 279-286.

Levy, E., F.J. Gough, D.K. Berlin, P.W. Guiana, and J.T. Smith. (1992). Inhibition of *Septoria tritici* and other phytopathogenic fungi and bacteria by *Pseudomonas fluorescens* and its antibiotics. *Plant Pathology* 41: 335-341.

Lim, H.S., Y.S. Kim, and S.D. Kim. (1991). *Pseudomonas stutzeri* YPL-1 genetic transformation and antifungal mechanism against *Fusarium solani*, an agent of plant root rot. *Applied and Environmental Microbiology* 57: 510-516.

Lively, D.H., M. Gormann, M.E. Haney, and J.A. Mabe. (1966). Metabolism of tryptophans by *Pseudomonas aureofaceins:* I. Biosynthesis of pyrrolnitrin. *Antimicrobial Agents Chemotherapy* 1: 462-469.

Loper, J.E. and J.S. Buyer. (1991). Siderophores in microbial interactions on plant surfaces. *Molecular Plant-Microbe Interactions* 4: 5-13.

Loper, J.E. and M.D. Henkels. (1999). Utilization of heterologous siderophores enhances levels of iron available to *Pseudomonas putida* in the rhizosphere. *Applied and Environmental Microbiology* 65: 5357-5363.

Lovic, B., C. Heck, J. Gallian, and J. Anderson. (1993). Inhibition of the sugarbeet pathogens *Phoma betae* and *Rhizoctonia solani* by bacteria associated with sugarbeet seeds and roots. *Journal Sugar Beet Research* 30: 169-184.

Maloy, O.C. (1993). *Plant Disease Control: Principles and Practice.* J Wiley & Sons, Chichester, UK.

Mathre, D.E., R.J. Cook, and N.W. Callan. (1999). From discovery to use: Traversing the world of commercializing biocontrol agents for plant disease control. *Plant Disease* 83: 972-983.

Maurhofer, M., C. Keel, D. Haas, and G. Défago. (1994). Pyoluteorin production by *Pseudomonas fluorescens* strain CHA0 is involved in the suppression of Pythium damping off of cress but not of cucumber. *European Journal of Plant Pathology* 100: 221-232.

Maurhofer, M., C. Keel, U. Schnider, C. Voisard, D. Haas, and G. Défago. (1992). Influence of enhanced antibiotic production in *Pseudomonas fluorescens* strain CHA0 on its disease suppression capacity. *Phytopathology* 82: 190-195.

Mavrodi, D.V., V.N. Ksenzenko. R.F. Bonsall, R.J. Cook, A.M. Boronin. and L.S. Thomashow. (1998). A seven gene locus for synthesis of phenazine-1-carboxylic acid by *Pseudomonas fluorescens* 2-79. *Journal of Bacteriology* 180: 2541-2548.

Mazzola, M., R.J. Cook, L.S. Thomashow. D.M. Weller, and L.S. Pierson III. (1992). Contribution of phenazine antibiotic biosynthesis to the ecological competence of fluorescent pseudomonads in soil habitats. *Applied and Environmental Microbiology* 58: 2616-2624.

Mazzola, M., D.K. Fujimoto, L.S. Thomashow, and R.J. Cook (1995). Variation in sensitivity of *Gaeumannomyces graminis* to antibiotics produced by fluorescent *Pseudomonas* spp. and effect on biological control of take-all of wheat. *Applied Environmental Microbiology* 61: 2554-2559.

Mercado-Blanco, J., K.M.G.M. Van der Drift, P. Olsson, J.E. Thomas-Oates, L.C. van Loon, and P.A.H.M. Bakker. (2001). Analysis of the pmsCEAB gene cluster involved in biosynthesis of salicylic acid and the siderophore pseudomonine in the biocontrol strain *Pseudomonas fluorescens* WCS374. *Journal of Bacteriology* 183: 1909-1920.

Métraux, J.P. (2001). Systemic acquired resistance and salicylic acid: Current state of knowledge. *Journal of Plant Pathology* 107: 13-18.

Métraux, J.P., H. Signer, J. Ryals, E. Ward, M. Wyss-Benz, J. Gaudin, K. Raschdorf, E. Schmid, W. Blum, and B. Inverardi. (1990). Increase in salicylic acid at the onset of systemic acquired resistance in cucumber. *Science* 250: 1004-1006.

Milner, J.L., L. Silo-Suh, J.C. Lee, H. He, J. Clardy, and J. Handelsman. (1996). Production of kanosamine by *Bacillus cereus* UW85. *Applied and Environmental Microbiology* 62: 3061-3065.

Mitchell, R. and M. Alexander. (1961). The mycolytic phenomenon and biological control of Fusarium in soil. *Nature* (London) 190: 104-110.

Mukerji, K.G. and K.L. Garg (eds.) (1988a). *Biocontrol of Plant Disease,* Vol. 1. CRC Press Inc., Boca Raton, FL.

Mukerji, K.G. and K.L. Garg (eds.) (1988b). *Biocontrol of Plant Diseases,* Vol. 2. CRC Press Inc., Boca Raton, FL.

Nakayama, T., Y. Homma, Y. Hashidoko, J. Mizutani, and S. Tahara. (1999). Possible role of xanthobaccins produced by *Stenotrophomonas* sp. strain SB-K88 in suppression of sugar beet damping-off disease. *Applied and Environmental Microbiology* 65: 4334-4339.

Neilands, J.B., K. Konopka, B. Schwyn, M. Coy. R.T. Francis, B.H. Paw, and A. Bagg. (1987). In *Iron Transport in Microbes. Plants and Animals.* eds. G. Winkelmann, D. van der Helm, and J.B. Neilands. VCH Press. Weinheim, pp. 3-33.

Nielsen, M.N., J. Sørensen, J. Fels, and H.C. Pedersen. (1998). Secondary metabolite- and endochitinase-dependent antagonism toward plant-pathogenic microfungi of *Pseudomonas fluorescens* isolates from sugar beet rhizosphere. *Applied and Environmental Microbiology* 64: 3563-3569.

Nelson, E.B., W.L. Chao, J.M. Norton, G.T. Nash, and G.E. Harman. (1986). Attachment of *Enterobacter cloacae* to hyphae of *Pythium ultimum*: Possible role in biological control of Pythium pre-emergence damping-off. *Phytopathology* 76: 327-335.

Notz, R., M. Maurhofer, H. Dubach, D. Haas, and G. Défago. (2002). Fusaric acid producing strains of *Fusarium oxysporum* alter 2,4-diacetylphloroglucinol bio-synthesis gene expression in *Pseudomonas fluorescens* CHA0 in vitro and in the rhizosphere of the wheat. *Applied and Environmental. Microbiology* 68: 2229-2235.

Notz, R., M. Maurhofer, U. Schnider-Keel, B. Duffy, D. Haas, and G. Défago. (2001). Biotic factors affecting expression of the 2,4-di-acetylpholoroglucinol biosynthesis gene phlA in *Pseudomonas fluorescens* biocontrol strain CHA0 in the rhizosphere. *Phytopathology* 91: 873-881.

Nowak-Thompson, B., N. Chaney, J.S. Wing, S.J. Gould, and J.E. Loper. (1999). Characterization of the pyoluteorin biosynthetic gene cluster of *Pseudomonas fluorescens* Pf-5. *Journal of Bacteriology* 181: 2166-2174.

Nowak-Thompson, B., S.J. Gould, J. Kraus, and J.E. Loper. (1994). Production of 2,4-diacetylphloroglucinol by the biocontrol agent *Pseudomonas fluorescens* Pf-5. *Canadian Journal of Microbiology* 40: 1064-1066.

Nowak-Thompson, B.. S.J. Gould, and J.E. Loper. (1997). Characterization of the pyoluteorin biosynthetic gene cluster of *Pseudomonas fluorescens* Pf-5. *Gene* 204: 17-24.

Ongena. M.. F. Daayf, P. Jacques, P. Thonart, N. Benhamou, T.C. Paulitz, P. Cornelis, N. Koedam, and R.R. Belanger (1999). Protection of cucumber against Pythium root rot by fluorescent pseudomonads: Predominant role of in-duced resistance over siderophores and antibiosis. *Plant Pathology* 48: 66-76.

Ordentlich, A., Y. Elad, and I. Chet. (1988). The role of chitinase of *Serratia marcescens* in biocontrol of *Sclerotium rolfsii*. *Phytopathology* 78: 84-88.

O'Sullivan, D.J. and F. O'Gara. (1992). Traits of fluorescent *Pseudomonas* spp. in-volved in suppression of plant root pathogens. *Microbiological Review* 56: 662-676.

Pierson, E.A. and D.M. Weller. (1994). Use of mixtures of fluorescent pseu-domonads to suppress take-all and improve the growth of wheat. *Phytopathology* 84: 940-947.

Pieterse, C.M.J., S.C.M. van Wees, E. Hoffland, J.A. van Pelt, and L.C. van Loon. (1996). Systemic resistance in *Arabidopsis* induced by biocontrol bacteria is in-dependent of salicylic acid accumulation and pathogenesis-related gene expres-sion. *Plant Cell* 8: 1225-1237.

Pieterse, C.M.J., S.C.M. van Wees, J. Ton, J.A. van Pelt, and L.C. van Loon. (2003). Signalling in rhizobacteria-induced systemic resistance in *Arabidopsis thaliana*. *Plant Biology* 4: 535-544.

Pieterse, C.M.J., S.C.M. van Wees, J.A. van Pelt, M. Knoester, R. Laan. H. Gerrits, P.J. Weisbeek, and L.C. van Loon. (1998). A novel signaling pathway controlling induced systemic resistance in *Arabidopsis*. *Plant Cell* 10: 1571-1580.

Raaijmakers, J.M., R.F. Bonsall, and D.M. Weller. (1999). Effect of population den-sity of *Pseudomonas fluorescens* on production of 2.4-diacetylphloroglucinol in the rhizosphere of wheat. *Phytopathology* 89: 470-475.

Raaijmakers, J.M., M. Leeman, M.M.P. van Oorschot, I. van der Sluis, B. Schippers, and P.A.H.M. Bakker. (1995). Dose-response relationships in biological control of Fusarium wilt of radish by *Pseudomonas* spp. *Phytopathology* 85: 1075-1081.

Raaijmakers, J.M., I. van der Sluis, M. Koster, P.A.H.M. Bakker, P.J. Weisbeek, and B. Schippers. (1995). Utilization of heterologous siderophores and rhizosphere competence of fluorescent *Pseudomonas* spp. *Canadian Journal of Microbiology* 41: 126-135.

Raaijmakers, J.M. and D.M. Weller. (1998). Natural plant protection by 2,4 diacetylphloroglucinol producing *Pseudomonas* spp. in take-all decline soils. *Molecular Plant-Microbe Interactions* 11: 144-152.

Raaijmakers, J.M., D.M. Weller, and L.S. Thomashow. (1997). Frequency of antibiotic-producing *Pseudomonas* spp. in natural environments. *Applied and Environmental Microbiology* 63: 881-887.

Raupach, G.S. and J.W. Kloepper. (1998). Mixtures of plant growth-promoting rhizobacteria enhance biological control of multiple cucumber pathogens. *Phytopathology* 88: 1159-1164.

Reddi, T. K., Y.P. Khudiakov, and A.V. Borovkov. (1969). *Pseudomonas fluorescens* strain 26.0, producing phytotoxic substances. *Microbiologiya* 38: 909-913.

Reitz, M., P. Oger, A. Meyer, K. Niehaus, S.K. Farrand, J. Hallman, and R.A. Sikora. (2002). Importance of the 0-antigen, core-region and lipid A of rhizobial lipopolysaccharides for the induction of systemic resistance in potato to *Globodera pallida*. *Nematology* 4: 73-79.

Reuveni, R. (1995). *Novel Approaches to Integrated Pest Management*. Lewis Publishers, Boca Raton, FL.

Rosas, S., F. Altamirano, E. Schröder, and N. Correa. (2001). In vitro biocontrol activity of *Pseudomonas aurantiaca*. *OYTON International Journal of Experimental Botany* 52: 203-209.

Rosas, S., M. Rovera, P. Emidi, M. Díaz, J. Andrés, and N. Correa. (2002). *Pseudomonas aurantiaca: su aporte a una agricultura sustentable*. XXI Reunión Latinoamericana de Rhizobiología-VI Congreso Nacional de la Fijación Biológica del Nitrógeno. 21-24 de Octubre 2002. Cocoyoc Morelos, México.

Ross, A.F. (1961). Systemic acquired resistance by localized virus infections in plants. *Virology* 14: 340-358.

Rovera, M., M. Correa, M. Reta, J. Andrés, S. Rosas, and N. Correa. (2000). Chemical identification of antifungal metabolites produced by *Pseudomonas aurantiaca* In Proceedings of the 5th International PGPR Workshop. http://www.ag.auburn.edu/argentina.

Ryals, J.A., U.H. Neuenschwander, M.G. Willits, A. Molina, H.Y,. Steiner, and M.D. Hunt. (1996). Systemic acquired resistance. *Plant Cell* 8: 1808-1819.

Schippers, B. (1992). Prospects for management of natural suppressiveness to control soilborne pathogens. In *Biological Control of Plant Diseases, Progress and Challenges for the Future*. NATO ASI Series A: Life Sciences. Vol. 230. eds. E. C. Tiamos. G. C. Panavizas, and R.J. Cook, Plenum Press, New York, pp. 21-34.

Schnider-Keel, U., A. Seematter, M. Maurhofer, C. Blumer. B. Duffy, G. Gigot-Bonnefoy, C. Reimmann, R. Notz. G. Défago, D. Haas, and C. Keel. (2000). Autoinduction of 2.4-diacetylphloroglucinol biosynthesis in the biocontrol agent *Pseudomonas fluorescens* CHA0 and repression by the bacterial metabolites salicylate and pyoluteorin. *Journal of Bacteriology* 182: 1215-1225.

Shanahan, P., D.J. O'Sullivan, P. Simpson, J.D. Glennon, and F. O'Gara. (1992). Isolation of 2,4-diacetylphloroglucinol from a fluorescent pseudomonad and investigation of physiological parameters influencing its production. *Applied Environmental Microbiology* 58: 353-358.

Smirnov, V.V. and E.A. Kiprianova. (1990). *Bacteria of Pseudomonas Genus.* Naukova Dumka, Kiev, Ukraine.

Stevans, A.M., K.M. Dolan, and E.P. Greenberg. (1994). Synergistic binding of the *Vibrio fischeri* LuxR transcriptional activator domain and RNA polymerase to the lux promoter region. *Proceedings of the National Academy of Science USA* 91: 12619-12623.

Sticher, L., B. Mauch-Mani, and J.P. Métraux. (1997). Systemic adquired resistance. *Annual Review of Phytopathology* 35: 235-270.

Sundhein, L., A.R. Poplawsky, and A.H. Ellingboe. (1988). Molecular cloning of two chitinase genes from *Serratia marcescens* and their expression in pseudomonas species. *Physiological and Molecular Plant Pathology* 33: 484-491.

Thomashow, L.S. and D.M. Weller. (1995). Current concepts in the use of introduced bacteria for biological disease control: Mechanisms and antifungal metabolites. In *Plant–Microbe Interactions,* eds. G. Stacey and N.T. Keen. Chapman and Hall, New York, pp. 187-235.

Thomashow, L.S., D.M. Weller, R.F. Bonsall, and L.S. Pierson III. (1990). Production of the antibiotic phenazine-1-carboxylic acid by fluorescent pseudomonas species in the rhizosphere of wheat. *Applied Environmental Microbiology* 56: 908-912.

Thrane, C., S. Olsson, T.H. Nielsen, and J. Sörensen. (1999). Vital fluorescent stains for detection of stress in *Pythium ultimum* and *Rhizoctonia solani* challenged with viscosinamide from *Pseudomonas fluorescens* DR54. *FEMS Microbiology Ecology* 30: 11-23.

Turner, J.M. and A.J. Messenger. (1986). Occurrence, biochemistry and physiology of phenazine pigment production. *Advances in Microbial Physiology* 27: 211-275.

Tuzun, S. and J. Kloepper. (1995). Practical application and implementation of induced resistance. In *Induced Resistance to Diseases in Plants,* eds. R. Hammerschmidt and J. Kuc. Kluwer Academic Press, Dordrecht, the Netherlands, pp. 152-168.

Uknes, S., B. Mauch-mani, M. Moyer, S. Potter, S. Willians, S. Dincher, D. Chandler, A. Slusarenko, E. Ward, and J. Ryals. (1992). Acquired resistance in Arabidopsis. *Plant Cell* 4: 645-656.

van Loon, L.C. (1997). Induced resistanse in plants and the role pathogenesis-related proteins. *European Journal of Plant Pathology* 103: 753-765.

van Loon, L.C., P.A.H.M. Bakker, and M.J. Pieterse. (1998). Systemic resistance induced by rhizosphere bacteria. *Annual Review of Phytopathology* 36: 753-765.

van Peer, R., G.J. Niemann, and B. Schippers. (1991). Induced resistance and phytoalexin accumulation in biological control of Fusarium wilt of carnation by *Pseudomonas* sp. strain WCS417r. *Phytopathology* 81: 728-733.

van Peer, R. and B. Schippers. (1992). Lipopolysaccharides of plant-growth promoting *Pseudomonas* sp. strain WCS417r induce resistance in carnation to fusarium wilt. *Netherlands Journal of Plant Pathology* 98: 129-139.

van Wees, S.C.M., E.A.M. De Swart, J.A. van Pelt, L.C. van Loon, and C.M.J. Pieterse. (2000). Enhancement of induced disease resistance by simultaneous activation of salicylate- and jasmonate-dependent defense pathways in *Arabidopsis thaliana gs. Proceedings of the National Academic of Sciences USA* 97: 8711-8716.

van Wees, S.C.M., M. Luijendijk, I. Smoorenburg, L.C. van Loon, and C.M.J. Pieterse. (1999). Rhizobacteria-mediated induced systemic resistance (ISR) in *Arabidopsis* is not associated with a direct effect on expression of known defense-related genes but stimulates the expression of the jasmonate-inducible gene Atvsp upon challenge. *Plant Molecular Biology* 41: 537-549.

Vincet, M.N., L.A. Harrison, J.M. Brackin, P.A. Kovacevich, P. Mukherji, D.M. Weller, and E.A. Pierson. (1991). Genetic analysis of the antifungal activity of a soilborne *Pseudomonas aureofaciens* strain. *Applied and Environmental Microbiology* 57: 2928-2934.

Voisard, C., C. Keel, D. Haas, and G. Defago. (1989). Cyanide production by *Pseudomonas fluorescens* helps suppress black root rot of tobacco under gnotobiotic conditions. *EMBO Journal* 8: 351-358.

Walsh, U.F., J.P. Morrisey, and F. O'Gara. (2001). Pseudomonas for biocontrol of phytopathogens: From functional genomics to commercial exploitation. *Current Opinion in Biotechnology* 12: 289-295.

Ward, E.R., S.J. Uknes, S.C. Willians, S.S. Dincher, D.L. Wiederhold, D.C. Alexander, P. Ahl-Goy, J.P. Métraux, and J.A. Ryals. (1991). Coordinate gene activity in response to agents that induce systemic acquired resistance. *Plant Cell* 3: 1085-1094.

Wei, G., J.W. Kloepper, and S. Tuzun. (1991). Induction of systemic resistance of cucumber to *Colletotrichum orbiculare* by select strains of plant growth-promoting rhizobacteria. *Phytopathology* 81: 1508-1512.

Wei, G., J.W. Kloepper, and S. Tuzun. (1996). Induced systemic resistance to cucumber diseases and increased plant growth by plant growth-promoting rhizobacteria under field conditions. *Phytopathology* 86: 221-224.

Whipps, J.M. (1997a). Developments in the biological control of soil-borne plant pathogens. *Advances in Botanical Research* 26: 1-134.

Whipps, J.M. (1997b). Ecological considerations involved in commercial development of biological control agents for soil-borne diseases. In *Modern Soil Microbiology,* eds. J.D. van Elsas, J.T. Trevors, and E.M.H. Marcel Dekker, Wellington, pp. 525-546

Whipps, J.M. (1997c). Interactions between fungi and plant pathogen in soil and the rhizosphere. In *Multithropic Interactions in Terrestrial Systems,* eds. A.C. Gange and B.K. Brown. Blackwell Science, Oxford, pp. 47-63.

Whistler, C.A. and L.S. Pierson III. (2003). Repression of phenazine antibiotic production in *Pseudomonas aureofasciens* strain 30-84 by RpeA. *Journal of Bacteriology* 185: 3718-3725.

Wienberg, E.D. (1970). Biosynthesis of secondary metabolites: Roles of trace elements. *Advances in Microbial Physiolgy* 4: 1-44.

Winkelmann, G. (ed). (1991). *Handbook of Microbial Iron Chelates,* CRC Press, Boca Raton, FL.

Winkelmann, G.. D. van der Helm. and J.B. Neilands (eds.). (1987). *Iron: Transport in Microbes, Plants and Animals.* VCH Press, Weinheim.

Yang, C.H. and D.E. Crowley. (2000). Rhizosphere microbial community structure in relation to root location and plant iron nutritional status. *Applied and Environmental Microbiology* 66: 345-351.

Yuan, Z.. S. Cang, M. Matsufuji. K. Nakata. Y. Nagamatsu, and A. Yoshimoto. (1999). High production of pyoluteorin and 2,4-diacetylphloroglucinol by environmental factors modulating antibiotic and siderophore biosynthesis by *Pseudomonas fluorescens* S272 grown on ethanol as a sole carbon source. *Journal of Fermentation in Bioengineering* 86: 559-563.

Zehnder, G.W., C. Yao, J.F. Murphy, E.R. Sikora, and J.W. Kloepper. (2000). Induction of resistance in tomato against cucumber mosaic cucumovirus by plant growth-promoting rhizobacteria. *Biocontrol* 45: 127-137.

Chapter 5

The Role of Competitive Root Tip Colonization in the Biological Control of Tomato Foot and Root Rot

Sandra de Weert
Irene Kuiper   Kevin Eijkemans
Faina D. Kamilova   Gail M. Preston
Ine H.M. Mulders   Paul Rainey
Guido V. Bloemberg   Igor Tikhonovich
Lev Kravchenko   André H.M. Wijfjes
Tanya Azarova   Ben J.J. Lugtenberg

INTRODUCTION

Rhizosphere colonization is one of the first steps in the pathogenesis of soilborne pathogens. Microbes are attracted to the root system because it secretes chemo-attractants (de Weert et al., 2002) and nutrients (Lugtenberg and Bloemberg, 2004). Roots can exude up to 30 percent of the fixed carbon as exudate metabolites. These components attract microbes to such an extent that the density of microbes in the rhizosphere can be 10 to 100 times higher than that in the bulk soil. This "rhizosphere effect" was discovered a century ago by the German microbiologist Hiltner (1904).

We thank Marco Simons for the isolation of strain PCL1085. The authors thank Rudy Scheffer and Bernadette Kroon from Syngenta Seeds B.V., Enkhuizen. the Netherlands, for supplying us with chemically untreated seeds of tomato cultivar Carmello. This work was supported by the NOW Earth and Life Sciences Council, project no. 809-39.003, and the Technology Foundation STW, project no. GBI 55.3868 (73-95), and EU proect QLK3-CT-200-31759 (ECO-SAFE) and QRLT-2002-00914 (Pseudomics).

The major soluble exudate components are organic acids, sugars, and amino acids. In tomato root exudates, the total amount of organic acid is five times higher than that of sugar (Lugtenberg, Kravchenko, and Simons, 1999), which, in turn, is double the amount of amino acid (Lugtenberg and Bloemberg, 2004). The major organic acids in tomato root exudate are citric, malic, and lactic acid. The major chemo-attractants for the cells of *Pseudomonas fluorescens* strain WCS365 in tomato exudates are the organic acids, malic and citric acids, as well as all major amino acids, whereas sugars do not play a role as chemo-attractants (de Weert et al., 2002).

Upon attraction toward exudate, some bacteria colonize the root, first as single cells, followed by the formation of microcolonies or biofilms (Bloemberg et al., 1997; Chin-A-Woeng et al., 1997; Bloemberg and Lugtenberg, 2004). These sites with high bacterial density are ideal sites for quorum sensing, a process in which bacteria secrete a sufficiently high concentration of AHLs (acyl homoserine lactones) to initiate processes required for many interactions with other organisms (Bull et al., 1991; Salmond et al., 1995; Bloemberg et al., 1997; Swift et al., 2001; Joint et al., 2002). It has been hypothesized that the high frequency of conjugation found in the rhizosphere (Van Elsas et al., 1988) is caused by the fact that bacteria in a biofilm can reach the required quorum of AHL molecules required for conjugation (Chin-A-Woeng et al., 1997). The area of the root surface that can be covered by microbes is limited to approximately 6 percent. The processes of root colonization and biofilm formation can be visualized best after labeling the microbes with auto-fluorescent proteins (Bloemberg et al., 1997, 2000; Stuurman et al., 2000; Chin-A-Woeng et al., 2004).

The trait of some bacteria to colonize the root can be used for applications such as biofertilization (Spaink et al., 1998), phytostimulation (Okon et al., 1997; Steenhoudt and Vanderleyden, 2000), phytoremediation (Kuiper, Bloemberg, and Lugtenberg, 2001; Kuiper et al., 2004), and biological control of plant diseases (Schippers et al., 1987; Weller, 1988; Thomashow and Weller, 1996; Bloemberg and Lugtenberg, 2001; Lugtenberg et al., 2001; Lugtenberg and Kamilova, 2004). The bacteria are usually applied by coating on seeds or by spraying in the furrow. Not all applications are successful; limited success is often correlated with poor root colonization (Schippers et al., 1987; Weller, 1988; Bull et al., 1991).

Because of the importance of competitive colonization for applications, we initiated a research program 15 years ago to study the mechanism of root colonization. In these studies, we initially use a gnotobiotic assay in which seedlings are inoculated with a 1:1 mixture of two bacteria. After the plant grows for one week, it is removed from its quartz sand-plant nutrient solution substrate, and the plant root or its parts are analyzed for the numbers and ratio of the two microbes. We analyze at least the root tip, because good competitive colonizers have to be able to reach the deeper root parts (Simons et al., 1996). After mutants were characterized for their colonization properties in the gnotobiotic system, their colonizing properties were tested in potting soil. With only a few exceptions, the mutants found initially in the gnotobiotic system also behaved as mutants in the potting soil. The advantage of the strategy to start with the gnotobiotic system is twofold: (1) the numbers of microbes under investigation reaching the root tip are 10 times higher than those in the gnotobiotic system; and (2) the assays in the gnotobiotic system are more reproducible (Lugtenberg et al., 2001). The gnotobiotic as well as the potting soil systems were used to compare competitive root tip-colonizing abilities of wild-type strains. It appeared that *P. fluorescens* strain WCS365 is the best colonizer of the biocontrol strains analyzed until recently. Therefore, we used this strain as the wild type from which competitive root tip-colonization mutants were isolated (Simons et al., 1997; Dekkers, van der Bij, et al., 1998; Lugtenberg et al., 2001; de Weert et al., 2002).

The analysis of colonization mutants has led to the discovery of many colonization traits so far, which are extensively reviewed by Lugtenberg et al. (2001). A few striking traits will be mentioned here. Flagellar motility was discovered as the first colonization trait (de Weger et al., 1987). It was discovered subsequently (de Weert et al., 2002) that motility is required for the chemotactic response of microbes toward exudate compounds as a first step in the process of root colonization. A well-functioning outer membrane was shown to be involved in colonization. Many colonization mutants have mutations affecting the structure of the LPS (lipopolysaccharide), a major outer membrane component (Lugtenberg and van Alphen, 1983). It appeared that for a good colonizer, part of the O-antigen side chain is required (Dekkers, van der Bij, et al., 1998). Good colonizers should also make their own vitamins and building blocks required for macromolecule synthesis. Apparently, the concentration of these components in the rhizosphere is very low. One of the most surprising findings was that a site-specific recombinase plays a role in colonization (Dekkers, Phoelich, et al., 1998). This enzyme promotes reciprocal recombination between short DNA sequences. This process can lead to inversion or deletion of the DNA fragment between the recognition sites (Sadowski, 1986; Dybvig, 1993). Apparently, one of the DNA configurations leads to a colonization-incompetent phenotype (Dekkers, Phoelich, et al., 1998). Characterization of colonization mutant PCL1206 led to the discovery that putrescine is a tomato exudate component (Kuiper, Bloemberg, Noreen, et al., 2001).

A recently developed enrichment procedure (Kuiper, Bloemberg, and Lugtenberg, 2001), applied to isolate mutants and wild-type strains enhanced in competitive colonization appeared to be successful. *mutY* mutants can accumulate combinations of mutations which enormously enhance a strain's colonizing ability (de Weert et al., 2004), whereas the procedure yields wild types, which compete much better for niches and/or nutrients on the root than *P. fluorescens* WCS365 (Lugtenberg et al., 2004).

MATERIALS AND METHODS

General

The methods used have been described previously in various primary publications (Dekkers, Bloemendaal, et al., 1998; Kuiper, Bloemberg, and Lugtenberg, et al., 2001; Lugtenberg et al., 2001). We will, therefore, only describe methods crucial for the understanding of the results by those unfamiliar with our colonization work.

Preparation of Tomato Exudates

Tomato (*Lycopersicon esculentum* Mill. cv. Carmello) seeds (S&G Seeds B.V., Enkhuizen, the Netherlands) were sterilized by gentle shaking for 3 min in a solution of 5 percent household sodium hypochlorite. The sterilized seeds were soaked six times for 30 min in sterile demineralized water. To synchronize the germination process, seeds were placed in petri dishes, containing a plant nutrient solution (PNS) (Hoffland et al., 1989), solidified with 1.5 percent Pronarose D1 (Hispanagar, Burgos, Spain) and incubated overnight upside down in the dark at 4°C, followed by incubation at 28°C for two days. Seed and seedling exudates were collected on wet filter paper in 10 petri dishes, containing 20 seeds each. Seed exudate was extracted from the filter paper after two days of incubation at 4°C in the dark. Seedling exudate was collected similarly after four days of incubation at 28°C in the dark. Root exudate was prepared in glass cylinders, each containing 5 ml PNS. Five sterile seedlings were placed on stainless-steel gauze at the bottom of the cylinder. Exudates from 25 plants were collected after 14 days of growth in a climate-controlled growth chamber (20°C, 70 percent relative humidity, 16 hours daylight), and checked for sterility. Exudates were evaporated to dryness at 45°C under vacuum, dissolved in 5 ml of water, and filtrated using membrane filter (pore size 45 μm).

HPLC Analysis of Tomato Exudate Organic Acids

The exudate samples of seed and root exudates were purified on a strong cation-exchanger (Dowex AG50W8, H^+ form). A sample of 5 ml of the redissolved exudate was loaded onto a column bed of 1.5×1.0 cm and passed at a rate of 1 ml/min. The eluent was collected (15 ml) and evaporated to dryness at 45°C under vacuum and redissolved in 200 µl deionized water. Quantitative analysis of organic acids was carried out, using a Jasco LC-900 series HPLC system (Jasco International Co.) consisting of a PU-980 pump, a DG-980-50 degassing module, a LG-980-02 ternary gradient unit, an UV detector UV-975 set to 210 nm, and a Rheodyne 20 µl valve loop injector. Data were collected and analyzed, using Borwin (JMBS Developments, France) chromatography software that was connected to the HPLC system Jasco LC-Net. The column used was an ion-exchange Supelcogel C-610-H (Supelco Gland, Switzerland, no. 213-4M23, 30 cm × 7.8 mm). The mobile phase was 10 mM H_3PO_4, the running rate was 0.6 ml/min. Of each sample, 20 µl was injected.

After optimizing the HPLC system using standard organic acids (Sigma, St. Louis, MO, USA; Fluka Chemie AG, Switzerland; Serva Feinbiochemica, Germany), exudate samples were analyzed. Organic acid peaks in the exudate chromatograms were identified by comparison of retention times of unknown peaks with retention times of standard compounds. For confirmation, exudate samples and known standards were coinjected to test whether the peaks overlap completely. The standards used are listed in Table 5.1. Quantification was done by establishing detector response calibration graphs and calculating the quantity of the acid in the exudate sample.

Competitive Tomato Root Tip Colonization

Competitive root tip-colonization assays were carried out after the inoculation of seeds or seedlings with a 1:1 mixture of wild-type and mutant cells, as described by Simons et al. (1996). The plants were grown for one week in a climate-controlled growth chamber, after which the 1 cm-long root tip was analyzed for the ratio of the two strains. Statistics on the results were performed, using the Wilcoxon-Mann-Whitney test (Sokal and Rohlf, 1981).

Competitive Growth in Exudates

To analyze competitive growth in root exudate, exudate was collected as described by Simons et al. (1997). Cells grown overnight in standard BM medium (Lugtenberg et al., 2001) were washed with PBS (phosphate buffered saline), and subsequently diluted to an OD 620nm value of 0.1. Subsequently, cells were diluted 1000-fold in a 1:1 ratio in 2 ml exudates and allowed to grow. Dilutions were plated on KB agar, supplemented with ×-gal at various time intervals, during four days. Plates were incubated at 28°C.

TABLE 5.1. Composition of organic acids in tomato seed, seedling, and root exudate.[a]

Organic acid[b]	Seed exudate[c] ng/seed (%)[d]	Seedling exudate[c] ng/seedling (%)	Root exudate[f] μg/plant (%)
Oxalic acid	300 ± 55 (50)	636 ± 77 (16.5)	1.4 ± 0.5 (5.7)
Ketoglutaric acid	105 ± 36 (17.5)	22 ± 14 (0.5)	nd[h]
Citric acid	nd	2060 ± 1020 (54)	13.6 ± 0.8 (55.2)
Pyruvic acid	113 ± 27 (19)	295 ± 17 (8)	1.0 ± 0.1 (4.2)
Malic	nd	180 ± 20 (5)	3.8 ± 2.3 (15.4)
t-Aconitic	nd	0.8 ± 0.3 (< 0.5)	0.02 ± 0.01 (0.1)
Succinic	tr[g]	102 ± 36 (2.5)	1.9 ± 0.3 (7.6)
Lactic	77 ± 12 (13)	475 ± 62 (12)	2.5 ± 1.2 (10.1)
Acetic	2 ± 1	1 ± 1	nd
Fumaric	3 ± 1 (0.5)	8 ± 1 (< 0.5)	nd
Propionic	nd	nd	0.07 ± 0.02 (0.3)
Pyruglutamic	tr	61 ± 1.5 (1.5)	0.35 ± 0.01 (1.4)
Total	600 (100)	3841 (100)	24.64 (100)

[a]Data are averages of analyses of three samples

[b]Mandelic and glyoxylic acids were not found in any of the exudate samples

[c]Seed exudate was collected after two days at 4°C

[d]Seedling exudate was collected after two days at 28°C

[e]Root exudate was collected after 14 days at 28°C

[f]Percentage of total organic acid

[g]tr, trace

[h]nd, not detected

Adhesion Experiments

Adhesion experiments were performed as described by Simoni et al. (1998). Briefly, a 10 ml syringe was filled with 12 g of PBS-moistened sand (7.5 ml sand column). The flow-through rate was 0.6 ml/min, and the dead volume of the tubing was 1 ml. The OD_{280} was set to approximately 1.0 in PBS, and subsequently the sample was loaded onto the column. The OD_{280} of 3 ml fractions was measured. When the OD_{280} did not change in subsequent three fractions, columns were washed with PBS solution. When the

OD_{280} did not change again in three subsequent fractions, the PBS was replaced by sterile demineralized water. Again, fractions of 3 ml were collected, and the OD_{280} was measured. When OD_{280} measurements were close to zero, the sampling was stopped.

RESULTS AND DISCUSSION

Nutrient Sources for Microbes in the Tomato Rhizosphere

Analysis and Identification of Tomato Seed, Seedling, and Root Exudate Organic Acids

HPLC analysis was used to analyze the organic acid composition of tomato seed, seedling, and root exudates. Exudate peaks were identified by the comparison of retention volumes with standard organic acids and subsequent injection of a mixture of the standard and exudate samples. Tomato seed, seedling, and root exudate were analyzed in three replicate samples. Results are shown in Table 5.1. The ratios of the various components were very different in the three exudates. Oxalic acid, pyruvic acid, and ketoglutaric acid are the major compounds in seed exudates but are minor compounds in root exudate. In contrast, citric acid was not detected in seed exudates but was by far the major compound in seedling and root exudates. Comparison of the amounts of soluble organic acids and sugars in the exudates showed that the amounts are similar in seed exudates, whereas the amount of organic acid exceeds that of sugars by factors of approximately two and five in seedling and root exudates, respectively. A comparison of the present results on organic acids with earlier data of our group on the other exudate components shows that exudate of seven-day-old tomato plants contains glutamic acid (4.4 nmole per plant) as the major amino acid (Simons et al., 1997), glucose (10 nmole per plant) as the major sugar, (Lugtenberg, Kravchenko, and Simons, 1999), and citric acid (60 nmole per plant) as the major organic acid (Table 5.1). It should be noted that the toxic polyamine putrescine is the major known N- and C-compound in tomato root exudate (175 nmole per plant; Kuiper, Bloemberg, Noreen, et al., 2001).

Nutritional Basis of Tomato Root Colonization

To test the hypothesis that growth on exudate organic acids is an important colonization trait, mutant PCL1085 was isolated. This is a Tn5*lacZ* mutant of *P. fluorescens* WCS365 which grows poorly on the two tomato exudate organic acids, malic and succinic acids, but well on glucose. The Tn5*lacZ* mutation appeared to be inserted in the gene *mqo,* which encodes malate:quinone oxidoreductase (malate dehydrogenase). Using double homologous recombination (Schafer et al., 1994), a second unmarked mutant in this gene, strain PCL 1463, was constructed.

Mutants PCL1085 and PCL1463 appeared to be such poor colonizers that they were completely outcompeted in competitive root-colonization assays by strain WCS365 or its Tn5*lacZ* mutant PCL1500. The latter strain has wild-type colonization characteristics. Its transposon was found to be located in a gene with unknown function. Mutant derivatives carrying the complete *mqo* gene complemented for both growth on malate as well as competitive tomato root tip colonization (Figure 5.1).

Screening of several wild-type pseudomonads, using the method of Glick (1995), showed that the efficient root colonizer and biocontrol strain *P. fluorescens* WCS365 was able to grow on BM-medium (Lugtenberg et al., 2001) with ACC (1-aminocyclopropane-1-carboxylic acid) as the sole nitrogen source. In an attempt to isolate mutants in the structural gene for ACC synthesis, Tn5*luxAB* mutant PCL1654 was isolated by screening a

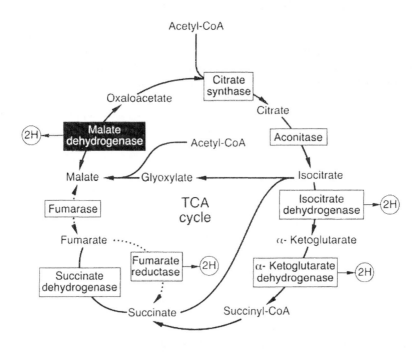

FIGURE 5.1. TCA cycle.

mutant bank of WCS365 for the inability to grow on ACC as the sole N-source. Cloning and sequence analysis of a DNA region flanking the transposon showed that the transposon of mutant PCL1654 had inserted in a gene with 84 percent amino acid identity (over a stretch of 195 amino acids) with *cis*-aconitate hydratase. In competitive tomato root tip-colonization assays, the mutant lost from the wild-type WCS365. We do not understand why the mutant was isolated in the used screening method. However, the results of the competitive colonization experiment and the poor growth in exudate confirm the conclusion drawn from work with mutant PCL1085 that mutants unable to utilize organic acids efficiently are poor competitive tomato root tip colonizers.

Since a mutant unable to grow on the major tomato root exudate's sugar glucose colonizes the root tip equally well as the parental *P. fluorescens* strain WCS365 (Lugtenberg, Kravchenko, and Simons, 1999), and since tomato root exudate contains five fold more organic acid than sugar, it is likely that organic acids are a/the major carbon source for efficient tomato root colonization by this strain.

It should be noted that exudate composition can depend on many factors, such as environmental factors and plant species' age and cultivar. For example, in the root exudate of grass, organic acids and sugars are equally well represented (Kuiper et al., 2002), in contrast to the root exudates of tomato (Table 5.1). Also, succinic acid is the major component of grass root exudates, representing 74 percent of the organic acids (Kuiper et al., 2002). In contrast, succinic acid is a minor component of tomato root exudate (Table 5.1).

The result suggests that growth on major exudate components can be used as an important parameter for inoculant quality, in applications for which root colonization is important, such as biocontrol, biofertilization, phytoremediation, and phytostimulation.

NOVEL COMPETITIVE ROOT
TIP-COLONIZATION MUTANTS

General Aspects

All mutants were first analyzed for (1) growth rate in the laboratory media King's B (King et al., 1954) and BM (Lugtenberg, Dekkers, et al., 1999) traits; (2) motility on swim plates; and (3) LPS ladder patterns on sodium dodecyl sulphate polyacrylamide gel electrophoresis, traits known to be involved in competitive root tip colonization. Only mutants not impaired in these traits but impaired in competitive root tip colonization in both sand and potting soil were further analyzed.

Mutant PCL1204 Is Impaired in an Operon with Unknown Function

This mutant has a 10-fold decreased competitive root tip-colonizing ability compared to wild-type *P. fluorescens* WCS365. It has its Tn5*lacZ* mutation in a gene homologous to PA3070 from *P. aeruginosa* PAO1 (Stover et al., 2000). PA3070 encodes a regulatory protein, and is part of a six-gene cluster in *P. aeruginosa* with unknown function. The operon of WCS365 appeared to contain five genes with strong similarity to those of *P. aeruginosa* (Figure 5.2), and is conserved among all sequenced pseudomonads. New WCS365 mutants were constructed by homologous recombination in genes PA3070 (mutant PCL1802) and PA3074 (mutant PCL1803). The new mutants were impaired in competitive root tip colonization. We, therefore, conclude that at least gene homolog PA3074 is required for efficient competitive tomato root tip colonization.

The *batI* (Bacteroides aerotolerance) system from *Bacteroides fragilis* (Tang et al., 1999) consists of a five-gene operon cluster *A, B, C, D,* and *E*. BatA and BatB show high similarity to one another. This is also the case for gene homolog PA3073 and PA3074 in WCS365. Genes *batA* and *batB* are involved in aerotolerance of *Bacteroides fragilis*. When *batD* is mutated, *B. fragilis* shows a decreased tolerance when exposed to atmospheric oxygen. By using the system described by Camacho-Carjaval et al. (2002), growth under different oxygen concentration (0.5 and 50 percent) was followed. However, no effect on growth rate could be observed for mutants PCL1204, PCL1802, and PCL1803 when compared to wild-type WCS365. Thus, no clear phenotypic function could be found for the gene impaired in mutant PCL1204 other than that it is impaired in competitive tomato root tip colonization.

Details about mutant PCL1204 are described by de Weert et al. (2004).

Mutant PCL1208 Has Its Insertion in a Rearrangement Hot Spot

This mutant has a 10-fold decreased competitive tomato root tip-colonizing ability, compared with its parental strain WCS365. It has its Tn5*lacZ* inserted in a gene with homology to an *rhs* family protein from *P. fluorescens* PfO-1. Several attempts to rescue more DNA sequences around the transposon insertion in the chromosome failed. It is known for *rhs* elements that they can insert themselves anywhere in the genomic

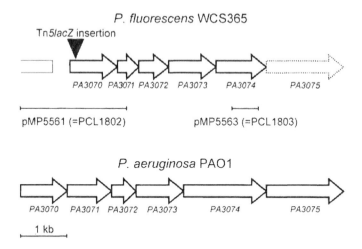

FIGURE 5.2. Map of gene cluster isolated from the genome of *P. fluorescens* WCS365 with the help of Tn*5lacZ*. The fragments used for single homologous recombination and complementation as well as the genomic organization in *P. aeruginosa* PAO1 are shown.

DNA. This is consistent with the insertion sequences present in the *hrtP* locus of the gene. The difficulties we have obtained for sequencing and cloning can be explained by this feature.

A rearrangements hot spot (RHS) is a repeated sequence in the genome of *E. coli*. These sequences are complex and contain up to 0.8 percent of the genome of *E. coli* K12 (Feulner et al., 1990). The number of *rhs* elements can vary per strain. *Rhs* elements are predicted to produce a set of roughly 160 kDa proteins with a long conserved N-terminus and a shorter C-terminus. An *rhs* element is composed of a G+C rich core ORF, an A+T rich core extension, and a H-rpt ORF containing insertion sequences. The high G+C content suggests an origin outside of *E.coli*. For *Bacillus subtilis* 168, it has been described that the *rhs* gene encodes a wall-associated protein (Foster, 1993). These binding proteins can provide the cell with an advantage in specific habitats, such as a mammalian host, or a soil/water environment (Foster, 1993). It can be hypothesized for mutant PCL1208 that this binding protein is involved in attachment to the root. The mutation could then lead to an impaired colonization phenotype of PCL1208. However, no differences between the parental strain and mutant were observed when attachment to synthetic surfaces, such as polystyrene or polyvinylchloride, was analyzed. It can be that this phenotype of attachment is not occurring on these synthetic surfaces. Therefore, attachment to seeds and roots was performed. Also in these cases, no differences were observed for PCL1208 in attachment to seeds when compared to its wild-type WCS365. Also, attachment to roots grown for seven days in the gnotobiotic system did not show differences when analyzed in a time course. Since the absence of an outer membrane protein could have an influence on the hydrophobicity of the cell surface, surface tension activity was determined by using the drop-collapsing assay (Jain et al., 1991). Again, no difference was observed between wild type and PCL1208 (data not shown). Finally, adhesion to sand particles was analyzed (see Materials and Methods section), since the mutant was screened on seedlings growing in a gnotobiotic sand system. No difference between WCS365 and PCL1208 could be found in their ability to adhere to sand particles.

Also for this mutant, no phenotype, except its defective competitive colonization ability, was found. Details about mutant PCL1208 are described by de Weert et al. (2004).

Mutant PCL1268 Is Impaired in Protein Secretion

Mutant PCL1268 is a Tn5*lacZ* derivative of *P. fluorescens* WCS365 impaired in competitive root tip colonization on the dicots potato and tomato, as well as on the monocot wheat. The transposon appeared to be inserted in an ORF homologous to the *secB* gene of *Vibrio cholerae* (Heidelberg et al., 2000). Upstream of *secB*, two ORFs were found (Figure 5.3). No other genes are present downstream of *secB*.

The mutant phenotype can be complemented by the introduction of the *secB* gene on the rhizosphere-stable plasmid pME6010 (Heeb et al., 2000). SecB of *E. coli* is part of a protein secretion pathway. Apparently, a defective protein-secretion pathway disturbs growth in exudate (see further on). For *E. coli*, it has been described that *SecB* is not essential for protein translocation, and that mutations in *secB* affect translocation of only some precursor proteins (Kumamoto and Beckwith, 1985). Seoh and Tai (1997) described that impaired synthesis of *SecB* in *E. coli* is causing most severe growth defects during growth in rich media when, at high growth rates, newly synthesized proteins are accumulating for transport. However, no difference in growth rate was observed when growth of PCL1268 was compared to that of the wild-type strain WCS365 in KB medium. Also growth in competition with the wild type in KB medium did not show any impaired growth of PCL1268. This implies that in rich medium, other proteins or systems can complement the *SecB* function, as is described by Santini et al. (1998) and Samuelson et al. (2000), who reported the existence of translocation pathways which are independent of the Sec system for *E. coli*.

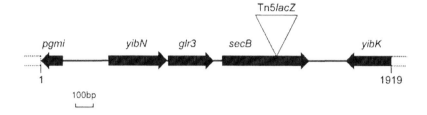

FIGURE 5.3. Chromosomal location of the Tn*5lacZ* in the competitive root tip-colonization mutant PCL1268. Arrows indicate the size and direction of transcription of the ORFs. The depicted genes are homologous to *PMGI* (2,3-bisphosphoglycerate-independent phosphoglycerate mutase), *yibN* (conserved hypothetical 15.6 KD protein), *glr3* (glutaredoxin 3), *secB* (protein-transport protein SecB), and *yibK* (hypothetical tRNA / rRNA methyltransferase).

When we analyzed the proteins present in cell envelope and periplasm after growth in KB and standard BM medium, no differences in protein patterns between mutant and wild-type strains were observed. However, since not all proteins can be detected using this procedure, it cannot be excluded that differences between wild type and mutant strains exist.

Analysis of competitive growth and survival between WCS365 and PCL1268 in tomato root exudate (Figure 5.4) showed that mutant PCL1268 is losing in competition to the wild-type strain. This indicates that during the utilization of exudate components, the *secB* mutation cannot be complemented by other systems, and that loss of the *secB* gene results in diminished rhizosphere fitness of the cells. This suggests that in tomato root exudate, and therefore probably also in the tomato rhizosphere, conditions are present under which *secB*-mediated transport is more important for cell growth and proliferation than in a rich medium. Certain exudate components can depend on binding proteins for their uptake, and the translocation of these binding proteins could be dependent on the *secB* function, as described previously for the maltose-binding protein (Kimsey et al., 1995), which has a role in the indirect uptake of nutrients for the SecB protein.

Mutants in the Type III Secretion System

Some bacterial plant and animal pathogens produce a so-called TTSS (type three secretion system), one of whose functions is to inject specific proteins into the cells of the eukaryotic partner. Evolutionarily, this system is de-

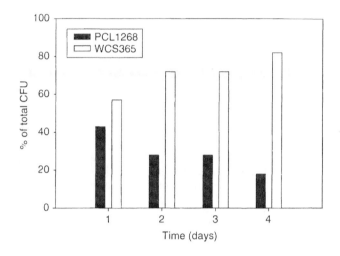

FIGURE 5.4. Growth and/or survival of *P. fluorescens* strain WCS365 in competition with competitive root tip-colonization mutant PCL1268 in tomato root exudate. Each bar represents the mean of two measurements. Cells grown overnight in standard BM medium were washed and diluted to an OD_{620} of 0.1. Subsequently, cells were diluted approximately 1000-fold and used to inoculate tomato root exudate in a 1:1 ratio. At the start of the experiment, and subsequently after each 24 hours, samples were taken, dilutions were plated on KB, and the number of CFU was determined.

rived from the flagellum. For the plant pathogen *P. syringae,* the present state of knowledge of TTSSs has recently been reviewed (He et al., 2004).

Rainey (1999) described many genes of the biocontrol strain *P. fluorescens* SBW25 that are specifically expressed in the rhizosphere. One of these genes is a homolog of *hrcC,* encoding a component of the TTSS (Preston et al., 2001), suggesting that the TTSS has functional significance in both pathogens and plant growth-promoting bacteria. Further analysis of the DNA surrounding *hrcC* led to the discovery of a 20 kb cluster that bears similarity to the TTSS from *P. syringae* (Preston et al., 2001). A strain of SBW25, containing a copy of *avrB* (from *P. syringae*), and engineered so that it overexpressed the TTSS pathway, caused a TTSS- and *AvrB*-dependent hypersensitive response on *Arabidopsis,* suggesting the *P. fluorescens*-TTSS pathway is functional. Moreover, Southern hybridization indicates that the cluster was widespread among saprophytic pseudomonads.

Rezzonico et al. (2002) found the *hrcV* gene in the *P. putida* biocontrol strain KD. The strain does not inhibit plant fungal pathogens in vitro, but protects cucumber against *Pythium ultimum* and tomato against *Fusarium oxysporum* in greenhouse experiments. To test whether the TTSS plays a role in the biocontrol activity of *P. putida* KD, a mutant with a W-cassette insert in *hrcC* was constructed. The biocontrol activity of this mutant against *Pythium ultimum* on cucumber was lower than that of the wild type. The authors suggested that the TTSS plays a role in disease suppression.

Following these findings, we compared the competitive root-colonizing abilities of *P. fluorescens* SBW25 TTSS mutants Rhi14 (*hrcD::Tn5*) and Rhi19 (*hrcR::Tn5*). It appeared that both mutants are 10-fold impaired in their competitive root tip-colonizing abilities, but were not impaired when analyzed alone on the root. This result could be explained by hypothesizing that the mutants are using the TTSS for either involvement in attachment to seed and/or root or for feeding on the root surface cells by means of the TTSS. Since no difference was found in attachment in competition with the wild type to seed and root, we hypothesize that the type three "injection" needle is not only used to inject proteins into plant/animal cells but also to suck nutrients from the plant cell. In fact, we find it likely that the latter function was developed earlier in evolution. It also explains why the TTSS is present in a wider group of microbes than pathogens only. Of course, other explanations cannot be excluded.

CONCLUSIONS

The results of our work (see references in the literature cited) and that of Rainey (1999) and Gal et al. (2003) suggest that the number of colonization genes may be several hundred. Therefore, a genomics approach, such as the one used in the EU programme Pseudomics (QRLT-2002-00914), is required to get a full view of the genes involved in colonization. In this article, we describe the analysis of the last colonization mutants isolated in the classical way.

REFERENCES

Bloemberg, G. V. and B. J. J. Lugtenberg. (2001). Molecular basis of plant growth promotion and biocontrol by rhizobacteria. *Current Opinions in Plant Biology* 4: 343-350.

Bloemberg, G. V. and B. J. J. Lugtenberg. (2004). Bacterial biofilms on plants: Relevance and phenotypic aspects. In *Microbial Biofilms*, eds. M. Ghannoum and G. A. O'Toole. ASM Press, Washington, DC.

Bloemberg, G. V., G. A. O'Toole, B. J. Lugtenberg, and R. Kolter. (1997). Green fluorescent protein as a marker for *Pseudomonas* spp. *Applied Environmental Microbiology* 63: 4543-4551.

Bloemberg, G. V., A. H. M. Wijfjes, G. E. M. Lamers, N. Stuurman, and B. J. J. Lugtenberg. (2000). Simultaneous imaging of *Pseudomonas fluorescens* WCS365 populations expressing three different autofluorescent proteins in the rhizosphere: New perspectives for studying microbial communities. *Molecular Plant-Microbe Interactions* 13: 1170-1176.

Bull, C. T., D. M. Weller, and L. S. Thomashow. (1991). Relationship between root colonization and suppression of *Gaeumannomyces graminis* var. *triciti* by *Pseudomonas fluorescens* strain 2-79. *Phytopathology* 81: 954-959.

Camacho-Carjaval, M. M., A. H. M. Wijfjes, I. H. M. Mulders, B. J. J. Lugtenberg, and G. V. Bloemberg. (2002). Characterization of NADH dehydrogenases of *Pseudomonas fluorescens* WCS365 and their role in competitive root colonization. *Molecular Plant-Microbe Interactions* 15: 662-671.

Chin-A-Woeng, T. F. C., W. de Priester, A. J. van der Bij, and B. J. J. Lugtenberg. (1997). Description of the colonization of a gnotobiotic tomato rhizosphere by *Pseudomonas fluorescens* biocontrol strain WCS365 using scanning electron microscopy. *Molecular Plant-Microbe Interactions* 10: 79-86.

Chin-A-Woeng, T. F. C., A. L. Lagopodi, I. H. M. Mulders, G. V. Bloemberg, and B. J. J. Lugtenberg. (2004). Visualisation of interactions of *Pseudomonas* and *Bacillus* biocontrol strains. In *Plant Surface Microbiology,* eds. A. Varma, L. Abott, D. Werner, and R. Hamps. Springer, Berlin, pp. 431-448.

Dekkers, L. C., C. J. Bloemendaal, L. A. de Weger, C. A. Wijffelman, H. P. Spaink, and B. J. J. Lugtenberg. (1998). A two-component system plays an important role in the root-colonizing ability of *Pseudomonas fluorescens* strain WCS365. *Molecular Plant-Microbe Interactions* 11: 45-56.

Dekkers, L. C., C. C. Phoelich, L. van der Fits, and B. J. J. Lugtenberg. (1998). A site-specific recombinase is required for competitive root colonization by *Pseudomonas fluorescens* WCS365. *Proceeding Natl. Acad. Sci. USA* 95: 7051-7056.

Dekkers, L. C., A. J. van der Bij, I. H. M. Mulders, C. C. Phoelich, R. A. Wentwoord, D. C. Glandorf, C. A. Wijffelman, and B. J. J. Lugtenberg. (1998). Role of the O-antigen of lipopolysaccharide, and possible roles of growth rate and of NADH: ubiquinone oxidoreductase *(nuo)* in competitive tomato root-tip colonization by *Pseudomonas fluorescens* WCS365. *Molecular Plant-Microbe Interactions* 11: 763-771.

de Weert, S., L. C. Dekkers, I. Kuiper, G. V. Bloemberg, and B. J. J. Lugtenberg. (2004). Generation of enhanced competitive root tip colonizing *Pseudomonas* bacteria through accelerated evolution. *Journal of Bacteriology* 186: 3153-3159.

de Weert, S., H. Vermeiren, I. H. M. Mulders, I. Kuiper, N. Hendrickx, G. V. Bloemberg, J. Vanderleyden, R. de Mot, and B. J. J. Lugtenberg. (2002). Flagella-driven chemotaxis towards exudate components is an important trait for tomato root colonization by *Pseudomonas fluorescens*. *Molecular Plant-Microbe Interactions* 15: 1173-1180.

de Weger, L. A., C. I. van der Vlugt, A. H. Wijfjes, P. A. Bakker, B. Schippers, and B. Lugtenberg. (1987). Flagella of a plant-growth-stimulating *Pseudomonas fluorescens* strain are required for colonization of potato roots. *Journal of Bacteriology* 169: 2769-2773.

Dybvig, K. (1993). DNA rearrangements and phenotypic switching in prokaryotes. *Molecular Microbiology* 10: 465-471.

Feulner, G., J. A. Gray, J. A. Kirschman, A. F. Lehner, A. B. Sadosky, D. A. Vlazny, J. Zhang, S. Zhao, and C. W. Hill. (1990). Structure of the *rhsA* locus from *Escherichia coli* K-12 and comparison of *rhsA* with other members of the *rhs* multigene family. *Journal of Bacteriology* 172: 446-456.

Foster, S. J. (1993). Molecular analysis of three major wall-associated proteins of *Bacillus subtilis* 168: Evidence for processing of the product of a gene encoding a 258 kDa precursor two-domain ligand-binding protein. *Molecular Microbiology* 8: 299-310.

Gal, M., G. M. Preston, R. C. Massey, A. J. Spiers, and P. B. Rainey. (2003). Genes encoding a cellulosic polymer contribute toward the ecological success of *Pseudomonas fluorescens* SBW25 on plant surfaces. *Molecular Ecology* 12: 3109-3121.

Glick, B. R. (1995). A novel procedure for rapid isolation of plant growth-promoting pseudomonads. *Canadian Journal of Microbiology* 41: 533-536.

He, S. Y., S. Bandyopadhyay, E. Bray, P. Hauck, Q.-L. Jin, O. Kolade, K. Nomura, R. Thilmony, W. Underwood, and J. Zwiesler-Vollick. (2004). Bacterial type III secretion in pathogenesis. In I. A. Tikhonovich, B. J. J. Lugtenberg, and N. A. Provorov (eds.). IS-MPMI, St-Petersburg, Russia.

Heeb, S., I. Yoshifumi, T. Nishijyo, U. Schnider, C. Keel, J. Wade, U. Walsh, F. O'Gara, and D. Aas. (2000). Small, stable shuttle vectors based on the minimal pVS1 replicon for use in gram-negative, plant-associated bacteria. *Molecular Plant-Microbe Interactions* 13: 232-237.

Heidelberg, J. F., J. A. Eisen, W. C. Nelson, R. A. Clayton, M. L. Gwinn, R. J. Dodson, D. H. Haft, E. K. Hickey, J. D. Peterson, L. A. Umayam, S. R. Gill, K. E. Nelson, T. D. Read, H. Tettelin, D. Richardson, M. D. Ermolaeva, J. Vamathevan, S. Bass, H. Qin, I. Dradoi, P. Sellers, L. McDonald, T. Utterback, R. D. Fleishmann, W. C. Noerman, O. White, S. L. Salzberg, H. O. Smith, R. R. Colwell, J. J. Mekalanos, J. C. Venter, and C. M. Fraser. (2000). DNA sequence of both chromosomes of the cholera pathogen *Vibrio cholerae*. *Nature* 406: 477-483.

Hiltner, L. (1904). Uber neue erfahrungen und probleme auf dem gebiet der bodenbacteriologie und unter besonderes berucksichtigung der grundugungen und brauche. *Arb Dtsch Landwirt Ges Berl* 98: 59-78.

Hoffland, E., G. R. Findenegg, and J. A. Nelemans. (1989). Solubilization of rock phosphate by rape. *Plant Soil* 113: 161-165.

Jain, D. K., D. K. C. Thomson, H. Lee, and J. T. Trevors. (1991). A drop-collapsing test for screening surfactant producing microorganisms. *Journal of Microbiology Methods* 13: 271-279.

Joint, I., K. Tait, M. E. Callow, J. A. Callow, D. Milton, P. Williams, and M. Camara. (2002). Cell-to-cell communication across the prokaryotic-eukaryotic boundary. *Science* 298: 1207.

Kimsey, H. H., M. D. Dagarag, and C. A. Kumamoto. (1995). Diverse effects of mutation on the activity of the *Escherichia coli* export chaperone SecB. *Journal of Biological Chemistry* 270: 22831-22835.

King, E. O., M. K. Ward, and D. E. Raney. (1954). Two simple media for the demonstration of pyocyanin and fluorescin. *Journal of Laboratory Medicine* 44: 301-307.

Kuiper, I., G. V. Bloemberg, and B. J. J. Lugtenberg. (2001). Selection of a plant-bacterium pair as a novel tool for rhizostimulation of polycyclic aromatic hydrocarbon-(PAH)-degrading bacteria. *Molecular Plant-Microbe Interactions* 14: 1197-1205.

Kuiper, I., G. V. Bloemberg, S. Noreen, J. E. Thomas-Oates, and B. J. J. Lugtenberg. (2001). Increased uptake of putrescine in the rhizosphere inhibits competitive root colonization by *Pseudomonas fluorescens* strain WCS365. *Molecular Plant-Microbe Interactions* 14: 1096-1104.

Kuiper, I., L. Kravchenko. G. V. Bloemberg, and B. J. J. Lugtenberg. (2002). *Pseudomonas putida* strain PCL1444, selected for efficient root colonization and naphthalene degradation, effectively utilizes root exudate components. *Molecular Plant-Microbe Interactions* 15: 734-741.

Kuiper, I., E. L. Lagendijk, G. V. Bloemberg, and B. J. J. Lugtenberg. (2004). Rhizoremediation: A beneficial plant-microbe interaction. *Molecular Plant-Microbe Interactions* 17: 6-15.

Kumamoto, C. and J. Beckwith. (1985). Evidence for specificity at an early step in protein export in *Escherichia coli*. *Journal of Bacteriology* 163: 267-274.

Lugtenberg, B. J. J. and G. V. Bloemberg. (2004). Life in the rhizosphere. In *Pseudomonas,* ed. J. L. Ramos. Kluwer Academic/Plenum Publishers, New York, pp. 403-430.

Lugtenberg, B. J. J., G. V. Bloemberg, A. Bolwerk, D. van den Broek, F. Cazorla-Lopez, T. F. C. Chin-A-Woeng, K. Eijkemans, F. D. Kamilova, I. Kuiper, I. H. M. Mulders, E. T. van Rij, and S. de Weert. (2004). In *Microbial Control of Tomato Foot and Root Rot,* eds. I. A. Tikhonovich, B. J. J. Lugtenberg, and N. A. Provorov. IS-MPMI, St. Petersburg. Russia.

Lugtenberg, B. J. J., L. C. Dekkers, M. Bansraj, G. V. Bloemberg, M. Camacho, T. F. C. Chin-A-Woeng, K. van den Hondel, L. Kravchenko, I. Kuiper, A. Lagopodi, I. Mulders, C. Phoelich, A. Ram, I. Tikhonovich, S. Tuinman, C. Wijffelman, and A. Wijfjes. (1999). *Pseudomonas* genes and traits involved in tomato root colonization. In *Biology of Plant-Microbe Interactions,* eds. P. J. G. M. de Wit, T. Bisseling, and W. J. Stiekema. International Society for Molecular Plant-Microbe Interactions, St. Paul, MN, pp. 324-330.

Lugtenberg, B. J. J., L. C. Dekkers, and G. V. Bloemberg. (2001). Molecular determinations of rhizosphere colonization by *Pseudomonas*. *Annual Review Phytopathology* 39: 461-490.

Lugtenberg, B. J. J. and F. D. Kamilova. (2004). Rhizosphere management: Microbial manipulation for biocontrol. *Encyclopedia of Plant and Crop Science* 1098-1101.

Lugtenberg, B. J. J., L. V. Kravchenko, and M. Simons. (1999). Tomato seed and root exudate sugars: Composition, utilization by *Pseudomonas* biocontrol strains and role in rhizosphere colonization. *Environmental Microbiology* 1: 439-446.

Lugtenberg, B. J. J. and L. van Alphen. (1983). Molecular architecture and functioning of the outer-membrane of *Escherichia coli* and other gram-negative bacteria. *Biochemistry Biophysics Acta* 737: 51-115.

Okon, Y., G. V. Bloemberg, and B. J. J. Lugtenberg. (1997). Biotechnology of biofertilization and phytostimulation. In *Agricultural Biotechnology*, ed. A. Altman. Dekker New York, New York, pp. 327-349.

Preston, G. M., N. Bertrand, and P. B. Rainey. (2001). Type III secretion in plant growth-promoting *Pseudomonas fluorescens* SBW25. *Molecular Microbiology* 41: 999-1014.

Rainey, P. B. (1999). Adaptation of *Pseudomonas fluorescens* to the plant rhizosphere. *Environmental Microbiology* 1: 243-257.

Rezzonico, F. G., L. Molina, C. Binder, and G. Defago. (2002). Type III protein secretion system in biocontrol strain *Pseudomonas putida* KD. In *Proceedings Interactions in the Microbial World,* ed. L. Molina. Amsterdam, the Netherlands.

Sadowski, P. (1986). Site specific recombinases: Changing partners and doing the twist. *Journal of Bacteriology* 165: 337-341.

Salmond, G. P., B. W. Bycroft, G. S. Stewart, and P. Williams. (1995). The bacterial "enigma": Cracking the code of cell-cell communication. *Molecular Microbiology* 16: 615-624.

Samuelson, J. C., M. Chen, F. Jiang, I. Möller, M. Wiedmann, A. Kuhn, G. J. Phillips, and R. E. Dalbey. (2000). YidC mediates membrane protein insertion in bacteria. *Nature* 406: 637-641.

Santini, C.-L., B. Ize, A. Chanal, M. Müller, G. Giordano, and L.-F. Wu. (1998). A novel Sec-independent periplasmic protein translocation pathway in *Escherichia coli. EMBO Journal* 17: 101-112.

Schafer, A., A. Tauch, W. Jager, J. Kalinowski, G. Thierbach, and A. Puhler. (1994). Small mobilizable multi-purpose cloning vectors derived from the *Escherichia coli* plasmids pK18 and pK19: Selection of defined deletions in the chromosome of *Corynebacterium glutamicum. Gene* 145: 69-73.

Schippers, B., A. W. Bakker, and P. A. H. M. Bakker. (1987). Interactions of deleterious and benificial rhizosphere microorganisms and the effect of cropping practices. *Annual Review of Phytopathology* 25: 339-358.

Seoh, H. K. and P. C. Tai. (1997). Carbon source-dependent synthesis of SecB, a cytosolic chaperone involved in protein translocation across *Escherichia coli* membranes. *Journal of Bacteriology* 179: 1077-1081.

Simoni, S. F., H. Harms, T. N. P. Bosma, and A. J. B. Zehnder. (1998). Population hetrogeneity affects transport of bacteria through sand columns at low flow rates. *Environmental Science Technology* 32: 2100-2105.

Simons, M., H. P. Permentier, L. A. de Weger, C. A. Wijffelman, and B. J. J. Lugtenberg. (1997). Amino acid synthesis is necessary for tomato root colonization by *Pseudomonas fluorescens* strain WCS365. *Molecular Plant-Microbe Interactions* 10: 102-106.

Simons, M., A. J. van der Bij, J. Brand, L. A. de Weger, C. A. Wijffelman, and B. J. J. Lugtenberg. (1996). Gnotobiotic system for studying rhizosphere colonization by plant growth-promoting *Pseudomonas* bacteria. *Molecular Plant-Microbe Interactions* 9: 600-607.

Sokal. R. R. and F. J. Rohlf. (1981). *Biometry.* Freeman, San Franscisco.

Spaink, H. P., A. Kondorosi, and P. J. J. Hooykaas. (1998). *The Rhizobiaceae.* Kluwer Academic Publishers, Dordrecht, the Netherlands.

Steenhoudt, O. and J. Vanderleyden. (2000). *Azospirillum*, a free-living nitrogen-fixing bacterium closely associated with grasses: Genetic, biochemical and ecological aspects. *FEMS Microbiology Reviews* 24: 487-506.

Stover, C. K., X. Q. Pham, A. L. Erwin. S. D. Mizoguchi, P. Warrener, M. J. Hickey, F. S. Brinkman. W. O. Hufnagle, D. J. Kowalik, M. Lagrou, R. L. Garber, L. Goltry, E. Tolentino, S. Westbrock-Wadman, Y. Yuan, L. L. Brody, S. N. Coulter, K. R. Folger, A. Kas, K. Larbig, R. Lim, K. Smith, D. Spencer, G. K. Wong, Z. Wu, I. T. Paulsen. J. Reizer, M. H. Saier, R. E. Hancock, S. Lory, and M. V. Olson. (2000). Complete genome sequence of *Pseudomonas aeruginosa* PAO1, an opportunistic pathogen. *Nature* 406: 959-964.

Stuurman, N., C. Pacios Bras. H. R. M. Schlaman, A. H. M. Wijfjes, G. V. Bloembergen, and H. P. Spaink. (2000). The use of GFP color variants expressed on stable broad-host range vectors to visualize *rhizobia* interacting with plants. *Molecular Plant-Microbe Interactions* 13: 1163-1169.

Swift, S., J. A. Downie, N. A. Whitehead, A. M. Barnard, G. P. Salmond, and P. Williams. (2001). Quorum sensing as a population-density-dependent determinant of bacterial physiology. *Advances in Microbial Physiology* 45:199-270.

Tang, Y. P., M. M. Dallas, and M. H. Malamy. (1999). Characterization of the Batl *(Bacteroides aerotolerance)* operon in *Bacteroides fragilis*: Isolation of a *B. fragilis* mutant with reduced aerotolerance and impaired growth in in vivo model systems. *Molecular Microbiology* 32: 139-149.

Thomashow, L. S. and D. M. Weller. (1996). Current concepts in the use of introduced bacteria for biological disease control: Mechanisms and antifungal metabolites. In *Plant-Microbe Interactions,* eds. G. Stacey and N. T. Keen. Chapman and Hall, New York, pp. 236-271.

Van Elsas, J. D., J. T. Trevors, and M. E. Starodub. (1988). Bacterial conjugation between pseudomonads in the rhizosphere of wheat. *FEMS Microbiology Ecology* 53: 299-306.

Weller, D. M. (1988). Biological control of soilborne plant pathogens in the rhizosphere with bacteria. *Annual Review Phytopathology* 26: 379-407.

Chapter 6

Biological Control
of Onion White Rot

A. Stewart
K.L. McLean

INTRODUCTION

Onion white rot is caused by the soilborne plant pathogen *Sclerotium cepivorum* Berk. It is one of the most important and destructive diseases of *Allium* species. Berkeley first described white rot of onions in England in 1841 (Walker, 1924). By 1923, the disease was well established in Europe, and initial observations had been made in America (Walker, 1924). In the South Pacific, white rot was first evident in Australia in 1911 (Royal Botanic Gardens Herbarium [DAR 56526], J. Walker) and in New Zealand in 1922 (Cunningham, 1922). Today, onion white rot is present in all *Allium*-growing countries.

The first indication of infection is a change in leaf color from bright green to blue green, followed by leaf-tip dieback and collapse of the leaves. Belowground, the roots are destroyed. A semiwatery decay of the scales also occurs, which is associated with the growth of superficial white mycelium on the affected areas (Tims, 1948; Fullerton et al., 1994). Within several days, the mycelium darkens and is transformed into sclerotia (Fullerton et al., 1994).

Sclerotia are the only reproductive structures of *S. cepivorum,* as no perfect stage has yet been described (Mordue, 1976) and no asexual spores are produced. The sclerotia are black, uniformly round, and usually measure 200 to 500 μm in diameter (Mordue, 1976). Occasionally larger, more irregular sclerotia have been produced (3 to 15 mm in length), which can be either unified structures or aggregations of the smaller sclerotia (Georgy and Coley-Smith, 1982; Backhouse and Stewart, 1988; Crowe, 1995; Metcalf et al., 1997). A sclerotium is composed of a narrow black rind that surrounds a medulla of compact interwoven hyphae. The interhyphal spaces of the sclerotium are filled with a gelatinous material that acts as a food reserve (Mordue, 1976). The protective nature of the rind also enables the sclerotia to remain viable in soil. Initial research, conducted in the United Kingdom, indicated that a high proportion of buried sclerotia survived for 20 years in soil (Coley-Smith, 1959; Coley-Smith, et al., 1990). However, other studies conducted in Canada, America, and New Zealand report a substantial decrease in populations shortly after burial, with only a small proportion of the sclerotia surviving from year to year and contributing to the buildup of inoculum over time to economically debilitating levels (Legget and Rahe, 1985; Alexander and Stewart, 1994; Harper et al., 2002; F. Crowe, personal communication). The variation in survival of the sclerotia is related to soil moisture, temperature, and microbial composition (Alexander and Stewart, 1994; McLean et al., 2002).

Sclerotia germinate only in response to root exudates produced by *Allium* species. The alk(en)yl cysteine sulphoxide compounds exuded by the roots are broken down by soil microorganisms to form thiols and sulphide compounds, which stimulate the sclerotia to germinate (Coley-Smith and King, 1969). Germination can occur in one of two ways, either by the projection of fine hyphal threads or the splitting of the rind, which causes a plug of hyphae to erupt (Coley-Smith, 1960). Hyphae directly penetrate the roots between or within the epidermal cell walls, whereas the basal stem tissue is infected via the production of infection cushions (Stewart et al., 1989). Fungal mycelium spreads beneath the cuticle (Abd-El-Razik et al., 1973), killing epidermal cells ahead of advancing hyphae (Stewart et al., 1989; Metcalf and Wilson, 1999). While fungal mycelium can spread from plant to plant, the mycelium cannot survive in the soil for any length of time, as *S. cepivorum* has no saprophytic ability (Scott, 1956a). However, whole planting rows can become infected via root contact (Scott, 1956b).

Chemicals have been relied upon to control onion white rot for the past 40 years. Initially, chemical compounds, including calomel, benomyl, dicloran, thiophanate-methyl, and botran were tested to control *S. cepivorum* (Booer, 1945, 1946; Croxall et al., 1953; Entwistle and Munasinghe, 1973; Maloy and Machtmes, 1974; Ryan and Kavanagh, 1976); however, insufficient concentrations, inappropriate application methods, inconsistent control, and phytotoxicity led to unsatisfactory levels of disease control (Entwistle and Munasinghe, 1973; Maloy and Machtmes, 1974).

Successful control of onion white rot was obtained after the introduction of systemic fungicides. The dicarboximide fungicides, iprodione, vinclozolin, and procymidone were initially highly successful against onion white rot in New Zealand and the United Kingdom (Wood, 1980; Entwistle and Munasinghe, 1980a,b; Fullerton and Stewart, 1991; Stewart and Fullerton, 1991), but less so in Australia (Porter et al., 1991). Unfortunately, in both New Zealand and the United Kingdom, the continued use of dicarboximides led to a decrease in the effectiveness that was due to enhanced degradation by soil microorganisms or, as yet, unaccounted-for factors (Entwistle, 1983; Walker et al., 1986; Slade et al., 1992; Tyson et al., 1999). However, Mexico and Australia are currently using iprodione and procymidone, respectively (Villalta et al., 2002; L. Perez Moreno, personal communication), and some in vitro fungicide tolerance has been reported with *S. cepivorum* isolates from the Bajio region of Mexico (Perez-Moreno et al., 1997).

More recently, the triazole fungicides, tebuconazole and triadimenol, have been used to control onion white rot in New Zealand, Mexico, and Australia (Fullerton et al., 1994, 1995; Tyson et al., 1999; Dennis, 2001; Duff et al., 2001; L. Perez Moreno, personal communication). However, a large variation in control between seasons has been reported with tebuconazole (Tyson et al., 1999). In New Zealand, tebuconazole provided 17 and 58 percent disease control in the 1997-1998 and 1998-1999 seasons, respectively (Tyson et al., 1999). The authors concluded that the timing of fungicide applications, changing soil conditions, and the interactions between soil temperature, soil moisture, and rainfall were the factors that contributed to the variation in fungicide efficacy. Presently, tebuconazole is approved for use by the Horticultural Development Council (available "off-label") in the United Kingdom but is not registered (Clarkson et al., 2002). In Canada and America, fungicides are not used to control onion white rot. Although dichloran is registered for use in Canada on onions, and iprodione and botran are registered for use on garlic in America, onion white rot control results have again been variable and growers no longer apply these chemicals (M.R. MacDonald, personal communication; F. Crowe, personal communication).

The reduction and variation in the efficacy of fungicides applied to control onion white rot indicates the need for alternative control measures. In addition, chemical residue buildup in the soil and high residue levels in consumable products are undesirable. Biological control is a viable alternative to fungicidal control. A number of life cycle stages of the pathogen can be targeted, including the overwintering sclerotia and mycelial penetration of root and stem tissue. In addition, resistance may be induced in the plant by the application of biological control agents.

STRATEGIES FOR BIOLOGICAL CONTROL

A number of fungi and bacteria have been identified which act as sclerotial mycoparasites, antibiotic producers, and effective saprophytic competitors.

Sclerotial Mycoparasites

Coniothyrium minitans Campbell has been reported to parasitize the sclerotia of *Sclerotium cepivorum* (Turner and Tribe, 1976; Harrison and Stewart, 1988; Stewart and Harrison, 1988) and produce pycnidia within them (Ghaffar, 1969a). *Coniothyrium minitans* hyphae penetrated the sclerotia through breaks in the surface of the rind and caused lysis of the medullary region, probably through the production of lytic enzymes (Stewart and Harrison, 1988). Sclerotia parasitized by *C. minitans* failed to germinate (Harrison and Stewart, 1988; Stewart and Harrison, 1988).

In glasshouse trials, *C. minitans* pycnidial dust and dressed seed provided disease control equivalent to the standard chemical control (Ahmed and Tribe, 1977). When prepared in a bulk carrier formulation, *C. minitans* reduced disease, compared with the pathogen control but not the chemical control (McLean and Stewart, 2000a). Other researchers have reported *C. minitans* to be ineffective at reducing white rot disease (Utkhede and Rahe, 1980a; De Oliveira et al., 1984). Biological control research with *C. minitans* is ongoing (Perez-Moreno and Rodriguez-Aguilera, 2002a,b). However, most researchers are pursuing other fungal genera, as *C. minitans* is a relatively slow-growing fungus, has minimal saprophytic ability (Entwistle, 1986), and is difficult to produce in bulk for field applications without specialized facilities. Also, commercial *C. minitans* products, such as Contans WG (Prophyta GmbH, Germany), are inferior compared with other preparations, as only low levels of antagonism toward *S. cepivorum* were observed (S. Bruckner, personal communication). Contans WG, however, successfully controls the related pathogen *Sclerotinia sclerotiorum* Lib. de Bary, which infects lettuce and oilseed rape (Jones and Whipps, 2001; Weber, 2002).

Sclerotial formation may influence the success of *C. minitans* against *S. sclerotiorum* compared with *S. cepivorum*. With *S. cepivorum* sclerotia, the rind forms late in the sclerotial production process and, once formed, the sclerotia does not increase in size (Willetts, 1971). However, with *S. sclerotiorum*, the sclerotia enlarges after rind formation, which may cause rind cells to rupture (Willetts and Bullock, 1992), and provide an opportune entry point for sclerotial mycoparasites. This could explain why a higher number of *S. sclerotiorum* sclerotia are parasitized compared with *S. cepivorum* when confronted with *C. minitans* (Turner and Tribe, 1976).

Sporidesmium sclerotivorum Uecker, Ayers & Adams is another fungus reported as a sclerotial parasite (Ayers and Adams, 1979; Adams and Ayers, 1981, 1983). This mycoparasite has the advantage that it can grow through soil from one sclerotium to the next, unlike *C. minitans* (Ayers and Adams, 1979; Adams and Ayers, 1981; Williams and Whipps, 1995). Although *S. sclerotivorum* is able to parasitize *S. cepivorum* sclerotia, it occurs at a much slower rate (10 percent after 4 weeks) compared with the larger *S. sclerotiorum* sclerotia (100 percent after 4 weeks) (Ayers and Adams, 1979). Additional studies indicated that the application of *S. sclerotivorum* weakened the sclerotia, but that death was related to the activity of other soil microorganisms, which colonized the damaged sclerotia (Adams and Ayers, 1983). Other research found that *S. sclerotivorum* did not establish well in all soil types (A. Stewart, personal communication). However, probably the greatest limiting factor for commercialization of *S. sclerotivorum* as a biological control agent is its very limited growth on most mycological media (Ayers and Adams, 1979), which would make bulk production difficult.

Trichoderma virens J. Miller, Giddens & Foster has been reported as a sclerotial parasite of *S. cepivorum* (Harrison and Stewart, 1988; Stewart and Harrison, 1988). The fungus grew over the surface of the sclerotia and penetrated through cracks in the rind created by collapsed rind cells (Figures 6.1 to 6.5). Conidiophores and conidial masses were produced on the surface of the parasitized sclerotia (Stewart and Harrison, 1988). However, in glasshouse trials, two other *Trichoderma* isolates, *T. harzianum* Rifai and *T. koningii* Oudemans, which were not reported as sclerotial parasites, were significantly better at reducing *S. cepivorum* infections (73 and 84 percent, respectively) than *T. virens* (56 percent) (Kay and Stewart, 1994).

Antibiotic Producers

Initial in vitro research indicated that *Penicillium* species had potential as biological control agents of onion white rot as *S. cepivorum* mycelial growth was inhibited at a distance in dual culture, thus indicating the presence of antibiotic substances (Ghaffar, 1969a; Moubasher et al., 1970; De Oliveira et al., 1984; Harrison and Stewart, 1988). Ghaffar (1969a) found *Penicillium nigricans* (Bain.) was the best *Penicillium* species due to the production of griseofulvin. Concentrations of 80 µg griseofulvin/g of soil resulted in 60 percent onion white rot disease after 15 days compared with lower concentrations of griseofulvin, which resulted in 100 percent disease after 15 days (Ghaffar, 1969b). *Pencillium godlewskii* Zaleski also gave promising results in in vitro assays; however, while disease reduction was significant compared with the untreated control, the level of disease was still greater than 88 percent (Abd-El-Razik et al., 1985). Harrison and Stewart

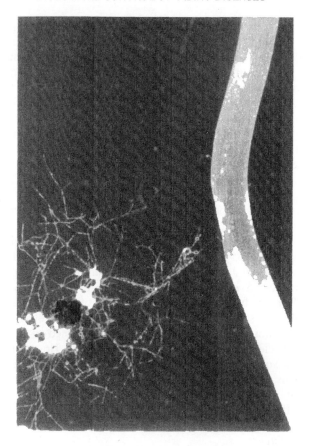

FIGURE 6.1. Healthy onion plants.

(1988) identified an isolate of *Penicillium expansum* (Link) with activity against white rot, but this species is known to cause a storage rot of onions (Dingley, 1969) and, therefore, it was not considered sensible to continue researching the commercial potential of this isolate (Harrison and Stewart, 1988). The low efficacy, short-term control of *S. cepivorum* and disease-inducing capabilities of *Penicillium* species indicate that this genus is likely to be unsuitable for commercial biological control development.

Chaetomium globosum Kunze. ex Steudel is another fungal species that has produced inhibition zones when challenged with *S. cepivorum* in dual culture (Harrison and Stewart, 1988). Culture filtrates of *C. globosum* A53 significantly reduced the growth rate of *S. cepivorum* (2.2 mm) compared

FIGURE 6.2. Germinating sclerotium of *Sclerotium cepivorum* infecting onion seedling.

with the control (7.6 mm) and caused increased hyphal branching and production of irregular sclerotia (Harrison and Stewart, 1988). In soil box trials, *C. globosum* A53 gave >80 percent disease control of onion white rot when applied in a sand and bran formulation (Kay and Stewart, 1994). However, in more recent soil box trials, *Trichoderma* species provided better control of onion white rot than *C. globosum* A53 (McLean and Stewart, 2000a). In vitro studies also determined that this isolate of *C. globosum* had lost the ability to sporulate and had irregular colony edges, which indicated poor and inconsistent growth (McLean, 1996). Further research with *C. globosum* A53 was not conducted.

FIGURE 6.3. Field symptoms of onion white rot.

Another antibiotic-producing genus that showed promise in early biological control investigations is *Bacillus*. Ghaffar (1969b) reported an increase in *Bacillus* populations just before maximum control of onion white
rot when *P. nigricans* was applied in a maize meal formulation. The author
reported that antagonistic bacterial species such as *Bacillus* may be responsible for control of onion white rot. Other research with *Bacillus* species
indicated that seed treatment with the *Bacillus subtilis* isolate BACT 2 significantly controlled onion white rot in both 'Autumn Spice' and 'Festival'
onion cultivars and BACT 4 and BACT 8 decreased bulb neck diameter,
which is desirable to speed maturity and curing and prolong keeping quality
(Utkhede and Rahe, 1983). In New Zealand, *Bacillus* species were able to
colonize the surface of *S. cepivorum* sclerotia, and bacterial cells were
found inside the sclerotia only when the rind was damaged (Backhouse and
Stewart, 1989). Researchers in Australia have isolated, screened, and identified bacterial species antagonistic to *Sclerotium cepivorum*. The most
common isolates were those of *Bacillus licheniformis* (Wong and Hughes,
1986); however, further work on these isolates has not been reported. *Bacillus* species have been described as opportunistic colonizers (Backhouse and
Stewart, 1989), and although antibiotics are produced, the antagonistic

FIGURE 6.4. Onion showing white rot infection.

effect appears to be minimal, which may explain why the research has not generally continued past the initial stages.

Trichoderma species have also been shown to produce antibiotics (Dennis and Webster, 1971a), and the antibiotics produced differ from species to species and among isolates of the same species (Dennis and Webster, 1971a,b). One of the most commonly produced antibiotic substances is 6-pentyl-α-pyrone (6PAP), which is responsible for the coconut aroma of some *Trichoderma* isolates. Six-pentyl-α-pyrone has been isolated from *T. harzianum, T. koningii, T. hamatum* Bonord. and *T. viride* Pers. ex Gray (Ghisalberti and Sivasithamparam, 1991). While control of *S. cepivorum*

FIGURE 6.5. Biological control of onion plot with *Trichoderma*.

has not been directly attributed to antibiotic production, 6PAP has been implicated in reducing the growth of *Rhizoctonia solani* Kuhn, and isolates of *Verticillium*, *Fusarium*, and *Aspergillus* (Dennis and Webster, 1971b; Claydon et al., 1987).

Competitive Saprophytes

Trichoderma Pers. ex Fr. species are ubiquitous, fast-growing saprophytes that can use a variety of substrates for growth (Klein and Eveleigh, 1998), which makes them excellent potential biological control agents. In glasshouse trials, De Oliveira et al. (1984) reported that conidial suspensions of *T. harzianum* gave 20 percent control of white rot in garlic. Kay and Stewart (1994) showed that sand:bran bulk carrier formulations of *T. atroviride* P. Karsten, *T. harzianum,* and *T. koningi* gave 51, 41, and 73 percent control of onion white rot, respectively. The sand:bran preparation was superior to coated seed and pellet formulations, as additional food was provided to assist *T. harzianum* with establishment and survival (Lewis and Papavizas, 1985). An additional glasshouse trial with the infested sand:bran formulations of *T. harzianum* (C52) and *T. koningii* (C62) significantly reduced disease from 39.8 percent in the untreated control to 7.7 and 5.4 percent, respectively (McLean and Stewart, 2000a). Research conducted in Australia indicated that *T. koningii* was able to colonize the rhizosphere of healthy onion roots (Metcalf and Wilson, 2001b), which indicated that competition was a likely mechanism of action.

Plant Growth Promotion

The application of some microorganisms has been reported to increase plant growth and make them more resistant to infection by plant diseases. Applications of *T. harzianum* and *T. viride* to soil significantly increased onion plant growth and controlled onion white rot (Payghami et al., 2001). Similarly, in New Zealand field trials, visible growth promotion of onion seedlings was observed after the application of *Trichoderma* species to soil (A. Stewart, unpublished data). *Bacillus subtilis* has shown growth promotion effects in a range of vegetable seedlings (Chang et al., 1986; Kleifeld and Chet, 1992). However, Utkhede and Rahe (1983) did not observe increased plant growth when the bacterium was applied to onion seeds.

Induced Host-Plant Resistance

Plant growth regulators have been used to induce resistance in onion seedlings to onion white rot infection. Abscisic acid, ATP, gibberellic acid, IAA, kinetin, and zeatin were tested (Galal and Bana, 2001). Gibberellic acid induced the greatest resistance with a 33, 58, and 70 percent reduction in *S. cepivorum* infection when two-month-old onion transplants were dipped in 25, 50, and 100 ppm solutions, respectively, before being inoculated with *S. cepivorum* and planted in the field (Galal and Bana, 2001).

Genetic Resistance

The development of stable onion lines exhibiting significant resistance to onion white rot has not yet been achieved through traditional plant breeding programs. Attempts to breed onion white rot–resistant onion lines have only produced onions with low levels of resistance compared with susceptible lines (Gabelman, 1986; Rahe, 1986), and resistance in these lines is highly variable (Utkhede and Rahe, 1980b; Utkhede et al., 1982; Utkhede and Rahe, 1984). Alternative methods to introduce resistance to existing onion cultivars using biotechnology are now being investigated. Although onions, like many other monocotyledonous species, are recalcitrant to transformation and regeneration in tissue culture (Eady, 1995; Eady et al., 1996), an onion transformation protocol recently developed and patented in New Zealand has enabled this technology to proceed (Eady, 2001) further. This protocol has been used to produce the world's first herbicide-resistant onion cultivars (Eady, 2002) and is now being used to develop onion cultivars with resistance to onion white rot. Three methods are being attempted to generate resistance: these include the insertion of genes encoding antifungal compounds, such as an oxalate; degrading protein or magainin peptides; and reducing the alk(en)yl cysteine sulphoxide production using gene silencing technology. To date, three types of transgenic plants utilizing each of these three methods to impart disease resistance have been generated in New Zealand (C.C. Eady, personal communication). Future work will evaluate the effect of expression of the antifungal compound *in planta* on resistance to the disease and what level of alk(en)yl cysteine sulphoxide reduction is required to prevent sclerotia from germinating. A system is being developed where gene expression is restricted to the roots and is not present in the edible bulb and leaves.

CURRENT BIOLOGICAL CONTROL RESEARCH

Current biological control research has mainly focused on the use of *Trichoderma* species. In the United Kingdom, a three-step screening process involving agar, sclerotial degradation, and onion seedling bioassays determined that *Trichoderma* species were the best potential biological control agents of *S. cepivorum* (Clarkson et al., 2001). Two isolates of *T. viride* (L4 and S17A) significantly reduced onion white rot incidence in pot trials from 75 percent (control) to 38 and 39 percent, respectively (Clarkson et al., 2002). Under field conditions, disease was again reduced with the application of L4 (5.5 percent) and S17A (8.5 percent) compared with the control (19 percent) (Clarkson et al., 2002).

In Australia, isolates of *T. koningii* showed potential as biological control agents. *Trichoderma koningii* (Td22), collected from a parasitized *S. cepivorum* sclerotia, reduced disease by 91.2 percent when applied in a millet formulation under commercial scale situations (Metcalf and Wilson,

2001a). Mode-of-action studies with another *T. koningii* isolate (Tr5) determined that the fungus colonized the epidermal mucilage of onion roots and, while healthy tissue was not penetrated, infected, or damaged tissue was invaded (Metcalf and Wilson, 2001b). *Trichoderma koningii* halted *S. cepivorum* infection, caused cell walls to dissolve, and caused hyphal tips to burst. In addition, contact between the two fungi was not critical as *T. koningii* produced two endochitinases (R_f 0.15 and 0.24) and two exo-acting chitinolytic enzymes (R_f 0.46 and 0.62) as part of the antagonism process (Metcalf and Wilson, 2001b). As *T. koningii* (Tr5) had low efficacy under commercial conditions, isolate Td22 became the focus of further study, as this isolate showed greater potential as a biological control agent under low-to-moderate disease pressure in commercial situations (Metcalf, 2002). The latest Australian research has investigated the use of paper mill waste as a growing medium for large-scale production of *T. koningii* (Td22) for further trials (Ramona and Line, 2002).

In New Zealand, an isolate of *T. atroviride* (C52), formally identified as *T. harzianum*, performed well in three separate glasshouse trials when formulated with bran flakes and sand (Kay and Stewart, 1994; McLean and Stewart, 2000a). In an initial trial, *T. atroviride* gave 51 percent disease control (Kay and Stewart, 1994), and in later trials, onion white rot disease was reduced from 39.8 and 66.5 percent in the untreated control to 7.7 percent and 34 percent, respectively (McLean and Stewart, 2000a). Based on these results, *T. atroviride* (C52) was prepared by Agrimm Technologies Ltd., Christchurch, New Zealand, into a range of formulations that were easier to prepare and disperse than the bran and sand delivery system. Pellets, seed coats, bulk carrier, and wettable powder formulations, all based on *T. atroviride* conidia, were also included.

All formulations were tested in varying combinations under commercial-scale field conditions in Canterbury, New Zealand. Results indicated that the pellet formulation was superior to the other formulations, providing control of onion white rot equivalent to the standard chemical applications under low (<25 percent) and moderate (26 to 60 percent) disease conditions (unpublished data). Intensive population dynamics studies with a *Trichoderma*-selective medium and a molecular marker for *T. atroviride* (C52) found the pellet preparation was able to maintain fungal concentrations at levels greater than 1×10^5 colony-forming units (cfu)/g of soil throughout the growing season (McLean, 2001). The bulk carrier formulation had a similar fungal concentration initially; however, the concentration gradually decreased over time and the fungus could not be detected after 19 weeks (McLean, 2001). The seed coat and wettable powder formulations also provided a high concentration initially; however, the fungal population was not maintained for the length of the trials (J. Hunt, personal communication).

Another benefit of the pellet formulation was that it could be easily dispersed using standard equipment, compared with the bulk carrier formulation that had to be separately delivered to the planting furrow, or the wettable powder formulation that needed to be applied after planting with a sprayer.

Complementary studies examined the infection sites of *S. cepivorum* on onion roots to enable optimization of the *T. atroviride* applications to better target *S. cepivorum* for reduced development of onion white rot (McLean and Stewart, 2000b). Sclerotia, placed 1 cm below the soil surface, caused the most severe infections with total plant collapse, compared with sclerotia at 10 and 20 cm depths which caused root infection only (McLean and Stewart, 2000b). In an additional experiment, sclerotia placed on the stem base caused significantly more severe infections, with total plant collapse, compared with sclerotia positioned along the length of a root or at the root tip (McLean and Stewart, 2000b). These studies indicated that applying *T. atroviride* to the planting furrow was the best strategy, as the stem base, where infections caused the most damage, would be protected.

Additional ecological studies indicated that *T. atroviride* was both rhizoplane- and rhizosphere-competent (McLean, 2001), which is uncommon for *Trichoderma* biological control agents, as most are reported to be rhizosphere-incompetent (Papavizas, 1981; Chao et al., 1986; Ahmad and Baker, 1987). By colonizing the rhizosphere, *T. atroviride* was able to outcompete *S. cepivorum* for infection sites on the root surface and stem base, which reduced the number of diseased plants.

These extensive New Zealand studies have led to the commercialization of *T. atroviride* C52 by Agrimm Technologies Ltd. Two formulations of *T. atroviride* C52 have been produced: Trichopel Ali 52, which is a pellet formulation that is added to the planting furrow following seed sowing, and Trichflow Ali 52, which is a wettable powder preparation applied later in the growing season. While presently these are only available in New Zealand, trials are underway to register the products in Australia (J. Hunt, personal communication).

INTEGRATED DISEASE MANAGEMENT

In New Zealand, *T. atroviride* controls onion white rot under low and moderate disease pressure. Under high disease pressure, however, an integrated approach is considered optimal to reduce infection. Integrated programs can combine biological control agents with sclerotial germination stimulants, organic amendments, soil partial sterilization, and/or reduced fungicide applications.

Germination Stimulants

The most common germination stimulant that has been investigated is diallyl disulfide (DADS), a synthetically produced form of onion or garlic oil. When applied to *S. cepivorum*-infested soil in the absence of a host, DADS causes the sclerotia to germinate, and the hyphae subsequently die as there is no host plant to infect. Initial studies indicated a large variation in disease control with DADS (Merriman et al., 1980; Coley-Smith and Parfitt, 1986). However, more recent trials have determined that DADS can successfully reduce *S. cepivorum* sclerotial viability (Crowe et al., 1994). In New Zealand, one and two applications of DADS increased onion yield by 33.5 and 79.5 percent, respectively (Tyson et al., 2000). Studies in Australia have shown applications of DADS to reduce disease by 45 percent and 64.4 to 84.5 percent (Dennis, 2001; MacManus et al., 2002), and in Mexico, DADS has also significantly reduced the number of sclerotia (Perez-Moreno and Sanchez-Pale, 1998). In Canada, both DADS and di-N-propyl disulphide (DPDS) were evaluated for control of *S. cepivorum* in soil (Hovius and McDonald, 2002). After a two-month exposure period to germination stimulants, DADS reduced sclerotial viability to 15 percent compared with DPDS (88.3 percent) and the control (99.2 percent) (Hovius and McDonald, 2002).

Diallyl disulphide was first registered in New Zealand in 1999 for use on onions (Alli-Up, Elliott Chemicals Ltd, Auckland, NZ). The cost of DADS applications is covered in the crop return and cover crops can be grown during treatment (Tyson et al., 2000). The possibility of registering Alli-Up in Australia and America is being investigated (G. Elliott, personal communication).

Organic Amendments

Organic matter applied to the soil can either cause the sclerotia to germinate and die as with germination stimulants (onion waste) or the sclerotial germination to be inhibited (eucalyptus leaves, brassica residues). Pot trials carried out with composted onion waste in the United Kingdom showed a 50 percent w/w rate to be the most effective at reducing sclerotial viability over a six-month period (50 percent) compared with 10 and 1 percent w/w rates (62 and 75 percent, respectively) (Coventry et al., 2002).

Eucalyptus leaves have also been applied to the soil as an amendment. Extensive studies in Egypt showed *S. cepivorum* to be controlled by the phenolic compounds derived from *Eucalyptus rostrata* Schlechtend. leaves when incorporated in soil (Salama et al., 1988; Ismail et al., 1989, 1990). In Australia, *E. rostrata* leaf mulch applied six weeks prior to sowing reduced

onion white rot incidence by 98 percent (Dennis, 2001). Detailed microscopic studies indicated that the presence of *Eucalyptus* leaves in the soil inhibited sclerotial germination even when onions were present (Dennis,
2001).

Soil incorporation of dried mustard (*Brassica juncea* L.) or rapeseed
(*Brassica napus* L.) plant material led to a decrease in *S. cepivorum* sclerotial viability from 48.9 percent (control) to 0 and 5.5 percent, respectively
(Smolinska, 2000). The recovery rate for the sclerotia retrieved from bras-
sica-amended soil was also low (13.3 percent for mustard and 27.8 percent
for rapeseed), compared with the control (74.4 percent) (Smolinska, 2000),
which implied that the sclerotia had degraded in response to the plant residues. However, the brassica residues were toxic to the onion seedlings with
poor emergence and abnormal seedling growth, compared with the control
(Smolinska, 2000).

Soil-Partial Sterilization

Soil-partial sterilization is also aimed at reducing pathogen inoculum in
the soil. This can be achieved by using chemicals or the sun's energy. In
1990, the U.S. Clean Air Act determined that the phasing out of the use of
methyl isothiocyanate-liberating compounds as soil sterilants would begin
in 2001 and be completed by 2010 (Whipps and Lumsden, 2001). Without
the option of methyl bromide, soil solarization is one of the only partial soil
sterilization methods available.

Soil solarization involves mulching or laying thin (50 μm thick) poly-
thene (polyethylene) sheeting over the soil during the hottest months of the
year. Soil solarization to control onion white rot has been undertaken in
the hot climate regions of Egypt (Satour et al., 1989, 1991) and Spain
(Basallote-Ureba and Melero-Vara, 1993; Prados et al., 2002). and the temperate regions of Australia (Porter and Merriman, 1983, 1985) and New
Zealand (McLean, Swaminathan, and Stewart, 2001). In hot regions, solarization is very effective at controlling onion white rot. In Egypt, there was
no onion white rot infection following a six-week solarization period, and
the onion plants were taller, heavier, and had a larger bulb size than onions
planted in nonsolarized soil (Satour et al., 1989). In Spain, a low disease incidence was evident following solarization, although no viable sclerotia
were detected (Basallote-Ureba and Melero-Vara, 1993). In the more temperate regions of the world, high soil moisture and longer solarization periods are required to reduce sclerotial viability, as the air temperatures and/or
sunshine hours are not as high as those in hot regions. In New Zealand, two
field trials testing a four-week solarization treatment resulted in a 40.2 and a
53.3 percent reduction in sclerotial viability. In a third trial, an eight-week

solarization treatment reduced viability to 8.7 percent (McLean, Swaminathan, and Stewart, 2001). While soil solarization was successful in New Zealand, the technique, for practical reasons, is better suited to hotter climates where the solarization period can be shorter.

Compatibility of Biological Control Agents with Fungicides

There are two aspects to the compatibility issue. First, a biological control agent needs to be compatible with fungicides, insecticides, and herbicides that are applied to control other diseases, pests, and weeds during the growing season, particularly those applied at the same time or in the same place as the biological control agent. Second, it would be advantageous for a biological control agent to be compatible with fungicides used to control onion white rot, since this would enable the integration of biological control agents and chemicals when disease pressure is high.

A study conducted in New Zealand examined the sensitivity of *T. atroviride* C52 conidia to all fungicides registered for use on onions in agar plate and soil pot trials (McLean, Hunt, and Stewart, 2001). In vitro, the conidia were least sensitive to procymidone (for control of onion white rot) and captan (for control of damping-off) and most sensitive to tebuconazole (for control of onion white rot), thiram (for control of damping-off), and mancozeb (for *Botrytis* and downy mildew control) (McLean, Hunt, and Stewart, 2001). In the pot trial, where the fungicides were applied to soil infested with *T. atroviride*, the effect of the fungicides was less extreme but followed a similar trend to the in vitro studies (McLean, Hunt, and Stewart, 2001). The pot trial results were considered more relevant to the field situation. Both thiram and captan are used as seed treatments in onion crops to control damping-off. It is advisable that captan is used instead of thiram when *T. atroviride* is also to be applied at planting. The other fungicides would probably not detrimentally affect *T. atroviride* populations in the soil as they are applied later in the season to foliage (McLean, Hunt, and Stewart, 2001).

Compatibility of Biological Control Agents with Organic Amendments

In Australia, spring-onion growers apply organic amendments to the soil prior to planting. Currently, this is poultry manure; however, alternatives are being investigated. Therefore, a biological control agent needs to be compatible with this practice. A collaborative project between researchers from the Department of Primary Industries, Victoria, Australia, and Lincoln

University, New Zealand, has studied the compatibility of *T. atroviride* (C52) as a pellet and a wettable powder formulation with organic amendments, including poultry manure, spent mushroom compost, and certified green compost. The results showed that *T. atroviride* was compatible with each organic amendment. The *T. atroviride* concentration was higher with the pellet formulation (10^4 to 10^7 cfu/g soil) compared with the wettable powder application (10^3 to 10^4 cfu/g soil), irrespective of the organic amendment type (K. McLean, unpublished data). Thus, *Trichoderma* biological control agents can be combined with organic amendments as part of an integrated control program.

A successful integrated disease management program is likely to involve a number of components. The use of germination stimulants, organic amendments, or soil solarization prior to planting can reduce the number of viable sclerotia in the soil. In some countries, soil solarization may be all that is required to control onion white rot. However, in most countries, additional control methods will be needed. In temperate climates, the use of germination stimulants or readily available organic amendments such as onion waste appears the most promising.

At planting, a *Trichoderma* biological control agent can be applied to the planting furrow with the seed, assuming that it is compatible with the fungicides and organic amendments used at that time. This type of integrated control regime should provide effective control of onion white rot, without heavy reliance on the fungicides which are becoming increasingly ineffective.

CONCLUSIONS AND FUTURE RESEARCH NEEDS

Large numbers of isolates have been identified with biological control activity against onion white rot, but few commercial products have been developed. The variability of biological control agents across sites and seasons is considered the major limiting factor to commercialization.

Improved efficacy can be achieved with more biological control research. The studies conducted in New Zealand can provide other countries with valuable information on the criteria needed for successful biological control. For example, fungal isolates such as *Trichoderma* need to be applied to the planting furrow at a concentration of at least 10^5 cfu/g soil in a food-base formulation in order for the biological control agent to be able to maintain a population threshold for the length of the growing season. Similarly, genetic studies on the genes implicated in mycoparasitism by *Trichoderma* species have shown that gene expression can be regulated by nitrogen catabolite repression (Steyaert, 2002). Thus, biological control activity could be reduced in the presence of inorganic nitrogen, and this has implications for appropriate timing of field application of a biological control agent.

Accepting that biological control agents used against *S. cepivorum* will probably give only satisfactory control under low-to-moderate disease pressures, and that an integrated control program will be required under high disease pressure will also enable biological control agent performance to be better managed. As the level of disease is difficult to predict from season to season, it would be wise to use an integrated approach routinely. Future research must focus on combining different control measures and optimizing their compatibility with one another to produce a multiple-step process to control *S. cepivorum*.

REFERENCES

Abd-El-Razik, A.A., A.M. El-Shabrawy, M.A. Sellam, and M.H. Abd-El-Rehim. (1985). Effectiveness of certain fungi and bacteria associated with sclerotia of *Sclerotium cepivorum* in Upper Egypt soil on controlling white rot of onion. *Egyptian Journal of Phytopathology* 17: 107-114.

Abd-El-Razik, A.A., M.N. Shatla, and M. Rushdi. (1973). Studies on the infection of onion plants by *Sclerotium cepivorum* Berk. *Phytopathologische Zeitschrift* 76: 108-116.

Adams, P.B. and W.A. Ayers. (1981). *Sporidesmium sclerotivorum*: Distribution and function in natural biological control of sclerotial fungi. *Phytopathology* 71: 90-93.

Adams, P.B. and W.A. Ayers. (1983). Histological and physiological aspects of infection of sclerotia of two *Sclerotinia* species by two mycoparasities. *Phytopathology* 73: 1072-1076.

Ahmad, J.S. and R. Baker. (1987). Rhizosphere competence of *Trichoderma harzianum*. *Phytopathology* 77: 182-189.

Ahmed, A.H.M. and H.T. Tribe. (1977). Biological control of white rot of onion *(Sclerotium cepivorum)* by *Coniothyrium minitans*. *Plant Pathology* 26: 75-78.

Alexander, B.J.R. and A. Stewart. (1994). Survival of sclerotia of *Sclerotinia* and *Sclerotium* spp. in New Zealand horticultural soil. *Soil Biology and Biochemistry* 26: 1323-1329.

Ayers, W.A. and P.B. Adams. (1979). Mycoparasitism of sclerotia of *Sclerotinia* and *Sclerotium* species by *Sporidesmium sclerotivorum*. *Canadian Journal of Microbiology* 25: 17-23.

Backhouse, D. and A. Stewart. (1988). Large sclerotia of *Sclerotium cepivorum*. *Transactions of the British Mycological Society* 91: 343-346.

Backhouse, D. and A. Stewart. (1989). Interactions between *Bacillus* species and sclerotia of *Sclerotium cepivorum*. *Soil Biology and Biochemistry* 91: 343-346.

Basallote-Ureba, M.J. and J.M. Melero-Vara. (1993). Control of garlic white rot by soil solarization. *Crop Protection* 12: 219-233.

Booer, J.R. (1945). Control of white rot in onions. *Nature* 155: 241-242.

Booer, J.R. (1946). Further experiments on the control of white rot (*Sclerotium cepivorum* Berk.) in onions, shallots and leeks. *Annals of Applied Biology* 33: 413-419.

Chang, Y.C., Y.C. Chang, R. Baker, O. Kliefield, and I. Chet. (1986). Increased growth of plants in the presence of the biological control agent *Trichoderma harzianum*. *Plant Disease* 70: 145-148.

Chao, W.L., E.B. Nelson, G.E. Harman, and H.C. Hoch. (1986). Colonization of the rhizosphere by biological control agents applied to seeds. *Phytopathology* 76: 60-65.

Clarkson, J.P., T. Payne, A. Mead, and J.M. Whipps. (2002). Selection of fungal biological control agents of *Sclerotium cepivorum* for control of white rot by sclerotial degradation in a UK soil. *Plant Pathology* 51: 735-745.

Clarkson, J., T. Payne, and J. Whipps. (2001). A screening system for identifying biological control agents of *Sclerotium cepivorum*. In *Biocontrol Agents: Mode of Action and Interaction with Other Means of Control*, eds. Y. Elad, S. Freeman, and E. Monte. IOBC/WRPS Bulletin, UK, 24, pp. 71-74.

Claydon, N., M. Allan, J.R. Hanson, and A.G. Avent. (1987). Antifungal alkyl pyrones of *Trichoderma harzianum*. *Transactions of the British Mycological Society* 88: 503-513.

Coley-Smith, J.R. (1959). Studies of the biology of *Sclerotium cepivorum* Berk: III. Host range, persistence and viability of sclerotia. *Annals of Applied Biology* 47: 511-518.

Coley-Smith, J.R. (1960). Studies of the biology of *Sclerotium cepivorum* Berk: IV. Germination of sclerotia. *Annals of Applied Biology* 48: 8-18.

Coley-Smith, J.R. and J.E. King. (1969). The production by species of *Allium* of alkyl sulphides and their effect on germination of sclerotia of *Sclerotium cepivorum* Berk. *Annals of Applied Biology* 64: 289-301.

Coley-Smith, J.R., C.M. Mitchell, and C.E. Sansford. (1990). Long-term survival of sclerotia of *Sclerotium cepivorum* and *Stromatinia gladioli*. *Plant Pathology* 39: 58-69.

Coley-Smith, J.R. and D. Parfitt. (1986). Some effects of diallyl disulphide on sclerotia of *Sclerotium cepivorum*: Possible novel control method for white rot disease of onions. *Pesticide Science* 37: 587-594.

Coventry, E., R. Noble, A. Mead, and J. Whipps. (2002). Control of *Allium* white rot (*Sclerotium cepivorum*) with composted onion waste. *Soil Biology and Biochemistry* 34: 1037-1045.

Crowe, F.J. (1995). First report of the large sclerotial form of *Sclerotium cepivorum* in North America. *Phytopathology* 85: 1038 (abstract).

Crowe, F.J., J. Debons, T. Darnell, M. Thornton, D. McGrath, P. Koepsell, J. Laborde, and E. Redondo Jaurez. (1994). Control of *Allium* white rot disease with DADS and related products. In *Proceedings Fifth International Workshop on Allium White Rot*, eds. A. R. Entwistle and J. M. Melero-Vara. Cordoba, Spain, pp. 14-28.

Croxall, H.E., R.W. Sidwell, and J.E.E. Jenkins. (1953). White rot (*Sclerotium cepivorum*) of onions in Worcestershire with special reference to control by seed treatment with calomel. *Annals of Applied Biology* 40: 166-175.

Cunningham, G.H. (1922). Some recent changes in the names of plant diseases. *New Zealand Journal of Agriculture* 24: 37-45.

Dennis, C. and J. Webster. (1971a). Antagonistic properties of species-groups of *Trichoderma:* I. Production of non-volatile antibiotics. *Transactions of the British Mycological Society* 57: 25-39.

Dennis, C. and J. Webster. (1971b). Antagonistic properties of species-groups of *Trichoderma:* II. Production of volatile antibiotics. *Transactions of the British Mycological Society* 57: 41-48.

Dennis, J.J. (2001). Progress towards an integrated control strategy for onion white root rot disease including the use of artificial germination stimulants. *Acta Horticulturae* 555: 117-121.

De Oliveira, V.L., M.D.M. Bellei, and A.C. Borges. (1984). Control of white rot of garlic by antagonistic fungi under controlled environmental conditions. *Canadian Journal of Microbiology* 30: 884-889.

Dingley, J.M. (1969). Records of plant diseases in New Zealand. In *Department of Scientific and Industrial Research Bulletin 192*. Government Printer, Wellington, New Zealand.

Duff, A.A., K.J. Jackson, and W.E. O'Donnell. (2001). Tebuconazole (Folicur) shows potential in the control of white rot *(Sclerotium cepivorum)* in garlic in subtropical Queensland, Australia. *Acta Horticulturae* 555: 247-250.

Eady, C.C. (1995). Towards the transformation of onions *(Allium cepa)*. *New Zealand Journal of Crop and Horticultural Science* 23: 239-250.

Eady, C.C. (2001). The transformation of onions and related *Alliums*. In *Transgenic Plants and Crops*, eds. Khachatourians, McHugen, Scorza, Nip, and Hui. Marcel Dekker, New York, USA. pp. 655-671.

Eady, C.C. (2002). Genetic transformation of onions. In *Allium Crop Science: Recent Advances*, eds. Rabinowitch and Currah. UK-CAB International, UK, pp. 119-144.

Eady, C.C., C.E. Lister, Y. Suo, and D. Schaper. (1996). Transient expression of *uida* constructs in in vitro onion *(Allium cepa* L.) cultures following particle bombardment and *Agrobacterium*-mediated DNA delivery. *Plant Cell Reports* 15: 958-962.

Entwistle, A.R. (1983). Changes in the effectiveness of iprodione for the control of *Allium* white rot. *Phytopathology* 73: 800 (abstract only).

Entwistle, A.R. (1986). Relationships between soil sclerotial populations of *Sclerotium cepivorum* and the incidence of *Allium* white rot. In *Proceedings Third International Workshop on Allium White Rot, Wellesbourne*, ed. A.R. Entwistle. UK, pp. 21-29.

Entwistle, A.R. and H.L. Munasinghe. (1973). Recent studies on the fungicidal control of white rot disease of salad onions. In *Proceedings 7th British Insecticide and Fungicide Conference, Brighton, England*, pp. 581-585.

Entwistle, A.R. and H.L. Munasinghe. (1980a). The effect of iprodione on sclerotium germination, root infection and mycelial spread of *Sclerotium cepivorum* in salad onions. *Annals of Applied Biology* 95: 329-339.

Entwistle, A.R. and H.L. Munasinghe. (1980b). The effect of seed, furrow and stem base spray treatment with iprodione on white rot disease *(Sclerotium cepivorum)* in spring-sown salad onions. *Annals of Applied Biology* 94: 215-224.

Fullerton, R.A. and A. Stewart. (1991). Chemical control of onion white rot *(Sclerotium cepivorum* Berk.) in the Pukekohe district of New Zealand. *New Zealand Journal of Crop and Horticultural Science* 19: 121-127.

Fullerton, R.A., A. Stewart and E.A. Slade. (1994). Onion white rot—it wont go away. *Commercial Grower,* pp. 17-23.

Fullerton, R.A., A. Stewart, and E.A. Slade. (1995). Use of demethylation inhibiting fungicides (DMIs) for the control of onion white rot *(Sclerotium cepivorum* Berk.) in New Zealand. *New Zealand Journal of Crop and Horticultural Science* 23: 121-125.

Gabelman, W.H. (1986). White rot resistance from *Allium cepa* cv. Zittauer Gelb. In *Proceedings Third International Workshop on Allium White Rot,* ed. A. R. Entwistle. Wellesbourne, UK, pp. 9-10.

Galal, A.A. and A.A. Bana. (2001). Influence of plant growth regulators on the interaction between onion plants and white rot pathogen *Sclerotium cepivorum* Berk. Assiut *Journal of Agricultural Sciences* 32: 1-14 (abstract only).

Georgy, N.I. and J.R. Coley-Smith. (1982). Variation in morphology of *Sclerotium cepivorum* sclerotia. *Transactions of the British Mycological Society* 79: 534-536.

Ghaffar, A. (1969a). Biological control of white rot of onion: I. Interactions of soil microorganisms with *Sclerotium cepivorum* Berk. *Mycopathologia et Mycologia Applicata* 38: 101-111.

Ghaffar, A. (1969b). Biological control of white rot of onion: II. Effectiveness of *Penicillium nigricans* (Bain.) Thom. *Mycopathologia et Mycologia Applicata* 38: 101-111.

Ghisalberti, E.L. and K. Sivasithamparam. (1991). Antifungal antibiotics produced by *Trichoderma* spp. *Soil Biology and Biochemistry* 23: 1011-1020.

Harper, G.E., C.M. Frampton, and A. Stewart. (2002). Factors influencing survival of sclerotia of *Sclerotium cepivorum* in New Zealand soils. *New Zealand Journal of Crop and Horticultural Science* 30: 29-35.

Harrison, Y.A. and A. Stewart. (1988). Selection of fungal antagonists for biological control of onion white rot in New Zealand. *New Zealand Journal of Experimental Agriculture* 16: 249-256.

Hovius, M.H.Y. and M.R. McDonald. (2002). Management of *Allium* white rot *(Sclerotium cepivorum)* in onions on organic soil with soil-applied diallyl disulphide and di-N-propyl disulfide. *Canadian Journal of Plant Pathology* 24: 281-286.

Ismail, I.M.K., A.A.M. Salama, M.I.A. Ali, and S.A.E. Ouf. (1989). Effect of amendment of non-autoclaved and autoclaved natural soil, with *Eucalyptus rostrata* leaves, on fungal population and total content of phenolic compounds. *Egyptian Journal of Microbiology* 1992: 299-320.

Ismail, I.M.K., A.A.M. Salama, M.I.A. Ali, and S.A.E. Ouf. (1990). Effect of soil amendment with *Eucalyptus rostrata* leaves on activity of hydrolytic enzymes, total phenol content and onion white rot disease incidence. *Zentralblatt fur Mikrobiologie* 1990: 219-228 (abstract only).

Jones, E.E. and J.M. Whipps. (2001). Biological control of *Sclerotinia sclerotiorum* in glasshouse lettuce. In *Biocontrol Agents: Mode of Action and Interaction with Other Means of Control,* eds. Y. Elad, S. Freeman, and E. Monte. IOBC/WPRS Bulletin, UK, 24: 83-87.

Kay, S.J. and A. Stewart. (1994). Evaluation of fungal antagonists for control of onion white rot in soil box trials. *Plant Pathology* 43: 371-377.

Kleifeld, O. and I. Chet. (1992). *Trichoderma harzianum*—interaction with plants and effect on growth response. *Plant and Soil* 144: 267-272.

Klein, D. and D.E. Eveleigh. (1998). Ecology of *Trichoderma*. In *Trichoderma* and *Gliocladium,* Vol. 1, eds. C. P. Kubicek and G. E. Harman. Taylor and Francis Ltd., London, pp. 57-74.

Legget, M.E. and J.E. Rahe. (1985). Factors affecting the survival of sclerotia of *Sclerotium cepivorum* in the Fraser Valley. *Annals of Applied Biology* 106: 255-263.

Lewis, J.A. and G.C. Papavizas. (1985). Characteristics of alginate pellets formulated with *Trichoderma* and *Gliocladium* and their effect on the proliferation of the fungi in soil. *Plant Pathology* 34: 571-577.

MacManus, G.P.V., R.D. Davis, K.L. Bell, R.A. Kopittke, and T. Napier. (2002). Sclerotial germination stimulant suppresses onion white rot. In *Proceedings Onions 2002 Conference,* Yanco, Australia, pp. 50-52.

Maloy, O.C. and R. Machtmes. (1974). Control of onion white rot by furrow and root-dip application of fungicides. *Plant Disease Reporter* 58: 6-9.

McLean, K.L. (1996). Control of onion white rot using beneficial microorganisms and soil solarisation. MSc thesis, Department of Plant Science, Lincoln University, Canterbury, New Zealand.

McLean, K.L. (2001). Biological control of onion white rot using *Trichoderma harzianum*. Ph D thesis, Soil, Plant and Ecological Sciences Division, Lincoln University, Canterbury, New Zealand.

McLean, K.L., G.E. Harper, C.M. Frampton, and A. Stewart. (2002). Biological aspects of *Sclerotium cepivorum* sclerotia. In *Proceedings 7th International Allium White Rot Workshop,* ed. M. R. Macdonald. Coalinga, California.

McLean, K.L., J. Hunt, and A. Stewart. (2001). Compatibility of the biocontrol agent *Trichoderma harzianum* (C52) with fungicides. *New Zealand Plant Protection* 54: 84-88.

McLean, K.L. and A. Stewart. (2000a). Application strategies for control of onion-white rot. *New Zealand Journal of Crop and Horticultural Science* 28: 115-122.

McLean, K.L. and A. Stewart. (2000b). Infection sites of *Sclerotium cepivorum* on onion. *New Zealand Plant Protection* 53: 118-121.

McLean, K.L., J. Swaminathan, and A. Stewart. (2001). Increasing soil temperature to reduce sclerotial viability of *Sclerotium cepivorum* in New Zealand soils. *Soil Biology and Biochemistry* 33: 137-143.

Merriman, P.R., S. Isaacs, R.R. Macgregor, and G.B. Towers. (1980). Control of white rot in dry bulb onions with artificial onion oil. *Annals of Applied Biology* 96: 163-168.

Metcalf, D.A. (2002). Development of biological control agents which control onion white root rot under commercial field conditions. In *Proceedings Onions 2002 Conference,* Yanco, Australia, pp. 63-68.

Metcalf. D.A., J.J.C. Dennis, and C.R. Wilson. (1997). First report of the large sclerotial form of *Sclerotium cepivorum* in Australia. *Australasian Plant Pathology*
26: 203.

Metcalf, D.A. and C.R. Wilson. (1999). Histology of *Sclerotium cepivorum* infection of onion roots and the spatial relationships of pectinases in the infection process. *Plant Pathology* 48: 445-452.

Metcalf, D.A. and C.R. Wilson. (2001a). The process of antagonism of *Sclerotium
cepivorum* in white rot affected onion roots by *Trichoderma koningii*. *Plant Pathology* 50: 249-257.

Metcalf. D.A. and C.R. Wilson. (2001b). Progress toward a biological control system for onion white root rot in Tasmania. *Acta Horticulturae* 555: 123-126.

Mordue. J.E.M. (1976). *Sclerotium cepivorum*. In *CMI Descriptions of Pathogenic
Fungi and Bacteria*. The Cambrian News Ltd., Aberystwyth, p. 512.

Moubasher, A.H., M.A. Elnaghy, and S.E. Megala. (1970). Fungi isolated from
sclerotia of *Sclerotium cepivorum* and from soil and their effects upon the pathogen. *Plant and Soil* 33: 305-312.

Papavizas, G.C. (1981). Survival of *Trichoderma harzianum* in soil and in pea and
bean rhizospheres. *Phytopathology* 71: 121-125.

Payghami, E., S. Massiha, B. Ahary, M. Valizadeh, and A. Motallebi. (2001). Enhancement of growth of onion (*Allium cepa* L.) by biological control agent
Trichoderma spp. *Acta Agronomica Hungarica* 49: 393-395.

Perez-Moreno, L., V.O. Portugal, J.R. Sanchez-Pale, C. Castaneda-Cabrera, and
A.R. Entwistle. (1997). Sensitivity in vitro of *Sclerotium cepivorum* Berk., to the
fungicides commonly used in its control. *Revista Mexicana de Fitopatologia* 15:
9-14 (abstract only).

Perez-Moreno, L. and A. Rodriguez-Aguilera. (2002a). Biological control of *Coniothyrium minitans*, on sclerotia of *Sclerotium cepivorum* Berk. In *Proceedings 7th
International Allium White Rot Workshop*. ed. M.R. Macdonald. Coalinga, California.

Perez-Moreno, L. and A. Rodriguez-Aguilera. (2002b). Biological effectiveness of
Coniothyrium minitans, on sclerotia of *Sclerotium cepivorum* Berk. In *Proceedings 7th International Allium White Rot Workshop*. Coalinga, California.

Perez-Moreno. L. and J.R. Sanchez-Pale. (1998). Control of soft rot (*Sclerotium
cepivorum* Berk.) of onion (*Allium sativum* L.) with sterilizers in the Bajio zone,
Mexico. *Revista Mexicana de Fitopatologia* 16: 72-78 (abstract only).

Porter, I.J., J.P. Maughan, and G.B. Towers. (1991). Evaluation of seed, stem and
soil applications of procymidone to control white rot (*Sclerotium cepivorum*
Berk.) in onions. *Australian Journal of Experimental Agriculture* 31: 401-406.

Porter, I.J. and P.R. Merriman. (1983). Effects of solarization of soil on nematode
and fungal pathogens at two sites in Victoria. *Soil Biology and Biochemistry* 15:
39-44.

Porter, I.J. and P.R. Merriman. (1985). Evaluation of soil solarization for control of
root diseases of row crops in Victoria. *Plant Pathology* 34: 108-118.

Prados, L.A.M., F.J. Bascon, P.C. Calvet, H.C. Corpas, R.A. Lara, J.M. Melero-
Vara, and M.J. Basallote-Ureba. (2002). Effect of different soil and clove treat-

ments in the control of white rot of garlic. *Annals of Applied Biology* 140: 247-253.

Rahe, J.E. (1986). Detection and selection for field resistance to onion white rot. In *Proceedings Third International Workshop on Allium White Rot*, ed. A.R. Entwistle. Wellesbourne. UK, pp. 11-17.

Ramona, Y. and M.A. Line. (2002). Potential for the large-scale production of a biocontrol fungus in raw and composted paper mill waste. *Compost Science and Utilization* 10: 57-62.

Ryan, E.W. and T. Kavanagh. (1976). White rot of onion *(Sclerotium cepivorum)*: 1. Control by fungicidal pelleting of onion seed. *Irish Journal of Agricultural Research* 15: 317-323.

Salama, A.A.M., I.M.K. Ismail, M.I.A. Ali, and S.A.F. Ouf. (1988). Possible control of white rot disease of onions caused by *Sclerotium cepivorum* through soil amendment with *Eucalyptus rostrata* leaves. *Revue d'Ecologie et de Biologie du Sol* 25: 305-314 (abstract only).

Satour, M.M., M.F. Abdel-Rahim, T. El-Yamani, A. Radwan, A. Grinstein, H.D. Rabinowitch, and J. Katan. (1989). Soil solarization in onion fields in Egypt and Israel: Short and long term effects. *Acta Horticulturae* 255: 151-159.

Satour, M.M., E.M. El-Sherif, L. El-Ghareeb, S.A. El-Hada, and H.R. El-Wakil. (1991). Achievements of soil solarization in Egypt. *FAO Plant Production and Protection Paper* 109: 200-212.

Scott, M.R. (1956a). Studies of the biology of *Sclerotium cepivorum* Berk.: I. Growth of the mycelium in the soil. *Annals of Applied Biology* 44: 576-583.

Scott, M.R. (1956b). Studies of the biology of *Sclerotium cepivorum* Berk.: II. The spread of white rot from plant to plant. *Annals of Applied Biology* 44: 584-589.

Slade, E.A., R.A. Fullerton, A. Stewart, and H. Young. (1992). Degradation of the dicarboximide fungicides iprodione, vinclozolin and procymidone in Patumahoe clay loam soil, New Zealand. *Pesticide Science* 35: 95-100.

Smolinska, U. (2000). Survival of *Sclerotium cepivorum* sclerotia and *Fusarium oxysporum* chlamydospores in soil amended with cruciferous residues. *Journal of Phytopathology* 148: 343-349.

Stewart, A., D. Backhouse, P.W. Sutherland, and R.A. Fullerton. (1989). The development of infection structures of *Sclerotium cepivorum* on onion. *Journal of Phytopathology* 126: 22-32.

Stewart, A. and R.A. Fullerton. (1991). Additional studies on the chemical control of onion white rot (*Sclerotium cepivorum* Berk.) in New Zealand. *New Zealand Journal of Crop and Horticultural Science* 19: 129-134.

Stewart, A. and Y.A. Harrison. (1988). Mycoparasitism of sclerotia of *Sclerotium cepivorum*. *Australasian Plant Pathology* 18: 10-14.

Steyaert, J. (2002). Studies on the genetic regulation of mycoparasitism in *Trichoderma hamatum*. MSc thesis, Soil, Plant and Ecological Sciences Division, Lincoln University, Canterbury, New Zealand.

Tims, E.C. (1948). White rot of shallot. *Phytopathology* 38: 378-394.

Turner, G.J. and H.T. Tribe. (1976). On *Coniothyrium minitans* and its parasitism of *Sclerotinia* species. *Transactions of the British Mycological Society* 66: 97-105.

Tyson, J.L., R.A. Fullerton. G.S. Elliott, and P.J. Reynolds. (2000). Use of diallyl disulphide for the commercial control of *Sclerotium cepivorum*. *New Zealand Plant Protection* 53: 393-397.

Tyson, J.L., R.A. Fullerton, and A. Stewart. (1999). Changes in the efficacy of fungicidal control of onion white rot. *Proceedings of the Fifty-second New Zealand Plant Protection Conference* 52: 171-175.

Utkhede, R.S. and J.E. Rahe. (1980a). Biological control of onion white rot. *Soil Biology and Biochemistry* 12: 101-104.

Utkhede, R.S. and J.E. Rahe. (1980b). Stability of cultivar resistance to onion white rot. *Canadian Journal of Plant Pathology* 2: 19-22.

Utkhede, R.S. and J.E. Rahe. (1983). Effect of *Bacillus subtilis* on growth and protection of onion against white rot. *Phytopathologische Zeitschrift* 106: 199-203.

Utkhede, R.S. and J.E. Rahe. (1984). Resistance to white rot infections in bulb onion seed lots. *Scientia Horticulturae* 22: 315-320.

Utkhede, R.S., J.E. Rahe, J.R. Coley-Smith, Q.P. van Deer Meer. J.G. Brewer, and V. Crisola. (1982). Genotype-environment interaction for resistance to onion white rot. *Canadian Journal of Plant Pathology* 4: 269-271.

Villalta, O., R. Crnov and I.J. Porter. (2002). Onion white rot—causing severe yield losses in spring onions. In *Proceedings Onions 2002*, ed. T. Napier Yanco. Australia, pp. 78-80.

Walker, A., P.A. Brown, and A.R. Entwistle. (1986). Enhanced degradation of iprodione and vinclozolin in soil. *Pesticide Science* 17: 183-193.

Walker, J.C. (1924). White rot of *Allium* in Europe and America. *Phytopathology* 14: 315-323.

Weber, Z. (2002). Efficacy of biopreparate Contans WG (*Coniothyrium minitans* Campb.) in winter oilseed rape protection against *Sclerotinia sclerotiorum* (Lib.) de Bary. *Rosliny-Oleiste* 23: 151-156 (abstract only).

Whipps, J.M. and R.D. Lumsden. (2001). Commercial use of fungi as plant disease biological control agents: Status and prospects. In *Fungi As Biocontrol Agents: Progress, Problems and Potential*, eds. T. M. Butt, C. Jackson, and N. Magan. CABI Publishing, Wallingford, UK, pp. 9-22.

Willetts, H.J. (1971). The survival of fungal sclerotia under adverse environmental conditions. *Biological Reviews* 46: 387-407.

Willetts, H.J. and S. Bullock. (1992). Developmental biology of sclerotia. *Mycological Research* 96: 801-816.

Williams, R.H. and J.M. Whipps. (1995). Growth and transmission of *Coniothyrium minitans* in soil. In *Biological Control of Sclerotium Forming Pathogens*, eds. J.M. Whipps and M. Gerlagh. IOBC/WPRS Bulletin, UK, 18: 39-43.

Wong, W.C. and I.K. Hughes. (1986). *Sclerotium cepivorum* Berk. in onion (*Allium cepa* L.) crops: Isolation and characterization of bacteria antagonistic to the fungus in Queensland. *Journal of Applied Bacteriology* 60: 57-60.

Wood, R.J. (1980). Control of onion white rot with iprodione. In *Proceedings 33rd New Zealand Weed and Pest Control Conference*, pp. 203-205.

Personal Communications

S. Bruckner, Prophyta GmbH, Inselstrasse 12, D-23999 Malchow/Poel. Germany.

F. Crowe, Botany and Plant Pathology, Oregon State University, 850 NW Dogwood Ln., Madras OR 97741.

C.C. Eady, Crop and Food Research, Canterbury Agriculture and Science Centre, Gerald St, Lincoln, Canterbury, New Zealand.

G. Elliott, Elliot Chemicals Ltd, 45 Kitchener Rd, Pukekohe, Auckland, New Zealand.

J. Hunt, Agrimm Technologies Ltd, 231 Fitzgerald Ave, Christchurch, New Zealand.

M.R. McDonald, Department of Plant Agriculture, University of Guelph, Ontario, Canada N1G 2W1.

L. Perez Moreno, Instituto de Ciencias Agricolas, Universidad de Guanajuato, Guanajuato, Mexico C. P. 36500.

Chapter 7

Biological Control of Fruit and Vegetable Diseases with Fungal and Bacterial Antagonists: *Trichoderma* and *Agrobacterium*

F. Scala
A. Raio
A. Zoina
M. Lorito

INTRODUCTION

Yields of food and fiber have increased substantially to meet the needs of an increasing population. The primary components of yield increase are improved pest control, primarily through the use of synthetic pesticides; application of large quantities of synthetic and natural fertilizers, especially nitrogen; and the development of plant varieties with higher-yield capacity and pest resistance. Unfortunately, these technologies are approaching their limits and are poorly sustainable from an environmental and human-safety point of view. Consequently, new tools and methodologies need to be developed. In addition, there is an increasing demand for higher quality and safety of the food chain and its end products, starting from the production

The authors are grateful to Claudia Capodilupo, Enrique Monte, Antonio Llobell, Maurizio Vurro, and Gary Harman for their contribution to the text of this paper. Research performed at our laboratory were supported by Project FIRB-MIUR 2002, Project EU "TRICHOEST" 2003, FAIR 98PL-4140 and 2E-BCAs, Project MIUR-MIPAF, project MIUR PRIN 2002 and 2003, project CNR-IPP, and project PON 2003.

phase of agricultural commodities. These are, in fact, the priorities of the agricultural-related research in Europe and the United States, where a reduction in production may even be considered acceptable if substantial increases in the quality and safety of the agricultural process are achieved (that is, by reducing the use of chemicals for disease control and fertilization). One way of lowering the use of chemicals, and hence the related side effects, is to apply beneficial microbes for crop protection. Plant-associated microorganisms, including eubacteria, actinomycetes, and fungi, are part of the natural ecosystem of healthy plants, occurring in the major habitats of the rhizosphere, leaf surfaces, and inside plant tissues. For biocontrol of plant diseases, a key point to be explored is the interaction between antagonistic organisms, host plants, and pathogens.

Soilborne plant pathogens represent a major problem of plant protection in many open field and greenhouse fruit and vegetable crops. Pathogens are often able to survive for several years in the soil as dormant resting structures until a susceptible crop is introduced again. These structures are able to withstand adverse environmental conditions and chemical applications, creating major control problems for agriculture. The fungal disease agents responsible for damping off, crown and root rots, and wilts are of utmost importance in all vegetable and fruit crops. Various *Pythium* spp. and *Phytophthora* spp. may damage the lower part of tomato, pepper, cucumber, and many other vegetable plants, both in soil and soil-less cultures. *Rhizoctonia solani* can also infect many species, causing symptoms resembling *Pythium* or *Phytophthora* rots. The rot disease caused by *Sclerotinia* spp. affects over 400 plant species, including most vegetable crops. The crown-rot- and wilt-inducing strains of *Fusarium oxysporum* are responsible for severe damage on many economically important plant species (tomato, cucumber, muskmelon, asparagus, radish, onion, flax, carnation, cyclamen). *Fusarium* wilt pathogens show a high level of host specificity and, based on the plant species and cultivars they can infect, they are classified into more than 1200 *formae speciales* and races. Parasitic weeds such as *Orobanche* species (broomrapes) attack nearly all vegetables, legumes, and sunflowers in southern Europe to the Balkans and Russia, the Middle East and North Africa. *Orobanche ramosa* together with *O. aegyptiaca* infest about 2.6 million hectares cropped to plants in the Solanaceae and grain, particularly tobacco, potato, tomato, and eggplant (Sauerborn, 1991). *Orobanche cumana* severely restricts and limits sunflower production in Spain and eastern Europe. The broomrapes interfere with water and mineral intake and divert considerable crop photosynthetic capacity to their own growth, and are responsible for both extensive qualitative and quantitative damage to these high value crops. Perennial weeds are among the most troublesome weeds to manage. For example, *Cirsium arvense* (Canada thistle) is considered one of the world's worst weeds (Holm et al., 1977), and the third most important weed in Europe (Schroeder et al., 1993). *Sonchus arvensis* is another perennial species. Although not as troublesome as *C. arvense*, it represents a considerable challenge, especially in organic farming. *Cyperus esculentus* is one of the most serious invasive alien species in the southern countries of Europe both in crops and noncrop areas. Perennials usually reproduce from vegetative buds in the root system and from seed and are very difficult to control because their extensive root system allows them to recover from control attempts.

Control strategies for the above pest problems as well as nematodes and arthropods (the latter two are pest problems beyond the scope of this chapter) include the application of soil fumigants such as methyl bromide, which is one of the most effective and widespread (but extremely expensive) practices used to control soil pests. Recent regulations have drastically reduced methyl bromide use, and it is to be phased out totally by 2005 due to negative environmental effects. The development of alternative strategies for safe and environmentally friendly bioconstraint management is specifically cited as one of the aims in research priorities. In many crops, no real alternatives to methyl bromide have been found. Other fumigants are expensive and generally less effective than methyl bromide for conventional agriculture and cannot be used in organic farming. Other control strategies such as soil solarization could be possible but have constraints: environmental (need high temperatures) and temporal (fields remain covered for much of a growing season). Seed treatments with conventional fungicides provide some initial protection to soil pathogens, but this is not effective for a long enough period in heavily infested soils. None of the fungicides allowable in organic agriculture are very effective on soilborne pathogens. Breeding high levels of human-toxic psoralens into vegetables, such as celery, to prevent damping-off fungi is used in organic agriculture in North America, but this may be unadvisable.

No traditional control methods have been effective for *Orobanche* spp., which are not usually amenable to control by persistent selective herbicides, as systemic herbicides cannot differentiate between the crop and the parasite, unless the crop is transgenically rendered resistant to herbicides (Joel et al., 1995). Furthermore, as these weeds attach to crop roots, they cannot be controlled mechanically, except by removing their flower stalks to reduce seed accumulation and dispersal. Perennial weeds are difficult to control using traditional methods, because they usually cannot be removed mechanically, due to the well-developed root systems or subterranean organs, and because they often require repetitive chemical treatments, which are expensive in conventional agriculture. None of the few herbicides allowable in organic agriculture control perennial or parasitic weeds.

Considerable effort during the past few decades has been dedicated to biological control of weeds and plant diseases, and many interesting and potential microorganisms have been found, but their use is still very limited. This is due to many constraints, including biological (virulence, stability, defence mechanisms of the target pest, interaction with other microorganisms); technological (scarcity of sporulation, lost of aggressiveness, special growth requirements); environmental (interaction with water, physical characteristics of the soil, physical and chemical barriers); and commercial (limited market, registration problems including secondary toxicity, and costs). Only a limited number of commercial products are available against a few diseases, and no commercial bioherbicides are available in the European market. Interestingly, cost has not been the major limiting factor for the adoption of biocontrol agents for the pest constraints discussed above. If effective, almost anything could compete with methyl bromide. The major problems come from consistency and lack of near-complete control activity when they are active. Many types of microbial agents have been employed in order to control plant diseases.

Coniothyrium minitans, for example, is an efficient mycoparasite of ascomycetous sclerotium-forming fungi, including important plant pathogenic species of *Sclerotinia* and *Sclerotium* (Whipps and Gerlagh, 1992). This organism has been used successfully in glasshouse and field experiments to control *Sclerotinia* diseases of a number of crop plants (Whipps and Lumsden, 2001), and a commercial product has been registered in seven European countries, the United States, and Mexico. However, pan-European registration has been held up due to concerns over the mode of action, and more work is needed in this area. The major constraints of its wider use in agricultural practice are the limited knowledge of its ecology and the scanty information on its physiology and genetics, preventing attempts at strain improvement. The concept of using nonpathogenic strains of *F. oxysporum* to control fusarium diseases came from the demonstration that suppression of the disease in suppressive soils results from interactions between pathogenic and nonpathogenic strains. Therefore, nonpathogenic strains were developed as biocontrol agents (Lemanceau and Alabouvette, 1991). The nonpathogenic *F. oxysporum* strains show several modes of action contributing to their biocontrol capacity (Couteaudier and Alabouvette, 1990; Lemanceau et al., 1993). They are able to compete for nutrients in the soil, affecting the rate of chlamydospore germination of the pathogen. They can also compete for infection sites on the root and can trigger plant defense reactions, inducing systemic resistance (Fuchs et al., 1997). Several strains of nonpathogenic *F. oxysporum* have good efficacy in many trials, but as with other biocontrol agents, there is a lack of consistency. Despite isolation of many promising pathogenic organisms that could be useful for control of parasitic weeds, none has received continual widespread use. Two very promising strains of *F. arthrosporioides* and *F. oxysporum* were isolated in Israel from juvenile *O. aegyptiaca* plants, and also attacked *O. ramosa* and *O. cernua* (Amsellem et al., 2001). Antagonist strains of *P. syringae* also have been identified that may be used in biocontrol against phytopathogenic fungi and bacteria. Strains of *P. syringae* have been reported to be effective as biocontrol agents against *Penicillium expansum* and *B. cinerea* on pears (Sugar and Spotts, 1999), *Monilinia fructicola* and *Rhizopus stolonifer* on peaches (Zhou et al., 1999). *P. digitatum* on citrus, *B. cinerea* on grape (Cirvilleri et al., 2000), *P. expansum* on apples (Conway et al., 1999), and *E. coli 0157:H7* on apple wounds (Janisiewicz et al., 1999) and as agent of induced resistance against *Plasmopara viticola* and *Uncinula necator* on grape (Kassemeyer and Busam, 1998).

Perennial weeds in arable farming are ideal targets for biological control. Biological control could replace one or more herbicide treatments. In organic farming systems, biological control of perennials, especially *Cirsium arvense,* would reduce the number of time-consuming and expensive mechanical treatments. *Phomopsis cirsii, Ramularia circii,* and *Septoria cirsii* were chosen as promising candidates in systematic field surveys of diseased *C. arvense* carried out in Denmark (Leth and Andreasen, 1999) and Russia (Berestetski, 1997).

A further application of fungal antagonists, and in particular yeasts, concerns the control of postharvest pathogens. This phytopathological issue is important not only for the production losses but also for the reduction of quality in food crops because of the presence of mycotoxins. Among the most promising microorganisms to be applied as biopesticides postharvest, antagonistic yeasts are probably the most studied and tested. The advantage of using these antagonists derives from their efficacy and the absence of accumulated antibiotics that may be harmful for the consumer (Droby and Chalutz, 1994). In the United States and Israel, yeast-based formulations for the control of fungal postharvest pathogens of citrus and pome fruits are already commercially available (Bio-Save and Aspire), and others are waiting to be registered. However, as in the case of other fungal antagonists, the level of efficacy of these formulations is not always comparable to one of the chemical fungicides. Different mechanisms of action have been suggested for antagonist yeasts, which could be complementary to each other: induction of host resistance, competition for space and nutrients, and secretion of lytic enzymes that depolymerize fungal walls (Castoria et al., 2001). A possible role of fungal cell wall–degrading enzymes (chitinases and glucanases) has been suggested for antagonist yeasts (Castoria et al., 1997), as well as in the case of *Trichoderma* (Woo et al., 1999).

In this chapter we focus our attention on the use of biocontrol strains belonging to the genera *Agrobacterium* and *Trichoderma*.

Biological control of crown gall by the nonpathogenic *Agrobacterium radiobacter* strain K84 was reported by New and Kerr (1972). They observed that the control of the disease was achieved when roots of young peach seedlings were dipped in a suspension containing the cells of the antagonist before being transplanted in a field heavily infested by tumorigenic agrobacteria. The antagonistic strain K84 was isolated in Australia from soil collected around a peach gall and was selected from among several nonpathogenic isolates because it was the only one that prevented the crown gall formation when coinoculated with a pathogenic strain in a ratio 1:1 on tomato plant roots.

According to the old classification based on pathogenicity, K84 is a biovar 2 strain belonging to the species *Agrobacterium radiobacter* (Keane et al., 1970; Kersters and De Ley, 1984). A next revision distinguished the *Agrobacterium* species on the basis of physiological and biochemical characteristics, and ascribed the biovar 2 agrobacteria to the species *Agrobacterium rhizogenes* independently from the pathogenic traits (Sawada et al., 1993). In 2001, Young et al., on the basis of 16S rDNA sequence analyses, proposed the inclusion of all *Agrobacterium* species in the genus *Rhizobium*. According to this study, the previous species *Agrobacterium rhizogenes* should be named *Rhizobium rhizogenes* (Young et al., 2001). This classification was recently discussed (Farrand et al., 2003), and retention of the genus *Agrobacterium* was recommended. Since the nomenclature and classification of the genus is still under discussion, the old epithets *A. radiobacter* and *A. tumefaciens* (Kersters and De Ley, 1984) will be adopted in this review to indicate nonpathogenic and pathogenic agrobacteria, respectively.

The fungi of the genus *Trichoderma* are among the most effective biopesticides and are applied against fungal diseases, both in conventional and organic farming. The fungi are commercially produced, and several patents protect their use (Harman et al., 1994, 1996). These fungi are commonly isolated from soil but are also present in other habitats. The genus *Trichoderma* includes more than 30 species, according to the latest nomenclature, with strains that mostly have an asexual life cycle. However, in some cases, the sexual form *Hypocrea* has been found. *Trichoderma* fungi show very high levels of genetic diversity and are able to perform a number of different biological functions. For this reason, they are considered of great interest for studying the biology of fungi and are very useful for application in agriculture and industry. For example, there are strains with a strong antagonistic potential toward phytopathogenic fungi, while others are effective soil colonizers, biodegraders, or producers of important metabolites.

It is known that many *Trichoderma* strains are "rhizosphere competent" and may also promote plant growth and root development. Recent studies about the mechanisms of the *Trichoderma*-plant-other-microbes interaction also pointed out that these fungi may induce systemic resistance in plants, thus increasing their defense responses against different pathogens (Harman, 2000; Harman et al., 2004). These new findings will significantly influence our capacities to manage agricultural ecosystems by providing a better knowledge to improve plant productivity while reducing or eliminating environmental degradation.

AGROBACTERIUM *SPP. AS A BIOCONTROL AGENT*

Biocontrol of plant diseases based on antagonists is a need for a modern and sustainable agriculture, because it reduces the environmental pollution and the toxic effects on humans due to the massive use of pesticides and fertilizers. Apart from these advantages, the use of antagonists may bring some other benefits that pesticides do not have. For instance, use of biocontrol strains can decrease or eliminate the development of resistance of pathogens observed when chemicals are applied, or antagonists that are rhizosphere competent can provide season-long protection against pathogens and improve root functions. However, much of the success of biocontrol depends on the preparation of the appropriate formulations of the antagonist and on the delivery systems for specific crops.

Crown Gall Disease

A. tumefaciens is a soilborne bacterium responsible for the induction of crown gall disease on many crops. More than 600 host species were described for this pathogen, mostly including dycotyledons and some monocotyledons and gymnosperms (De Cleene and DeLey, 1976). The disease is widespread all over the world, mainly in temperate areas (Moore, 1988). Although the disease can damage orchard and landscape planting, it represents a serious economical problem in the nurseries that grow stone fruit trees, cane berries, grapes, roses (Dye et al., 1975), and other ornamentals, since the infected plants must be culled and discarded. Annual cullage is usually less than 5 percent but can reach 100 percent in epidemic years (Moore, 1988).

The tumor-inducing ability of *A. tumefaciens* is due to the presence in the bacterial cell of a 200 kb plasmid, named pTi (tumor inducing plasmid). A specific fragment of this plasmid called T-DNA (transferred DNA) containing all the information needed for tumorigenesis is delivered and integrated into host genome (Chilton et al., 1977). T-DNA genes direct the biosynthesis of auxins and cytokinins responsible for uncontrolled plant cell proliferation and the synthesis of opines, tumor-specific compounds that can be utilized by the inducing *Agrobacterium* strain as carbon and nitrogen sources (Clarke et al., 1992). Opine production by tumor tissue provides the inciting bacterium strain with a selective growth substrate favoring its propagation. Thus, opines are thought to play a major role in the epidemiology of crown gall and the ecology of *Agrobacterium* spp. The kind of opine synthesized and catabolized varies between strains, and that in fact can be described by their opine characteristics (Tempè and Petit, 1982). Opines not only serve as carbon and nitrogen sources for the tumor-inducing bacterium, but a few of the opines can also induce conjugal transfer and exchange of genetic material among agrobacteria (Moore et al., 1997).

Tumors generally develop at the crown or on the roots of the host, but in some instances, the bacterium is able to induce tissue proliferation on epigeous parts of the plants (Miller, 1975; Bouzar et al., 1995; Burr et al., 1998). Endophytic and systemic behavior of agrobacteria has been observed in several hosts, and it is a key factor in the epidemiology of the disease (Burr et al., 1998; Martì et al., 1999; Zoina et al., 2001). This behavior has great relevance for crown gall disease management, since the pathogen may be transmitted through plant vegetative propagation and micropropagation systems (Cooke et al., 1992; Poppenberger et al., 2002).

Disease Control by Conventional Methods

The use of physical and chemical methods for controlling crown gall has been attempted, but only incomplete control has been achieved. Heat therapy and hot water treatments reduced disease incidence on *Prunus* spp. seedlings and grape cuttings, respectively (Moore and Allen, 1986; Ophel et al., 1990). Chemical control of crown gall on apple rootstocks by treatments with Terramycin and Copac E partially controlled the disease, but phytotoxic effects were also observed (Canfield et al., 1992).

Soil solarization reduced the level of agrobacteria population in sandy soil but was less effective in silty clay soil (Raio et al., 1997). Moreover, a residual population of agrobacteria was detected in the rhizosphere of weed plants growing at the edge of solarized plots, showing that weeds may represent a dangerous reservoir of the pathogen in the soil (Raio et al., 1997).

Biological Control

So far, the most effective method for preventing crown gall has been the use of the biocontrol strain K84. It has demonstrated remarkable and widespread success in controlling the disease in different hosts and countries (Du Plessis et al., 1985; Lopez et al., 1987; Psallidas, 1988; Bouzar et al., 1991; Serfontein and Staphorst, 1994; Moore and Canfield, 1996; Zoina et al., 2003). The main mechanism involved in biological control is the production of agrocin 84, an antibiotic highly specific against agrobacteria (Kerr and Htay, 1974). This agrocin is a di-substituted adenine nucleoside analog which is active against agrocinopine-type pathogenic *Agrobacterium,* in which it inhibits DNA replication (Roberts et al., 1977; Kerr and Tate, 1984). Sensitivity to agrocin 84 is related to agrocinopine uptake, and there is evidence that both molecules are transported by the same Ti plasmid-encoded permease (Murphy and Roberts, 1979; Ryder et al., 1987). Nopaline-type pathogenic agrobacteria that are agrocin 84–sensitive are able to synthesize both nopaline and agrocinopine A. The antagonist has genes for the catabolism of both opines and therefore competes with the pathogenic strains for these substrates; in addition, it produces agrocin 84 which is actively transported into the pathogens via the permease designed for the uptake of agrocinopine A, determining cell death (Kerr, 1989). Synthesis of agrocin 84 is coded by genes harbored on a 47.7 kb plasmid (pAgK84), which also carries genes that confer to the antagonist the agrocin 84 immunity and conjugal transfer capacity (Farrand et al., 1985; Ryder et al., 1987). The genes involved in biosynthesis of agrocin 84 have been characterized and localized to a 21 Kb segment of pAgK84 (Wang et al., 1994). Even though antibiosis by agrocin 84 plays a leading role in the control of sensitive pathogenic strains, several experiments have demonstrated that strains of *A. tumefaciens* resistant to agrocin 84 can be partially controlled by the antagonist (Cooksey and Moore, 1982; Du Plessis et al., 1985; van Zyl et al., 1986; Lopez et al., 1989; Vicedo et al., 1993). This finding shows that biocontrol by strain K84 is a complex phenomenon including different mechanisms. The antagonist is strongly competitive for the colonization of the rhizosphere of host and nonhost plants (Macrae et al., 1988; Vicedo et al., 1993); a study performed in Oregon showed that K84 is able to persist for up to two years in a field environment as a rhizosphere inhabitant or in association with crown gall tissue (Stockwell et al., 1993). Field studies performed by Lopez et al. (1989) showed that both strain K84 and a mutant unable to produce agrocin 84 were able to reduce infection by some pathogenic agrobacteria that were insensitive in vitro to agrocin 84. This behavior could in part be due to the high efficiency of the antagonist for root colonization ability and competition for occupation of the infection sites, but the involvement of other factors as production of agrocins other than agrocin 84 was also hypothesized (Farrand and Wang, 1992). Later, it was found that K84 also produces agrocin 434 (Donner et al., 1993). Synthesis of agrocin 434 is coded by genes harbored by the 300 to 400 kb cryptic plasmid (pAgK434 or pAtK84a) of strain K84. This compound is a di-substituted cytidine nucleoside and is active against biovar 2 agrobacteria (Donner et al., 1993; McClure et al., 1998). Penyalver et al. (1994) showed that K84 also produces a third antibiotic-like substance named ALS84 that in vitro inhibited tumorigenic agrobacteria and a range of phytopathogenic bacteria. The inhibitory activity of ALS84 is correlated with the production of siderophores by K84 under iron-limiting conditions (Penyalver et al., 2001). So far, the genetic bases related to the ALS84 production and its role in the biocontrol activity of K84 have not yet been investigated.

Strain K84 harbors a third plasmid of 173 kb in size (pNoc or pAtK84b) that is thought to be a derivative of a nopaline-type Ti plasmid disarmed in the oncogenic T-DNA and Vir regions but retaining genes involved in catabolism of nopaline (Clare et al., 1990). In a field experiment, a derivative of strain K84 harboring only pAtK84b partially inhibits gall formation on almond seedlings inoculated with a tumorigenic *Agrobacterium* strain (McClure et al., 1998), evidencing the involvement of pAtK84b in K84 biocontrol activity. Probably pTi genes that are important in the early stages of the pathogenic process (chemotaxis, attachment) are carried on pAtK84b and may enhance the potential for competition of K84 at the wound site of infection.

It is evident that biological control of crown gall by strain K84 is a complex process including several factors acting synergistically and playing roles in the inhibition process. The relative contribution of each factor will depend on the kind of pathogen present, the ratio of pathogens to biocontrol strains, the host plant, and the method of application of the biocontrol agent (McClure et al., 1998).

Strain K84 has been used commercially for the control of crown gall in Australia since 1973, and it is now used in many other countries mostly for protecting stone fruit trees. Application of K84 is easy to perform, and no risks for human health have been reported (Moore, 1988). Treatment consists of preparing a suspension of strain K84 (about 10^7 cells/ml) and dipping the plant material (seeds, cuttings, or roots of young plants) into it before transplanting in soil (Kerr, 1989).

Failure of Biocontrol by K84 Strain

Since *A. tumefaciens* can be transmitted through plant vegetative propagation and micropropagation systems (Cooke et al., 1992; Poppenberger et al., 2002), the endophytic and systemic behavior of agrobacteria has great relevance in plant disease management. So far effective methods for the detection of *A. tumefaciens* in asymptomatic infected plants are not available, and since the antagonist is effective only to prevent pathogen infection, one cause of failure of the application of the biological control method is the use of plant propagation material latently infected (Cubero et al., 1999).

The reported failures of K84 as a biocontrol agent have been mainly due to the occurrence of indigenous populations of agrocin 84–insensitive strains of pathogenic agrobacteria (Table 7.1). This is the case of *Agrobac-*

TABLE 7.1. Biocontrol efficacy of strain K84 on different crops.

Host	K84 efficacy	Country	Source
Almond	High	California (USA); Spain	Schroth and Moller, 1976; Lopez et al., 1987
Apple	Null	Hungary; Italy	Grimm and Sule, 1981; Bazzi et al., 1980
Apricot	High	California (USA); Hungary	Schroth and Moller, 1976; Grimm and Sule, 1981
Cherry	High	Hungary; Italy, Oregon (USA); Poland	Grimm and Sule, 1981; Bazzi and Mazzucchi, 1978; Moore and Canfield, 1996; Sobiczewski and Piotrowski, 1983
Chrysanthemum	Null	Italy	Bazzi and Rosciglione, 1982; Bush and Pueppke, 1991
Euonymous	Variable	Oregon (USA)	Moore and Canfield, 1996
Ficus benjamina	Variable	Italy	Zoina et al., 2001
Grape	Null	Italy; Greece	Bazzi and Burr, 1986; Panagopoulos et al., 1979
Hop	High	South Africa	Serfontein and Staphorst, 1994
Peach	High	Algeria; California (USA); Canada, Greece; Hungary; Italy; Spain	Bouzar et al., 1991; Schroth and Moller, 1976; Dhanvantari, 1976; Psallidas, 1988; Grimm and Sule, 1981; Zoina et al., 1994; Lopez et al., 1987
Pear	Null	Hungary	Grimm and Sule, 1981
Plum	High	California (USA); Italy; Spain	Schroth and Moller, 1976; Bazzi and Mazzucchi, 1978; Lopez et al., 1987
Raspberry	High	Hungary	Sule, 1978
Rose	High	New Zealand; Spain	Dye et al., 1975; Lopez et al., 1981
Walnut	Variable	Oregon (USA)	Moore and Canfield, 1996
Willow	High	New Zealand	Spiers, 1980

terium strains infecting *Crysanthemum* spp., *Ficus benjamina, Rubus* spp., and *Vitis* spp. that harbor Ti plasmids other than the nopaline type (Perry and Kado, 1982; Vadequin-Dransart et al., 1995; Burr et al., 1998).

A real threat to biological control of crown gall by the K84 strain is the selection of pathogenic recombinants resistant to agrocin 84. This event was observed for the first time in a field experiment performed in Greece (Panagopoulos et al., 1979), where almond plants were inoculated with an agrocine 84–sensitive *A. tumefaciens* strain and protected with K84. Although significant reduction of disease was achieved, a conspicuous number of plants developed tumors on roots. Since *A. tumefaciens* isolate producers of agrocin 84 were reisolated from those tumors, it was hypothesized that the incomplete control was due to the transfer of agrocin-controlling genes of strain K84 to the tumorigenic strain (Panagopoulos et al., 1979), which also became resistant to agrocin 84. The occurrence of this transfer was probably due to the high densities of virulent bacteria used for plant inoculation that were not completely controlled by the antagonist and that induced gall formation. The nopaline synthesized in tumors induced bacterial conjugation and the agrocin-controlling genes transfer (Ellis and Kerr, 1979).

This event could happen also as a consequence of improper application of the biological control agent. For example, if K84 is applied at populations too low to control disease, then agrocin 84–sensitive isolates can infect the plant and develop galls. K84 could subsequently colonize the developing galls, and plasmid transfer could occur in the gall tissue (Stockwell et al., 1996).

Transconjugants can originate from different plasmid exchanges (Figure 7.1); however, the most dangerous are those harboring both the oncogene traits of the Ti plasmid and the genes for agrocine production and immunity. These recombinants are most frequently originated by pAgK84 transfer to pathogenic strains and are no longer subject to biological control. After the first report by Panagopoulos et al. (1979), the appearance of dangerous transconjugants agrobacteria has been observed in experimental fields in Spain and the United States (Vicedo et al., 1993; Stockwell et al., 1996), under conditions very similar to those of natural infection in K84-treated plants in the nursery. These experiments indicated that the selection of recombinant populations of agrobacteria following the use of the biological control method may represent a real threat to the sustained efficacy of K84 as a biocontrol agent.

Dangerous transconjugants, originated by the transfer of pAgK84 to tumorigenic agrobacteria, were recovered from galls developing on cherry and raspberry K84-treated plants grown in commercial nurseries in Oregon and California (Lu, 1994). Transconjugants were also obtained from galled peach rootstocks collected from an Italian nursery (Peluso et al., 2001).

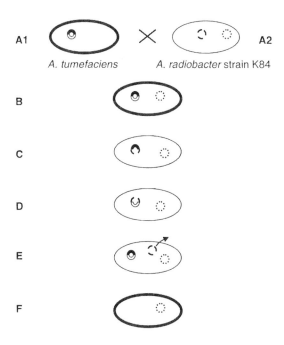

FIGURE 7.1. Plasmid exchanges between the antagonistic strain K84 and *A. tumefaciens* strains. *(A1) A. tumefaciens* cell with the pTi. The black part of the pTi represents the T-DNA. Chromosome is not represented. *(A2)* K84 cell with the pAtK84b (broken line) and pAgK84 (dotted line). Chromosome and pAgK434 are not represented. *(B)* Transfer of pAgK84 from K84 to *A. tumefaciens. (C)* Transfer of pTi from *A. tumefaciens* to K84 and recombination with pAtK84b. The recombinant plasmid harbors the oncogenic traits (T-DNA). *(D)* Transfer of pTi from *A. tumefaciens* to K84 and recombination with pAtK84b. The recombinant plasmid does not harbor the oncogenic traits (T-DNA). *(E)* Transfer of pTi from *A. tumefaciens* to K84 and loss of the pAtK84b. *(F)* Transfer of pAgK84 to avirulent *A. tumefaciens* strains. Transconjugants B, C, and E combine pathogenicity with immunity to agrocin 84. Such strains are no longer subject to biological control by strain K84. *Source:* Modified from Kerr, 1989.

Field biocontrol trials showed that these transconjugants were able to induce high disease incidence on peach rootstocks and were not controlled by treating the plants with K84 (Raio et al., 2003).

Both in the American and Italian nurseries, a breakdown in the control of crown gall was reported (Lu, 1994; Peluso et al., 2001), showing in a definitive way that the commercial use of K84 may represent a serious risk.

The tra⁻ Deletion Mutant Strain K1026

In order to reduce the risk of acquisition of pAgK84 by *A. tumefaciens,* a derivative of strain K84, unable to transfer its a grocin plasmid, has been developed by using recombinant DNA techniques (Jones et al., 1988). Transfer of pAgK84 is controlled by a defined region *(tra)* located on the same plasmid. In order to obtain the *tra*⁻ mutant of K84, a map of the pAgK84 plasmid was constructed (Kerr, 1989). Fragment B1 in the outer circle of the map covers most of the *tra* region. It was cloned, then cut with restriction enzyme *Eco* RI (inner circle) and recloned. One clone was identified which lacked *Eco* RI fragments D1 and H. After several more manipulations, this deleted fragment replaced the intact fragment in pAgK84 in strain K84. The new strain was called K1026. It has the same chromosomal background and plasmid complement as strain K84 from which it differs only for a 5.9 kb deletion covering the *tra* region of pAgK84 (Jones et al., 1988; Kerr, 1989).

The in vivo efficacy of strain K1026 was the same as K84 in different experimental conditions, hosts, and countries (Penyalver et al., 2000). Both strains were able to control disease induction by both agrocin 84–sensitive and resistant pathogens, showing that strain K1026 was also able to use other mechanisms not related to agrocin 84 susceptibility in the same way as strain K84 (Penyalver et al., 2000). In a comparative study, pAgK84 transfer from K84 to the recipient pathogen was detected, but no plasmid transfer was evidenced when the K1026 strain was used as a biocontrol agent (Vicedo et al., 1993). The same study showed that root colonization ability and persistence in the rhizosphere of K1026 strain were identical to K84.

Strain K1026 was the first genetically engineered organism to be released for testing in Australia (Kerr, 1989) and has been patented by Luminis Pty. Ltd., Adelaide University's intellectual property company. It is commercialized by Bio-Care Technology Pty. Ltd., Sommersby, Australia, which successfully applied for the registration of K1026 as a pesticide. K1026 is sold as peat cultures under the trade name "No Gall" for use on stone fruits, almond, pecan, walnut, and rose (Kerr, 1989).

Strain K1026 is a genetically engineered microorganism, and it is harmless to plants, animals, and humans, as demonstrated by the following remarks: (1) strain K84, the progenitor of K1026, has been registered as a biopesticide and been used commercially in several countries for many years where there have been no reports of harm; (2) strain K1026 is identical to K84, except that it lacks a 5.9 kb portion of the agrocin 84 plasmid, therefore preventing pAgK84 transfer; (3) no foreign DNA remains in

K1026; (4) K1026 cannot grow at 37°C (human body temperature); and (5) agrocin 84 is specific for agrocinopine-catabolizing agrobacteria, most of which are plant pathogens, while other organisms are unaffected (Kerr, 1989).

The use of *A. radiobacter* K1026 should prevent the risk of pAgK84 transfer into pathogenic *Agrobacterium* strains and safeguard the efficacy of *A. radiobacter* for biological control of crown gall.

Is the K1026 Strain Really Safe?

The presence of pAtK84b in K84 and K1026 strains has been thought to safeguard biological control by preventing acquisition of Ti plasmid because incompatible plasmids are unable to replicate in the same cell (Kerr and Ellis, 1982). However, some experimental data show that such protection may not be as effective as once thought. The transfer of a transposon-tagged Ti plasmid from *A. tumefaciens* strain to *A. radiobacter* strain K84 was detected after coinoculation of the two bacteria on tomato plants (Stockwell et al., 1990). Furthermore, a spontaneous transconjugant strain, originated by the conjugal transfer of Ti plasmid from a tumorigenic strain to K84, was detected in a tumor from a K84-treated plant grown in soil inoculated with a nopaline strain of *A. tumefaciens* sensitive to agrocin 84 (Vicedo et al., 1996). Recombination between pAtK84b of K84 and nopaline Ti plasmid of *A. tumefaciens* took place, originating a new Ti plasmid, and K84 was converted to a virulent form (Vicedo et al., 1996). This transconjugant showed the same host range and tumor-inducing ability of the pathogenic-inoculated strains was not controlled by K84, and survived in the root system at larger population densities than the wild-type Ti plasmid donor strain (Lopez-Lopez et al., 1999). This is the only reported case regarding the appearance of recombinant strains harboring Ti plasmid associated with chromosomal and plasmid backgrounds of K84 under field conditions. The frequency at which Ti plasmid is transferred by conjugation to strain K84 and the consequences on biocontrol efficacy are not yet known. However, experimental data collected over the years show that it seems to be a rare event. The occurrence of this type of recombination may represent a risk to the biocontrol of crown gall using strain K1026. Available data from several biocontrol experiments show that so far the transfer of Ti plasmid to K1026 has never been detected (Lopez-Lopez et al., 1999); the use of this biocontrol strain should be safe and should be promoted wherever strain K84 is used to control crown gall.

TRICHODERMA *SPP. AS A BIOCONTROL AGENT*

Plant diseases due to infections of pathogenic fungi represent a limiting factor for the cultivation and conservation of plant crops of major interest. The consequences of parasite attacks are a quantitative and qualitative reduction of production, and represent a source of risk for human and animal health because of the accumulation of cancerogenic mycotoxins in contaminated food. The chemical control on phytopathogenic fungi is limited not only by the economical cost but also by concerns on its impact on consumer health and the environment. As an alternative, new technologies for plant defense can be set up based on the utilization of transgenic plants and microorganisms or the application of antagonist microbes (fungi or bacteria) capable of controlling important diseases. Biopesticides based on antagonistic fungi are produced and commercialized for the biocontrol of phytopathogenic fungi causing root and crown rot of vegetables, seedlings, damping-off, vascular diseases, take-all of cereals, root infection of fruit trees, mold of grapes and other crops, and monilia disease and soft rot of pome and stone fruit (Harman and Björkman, 1998). The size of the market for biopesticides based on antagonistic fungi is modest compared with chemicals, but it is steadily growing and is expected to increase significantly with the introduction of these products in India, China, Africa, and Central Asia. In addition to the registered products, a large variety of bioprotectants, biostimulators, and biofertilizers nonregistered and based on *Trichoderma* or other antagonistic fungi are being sold everywhere. For instance, commercial products based on the strains of *Trichoderma* spp. include Binab T (*T. harzianum* and *T. polysporum* registered in Sweden, the United Kingdom, and the United States), Bio Fungus (*Trichoderma* spp. in Belgium), EcoSOM (*T. viride* in India), RootShield T-22G, T-22 Planter Box, Bio-Trek (*T. harzianum* in the United States and northern Europe), Root Pro (*T. harzianum* in Israel), Supresivit (*T. harzianum* in the Czech Republic), Trichodex (*T. harzianum* in Italy, Israel, and the United States), Trieco (*T. viride* in India), Trichoderma 2000 (*Trichoderma* spp. in Israel), Trichopel Trichojet, Trichodowels, and Trichoseal (*T. harzianum* and *T. viride* in New Zealand). The availability and diffusion of these products is much wider than people normally think, also because of the need to reduce the application of chemical pesticides, or to obtain the certification of organic farming. In fact, *Trichoderma* is listed, both in Europe and the United States, as an active principle allowed in organic farming. Regardless of the obvious potential, some problems limit the development and application of these biopesticides. In addition to the lack of very effective and correctly formulated strains, often there is a limited availability of (1) basic information needed for product registration, including a sufficient knowledge of the mechanisms of action; (2) enough efficacy tests for the geographic areas in the countries where the product has to be registered; and (3) methods for monitoring the production of eventual toxic metabolites produced by these fungi, allowing an evaluation of possible risks derived from a large-scale application. A deep understanding of the mechanism of action of the *Trichoderma* spp. today in use is clearly required to expand the application of these biopesticides, and it will help in the selection of strains more adapted to the various crop conditions. However, the mechanisms of the *Trichoderma*-plant-other-microbes interaction are very complex and include mycoparasitism (necrotrophic direct fungus-fungus interaction), competition for nutrients, antibiosis, inhibition of pathogen enzymes, antagonism against nematodes, colonization of rhizosphere, phyllosphere, and spermosphere, promotion of plant growth and root development, and induction of systemic resistance against different pathogens (Harman et al., 2004). Several cell wall–degrading enzymes and antibiotics have been included among the factors that play a main role in this complex biological process (Harman and Kubicek, 1998). Some chitinases and glucanases have been found to be critical for biocontrol (Limón et al., 1996; Lorito, 1998; Woo et al., 1999) and are being applied to develop new defense strategies (transgenic and not) against fungal phytopathogens (Lorito et al., 1996, 2001; Esposito et al., 2000). The use of mixtures of these enzymes (chitinases and glucanases) capable of degrading the fungal cell walls appears especially promising, since they are active on a wide spectrum of fungi (no resistant phytopathogenic fungi have been found so far), produced in large amount by *Trichoderma* spp., stable at room temperature, capable of reaching efficacy levels similar to that of chemical fungicides (they are considered the most potent antifungal enzymes known), and capable of acting even if applied ectopically to crops in greenhouse or postharvest (Harman et al., 1993; Lorito, Hayes et al., 1994; Lorito, Peterbauerm et al., 1944, 1996). Moreover, these enzymes are strongly synergistic with synthetic fungicides of different classes, and in particular, azole and other compounds that affect cell membranes (Lorito, Hayes, et al., 1994; Lorito, Peterbauer, et al., 1994; Lorito et al., 1996). However, there are still many aspects of molecular interactions that these fungi establish with plants and other microbes that are obscure. This is not only because of the complexity of the processes involved but also because some of the most powerful and recent methods of investigation based on molecular genetics and biochemistry techniques (that is, EST sequencing, application of DNA array and miniarray, and proteome analysis) have not yet been applied to study the mechanisms of biocontrol.

Trichoderma *in Biocontrol*

Trichoderma strains can be useful in biocontrol against a wide range of pathogens including *S. rolfsii, Pythium* spp., *R. solani, B. cinerea, Armillaria mellea, Heterobasidium annosum,* and *Phytophthora* spp. (Cook and Baker, 1983; Gullino and Garibaldi, 1988; Harman et al., 1989; Smith et al., 1990). Some strains have been selected for commercial use because of their remarkable characteristics for biocontrol activity. Strain T-22, for example, has a very high rhizosphere competence, is active against many pathogens, can be applied to a large number of crops, and improves average yields of crops, besides other things. Application of antagonists can be carried out in different ways. Once large amounts of cells, mycelium, and spores of the antagonist have been produced by fermentation, the biomass is processed and incorporated into different substrates to prepare formulations in granules, powders, pellets, liquids, and so forth. Many studies have been done to explore the efficacy of different *Trichoderma* formulations and delivery systems. To control pathogens in the soil, for example, *Trichoderma* has been added to it as a mixture of bran and sand containing the germling antagonist (Lewis and Papavizas, 1984), or in the form of alginate pellets (Knudsen et al., 1991). *T. harzianum* has been applied also directly on seeds as gels by the use of fluid drilling (Fisher et al., 1983; Conway, 1986), while *T. virens* has been applied to cotton seeds using a latex sticker to give protection against *Pythium* damping-off (Howell, 1991). Seed-coating systems have also been attempted. A liquid-coating formulation consisting of a mixture containing *T. harzianum,* combined with a suspension of aqueous binder and solid particulates, has been applied by spraying onto cucumber seeds to control *Pythium* damping-off (Taylor et al., 1991). It appears very useful for the success of biocontrol to include in the coating mixture substances that may favor the growth of the antagonist. The inclusion of some polysaccharides and alcohols to the coating mixture of *T. harzianum* used onto pea seeds significantly increased the biocontrol activity against *Pythium* damping-off (Nelson et al., 1988).

Moreover, a strong support to the use of *Trichoderma*-based biofungicides comes from the possibility of integrating their application with other control methods, thus reducing the impact on the environment. For instance, the association of these biopesticides with alternative methods of soil sterilization is very appealing, given the sometime limited level of efficacy of these techniques applied singly as compared with chemicals. The fact that the application of methyl bromide, the most used soil fumigant and also the most effective at reducing the ozone level in the atmosphere, will be completely banned by the year 2005 (if this deadline is maintained) has provided a strong impulse to the development of "biological" methods based on soil solarization and antagonistic microbes (Katan, 2000). The efficacy of soil solarization is related to different heat sensitivity of soilborne agents and to different increase of temperature levels in soils. The effects of solarization can be enhanced by adding organic substances to the soil during the process. The combination of organic amendments and solarization offers a different approach and additional options for pest management in soil, but it is closely related to the disposable organic amendments. Recently, the role of a greenhouse cover film in ethylene tetrafluoroethylene (ETFE) in improving soil thermal regimes has been investigated (Cascone et al., 2001; Polizzi et al., 2001). The results showed that the ETFE film induces a greater greenhouse effect and determines higher soil thermal regimes than those monitored under ethylene vinyl-acetate (EVA) greenhouse cover and reported in literature. The high increase of temperature (about 5 to 6°C at 30 cm soil depth) should be repeatable in the greenhouse soil by using an ETFE film as greenhouse cover, improving the efficacy of soil solarization. The use of this film can reduce the period of soil solarization or can extend the treatment in country with lower solar radiation intensity. In another work, the effectiveness of soil solarization with different plastic films as a control of corky root and root-knot on tomato plants was tested (Cascone et al., 2001). The results showed the possibility of the use of a colored film (green coextruded), which can also be left in place after transplanting to control weeds and reduce reinfection risks. In addition to these improvements applicable to solarization, the integration of this technique with the use of microbe-based formulations, and in particular *Trichoderma* spp., should overcome the downside of both methods. In fact, *Trichoderma* spp. are among the most resistant fungi to high temperatures and are able to rapidly colonize sterilized soils. It is, therefore, conceivable that the wide-spectrum efficacy of solarization integrated with the easiness of application of formulated biofungicide should provide a protection comparable to the one obtained with chemical pesticides.

Antagonistic Mechanisms of Action of Trichoderma spp. in Biocontrol

Since Weindling (1932, 1934, 1941; Weindling and Fawcett, 1936) first discovered the biocontrol capacities of *Trichoderma*, a tremendous amount of work has been carried out in order to understand the mechanisms of action of the strains belonging to this genus, and to evaluate their possible application in agriculture. It is known that *Trichoderma*, as well as other antagonistic fungi, are able to attack phytopathogenic fungi in several ways (Harman, 2000). Production of antibiotic compounds, competition for nutrients and space, mycoparasitism, or some combinations of these have been reported as effective mechanisms of antagonism in *Trichoderma*. Antibiotics, alone or in combination with other substances, may inhibit growth of microorganisms. The evidence for the involvement of antibiotics in biocontrol has been pursued in different ways (Weller and Thomashow, 1993). Isolation of antibiotics and identification of their coding genes, together with the use antibiotic- deficient mutants obtained, for example, through site-directed mutagenesis, provided convincing evidence for the role of these substances in biocontrol. *Trichoderma* spp. produce a number of antibiotic compounds, including pyrones, pyriodones, butenolides, antraquinones, and piperazines. Gliotoxin and gliovirin, two derivatives of piperazine produced by *T. virens,* seem to play an important role in the control of *Pythium* and *Rhizoctonia,* causal agents of root rots (Wilhite et al., 1994). These substances probably act in synergy with cell wall–degrading enymes (chitinases, β-1,3-glucanases, etc.) or the other lytic enzymes (proteases and lipases), also produced by the antagonist during biocontrol. Similarly, peptaibol antibiotics and hydrolytic enzymes produced in parallel by *T. harzianum* are synergically involved in the mechanism of action of this fungus against phytopathogenic fungi (Schirmböck et al., 1994). Competition for space or specific infection sites, nutrients, and other factors necessary for growth in the rhizosphere may also be the way through which the antagonist controls plant pathogens. Competition for space has not been clearly demonstrated. In the case of *T. harzianum,* there is evidence showing that the antagonist plays a role in biocontrol when it establishes its dominance in a specific environmental niche. *T. harzianum* is able to control the attack of *Botrytis cinerea* on grapes by colonizing blossom tissue and excluding the pathogen from its infection site (Gullino, 1992). The mechanisms of the competition for nutrients like carbon, nitrogen and iron, root, and seed exudates have been better defined. Competition for carbon and nitrogen between *T. harzianum* and *F. oxysporum* f. sp. *melonis* on rhizosphere colonization has been identified as the most important mechanism involved in the biocontrol of this vascular disease. It was demonstrated that all the other possible modes of action played little or no role in this case (Sivan and Chet, 1989). Mechanisms by which antagonists parasitize other fungi have been thoroughly studied by using *Trichoderma* spp. Many studies report the presence of propagules of the antagonist in fungal structures of the pathogen. *Trichoderma* hyphae have been isolated from sclerotia of *Rhizoctonia* spp., *Sclerotinia* spp., *Sclerotium* spp., and so forth (Gladders and Coley-Smith, 1980; Artigues and Davet, 1984; Zazzerini and Tosi, 1985; Howell, 1987). The mechanisms of mycoparasitism are complex and include several steps. The antagonist constitutively produces small amounts of the cell wall–degrading enzyme exochitinase that allows the release of oligosaccharides from the host fungus cell wall. These molecules trigger in

Trichoderma the induction of endochitinases, which are very powerful degraders of the cell wall of the host fungus before the antagonist and pathogen come into contact. When the contact occurs, *Trichoderma* hyphae grow over or along the host hyphae forming appressoria, through which the lumen of the host is directly reached. Recognition probably is due to lectin binding present on the hyphae surface. Successively, *Trichoderma* produce more fungitoxic cell wall–degrading enzymes, and probably also peptaibol antibiotics: these substances act synergistically to complete the degradation of the host cell wall (Harman et al., 2004).

Induced Resistance in Plants

Antagonists do not perform only by attacking the pathogens, as previously described, but they can also control diseases by interacting directly with plants. Recently, a large body of evidence showing the induction of localized and systemic resistance in plants treated with fungal antagonists, including *Trichoderma* spp., has been obtained. These defense responses of plants to pathogen attacks or other factors (for instance, physical or insect-caused injuries, and application of chemicals) are very complex and involve signal molecules such as salicylic acid, jasmonic acid, ethylene, and nitrous oxide. Signals trigger a cascade of reactions, leading to the production of metabolites (for example, phytoalexins) and proteins (pathogenesis-related proteins) that support plant disease resistance. More recently, the mechanisms of a type of induced resistance phenotypically similar to induced systemic resistance (ISR) activated by the action of rhizobacteria has been partially clarified (Pieterse and van Loon, 1999). In this case, however, PR proteins are not expressed, and plants react more strongly to the attack of pathogens, because of an increased sensitivity to jasmonic acid and ethylene (Kuc, 2001). Some fungi, like *Fusarium, Rhizoctonia,* and symbiontic mycorrhizal fungi as well as *Trichoderma* spp., also have the ability to promote the enhancement of plant growth and induce resistance to pathogens. In the case of *Trichoderma* spp., one of the first reports published by De Meyer et al. (1998) observed an induced systemic resistance in *T. harzianum* biocontrol of *B. cinerea.* Since then, a number of papers reported the induction of systemic resistance in a variety of plants by the application of different *Trichoderma* species (Harman et al., 2004). The ability of these fungi to stimulate in host plant defense responses has been definitely established: in cotton with some *T. virens* strains (Howell et al., 2000), in cucumber with some *T. harzianum* strains, in pepper (Sid Ahmed et al., 2000), and in tobacco (Chang et al., 1997) against various pathogens as well as *R. solani* and some *Phytophthora* spp. In these cases, the treatment with the biocontrol agent causes an increase or an exnovo synthesis of substances involved in active defense mechanisms as, for example, terpenoids (Howell et al., 2000), capsidiol (Sid Ahmed et al., 2000), and the increase of peroxidases, beta-glucosidases, cellulases, and chitinases activities (Yedidia et al., 2000). In grapevine cell culture, an elicitor produced by *T. viride* was able to activate defense responses, that is, cellular necroses and accumulation of resveratrol, one of the most important phytoalexins produced in response to infections by *B. cinerea* and *Plasmopara viticola* (Calderon et al., 1993; Dai et al., 1995). In case of soilborne diseases, to discriminate if the control is due to the effects of localized or systemic acquired resistances versus direct action of *Trichoderma* on the pathogen required the use of mutants. It has been shown that by using *T. virens* mutants with different antibiotic or mycoparasitic capabilities, biocontrol of *R. solani* on cotton seedlings was mostly depending on the ability of the strain to induce in the plant the production of phytoalexin such as hemoglossypol, deoxyhemoglossypol, and gossypol (Howell et al., 2000). This result is of great interest, because it provides a new perspective of the mechanisms involved in biocontrol, and suggests that induced resistance in some cases may play a role more important than the direct action of the antagonist on pathogens (Harman et al., 2004). In addition, the induced resistance may confer protection for a long time and against a variety of pathogens, including viruses (Harman et al., 2004). For instance, the application of *T. harzianum* about three months earlier to tomato roots reduced significantly the infection on leaves caused by *Alternaria solani*.

The resistance in plants induced by *Trichoderma* strains may be caused by different molecules. These fungi produce an enzyme xylanase of 22 kDa, that is able to induce a localized reaction of resistance (Lotan and Fluhr, 1990). Several peptides and proteins produced by *T. virens* have been found to have elicitor activity and to induce the biosynthesis of terpenoid phytoalexin and peroxidase activity in cotton (Howell et al., 2000). Unpublished results show that *T. harzianum* produces proteins that are homologous to the products of the avirulence genes *Avr4* and *Avr9* found in the tomato pathogen *Cladosporium fulvum*. These proteins are race- or pathovar-specific elicitors that induce resistance responses in plants carrying the matching resistance genes (Kloepper et al., 1993). Another class of molecules able to induce resistance may be produced by the interaction that *Trichoderma* has with pathogens and plants. The action of cell wall–degrading enzymes, both on the cell walls of fungal pathogens and plants, releases oligosaccharides that may trigger both the mechanisms of antagonism in *Trichoderma* and defense responses in plants. Rhizosphere-competent strains colonize plant roots and are able to penetrate the root epidermis, causing a change in the plant metabolism (Harman et al., 2004). An increase in the levels of the plant enzymes involved in the defense mechanisms, such as peroxidases, chitinases, and β-1,3 glucanases, has been observed (Harman et al., 2004). Production and accumulation of phytoalexins have also been reported. Root colonization of cucumber induced an increase accumulation in leaves of phenolic compounds that are strongly inhibitory toward *Pseudomonas syringae* pv. *lachrymans, A. tumefaciens, F. oxysporum* f. sp. *melonis, B. cinerea,* and other fungi (Harman et al., 2004).

Different techniques of genetic engineering have been set up for *Trichoderma* spp. (homologous recombination, use of reporter genes, secretion of heterologous proteins) (Lorito et al., 1996; Woo et al., 1999; Kullnig et al., 2000). Nevertheless, many of the molecular factors that regulate the interaction between the antagonist, plant, and/or pathogen are still largely unknown. On the other hand, the mechanisms involved in the plant–pathogen interaction are better understood. In fact, different genes and factors involved in the microorganism avirulence and pathogenicity, or in the plant resistance process have been identified. Cross-talk between pathogen elicitors and plant receptors is very well studied (Dangl and Jones, 2001). It is clear that to identify the genetic and molecular factors that regulate the plant–pathogen–antagonist interaction a holistic approach is needed, aiming to study the changes of gene expression occurring in all the partners as consequence of this interaction. This strategy is close to the natural crop conditions, where plants are exposed at the same time to pathogens and antagonists, and the development of the disease is the result of such a complex, multifarious interaction. This approach can take advantage of the information already available on the mechanisms of plant–pathogen (Del Sorbo et al., 2000; Dangl and Jones, 2001) and plant–antagonist interactions, that is, of the role of lytic enzymes and antibiotics in biocontrol (Lorito and Woo, 1998) in developing new genetic and molecular procedures not yet applied to study this multicomponent interaction. In particular, postgenomic analysis with the developing of EST sequences, gene libraries, arrays, proteomic maps, and transcriptomic analysis will be particularly useful to identify and characterize new genes, proteins, and signal molecules involved in the interaction, which can be readily utilized for plant protection strategies (Lorito et al., 1998).

Plant Growth Improvement

The beneficial effects of some microorganisms that colonize roots on growth and yields of plants are known, as in the case of rhizobacteria. Rhizosphere-competent strains of *Trichoderma* are able to promote both root development and growth in a wide range of plants (Harman, 2000). The effects of strain T-22 of *Trichoderma* have been studied, especially on maize plants. Seeds treated with the antagonist produced plants with a much deeper rooting, compared with the untreated plants, and also more tolerant to drought (Harman and Björkman, 1998). In addition, it has been shown that this effect becomes even stronger when other root-colonizing microorganisms, like mycorrhyzal fungi, are applied. A better development of root caused by the presence of the antagonists may also bring about a significant increase in yield. In trials carried out on maize, treated plants showed an average yield increase of about 5 percent (Harman and Björkman, 1998). Thus, *Trichoderma* can be useful in this perspective, especially in fields where crops, not being cultivated in optimal agricultural regimes, have wide margins of improving their yields. However, it should be underlined that with some lines of maize, *Trichoderma* caused a diminution both in plant growth and yield in comparison to untreated plants (Harman and Björkman, 1998). These results suggest that more studies are needed to clarify all the mechanisms involved in the plant–antagonist interactions in order to identify the most useful strains that can symbiotically behave in specific ecosystems. The mechanisms through which *Trichoderma* increases root and shoot growth and yields may depend on several physiological interactions. *Tricoderma* strains, for instance, are able to degrade HCN present in soil and produced by some deleterious rhizobacteria (Lynch, 1996). *Trichoderma* also produces a diffusible metabolite that has plant hormone characteristics (Windham et al., 1986). Further, this fungus is able to solubilize minor nutrients like Mn^{4+}, Cu^{2+}, and Fe^{3+} by reducing them to the soluble forms Mn^{2+}, Cu^{1+}, and Fe^{2+}, which makes these minerals available for plant growth (Altomare et al., 1996). Similar results have also been observed with nitrogen. Plants grown from seeds treated with *Trichoderma* produced the maximum yield even when about 40 percent less nitrogen-containing fertilizer was supplied in comparison with plants from untreated seeds. This means that *Trichoderma* may improve the nitrogen-uptake efficiency of plants, and consequently reduce nitrogen pollution of soils and waters (Harman et al., 2004).

Trichoderma *spp. in the Biotechnology and "-Omics" Era*

Trichoderma is a fungal genus of great and demonstrable biotechnological value, but its genome is poorly surveyed compared with other model microorganisms, due to the high diversity of its species, the absence of optimized systems for its exploration, and the variety of genes expressed under a wide range of ambient conditions. Because of their ubiquity and rapid substrate colonization, *Trichoderma* species have been widely used as biocontrol organisms for agriculture, and their enzyme systems are applied in industry (Harman and Kubicek, 1998). However, there is a clear need to explore beyond the phenotype to exploit the underlying genetic systems.

Many *Trichoderma* strains, traditionally classified as *T. harzianum,* act as parasites of economically important plant pathogens and were considered both as mycoparasites and as "biocontrol agents." There is, however, an immense genetic variability within each aggregate of this "species." For instance, the presence of at least four distinct species within the *T. harzianum* aggregate has been demonstrated, and the existence of biotypes within *T. harzianum s.str.* linked to biocontrol and mycoparasitic activity also confirmed. There are many more species and genotypes which have not been analyzed to the same degree, but which may show further important attributes and activities, and other genetically diverse aggregates are likely to exist (Grondona et al., 2002). It has been recently shown that as many as 76 percent of all fungal isoenzymes known to be cell wall–degrading enzymes are present in only four *Trichoderma* species, that is, *T. harzianum s.str., T. longibrachiatum, T. atroviride,* and *T. asperellum* (Sanz, 2001). The great diversity and highly active nature of *Trichoderma* enzymatic systems has led to their successful use in environmental and industrial biodegradation, in composting, textiles. food and feed production, and pulp and paper treatment (Harman and Kubicek, 1998). These multiple applications eloquently demonstrate the large genetic and phenotypic diversity within the genus.

A fundamental part of the *Trichoderma* antifungal system consists of a series of genes coding for a surprising variety of extracellular lytic enzymes, including endochitinases, N-acetyl-β-glucosaminidases, chitin-1,4-β-chitobiosidases, proteases. endo- and exo-β-1.3-glucanases, endo β-1,6-glucanases, lipases. xylanases. mananases, pectinases. pectin lyases, amylases, phospholipases, RNAses, DNAses, etc. (Benítez et al., 1998; Lorito, 1998). The chitinolytic and glucanolytic enzymes are especially valuable for their CWDE activity on fungal plant pathogens, by hydrolyzing polymers not present in plant tissues. Each of these classes of enzymes contains diverse proteins with distinct enzymatic activity, and some have been purified and characterized, and their genes cloned (see Harman and Kubicek, 1998, for a review). Once purified, most have been shown to have strong antifungal activity. especially in combination. Much research over the past ten years has demonstrated that CWDEs from *Trichoderma* have a potential in agriculture as active components of novel fungicides (Harman, 2000). *Trichoderma* enzymes have numerous structural and kinetic properties which increase the probability of evading inhibitory mechanisms (Ham et al., 1997), and have been demonstrated to act synergistically with PR proteins of plants (Lorito et al., 1996). *Trichoderma* CWDEs do not act on humans and animals, as confirmed by EPA tests during registration of strains as biocontrol agents in the EU, and degrade innocuously in the environment. Single or mixed CWDEs with elevated antifungal effects, obtained from culture filtrates or through heterologous expression, can be included in commercial formulations as they are easily characterized, resist desiccation, are solid at temperatures up to 60°C, and are active over wide ranges of pH and temperatures. The inhibitory effect of chemical fungicides can be substantially improved by the addition of minute quantities (10 to 20 ppm) of *Trichoderma* CWDEs (Lorito, Peterbauer, et al., 1994), and CWDE genes can be expressed in plants (Lorito et al., 1998).

In addition to that, *Trichoderma* strains seem an inexhaustible source of antibiotics, from the acetaldehydes gliotoxin and viridin (Dennis and Webster, 1971), to alpha-pyrones (Keszler et al., 2000), terpenes, polyketides, isocyanide derivatives, piperacines, and complex families of peptaibols (Sivasithamparam and Ghisalberti, 1998). These produce synergistic effects in combination with CWDEs, with strong inhibitory activity on many fungal plant pathogens (Schirmböck et al., 1994; Lorito et al., 1996). The potential of genes coding for biosynthetic pathways of antibiotics (e.g., polyketides) with applications in human and veterinary medicine must be explored.

Molecules produced by *Trichoderma* also have potential for promoting plant growth (Inbar et al., 1994). Application of *T. harzianum* to plants led to improved seed germination, increased plant size, and augmentation of leaf area and weight (Linko et al., 1998; Altomare et al., 1999). The scenario of combined systemic biofungicides and plant growth promoters has great market potential if the molecular basis of the activities can be identified.

Many purified *Trichoderma* CWDEs are of interest to agri-food business. The enormous potential of the β-(1,4)-endoglucanase produced by *T. longibrachiatum* and *T. reesei* has been used to solve filtration problems associated with the presence of β-glucans during beer production. Addition of this enzyme is a frequent practice in this industrial sector, and genes coding for its production has been incorporated in transgenic yeasts for beer production (Pérez-González et al., 1993). β-(1,4)-endoglucanase produced by *T. longibrachiatum* promotes the liberation of aromatic terpene precursors in grape that must lead to the final fruity aroma of wines (Celis et al., 2000). Finally, a third very distinct example: for years, *Trichoderma* cellulases and hemicellulases have been added to chicken feed formulations to improve digestibility and fecal production.

The information above demonstrates the importance of *Trichoderma* in biotechnology. Further evidence of the breadth of interest in this genus as a means of wealth creation is demonstrated by the International *Trichoderma* Consortium, a collaborative European-led and promoted program between more than a dozen laboratories in ten countries for sequencing and functional analysis of the *Trichoderma* genome. While sequencing of a complete genome is a major step forward in knowledge, the gene sequences alone are of limited value for understanding gene function or cellular physiology. To determine gene function, it is essential to look at gene and protein responses to biological and environmental stimuli. The sequencing of expressed genes from cDNA libraries allows resource optimization by avoiding the sequencing of noncoding regions of genomic DNA and of genes not related to the functional status targeted. The expression profiles analyzed using sets of gene sequences arrayed in membranes or other supports, that can be screened with probes from defined functional situations, are powerful tools to assign functions to specific groups of gene products in different physiological scenarios. Proteomics can complement this functional genomic approach. In contrast to the expressome (set of genes expressed in specific conditions) the proteome is dynamic and constantly changing in response to both environment and genotype. It allows modifications to the final protein product (e.g., glycosylation, phosphorylation. acetylation, truncation), and to the way these molecules interact in the living cell. The project of the International *Trichoderma* Consortium integrates all these innovative approaches for the selection of sets of genes for which their putative function could lead to commercial applications. The sequence, array expression, and proteomic data will be jointly analyzed using bioinformatic tools developed to identify gene functions and, therefore, potential applications.

CONCLUSIONS

Growing concern over the presence of chemical residues in the food chain, the occurrence of fungicide-resistant strains and herbicide-resistant weeds, and the banning of some effective pesticides have generated a strong interest in the development of alternatives to synthetic agro-chemicals that are both effective and economically feasible. Sales of organic products have increased dramatically in recent years (for instance, in Europe), and organic farming is the fastest-growing sector of agriculture and an important point in the EU agri-food policy. There is an increasing interest in the biological control of plant diseases, pests, and weeds as an environment-friendly practice to be used in conventional, low-input agriculture and organic farming.

Consumers ask for healthy food, organic farmers ask for alternative to agro-chemicals. conventional farmers require friendly techniques, and companies, especially SMEs, develop or market biocontrol agents and their applications. Considerable attention is devoted to overcome the lack of knowledge on methods of production. formulation, or application of microbial control agents. New fermentation technologies. production of chopped mycelia, increasing shelf life of formulations, compatibility and efficacy of irrigation methods, and other methods of microbial application have been developed.

The public, even if its mind is open to organic farming practices and safe methods for food production, may have concerns about the risk associated with the release in the environment of microbes and eventual undesirable effects on nontarget organisms. Therefore, it is important, for instance, to develop methodologies for the assessment of the environmental impact of microbial biocontrol agents such as *Trichoderma* and *Agrobacterium* on natural microbial populations, and track their movement by constructing labeling methods.

Moreover, there are some other problems for the use of biological over chemical control that must be overcome to increase its usefulness in diverse habitats on horticultural, fruit, forest, and field crops. Results of biological control are sometime inconsistent and lower than chemical control. In many cases, biocontrol agents are too specific or, under some environmental conditions, slow acting. A better assessment of the efficacy in the field, by also evaluating the best methods of application in different environmental conditions, is needed. These trials should be carried out to verify the control of diseases on different crops in many different countries. The biopesticides should be supplied to producer associations that should organize their own on-site demonstration trials. This could enlarge the knowledge on the efficacy of biocontrol treatments, and enlarge the consumers' audience and their confidence in biopesticides.

The understanding of the genetic bases of the mechanisms of interaction between antagonistic microbes and other microorganisms or plants will deliver knowledge of utmost importance and will open the possibility to engineer more effective and safe strains. The other important issue is the understanding of the parameters that control the ecological fitness of biocontrol agents, which can help to increase the compatibility of beneficial microbes, for instance, to the rhizosphere and similar habitats. Finally, commercialization of biocontrol agents has been slow mainly due to the lack of consistency and efficacy of the microorganisms used. Further research must be carried out, for example, to explore the synergy of integrated use of more than one microorganism, or of microbes and bioactive metabolites. For instance, mixtures of cell wall–degrading enzymes produced by *Trichoderma* strains optimized for synergistic antimicrobial activity could be applied in combination with living biocontrol agents and chemical fungicides, or pathogen-toxin mixtures with enhanced biocontrol activity against weeds could be developed and increased.

REFERENCES

Altomare, C., T. Björkman, W.A. Norvell, and G.E. Harman. (1996). Solubilization of manganese dioxide by the biocontrol fungus *Trichoderma harzianum* strain 1295-22. Abstracts Internet. Union Microbiology Society, Jerusalem, p. 171.

Altomare, C., W.A. Norvell, T. Björkman, and G.E. Harman. (1999). Solubilization of phosphates and micronutrients by the plant-growth promoting and biocontrol fungus *Trichoderma harzianum* Rifai strain 1295-22. *Applied Environmental Microbiology* 65: 2926-2933.

Amsellem, Z., Y. Kleifeld, Z. Kerenyi. L. Hornok, Y. Goldwasser, and J. Gressel. (2001). Isolation, identification, and activity of mycoherbicidal pathogens from juvenile broomrape plants. *Biological Control* 21: 274-284.

Artigues, M. and P. Davet. (1984). Comparaison des aptitudes parasitaires de clones de *Trichoderma* vis-à-vis de quelques champignons á sclérotes. *Soil Biology and Biochemistry* 16: 413-417.

Bazzi, C. and T.J. Burr. (1986). La rogna della vite. *Informatore Fitopatologico* 3: 11-14.

Bazzi, C. and U. Mazzucchi. (1978). Biological control of crown gall on sweet cherry, myrobalan and peach seedlings in Italy. In *Proceedings 4th International Conference on Plant Pathogenic Bacteria*. I.N.R.A.. Angers, France, pp. 251-253.

Bazzi, C.. U. Mazzucchi, and S. Boschieri. (1980). Prova di lotta biologica al tumore batterico del melo in Alto Adige. *Informatore Fitopatologico* 6: 3-6.

Bazzi, C. and B. Rosciglione. (1982). *Agrobacterium tumefaciens* biotype 3, causal agent of crown gall on *Chrysanthemum* in Italy. *Phytopathologische Zeitschrift* 103: 280-284.

Benítez. T., C. Limón, J. Delgado-Jarana, and M. Rey. (1998). Glucanolytic and other enzymes and their genes. In *Trichoderma* and *Gliocladium,* Vol. 2, eds. G.E. Harman and C.P. Kubicek. Taylor and Francis, London.

Berestetski, A.O. (1997). Study of the mycobiota of *Cirsium arvense* for developing a bioherbicide. 10th EWRS Symposium, Poznan. Poland.

Bouzar, H., W.S. Chilton, X. Nesme, Y. Dessaux, V. Vaudequin, A. Petit. J.B. Jones, and N.C. Hodge. (1995). A new *Agrobacterium* strain isolated from aerial tumors on *Ficus benjamina* L. *Applied and Environmental Microbiololgy* 61: 65-73.

Bouzar, H., N. Daouzli. Z. Krimi. A. Alim, and E. Khemici. (1991). Crown gall incidence in plant nurseries of Algeria, characteristics of *Agrobacterium tumefaciens* strains, and biological control of strains sensitive and resistant to agrocin 84. *Agronomie* 11: 901-908.

Burr, T.J., C. Bazzi, S. Sule. and L. Otten. (1998). Biology of *Agrobacterium vitis* and the development of disease control strategies. *Plant Disease* 82: 1288-1297.

Bush, A.L. and S.G. Pueppke. (1991). Cultivar-strain specificity between *Chrysanthemum morifolium* and *Agrobacterium tumefaciens. Physiological and Molecular Plant Pathology* 39: 309-323.

Calderon, A.A., J.M. Zapata, R. Munoz, M.A. Pedreno, and A.R. Barcelo. (1993). Resveratrol production as a part of the hypersensitive-like response of grapevine cells to an elicitor from *Trichoderma viride. New Phytologist* 124: 455-463.

Canfield, M.L., C. Pereira, and L.W. Moore. (1992). Control of crown gall in apple (malus) rootstocks using Copac E and Terramicin. *Phytopathology* 82: 1153.

Cascone. G., G. Polizzi. C. Arcidiacono, and A. D'Emilio. (2001). Influenza sulla solarizzazione sotto serra di un film di copertura in ETFE. *Colture Protette* 3: 77-86.

Castoria. R.. F. De Curtis, G. Lima, L. Caputo, S. Pacifico, and V. De Cicco. (2001). *Aureobasidium pullulans* (LS-30) an antagonist of postharvest pathogens of fruits: Study on its modes of action. *Postharvest Biology and Technology* 22: 7-17.

Castoria, R., F. De Curtis, G. Lima, and V. De Cicco. (1997). β-1,3-glucanase activity of two saprophytic yeasts and possible mode of action as biocontrol agents against postharvest diseases. *Postharvest Biology and Technology* 12: 293-300.

Celis, J.E., M. Kruhoffer, I. Gromova, C. Frederiksen, M. Ostergaard, T. Thykjaer, P. Gromov. J. Yu, et al. (2000). Gene expression profiling: Monitoring transcription and translation products using DNA microarrays and proteomics. *FEBS Letters* 480: 2-16.

Chang, P.F.L., Y. Xu, M.L. Narasimhan, K.T. Cheah, M.P. D'Urzo, B. Damsz, A.K. Kononowicz, L. Abad, et al. (1997). Induction of pathogen resistance and pathogenesis-related genes in tobacco by a heat-stable *Trichoderma* mycelial extract and plant signal messengers. *Physiologia Plantarum* 100: 341-352.

Chilton, M.D., M.H. Drummond, D.J. Merlo, D. Sciaky, A.L. Montoya, M.P. Gordon, and E.W. Nester. (1977). Stable incorporation of plasmid DNA into higher plant cells: The molecular basis of crown gall tumorigenesis. *Cell* 11: 236-271.

Cirvilleri G., P. Bella, and V. Catara. (2000). Molecular detection and biological control activity of *Pseudomonas* strains. 5th Congress of the European Foundation for Plant Pathology. Taormina-Giardini Naxos, Italy, p. 85.

Clare, B.G., A. Kerr. and D.A. Jones. (1990). Characteristics of the nopaline catabolic plasmid in *Agrobacterium* strains K84 and K1026 used for biological control of crown gall disease. *Plasmid* 23: 126-137.

Clarke. H.R.G.. J.A. Leigh, and C.J. Douglas. (1992). Molecular signals in the interaction between plants and microbes. *Cell* 71: 191-199.

Conway, K.E. (1986). Use of fluid-drilling gels to deliver biological control agents to soil. *Plant Disease* 70: 835-839.

Conway, W.S., W.J. Janisiewicz, J.D. Klein, and C.E. Sams. (1999). Strategy for combining heat treatment, calcium infiltration, and biological control to reduce postharvest decay of Gala apples. *HortScience* 34: 700-704.

Cook, R.J. and K.R. Baker. (1983). *The Nature and Practice of Biological control of Plant Pathogens.* American Phytopathological Society. St. Paul, MN.

Cooke, D.L., W.M. Waites, and C. Leifert. (1992). Effects of *Agrobacterium tumefaciens, Erwinia carotovora, Pseudomonas syringae* and *Xanthomonas campestris* on plant tissue cultures of *Aster, Cheiranthus, Delphinium,* Iris and Rosa; disease development in vivo as a result of latent infection in vitro. *Journal of Plant Disease and Protection* 99: 469-481.

Cooksey, D.A. and L.W. Moore. (1982). Biological control of crown gall with an agrocin mutant of *Agrobacterium radiobacter. Phytopathology* 72: 919-921.

Couteaudier. Y. and C. Alabouvette. (1990). Quantitative comparison of *Fusarium oxysporum* competitiveness in relation with carbon utilization. *FEMS Microbiology Ecology* 74: 261-268.

Cubero, J., M.C. Martinez, P. Llop, and M.M. Lopez. (1999). A simple and efficient PCR method for the detection of *Agrobacterium tumefaciens* in plant tumours. *Journal of Applied Microbiology* 86: 591-602.

Dai, G.H., C. Andary, L. Mondolot-Cosson, and D. Boubals. (1995). Involvement of phenolic compounds in the resistance of grapevine callus to downy mildew *(Plasmopara viticola)*. *European Journal of Plant Pathology* 101: 541-547.

Dangl, J.L. and J.D. Jones. (2001). Plant pathogens and integrated defense responses to infection. *Nature* 411: 826-833.

De Cleene, M. and J. De Ley. (1976). The host range of crown gall. *Botanical Gazette* 42: 389-466.

Del Sorbo, G., F. Scala, G. Parrella, M. Lorito, C. Comparini, M. Ruocco, and A. Scala. (2000). Functional expression of the gene *cu*, encoding the phytotoxic hydrophobin cerato-ulmin, enables *Ophiostoma quercus,* a non-pathogen on elm, to cause symptoms of Dutch elm disease. *Molecular Plant-Microbe Interactions,* St. Paul 13: 43-53.

De Meyer, G., J. Bigirimana, Y. Elad, and M. Hofte. (1998). Induced systemic resistance in *Trichoderma harzianum* T 39 biocontrol of *Botrytis cinerea*. *European Journal of Plant Pathology* 104: 279-286.

Dennis, C. and J. Webster. (1971). Antagonistic properties of species groups of *Trichoderma*: III. Hyphal interactions. *Transactions of British Mycological Society* 57: 363-369.

Dhanvantari, B.N. (1976). Biological control of crown gall of peach in southwestern Ontario. *Plant Disease Reporter* 60: 549-551.

Donner, S.C., D.A. Jones, N.C. McClure, G.M. Rosewarne, M.E. Tate, A. Kerr, N.N. Fajardo, and B.G. Clare. (1993). Agrocin 434, a new plasmid encoded agrocin from the biocontrol *Agrobacterium* strains K84 and K1026, which inhibits biovar 2 agrobacteria. *Physiological and Molecular Plant Pathology* 42: 185-194.

Droby, S. and E. Chalutz. (1994). Mode of action of biocontrol agents of postharvest diseases. In *Biological Control of Postharvest Diseases: Theory and Practice,* eds. C.L. Wilson and M.E. Wisniewski. CRC Press, Boca Raton, FL, pp. 63-75.

Du Plessis, H.J., M.J. Hatting, and J.J. Van Vuuren. (1985). Biological control of crown gall in South Africa by *Agrobacterium radiobacter* strain K84. *Plant Disease* 69: 302-305.

Dye, E.W., W.J. Kemp, M.J. Amos, and W.C. Parker. (1975). Crown gall in roses can be controlled. *Commercial Horticulture* 7: 5-7.

Ellis, J.G. and A. Kerr. (1979). Transfer of agrocin 84 production from strain 84 to pathogenic recipients: A comment on the previous paper. In *Soil-borne Plant Pathogens,* eds. B. Schippers and W. Gams. Academic Press, London, pp. 579-583.

Esposito, S., M.G. Colucci, L. Frusciante, E. Filippone, M. Lorito, and R.A. Bressan. (2000). Antifungal transgene expression in *Petunia hybrida*. *Acta Horticulturae* 508: 157-161.

Farrand, S.K., J.E. Slota, J.S. Shim, and A. Kerr. (1985). Tn5 insertion in the agrocin 84 plasmid: The conjugal nature of pAgK84 and the location of determinants for transfer and agrocin 84 production. *Plasmid* 13: 106-117.

Farrand, S.K., P.B. Van Berkum, and P. Oger. (2003). *Agrobacterium* is a definable genus of the family *Rhizobiaceae*. *International Journal of Systematic and Evolutionary Microbiology* 53: 1681-1687.

Farrand. S.K. and C. Wang. (1992). Do we really understand crown gall control by *Agrobacterium radiobacter* strain K84? In, *Biological Control of Plant Diseases,* ed. E.S. Tjamos. Plenum Press, New York. pp. 287-293.

Fisher. C.G.. K. Conway and J.E. Motes. (1983). Fluid drilling: a potential delivery system for fungal biological control agents with small-seeded vegetables. *Proceedings of the Oklahoma Academy of Science* 63: 100-101.

Fuchs, J.G., Y. Moënne-Loccoz. and G. Défago. (1997). Non-pathogenic *Fusarium oxysporum* strain Fo47 induces resistance to Fusarium wilt in tomato. *Plant Disease* 81: 492-496.

Gladders, P. and J.R. Coley-Smith. (1980). Interactions between *Rhizoctonia tuliparum* sclerotia and soil micro-organism. *Transactions of the British Mycological Society* 74: 579-586.

Grimm, R. and S. Sule. (1981). Control of crown gall *(Agrobacterium tumefaciens)* in nurseries. In *Proceedings of the Fifth International Conference on Plant Pathogenic Bacteria.* Cali-Colombia. Centro Internacional de Agricultura Tropical (CIAT), pp. 531-537.

Grondona, I., A. Llobell, P.F. Cannon, M. Lorito, Y. Elad, S. Freeman, J. Katan, M. Rey, and E. Monte. (2002). Case study: *Trichoderma* as an alternative to methyl bromide in strawberries. In *Methyl Bromide Alternatives: The Remaining Challenges,* eds. T.A. Batchelor and J.M. Bolivar. European Commission/Spanish Ministry of Science and Technology. Brussels, pp. 127-129.

Gullino, M.L. (1992). Control of Botrytis rot of grapes and vegetables with *Trichoderma* spp. In *Biological Control of Plant Diseases: Progress and Challenges for the Future,* eds. E.C. Tjamos, G.C. Papavizas, and R.J. Cook. Plenum Press, New York. pp. 125-132.

Gullino, M.L. and A. Garibaldi. (1988). Biological and integrated control of grey mould of grapevine results in Italy. *Bulletin Oeppital* 18: 9-12.

Ham, K.S., S.C. Wu, A.G. Darvill, and P. Albersheim. (1997). Fungal pathogens secrete an inhibitor protein that distinguishes isoforms of plant pathogenesis-related endo-1,3-glucanases. *Plant Journal* 11: 169-179.

Harman, G.E. (2000). The dogmas and myths of biocontrol: Changes in perceptions based on research with *Trichoderma harzianum* T-22. *Plant Disease* 84: 377-393.

Harman, G.E. and T. Björkman. (1998). Potential and existing uses of *Trichoderma* and *Gliocladium* for plant disease control and plant growth enhancement. In *Trichoderma* and *Gliocladium,* Vol. 2, eds. Harman, G.E. and C.P. Kubicek. Taylor and Francis Ltd., London, pp. 229-265.

Harman, G.E., C.K. Hayes, M. Lorito, R.M. Broadway, A. Di Pietro, and A. Tronsmo. (1993). Chitinolytic enzymes of *Trichoderma harzianum*: Purification of chitobiosidase and endochitinase. *Phytopathology* 83: 313-318.

Harman, G.E., C.R. Howell, A. Viterbo, I. Chet, and M. Lorito. (2004). *Trichoderma* species: Opportunistic, avirulent plant symbionts. *Nature Reviews Microbiology* 2: 43-56.

Harman. G.E. and C.P. Kubicek. (1998). *Trichoderma* and *Gliocladium,* Vols. 1 and 2. Taylor and Francis Ltd.. London.

Harman. G.E., M. Lorito, A. Di Pietro, and C.K. Hayes. (1994). Antifungal synergistic combination of enzyme fungicide and non-enzymatic fungicide and use thereof. U.S. Patent No. 5,326,561, Issued July 5, 1994.

Harman, G.E., M. Lorito, A. Di Pietro. C.K. Hayes, F. Scala, and C.P. Kubicek. (1996). Combinations of fungal cell wall degrading enzyme and fungal cell membrane affecting compound. U.S. Patent Serial Number 611.504, filed March 5. 1996.

Harman, G.E., A.G. Taylor, and T.E. Stasz. (1989). Combining effective strains of *Trichoderma harzianum* and solid matrix priming to improve biological seed treatment. *Plant Disease* 73: 631-637.

Holm, L.G., D.L. Plunkett, J.V. Pancho. and J.P. Herberger. (1977). *The World's Worst Weeds: Distribution and Biology.* University Press of Hawaii, Honolulu.

Howell, C.R. (1987). Relevance of mycoparasitism in the biological control of *Rhizoctonia solani* by *Gliocladium virens. Phytopathology* 77: 992-994.

Howell, C.R. (1991). Biological control of *Pythium* damping-off of cotton with seed-coating preparations of *Gliocladium virens. Phytopathology* 81: 738-741.

Howell, C.R.. L.E. Hanson, R.D. Stipanovic, and L.S. Puckhaber. (2000). Induction of terpenoid synthesis in cotton roots and control of *Rhizoctonia solani* by seed treatment with *Trichoderma virens. Phytopathology* 90: 248-252.

Inbar, J., M. Abramsky, D. Coen, and I. Chet. (1994). Plant growth enhancement and disease control by *Trichoderma harzianum* in vegetable seedlings grown under commercial conditions. *European Journal of Plant Pathology* 100: 337-346.

Janisiewicz, W.J., W.S. Conway, and B. Leverentz. (1999). Biological control of postharvest decays of apples can prevent growth of *Escherichia coli* 0157:H7 in apple wounds. *Journal of Food Protection* 62: 1372-1375.

Joel, D.M, Y. Kleifeld, D. Losner-Goshen, G. Herzlinger, and J. Gressel. (1995). Transgenic crops against parasites. *Nature* 374: 220-221.

Jones, D.A., M.H. Ryder, B.G. Clare, S.K. Farrand. and A. Kerr. (1988). Construction of a Tra-deletion mutant of pAgK84 to safeguard the biological control of crown gall. *Molecular and General Genetics* 212: 207-214.

Kassemeyer, H.H. and G. Busam. (1998). Induced resistance of grapevine— Perspectives of biological control of grapevine diseases. *Bulletin-OILB-SROP* 21: 43-45.

Katan, J. (2000). Soil and substrate disinfestations as influenced by new technologies and constraints. In *Proceedings IS Chemical and Non-Chemical Soil and Substrate Disinfestation,* eds. M.L. Gullino. J. Katan, and A. Matta. *Acta Horticolturae* 532: 29-35.

Keane, P.J., A. Kerr, and P.B. New. (1970). Crown gall of stone fruit: II. Identification and nomenclature of *Agrobacterium* isolates. *Australian Journal of Biological Sciences* 23: 585-595.

Kerr, A. (1989). Commercial release of a genetically engineered bacterium for the control of crown gall. *Agricultural Science* 11: 41-44.

Kerr. A. and J.G. Ellis. (1982). Conjugation and transfer of Ti plasmid in *Agrobacterium tumefaciens.* In *Molecular Biology of Plant Tumors,* eds. G. Kahl and J.S. Schell. Academic Press, New York, pp. 321-344.

Kerr, A. and K. Htay. (1974). Biological control of crown gall through bacteriocin production. *Physiology and Plant Pathology* 4: 37-44.

Kerr, A. and M.E. Tate. (1984). Agrocins and the biological control of crown gall. *Microbiological Science* 1: 1-4.

Kersters, K. and J. De Ley. (1984). Genus III. *Agrobacterium* Conn 1942. In *Bergey's Manual of Systematic Bacteriology,* Vol. I, eds. N.R. Krieg and J. G. Holt. The William & Wilkins Co., Baltimore, MD, pp. 244-254.

Keszler, A., E. Forgacs, L. Kotai, J.A. Vizcaíno, E. Monte, and I. García-Acha. (2000). Separation and identification of volatile components in the fermentation broth of *Trichoderma atroviride* by solid-phase extraction and gas chromatography-mass spectroscopy. *Journal of Chromatography Sciences* 38: 421-424.

Kloepper, J.W., S. Tuzun, L. Liu, and G. Wei. (1993). Plant growth-promoting rhizobacteria as inducers of systemic disease resistance. In *Pest Management: Biologically Based Technologies,* eds. R.D. Lumsden and J.L. Vaughn. American Chemical Society Books, Washington, DC, pp.156-165.

Knudsen, G.R., D.J. Eschen, L.M. Dandurand, and L. Bin. (1991). Potential for biocontrol of *Sclerotinia sclerotiorum* through colonization of sclerotia by *Trichoderma harzianum. Plant Disease* 75: 466-470.

Kuc, J. (2001). Concepts and direction of induced systemic resistance in plants and its application. *European Journal of Plant Pathology* 107: 7-12.

Kullnig, C., R.L. Mach, M. Lorito, and C.P. Kubicek. (2000). Enzyme diffusion from *Trichoderma atroviride* (*T. harzianum* P1) to *Rhizoctonia solani* is a prerequisite for triggering of *Trichoderma ech42* gene expression before mycoparasitic-contact. *Applied and Environmental Microbiology* 66: 2232-2234.

Lemanceau, P. and C. Alabouvette. (1991). Biological control of fusarium diseases by fluorescent *Pseudomonas* and non-pathogenic *Fusarium. Crop Protection* 10: 279-286.

Lemanceau, P., P.A.H.M. Bakker. W. Jan De Kogel, C. Alabouvette, and B. Schippers. (1993). Antagonistic effect on non-pathogenic *Fusarium oxysporum* strain Fo47 and pseudobactin 358 upon pathogenic *Fusarium oxysporum* f. sp. *dianthi. Applied and Environmental Microbiology* 59: 74-82.

Leth, V. and C. Andreasen. (1999). *Phomopsis cirsii:* A promising control agent for *Cirsium arvense.* In *Program Abstracts, X International Symposium on Biological Control of Weeds.* USDA-ARS and Montana State University. Bozeman, MT, p. 116.

Lewis, J.A. and G.C. Papavizas. (1984). A new approach to stimulate population proliferation of *Trichoderma* species and other potential biocontrol fungi introduced into natural soils. *Phytopathology* 74: 1240-1244.

Limón. M.C., A. Llobel. J.A. Pintor Toro, and T. Benítez. (1996). Overexpression of chitinase by *Trichoderma harzianum* strains used as biocontrol fungi. In *Chitin Enzymology,* Vol. 2, ed. R.A.A. Muzzarelli. AtecEdizioni, Italy.

Linko, M., A. Haikara, A. Ritala, and M. Penttilla. (1998). Recent advances in the malting and brewing industry. *Journal of Biotechnology* 65: 85-98.

Lopez, M.M., M.T. Gorris, C.I. Salcedo, A.M. Montojo, and M. Mirò. (1989). Evidence of biological control of *Agrobacterium tumefaciens* strains sensitive and resistant to agrocin 84 by different *Agrobacterium radiobacter* strains on stone fruit trees. *Applied and Environmental Microbiology* 55: 741-746.

Lopez, M.M., M.T. Gorris, F.J. Temprano, and R.J. Orive. (1987). Results of seven years of biological control of *Agrobacterium tumefaciens* in Spain. *Eppo Bulletin* 17: 273-279.

Lopez, M.M., M. Miró, R. Orive, F. Temprano, and M. Poli. (1981). Biological control of crown gall on rose in Spain. In *Proceedings of the Fifth International Conference on Plant Pathogenic Bacteria. Cali-Colombia.* Centro Internacional de Agricultura Tropical (CIAT), pp. 538-547.

Lopez-Lopez, M.J., B. Vicedo, N. Orellana, J. Piquer, and M.M. Lopez. (1999). Behavior of a virulent strain derived from *Agrobacterium radiobacter* strain K84 after spontaneous ti plasmid acquisition. *Phytopathology* 89: 286-292.

Lorito, M. (1998). Chitinolytic enzymes and their genes. In *Trichoderma* and *Gliocladium*, Vol. 2, eds. G.E. Harman and C.P. Kubicek. Taylor and Francis, London pp. 73-99.

Lorito, M., C.K. Hayes, A. Di Pietro, S.L. Woo, and G.E. Harman. (1994). Purification, characterization and synergistic activity of a glucan 1.3-β-glucosidase and an N-acetyl-β-glucosaminidase from *Trichoderma harzianum*. *Phytopathology* 84: 398-405.

Lorito, M., C. Peterbauer, C.K. Hayes, and G.E. Harman. (1994). Synergistic interaction between fungal cell wall degrading enzymes and different antifungal compounds enhances inhibition of spore germination. *Microbiology* 140: 623-629.

Lorito, M., F. Scala, A. Zoina, and S.L. Woo. (2001). Enhancing biocontrol of fungal pests by exploiting the Trichoderma genome. In *Enhancing Biocontrol Agents and Handling Risks,* eds. J. Gressel and M. Vurro. IOS Press, Amsterdam, Chapter 22.

Lorito, M. and S.L. Woo. (1998). Advances in understanding the antifungal mechanism(s) of *Trichoderma* and new applications for biological control. In *Molecular Approaches in Biological Control,* Vol. 21(9), eds. B. Duffy, U. Rosenberger, and G. Défago. IOBC wprs Bulletin/Bulletin OILB srop, Dijon, France, pp. 73-80.

Lorito, M., S.L. Woo, M. D'Ambrosio, G.E. Harman, C.K. Hayes, C.P. Kubicek, and F. Scala. (1996). Synergistic interaction between cell wall degrading enzymes and membrane affecting compounds. *Molecular Plant-Microbe Interactions* 9: 206-213.

Lorito, M., S.L. Woo, I.G. Fernandez, G. Colucci, G.E. Harman, J.A. Pintor-Toro, E. Filippine, et al. (1998). Genes from mycoparasitic fungi as a source for improving plant resistance to fungal pathogens. *Proceedings of the National Academy of Sciences of USA* 95: 7860-7865.

Lotan, T. and R. Fluhr. (1990). Xylanase, a novel elicitor of pathogenesis-related proteins in tobacco, uses a non-ethylene pathway for induction. *Plant Physiology* 93: 811-817.

Lu, S.F. (1994). Isolation of putative pAgK84 transconjugants from commercial cherry and raspberry plants treated with *Agrobacterium radiobacter* strain K84. MS thesis, Oregon State University, Corvallis.

Lynch, J.M. (1996). *Trichoderma:* Potential in regulatory soil function. Abstracts Internat. Union Microbiology Society, Jerusalem, p. 12.

Macrae, S.. J.A. Thomson, and J. Van Staden. (1988). Colonization of tomato plants by two agrocin-producing strains of *Agrobacterium tumefaciens*. *Applied and Environmental Microbiology* 54: 3133-3137.

Martì. R.. J. Cubero, A. Daza, J. Piquer, C.I. Salcedo. C. Morente, and M.M. Lopez. (1999). Evidence of migration and endophytic presence of *Agrobacterium tumefaciens* in rose plants. *European Journal of Plant Pathology* 105: 39-50.

McClure, N.C., A. Ahmadi, and B.G. Clare. (1998). Construction of a range of derivatives of the biological control strain *Agrobacterium rhizogenes* K84: A study of factors involved in biological control of crown gall disease. *Applied and Environmental Microbiology* 64(10): 3977-3982.

Miller, H.N. (1975). Leaf, stem. crown and root galls induced in *Chrysanthemum* by *Agrobacterium tumefaciens*. *Phytopathology* 65: 805-811.

Moore, L.W. (1988). Use of *Agrobacterium radiobacter* in agricultural ecosystems. *Microbiological Sciences* 5: 92-95.

Moore. L.W and J. Allen. (1986). Controlled heating of root-pruned dormant *Prunus* spp. seedlings before transplanting to prevent crown gall. *Plant Disease* 70: 532-536.

Moore, L.W. and M. Canfield. (1996). Biology of *Agrobacterium* and management of crown gall disease. In *Principles and Practice of Managing Soilborne Plant Pathogens*, ed. Robert Hall. APS Press, St. Paul, MN, pp.153-191.

Moore. L.W., W.S. Chilton, and M.L. Canfield. (1997). Diversity of opines and opine-catabolizing bacteria isolated from naturally occurring crown gall tumors. *Applied and Environmental Microbiology* 63: 201-207.

Murphy, P.J. and W.P. Roberts. (1979). A basis for agrocin 84 sensitivity in *Agrobacterium radiobacter*. *Journal of General Microbiology* 114: 207-213.

Nelson, E.B.. G.E. Harman, and G.T. Nash. (1988). Enhancement of *Trichoderma*-induced biological control of *Pythium* seed rot and pre-emergence damping-off of peas. *Soil Biology and Biochemistry* 20: 145-150.

New, P.B. and A. Kerr. (1972). Biological control of crown gall: Field measurements and glasshouse experiments. *Journal of Applied Bacteriology* 35: 279-287.

Ophel, K., P.R. Nicholas, P.A. Magarey, and A.W. Bass. (1990). Hot water treatment of dormant grape cuttings reduces crown gall incidence in a field nursery. *American Journal of Enology and Viticulture* 41: 325-329.

Panagopoulos, C.G.. P.G. Psallidas, and A.S. Alivizatos. (1979). Evidence of a breakdown in the effectiveness of biological control of crown gall. In *Soil-borne Plant Pathogens,* eds. B. Schippers and W. Gams. Academic Press. London, pp. 569-578.

Peluso, R., A. Zoina. F. Morra, L. Sigillo, and A. Raio. (2001). Outbreak of dangerous transconjugant strains in a stone fruit nursery field. In *Proceedings of VIII Convegno Nazionale S.I.Pa.V.: Le conoscenze delle Interazioni Biologiche in Patologia Vegetale per una Agricoltura Ecocompatibile,* ed. Università degli Studi della Basilicata. Potenza, Italy.

Penyalver, R., P. Oger, M.M. Lopez, and S.K. Farrand. (2001). Iron-binding compounds in *Agrobacterium* spp.: The biological control strain K84 produces a hydroxamate siderophore. *Applied and Environmental Microbiology* 67: 654-664.

Penyalver, R., B. Vicedo, and M.M. Lopez. (2000). Use of genetically engineered *Agrobacterium* strain K1026 for biological control of crown gall. *European Journal Plant Pathology* 106: 801-810.

Penyalver, R., B. Vicedo, C.I. Salcedo, and M.M. Lopez. (1994). *Agrobacterium radiobacter* strain K84, K1026 and K84 Agr⁻ produce an antibiotic-like substance, active in vitro against *A. tumefaciens* and phytopathogenic *Erwinia* and *Pseudomonas*. *Biocontrol Sciences and Technology* 4: 259-267.

Pérez-González, J.A., R. González, A. Querol, J. Sendra, and D. Ramón. (1993). Construction of a recombinant wine yeast strain expressing (1,4) endoglucanase and its use in microvinification processes. *Applied Environmental Microbiology* 59: 2801-2806.

Perry, K.L. and C.I. Kado. (1982). Characteristics of Ti plasmids from broad host-range and ecologically specific biotype 2 and 3 strains of *Agrobacterium tumefaciens*. *Journal of Bacteriology* 151: 343-350.

Pieterse, C.M.J. and L.C. van Loon. (1999). Salicylic acid-independent plant defense pathways. *Trends in Plant Science* 4: 52-58.

Polizzi, G., G. Cascone, G. Polizzi, K. Sciortino Jr., and I. Castello. (2001). Recovery of microfungi from greenhouse soil after exposure to different treatments of solarization. In *Proceedings 5th European Foundation for Plant Pathology Congress: Biodiversity in Plant Pathology*. Taormina Giardini-Naxos, Italy, pp. 585-588.

Poppenberger, B., W. Leonhardt, and H. Redi. (2002). Latent persistence of *Agrobacterium vitis* in micropropagated *Vitis vinifera*. *Vitis* 41: 113-114.

Psallidas, P.G. (1988). Large-scale application of biological control of crown gall in Greece. *Eppo Bulletin* 18: 61-66.

Raio, A., R. Peluso, G. Puopolo, and A. Zoina. (2003). Root colonization ability and virulence of spontaneous transconjugant strains originated by plasmid transfer from *Rhizobium rhizogenes* K84 biocontrol agent to tumorigenic *Rhizobium* strains. In *Proceedings of International Symposium: Structure and Function of Soil Microbiota*, ed. A. Raio. Marburg, Germany, p. 145.

Raio, A., A. Zoina, and L.W. Moore. (1997). The effect of solar heating of soil on natural and inoculated agrobacteria. *Plant Pathology* 46: 320-328.

Roberts, W.P., M.E. Tate, and A. Kerr. (1977). Agrocin 84 is a 6-N-phosphoramidate of an adenine nucleotide analogue. *Nature* 265: 379-381.

Ryder, M.H., J.E. Slota, A. Scarim, and S.K. Farrand. (1987). Genetic analysis of Agrocin 84 production and immunity in *Agrobacterium* spp. *Journal of Bacteriology* 9: 4184-4189.

Sanz. (2001). PhD thesis, Universisity of Salamanca.

Sauerborn, J. (1991). The economic importance of the phytoparasites *Orobanche* and *Striga*. In *Fifth International Symposium on Parasitic Weeds,* eds. J. Ransom, L.J. Musselman, A.D. Worsham, and C. Parker. CIMMYT, Nairobi, Kenya.

Sawada, H., H. Ieki, H. Oyaizu, and S. Matsumoto. (1993). Proposal for rejection of *Agrobacterium tumefaciens* and revised descriptions for the genus *Agrobacterium* and for *Agrobacterium radiobacter* and *Agrobacterium rhizogenes*. *International Journal of Systematic Bacteriology* 43: 694-702.

Schirmböck, M., M. Lorito, Y.L. Wang, C.K. Hayes, I. Arisan-Atac. F. Scala, G.E. Harman, and C.P. Kubicek. (1994). Parallel formation and synergism of hydrolytic enzymes and peptaibol antibiotics, molecular mechanisms involved in the antagonistic action of *Trichoderma harzianum* against phytopathogenic fungi. *Applied and Environmental Microbiology* 60: 4364-4370.

Schroeder, D., H. Müller-Schärer, and C.S.A. Stinson. (1993). A European weed survey in 10 major crop systems to identify targets for biological control. *Weed Research* 33: 449-458.

Schroth, M.N. and W.J. Moller. (1976). Crown gall controlled in the field with a non-pathogenic bacterium. *Plant Disease Reporter* 60: 275-278.

Serfontein, S. and J.L. Staphorst. (1994). Crown gall of hop caused by *Agrobacterium tumefaciens* biovar 1 in South Africa. *Plant Pathology* 43: 1028-1030.

Sid Ahmed, A., C.P. Sánchez, and M.E. Candela. (2000). Evaluation of induction of systemic resistance in pepper plants *(Capsicum annuum)* to *Phytophthora capsici* using *Trichoderma harzianum* and its relation with capsidiol accumulation. *European Journal of Plant Pathology* 106: 817-824.

Sivan, A. and I. Chet. (1989). The possible role of competition between *Trichoderma harzianum* and *Fusarium oxysporum* on rhizosphere colonization. *Phytopathology* 79: 198-203.

Sivasithamparam, K. and E.L. Ghisalberti. (1998). Secondary metabolism in *Trichoderma* and *Gliocladium*. In *Trichoderma* and *Gliocladium*, Vol. 2, eds. G.E. Harman and C.P. Kubicek. Taylor and Francis Ltd., London, pp. 139-191.

Smith, V.L., W.F. Wilcox, and G.E. Harman. (1990). Potential for biological control of *Phytophthora* root and crown rots of apple by *Trichoderma* and *Gliocladium* spp. *Phytopathology* 80: 880-885.

Sobiczewski, P. and A. Piotrowski. (1983). Preliminary investigation on the biological control of crown gall. *Fruit Science Reports* 10: 189-194.

Spiers, A.G. (1980). Biological control of *Agrobacterium* species in vitro. *New Zealand Journal of Agricultural Research* 23: 133-137.

Stockwell, V.O., M.D. Kawalek, L.W. Moore, and J.E. Loper. (1990). Plasmid transfer between *Agrobacterium radiobacter* K84 and *A. tumefaciens* in crown gall tissue. *Phytopathology* 80: 1001-1009.

Stockwell, V.O., M.D. Kawalek, L.W. Moore, and J.E. Loper. (1996). Transfer of pAgK84 from the biocontrol agent *Agrobacterium radiobacter* K84 to *A. tumefaciens* under field conditions. *Phytopathology* 86: 31-37.

Stockwell, V.O., L.W. Moore, and J.E. Loper. (1993). Fate of *Agrobacterium radiobacter* K84 in the environment. *Applied and Environmental Microbiology* 59: 2112-2120.

Sugar, D. and R.A. Spotts. (1999). Control of postharvest decay in pear by four laboratory-grown yeasts and two registered biocontrol products. *Plant Disease* 83: 155-158.

Sule, S. (1978). Biological control of crown gall with a peat cultured antagonist. *Phytopathologische Zeitschrift* 91: 273-275.

Taylor, A.G., T.G. Min, G.E. Harman, and X. Jin. (1991). Liquid coating formulation for the application of biological seed treatments of *Trichoderma harzianum*. *Biological Control* 1: 16-22.

Tempè, J. and A. Petit. (1982). Opine utilization by *Agrobacterium*. In *Molecular Biology of Plant Tumors,* eds. G. Kahl and J.S. Shell. Academic Press, New York, pp. 451-459.

Vadequin-Dransart, V., A. Petit, C. Poncet, C. Ponsonnet, X. Nesme, J.B. Jones, H. Bouzar, W.S. Chilton, and I. Dessaux. (1995). Novel Ti plasmids in *Agrobacterium* strains isolated from fig tree and chrysanthemum tumors and their opinelike molecules. *Molecular Plant-Microbe Interactions* 8: 311-321.

Van Zyl, F.G.H., B.W. Strijdom, and J.L. Staphorst. (1986). Susceptibility of *Agrobacterium tumefaciens* strains to two agrocin-producing *Agrobacterium* strains. *Applied and Environmental Microbiology* 52: 234-238.

Vicedo, B., M.J. Lopez, M.J. Asìns, and M.M. Lopez. (1996). Spontaneous transfer of the Ti plasmid of *Agrobacterium tumefaciens* and the nopaline catabolism plasmid of *A. radiobacter* strain K84 in crown gall tissue. *Phytopathology* 86: 528-534.

Vicedo, B., R. Penalver, M.J. Asìns, and M.M. Lopez. (1993). Biological control of *Agrobacterium tumefaciens*, colonization and pAgK84 transfer with *Agrobacterium radiobacter* K84 and the Tra-mutant strain K1026. *Applied and Environmental Microbiology* 59: 309-315.

Wang, C.L., S.K. Farrand, and I. Hwang. (1994). Organization and expression of the genes on pAgK84 that encode production of agrocin 84. *Molecular Plant-Microbe Interactions* 7: 472-481.

Weindling, R. (1932). *Trichoderma lignorum* as a parasite of other soil fungi. *Phytopathology* 22: 837-845.

Weindling, R. (1934). Studies on a lethal principle effective in the parasitic action of *Trichoderma lignorum* on *Rhizoctonia solani* and other soil fungi. *Phytopathology* 24: 1153-1179.

Weindling, R. (1941). Experimental consideration of the mold toxins of *Gliocladium* and *Trichoderma*. *Phytopathology* 31: 991-1003.

Weindling, R. and H.S. Fawcett. (1936). Experiments in the controls of Rhizoctonia damping-off of citrus seedlings. *Journal of Agriculture Sciences, CA Agric. Exp. Station* 10: 1-16.

Weller, D.M. and L.S. Thomashow. (1993). Microbial metabolites with biological activity against plant pathogens. In *Pest Management: Biologically Based Strategies,* eds. R.D. Lumsden and J.L. Vaughn. American Chemical Society, Washington, DC, pp.172-180.

Whipps, J.M. and M. Gerlagh. (1992). Biology of *Coniothyrium minitans* and its potential for use in disease biocontrol. *Mycological Research* 96: 897-907.

Whipps, J.M. and R.D. Lumsden. (2001). Commercial use of fungi as plant disease biological control agents: Status and prospects. In *Fungal Biocontrol Agents—Progress, Problems and Potential,* eds. T. Butt, C. Jackson, and N. Magan. CAB International, Wallingford, UK, pp. 9-22.

Wilhite, S.E., R.D. Lumdsen, and D.C. Straney. (1994). Mutational analysis of gliotoxin production by the biocontrol fungus *Gliocladium virens* in relation to suppression of *Pythium* damping-off. *Phytopathology* 84: 816-821.

Windham, M.T., Y. Elad, and R. Baker. (1986). A mechanism for increased plant growth induced by *Trichoderma* spp. *Phytopathology* 76: 518-521.

Woo, S.L., B. Donzelli, F. Scala, R. Mach, G.E. Harman, C.P. Kubicek, G. Del Sorbo, and M. Lorito. (1999). Disruption of the *ech42* (endochitinase-encoding) gene affects biocontrol activity in *Trichoderma harzianum* P1. *Molecular Plant-Microbe Interactions* 12: 419-429.

Yedidia, I., N. Behamou, Y. Kapulnik, and I. Chet. (2000). Induction and accumulation of PR proteins activity during early stages of root colonization by the mycoparasite *Trichoderma harzianum* strain T-203. *Plant Physiology and Biochemistry* 38: 863-873.

Young, J.M., L.D. Kuykendall, E. Martinez-Romero, A. Kerr, and H. Sawada. (2001). A revision of *Rhizobium* Frank 1889, with an emended description of the genus, and the inclusion of all species of *Agrobacterium* Conn 1942 and *Allorhizobium undicola* de Lajudie et al.. 1998 as new combinations: *Rhizobium radiobacter, R. rhizogenes, R rubi, R. undicola* and *R. vitis*. *International Journal of Systematic and Evolutionary Microbiology* 51: 89-103.

Zazzerini, A. and I. Tosi. (1985). Antagonistic activity of fungi isolated from sclerotia of *Sclerotinia sclerotium*. *Plant Pathology* 34: 415-421.

Zhou, T., J. Northover, and K.E. Schneider. (1999). Biological control of porthawest diseases of peach with phyllosphere isolates of *Preudomonas syringae*. *Canadian Journal of Plant Pathology* 21: 375-381.

Zoina, A., R. Peluso, L. Sigillo, and A. Raio. (2003). Efficacia di diverse modalità di lotta al tumore batterico dei fruttiferi. *Informatore Agrario* 16: 69-71.

Zoina, A., A. Raio, M. Lorito, G. Del Sorbo, B. Aloj, and F. Scala. (1994). Lotta biologica al tumore radicale del pesco: esiti di una prova quinguenaal. *Atti Giornate Fitopathologische Bolagna* 3: 359-364.

Zoina, A., A. Raio, R. Peluso, and A. Spasiano. (2001). Charaterization of agrobacteria from weeping fig *(ficus benjamina)*. *Plant Pathology* 50: 620-627.

Chapter 8

An Overview of Biological Control of Fruit and Vegetable Diseases

Ram Gopal Kapooria

INTRODUCTION

Fruits and vegetables play an important role in human nutrition. In fact, no meal is complete without vegetables. A diet rich in fresh fruits and vegetables is key to good health, especially in our times, when diseases have caused a heavy-handed effect on human health.

It is a paradox that fruits and vegetables protect humans against diseases but are themselves attacked by numerous pathogens and suffer from a variety of diseases that reduce their quality and yield. Losses from fruit and vegetable diseases vary considerably, but on an average are responsible for 17 to 23 percent loss.

In the past 100 years, agriculture has evolved rapidly through the mechanical age into the current chemical age. This transition has brought in many problems, and agriculture is moving toward a biological age, which ensures an efficient management of biological systems, with emphasis on renewal of resources. Continued dependence on chemicals for greater productivity and protection of agricultural crops has caused too much damage to the ecosystem. It has also been responsible for fungicide resistance in plant pathogens (Brent, 1995). These happenings have given birth to new and innovative strategies of plant disease management, such as integrated

The encouragement received from friends and colleagues during this research is sincerely acknowledged. I am grateful to Prof. K. Mwauluka for reading through the manuscript and for making valuable suggestions.

pest management (IPM), biocontrol of plant diseases, and improved plant nutrition and protection through vesicular-arbuscular mycorrhizae. In recent years, the idea of sustainable agriculture has become fashionable, and approaches to crop production and disease management have seen enhanced modification of some common practices.

SOURCES OF BIOCONTROL AGENTS

Microbes occur in every kind of habitat. On plants, they exist on surfaces of roots, stem, leaves, flowers, and fruits. Various names have been used to describe microbes occurring on specific plant surfaces and capable of inducing many types of interactions in plant pathogens. Such terms as phyllosphere and phylloplane, rhizosphere, and rhizoplane signify the microflora of these habitats. Isolation, identification, and characterization of these microbes, and the evaluation of their antagonistic properties against plant pathogens may provide a large number of agents of biocontrol. Several bacteria, fungi, and yeasts have been obtained from plant surfaces, and they have shown encouraging results for the management of plant diseases. Exhibit 8.1 and Table 8.1 give names of several fungi and bacteria that have been extensively used as biocontrol agents in plant disease management.

Phylloplane Antagonists

Epiphytic filamentous fungi, yeasts, and bacteria were found to control a number of fungal diseases. Wood (1950) showed their effectiveness against gray mold of lettuce, while Newhook (1957) demonstrated the control of gray mold of tomato. In these experiments, bacteria and yeasts (Wood, 1950), or fungi (*Cladosporium herbarum* and *Penicillium* spp.) featured as agents of biocontrol (Newhook, 1957). Several bacteria and streptomyces from the phylloplane can successfully reduce the incidence of gray mold on beetroot, onion, and lettuce (Clark and Lorbeer, 1977; Pennock-Vos et al., 1990; White et al., 1990). Bhatt and Vaughan (1962) showed the control of blossom blight and green fruit rot of strawberries with *Cladosporium herabrum*. Sufficient control of *B. cinerea* Pers.:Fr. by isolates of *Trichoderma* and *Gliocladium* spp. was reported by Dubos and Bulit (1981), Bisiach et al. (1985), and Guillino and Garibaldi (1988); on grapes by Tronsmo and Dennis (1974); on strawberries by Peng and Sutton (1990); and on apple by Tronsmo and Raa (1977). Elad and Zimand (1993) also reported control of gray mold of cucumber by *Gliocladium* and *Trichoderma* species. Blakeman and Fokkema (1982) confirmed that phylloplane microbes possess the potential to control many foliar diseases of pants. Mi-

EXHIBIT 8.1. Commonly used species of actinomycetes, fungi, and yeasts in biocontrol of plant diseases.

Microbe Species

Aspergillus niger Tieghem
Candida saitoana Nakase and Suzuki
C. oleophila Montrocher
Chaetomium globosum Kuntze:Fries
Cladosporium herbarum Link
Coniothyrium minitans Cambell
Cryptococcus lorentii (Kufferath) Skinner
Debaromyces hansenii (Zoph) Lodder et Krejer-Van Rij
Epicoccum nigrum Link
E. purpurascens Ehrenb
Fusarium oxysporum Schlecht (nonpathogenic)
Fusarium proliferatum (Matsushima) Nirenberg
Gliocladium virens J.H. Miller
Heteroconium chaetospira (Grove) M.B. Ellis
Paceilomyces lilacinus (Thom) Samson
Pencillium frequentans Westling
P. funiculosum Thom
P. oxalicum Currie &Thom
Penicillium spp.
Peniophora gigantea (Fries:Fries) Massee
Pichia anomala (Hansen) Kurtzman
P. guilliermondii Wickerham
Pythium oligandrum Drechsler
Sporidesmium sclerotivorum Uecker, Ayers & Adams
Sterile hyphomycete fungus
Streptomyces griseoviridis
Talaromyces flavus (Klöcker) Stolk & Samson
Trichoderma aureoviride Rifai aggr.
T. harzianum Rifai
T. koningi Oudemans aggr.
T. pseudokoningii Rifar aggr.
T. viride Pers:Fr.

TABLE 8.1. Bacterial species used as biocontrol agents for plant disease control.

Bacterial Species	Strain/Isolate
Bacillus cereus	
B. subtilis	
Enterobacter aerogens	
E. cloacae	
Erwinia herbicola	EPS528, EPS482, Cu135
Pasteuria penetrans	
P. cepacia	
Pseudomonas florescens	A506, ATB, CL42, CL 66, CL82 EPS 288, EPS 375, EPS 381, Pfl, 63-63-28, 89B-27, WCS 374, WCS 417
P. putida	WCS 358
Serratia marcescens	90-166
Xanthomonas campestris pv. *diffenbachiae*	

crobial antagonists isolated from blossom, leaves, and fruits of mango yielded 121 biocontrol agents from a total of 648 microbes, which were effective against anthracnose of mango caused by *Colletotrichum gloeosporioides* (Penz.) Penz. and Sacc. (Koomen and Jeffries, 1993). Although several of these bacteria and yeasts could inhibit conidial germination of *C. gleosporioides,* only *Pseudomonas fluorescens* showed reduction of anthracnose development.

Rhizosphere and Rhizobacteria

The significance of rhizoplane and rhizosphere as a rich and diverse site has been known for a long time, and the microflora from this site has been exploited extensively for the control of soilborne diseases (Mukerji, 2002).

Pandey and Upadhyay (2000) compared microbial populations from the rhizosphere and nonrhizosphere soil of pigeon-pea and showed their antagonistic activity against several diseases. Isolates of *Trichoderma* spp. from the rhizosphere of ginger (*Zingiber officinale* Roscoe.) were shown to be antagonistic to dry root rot of chilli caused by *Rhizoctonia solani* Kühn (Bunker and Mathur, 2001). They also reported that *Gliocladium virens* Miller and *Trichoderma aureoviride* Rifai aggr. were also quite effective.

Montesinos et al. (1996) prepared several isolates of *Erwinia herbicola* and *Pseudomonas fluorescence* from the aerial parts and roots of several plants which reduced the brown spot of pear caused by *Stemphillium vesicarium* (Wallr.) Simmons by 57 percent. Specific rhizosphere bacteria have been shown to be effective against many diseases such as anthracnose of cucumber (Liu et al., 1995).

Among many bacteria, actinomycetes, and fungi isolated from watermelon roots, a nonpathogenic strain of *Fusarium oxysporum* Schlecht recovered from a suppressive soil showed 35 to 75 percent control of *Fusarium* wilt of watermelon (Larkin et al., 1996). Park et al. (1988), Lemanceau and Alabouvette (1991), and Mandeed and Baker (1991) reported biocontrol of Fusarial diseases by nonpathogenic *Fusarium* and fluorescent pseudomonads. The rhizosphere microbes have been generally found to control several diseases like damping-off, root rots, and wilts of plants. Several authors reported (Broadbent et al., 1971; Halder et al., 1983; Ahmad and Baker, 1987; Nelson, 1988; Weller, 1988; Raaijmakers et al., 1995) that rhizosphere bacteria could also control soilborne diseases.

Since the 1970s, rhizobacteria have been found particularly beneficial because of their different effects on the activities of plants and pathogens. The term *rhizobacteria* was applied to those bacteria of the rhizosphere that colonize the roots of plants (Kloepper and Schroth, 1978; Suslow et al., 1979). Two types of rhizobacteria occur naturally and may be (1) deleterious rhizobacteria (DRB) or (2) plant growth-promoting rhizobacteria (PGPR).

1. The DRB have been classified as exo-minor pathogens (Burr and Caesor, 1984; Schippers et al., 1987) and are known to reduce crop yields, just like the major plant pathogens. They include species of *Achromobacter, Arthrobacter, Citrobacter, Enterobacter, Pseudomonas,* and *Serratia* (Dube, 2001).

2. The PGPR are generally beneficial and enhance growth and yield of plants as well as control plant diseases. These bacteria are commonly isolated from disease-suppressive soils and include fluorescent pseudomonads *(P. fluorescens* and *P. putida)* and nonfluorescent *Pseudomonas* sp., *Bacillus subtilis,* and *Serratia* spp. (Dube, 2001). Their role as agents of biocontrol of plant diseases is revealed from the following examples.

Isolates A-13 and AF1 of *Bacillus subtilis* have been extensively studied, and Uthkhede and Rahe (1980) showed its beneficial effect on the control of white rot of onion and an increase of carrot yield by 48 percent reported by Merriman et al. (1974). Yuen et al. (1985) also reported yield increases of many other plants. The most common method of application of *B. subtilis* strains is by treating seeds with bacterial suspension.

A strain of *Enterobacter cloacae* has been shown to be quite effective against several such diseases as pre- and postemergence rots and wilts caused by *Pythium, Rhizopus,* and *Fusarium* species in cucumbers, pea, and peach (Sneh et al., 1984; Chao et al., 1986; Wilson et al., 1987; Howell et al., 1988).

Pseudomonas fluorescens and *P. putida* are known to suppress a number of important diseases of plants. They have been tested against damping-off of cucumber and also for the control of diseases caused by *Gaeumannomyces graminis* f. sp. *tritici* Walker, *Rhizoctonia solani, Pythium* spp., *Sclerotium rolfsii* Sacc., and *Thielaviopsis basicola* (Berk. & Broom) Ferraris (Cook and Rovira, 1976; Savitry and Gananamanickam, 1987; Défago et al., 1990; Paulitz and Loper, 1991; Thomashow et al., 1993). Pseudomonads have also been found inhibitory to *Xanthomonas citri* and *Sarocladium oryzae* (Sawade) W. Gams (Unnamalai and Gananamanickam, 1984; Sakthivel and Gananamanickam, 1986). It may be noted that some strains of pseudomonads can also control certain foliar diseases of wheat, such as *Septoria tritici* Roberge and *Puccinia graminis* f. sp. *tritici* Cummins (Voisard et al., 1989; Flaishman et al., 1996). The most commonly used strains of *Pseudomonas* are 2-79, CHAO, PN3 and RBT13.

Manure-Based Compost Extract Formulations

Various manure-based compost extract formulations have been found effective in comparison with extracts prepared from composts lacking manure. Cattle manure-based compost showed excellent control of gray mold (*Botrytis cinerea* Pers.) on strawberries and various powdery mildews (Wettzien, 1991). Elad and Shteinberg (1994) reported partial control of *B. cinerea* on tomato, pepper foliage, and grape berries.

Yohalem et al. (1996) and Ketterer et al. (1992) reported that water extracts of spent-mushroom-compost could also control apple scab. Zhang et al. (1998) demonstrated that if biocontrol agents are fortified with compost mixtures they could also reduce several soilborne diseases.

Biocontrol Agents from Guttation Fluid and Nematodes

A recent development in the biocontrol of plant diseases has been found in bacteria isolated from the guttation fluid of susceptible host varieties. Such bacteria were investigated for their potential as biocontrol agents by Fukeri et al. (1999), who reported encouraging results for the biocontrol of plant diseases.

Biocontrol agents have also been isolated from the cysts of *Heterodera glycines* Ichinohe, from which an unidentified sterile hyphomycete nematophagous fungus was isolated (Kun et al., 1998) as an effective biocontrol agent.

Mycorrhiza in Plant Disease Control

Arbuscular mycorrhizal fungi (AMF) are an economically and ecologically important group of symbiotic fungi, which colonize roots of over 80 percent of plant species. They are almost ubiquitous in natural and agricultural terrestrial ecosystems, and their role in controlling plant diseases has been found very valuable. Mark and Cassels (1996), Trotta et al. (1996), Cordier et al. (1998), and Norman et al. (1996) reported that diseases caused by *Aphanomyces, Fusarium, Phytophthora,* and *Sclerotium* could be successfully controlled by AMF. The subject has been reviewed by many workers, among which the reviews of Hooker et al. (1994) and Azcón-Aguilar and Barea (1996) are particularly useful.

Azcón-Aguilar and Barea (1996) proposed a number of mechanisms by which AMF can control soilborne diseases. These mechanisms include changes in the nutritional status of the host plant, biochemical and anatomical changes in plant tissues, stress alleviation of plants, microbial changes and their interactions in rhizosphere, and the morphological changes in the root system.

Plant Extracts and Disease Control

The fungicide resistance of pathogens, excessive use of chemicals in agriculture, and the resulting damage to the ecosystems have been quite alarming since around 1970. Plant pathologists have recognized these problems, and current studies have been focused on evaluating plants that produce inhibitory substances that could be used in the control of plant diseases. Ahmad and Prasad (1995) used leaf extract of *Lawsonia alba* and root extract of *Datura stramonium* (L.) against anthracnose of papaya caused by *Colletotrichum papaya* and showed their control.

Singh and Majumdar (2001) tested extracts of garlic (*Allium sativum* L.), onion (*A. cepa* L.), neem (*Azadirachta indica* A. Juss), turmeric (*Curcuma longa* L.), datura (*Datura stramonium* L.), basil (*Ocimum sanctum* L.), and ginger (*Zingiber officinale* Roscoe.) for the control of *Aletrnaria alternata* (Fr.) Keissler on pomegranate (*Punica granatum* L.), and reported that concentrations of all extract types showed significant disease reduction, but the most effective extract was from garlic, which was followed by turmeric.

Rawal and Thakore (2003) tested the efficacy of extracts from *Azadirachta indica* A. Juss., *Brasisca alba* L., *Cassia siamia* L., *Datura stramonium, Diospyros cordifolia* L., *Lantana camara* L., *Madhuca indica* L., and *Ocimum sanctum* against the fruit rot of sponge gourd (*Luffa cylindrica* Roem) caused by *Fusarium solani* (Mart.) Sacc., and found that extracts of *A. indica, D. stramonium, D. cordifolia, L. camara,* and *O. sanctum* were inhibitory to the pathogen.

The antibacterial activity of plant extracts to phytopathogenic bacteria like *Xanthomonas campestris* pv. *campestris,* the cause of black rot of cabbage, was shown by Sharma and Mehta (2001). Mistry et al. (2001) evaluated various plant extracts for their in vitro effect against leaf spot of papaya caused by *Alternaria alternata.*

Plant Residues and Biocontrol of Diseases

Cruciferous plant residues can reduce the incidence of *Aphanomyces* root rot of pear when used as a soil amendment. When hydrolyzed, these residues release numerous compounds such as isothiocyanates, thiocynate ion, nitrites, and epithionitrites. Volatile substances degrade the glucosinolates released from cruciferous plants, and the products of their degradation are toxic to *Aphanomyces euteiches* Drechsler (Smolinska et al., 1997).

Soil amendment with hairy vetch (*Vicia pillosa* Roth) could suppress soilborne pathogens. In addition to hairy vetch, many other legumes also give similar results and are effective in the control of soilborne diseases of plants. *Medicago sativa* L. soil amendment can effectively reduce soil populations of *Thielaviopsis basicola* (Condole and Rothrock, 1997). Soil amendment with legumes has been shown to be particularly effective against infections of *Thielaviopsis basicola, R. solani,* and *Pythium* spp. It has been suggested that the control of soilborne diseases is due to the presence of inhibitory volatile compounds such as ammonia, which can penetrate and disrupt cell membranes.

BIOCONTROL OF VEGETABLE DISEASES

Many workers using different antagonistic microbes have investigated the biocontrol of several foliar diseases as shown below.

Lettuce

Blakeman and Fokkema (1982) reported biocontrol of *B. cinerea* on lettuce with epiphytic filamentous fungi, while Wood (1950) reported its control with bacteria and actinomycetes. Clark and Lorbeer (1977), Pennock-Vos et al. (1990), and White et al. (1990) also demonstrated that certain bacteria and S*treptomyces* could control *B. cinerea* infections of lettuce. El-Tarabily et al. (2002) showed biocontrol of *Sclerotinia minor* Jagger on lettuce by certain organisms isolated from a lettuce field in Al-Ain, UAE. They demonstrated that *Serratia marcescens, Streptomyces viridodiasticus,* and *Micromonopora carbonacea* were the most promising agents of biocontrol from among a total of 85 bacteria, 94 streptomyces, and 35 nonstreptomyces isolates. Adams and Ayers (1982) showed biocontrol of Sclerotinia lettuce drop in the field by *Sporidesmium sclerotivorum* Ucker, Ayers & Adams. Budge and Whipps (1991) showed that greenhouse grown lettuce and celery could be protected from *S. sclerotivorum* by *Coniothyrium minitans* and *Trichoderma* spp. In a separate study, Budge et al. (1995) found biocontrol of *S. sclerotivorum* with *C. miniatans* and *Gliocladium virens.*

Chillies and Pepper

Phytophthora capsici Leonian causes rot in the aerial and subterranean parts of pepper and also many other plants. Its control with resistant varieties and fungicides showed little success, but it was controlled with *Trichoderma harzianum,* which also controlled other chilli diseases caused by *Rhizoctonia* and *Pythium* (Sid Ahmed et al., 1999). Manoranjitham et al. (2000) reported that a talc-based formulation of *Trichoderma viride* and *Pseudomonas fluorescens* was effective in the control of damping-off in chillies. Many workers have reported efficacy of *P. fluorescens* isolates against many soilborne and foliar diseases of plants. Ramamoorthy and Samiyappam (2001) reported that isolates Pf1 and ATB of *P. fluorescens* gave increased growth of chilli plants, produced maximum amount of indole acetic acid in vitro, and decreased incidence of *Colletotrichum capsici* (Sydow) E. Butler fruit rot under greenhouse conditions. Bunker and Mathur (2001) reported that *Trichoderma harzianum, T. aureoviride,* and *Gliocladium virens* could protect chilli against dry root rot caused by *Rhizoctonia solani.*

Cabbage and Chinese Cabbage

Leifert et al. (1993) reported biocontrol of *Botrytis cinerea* and *Alternaria brassicicola* (Schwein.) Wiltshire on Dutch white cabbage by *Pseudomonas fluorescens* isolates CL42, CL66, CL82 and *Serratia plymuthica* isolate CL43.

Club root of cultivated *Brassicas* spp., caused by soilborne *Plasmodiophora brassicae* Woronin, could be controlled by treatment of cabbage seeds with isolates of *Heteroconium chaetospira* (Grove) M.B. Ellis (Nara-siwa et al., 1998).

Parwal and Okra

The parwal (*Trichosanthes dioica* Roxb) is a popular vegetable in India. Its vine and fruits suffer from diseases caused by *Phytophthora cinnamoni* Rands. Singh et al. (2002) reported its control with certain organic amend-ments and *Trichoderma harzianum*. Khan and Verma (2002) reported that root-knot nematode [*Meloidogyne incognita* (Kofold & White) Chitwood] on okra (*Abelmoschus esculentus* L.) could be achieved with the application of *Paecilomyces lilacinus*.

Eggplant and Tomato

Marois et al. (1982) reported that *Talaromyces flavus* (Klöcker) Stolk et Samson is a promising biocontrol agent of *Verticillium* wilt of eggplant. Tomato wilt caused by *Fusarium oxysporum* f. sp. *lycopersici* (Sacc.) Snyder & Hansen, which is a widespread disease of general concern, could be effectively controlled using biocontrol agent *Penicillium oxalicum* (De Cal et al., 1995). Suppression of tomato wilt has also been reported by fluo-rescent pseudomonads (Lemanceau and Alabouvette, 1993) and nonpatho-genic strains of *Fusarium* (Davis, 1968; Abbatista et al., 1988; Tamietti et al., 1993). Fuchs and Défago (1991) showed that tomato plants could be protected from its wilt disease by a combination of a nonpathogenic *Fusarium* and different bacteria even in untreated soil. Many different biocontrol agents have been found effective against *Fusarium* wilt of tomato and include nonpathogenic species of *Fusarium*. Some other bio-control agents are effective against *Fusarium* wilt of tomato, namely, *Cepahalosporium* sp. (Phillips et al., 1967), *Macrophonospora* sp. (Smith, 1957), and *Penicillium oxalicum* (De Cal et al., 1999), M'Piga et al. (1997) reported that tomato plants develop increased resistance to *Fusarium oxy-sporum* f. sp. *lycopersici* if they have been treated with the endophytic strain 63-28 of *Pseudomonas fluorescens*.

The biocontrol of southern blight of tomato, caused by *Sclerotium rolfsii* Curzi, could be achieved by a simplified formulation of *Trichoderma koningii* (Latunde-Dada, 1993).

Carrot, Radish, Sugarbeet, and Onion

Species of streptomycetes and nonstreptomycete actinomyces can con-trol cavity spot of carrots caused by *Pythium coloratum* (El-Tarabily et al.,

1997). *Fusarium* wilt of radish could be controlled by fluorescent pseudomonads (Scher and Baker, 1982; Raaijmakers et al., 1995). Seed and seedling diseases of radish caused by *Pythium* spp. and *Rhizoctonia solani* have been controlled with *Trichoderma harzianum* (Hartman et al., 1980).

Holmes et al. (1998) studied damping-off of sugarbeet caused by *Pythium ultimum* Trow and showed its control with formulations of *Pythium oligandrum*. Cordoso and Echandi (1987) showed biocontrol of *Rhizoctonia* root rot of snapbean with binucleate *Rhizoctonia*-like fungi.

Phyllosphere bacteria controls onion diseases due to *Botrytis squamosa* Viennot-Bourgin and *B. cinerea* (Clark and Lorbeer, 1977). *Chaetomium globosum*, *Trichoderma viride*, and *T. harzianum* were effective biocontrol agents against the white rot of onion (Kay and Stewart, 1994). When these antagonists were applied as a soil additive of sand and bran fungal homogenate in a ratio of 1:2:1, they showed control of white rot of onion equivalent to the control achieved by the fungicide procymidione. *Cladosprium herbarum* controls *B. cinerea* on sugarbeet and onion (Clark and Lorbeer, 1977; Pennock-Vos et al., 1990; White et al., 1990).

Bean and Pea

Papavizas and Lewis (1989) observed that chemical control of southern blight and damping-off of snapbean caused by *Sclerotium rolfsii* is impractical as well as uneconomical. They conducted greenhouse trials on the control of these diseases and demonstrated that *Gliocladium* and *Trichoderma* spp. could control them. Zhou and Recleder (1989) controlled white mold by applying a suspension of spores of *Epicoccum purpurascens* to snapbean plants.

Gliocladium virens alone caused better control of wilt complex of French bean (*Phaseolus vulgarius* L.) caused by *Sclerotium rolfsii, R. solani,* and *F. oxysporum* f. sp. *phaseoli* Kendr. & Snyder than a mixture of fungicides (Mukherjee and Tripathi, 2000).

Callan et al. (1990) reported that warm-season crops like snap bean (*Phaseolus vulgaris*), lima bean (*P. limensis*), and other similar crops are subject to imbibitional chilling injury and become susceptible to pre-emergence damping-off from *Pythium* spp. This susceptibility occurs when seeds are planted in cold soil, where they exude soluble carbohydrates, potassium ions, and proteins, which predispose seeds of pea and bean to damping-off. The disease could be controlled by biopriming of dry seeds with *Pseuodmonas fluorescens* strain AB254.

Root rot of pea and bean caused by *Aphanomyces eutiches* contributes to 15 percent annual losses in the Great Lakes states. Muehlchen et al. (1990) reported that the incidence of root rot of pea and bean could be controlled if white mustard (*Sinapsis alba* L.) is planted after two successive crops of pea and bean. Smolinska et al. (1997) reported that the addition of seed meal of *Brassica napus* L. to the soil as a soil amendment also caused a reduction in the incidence of *Aphanomyces* root rot.

Cucumber

The consumption of cucumbers in salads and pickles has become a popular practice in most countries due to their low caloric value, which reduces the risk of obesity among its consumers. In addition, it also acts as a taste enhancer of the meal. Although it is a tropical crop of warm climates, it can be grown in cool temperate climates in greenhouses. Under such conditions, the plants are very susceptible to diseases, such as downy mildew (*Pseudoperonospora cubensis* Berk. & Curtis Rostov.). Fusarium wilt (*Fusarium oxysporum* f. sp. *niveum* E.M. Smith) Snyder & Hansen, anthracnose (*Colletotrichum orbiculare* Berk. & Mont.) v. Arx, powdery mildew (*Erysiphe cichoraceum* DC), and gray mold *(Botrytis cinerea)*. Gray mold, caused by *Botrytis cinerea* in greenhouse cucumbers, could be controlled by the use of *Trichoderma harzianum* alone or in combination with fungicides (Elad and Zimand, 1993). *Pythium* root rot of long English cucumber grown in hydroponic culture has been controlled by *Pseudomonas fluorescens* strain BTP1 and its siderophore negative mutant M3 (Onega et al., 1999). Singh et al. (1999) reported that chinolytic bacteria could control *Fusarium* wilt of cucumber. Liu et al. (1995) reported that cucumber plants treated with *P. putida* strain 89B-27 and *Serratia marcescens* strain 90-166 develop resistance against anthracnose *(Colletotrichum orbiculare)*. Nonpathogenic strains of *F. oxysporum* combined with *P. putida* have also been shown to control *Fusarium* wilt of cucumber (Park et al., 1988). The damping-off of cucumbers, caused by *Rhizoctonia* spp., could be controlled by nonpathogenic binucleate rhizobacteria (Villajuan-Abgona et al., 1996).

Raupach and Kloepper (1988) showed that several rhizobacteria could control anthracnose *(Colletotrichum orbiculare)*, angular leaf spot [*Pseudomonas syringae* pv. *lachrymans* (Smith & Bryan)], and cucumber wilt [*Erwinia tracheiphila* (Smith), Bergey et al.]. *Bacillus pumilus* strain INR7, *B. subtilis* strain GBO3 and *Curobacterium flaccumfaciens* strain ME1, when applied as seed treatment or foliar sprays, could control several diseases of cucumber to a level equivalent to synthetic fungicides (Raupach and Kloepper, 1988).

BIOCONTROL OF FRUIT DISEASES

During the past several years there has been an upsurge of research activities on the biocontrol of fruit diseases. Although considerable work has been done on the postharvest diseases of fruits, some work has also been done on the fruit diseases in the field. The following examples illustrate the research done in this area.

Apple and Pear

Gray mold, caused by *Botrytis cinerea,* is a major threat to many fruits and vegetables, and a lot of work has been done on the biocontrol of *B. cinerea.* Tronsmo and Raa (1977) showed the antagonistic action of *Trichoderma koningii* and *Gliocladium* sp. against gray mold of apple. Roberts (1990) reported that *Cryptococcus laurentii* could also control apple gray mold. El-Ghaouth et al. (1998) reported that *Candida saitoana* reduced diseases caused by *B. cinerea* and *Penicillium expansum* in apple. Teixidó et al. (1998) showed that *Candida sake* was also effective in controlling gray mold of apples. Leibinger et al. (1997) demonstrated the effectiveness of mixtures of yeasts and bacteria, and reported that they could control apple diseases caused by *B. cinerea* and *Penicillium expansum.* Roiger and Jeffers (1991) studied crown and root rot of apple seedlings caused by *Phytophthora* sp. and reported their control with *Trichoderma* spp. Culleen et al. (1984) reported the antagonistic effect of *Chaetomium globosum* against apple scab, caused by *Venturia inaequalis* (Cooke) Winter in Thüm, under field conditions (Nicholson and Rahe, 2004).

Lindow et al. (1996) reported results of field trials conducted over a period of 16 years on the control of fire blight of pear. These workers reported that use of *Pseudomonas fluorescens* strain A506 in combination with streptomycin and oxytetracycline could reduce pear fire blight by 40 to 50 percent.

Montesinos et al. (1996) isolated many strains of *Erwinia herbicola* and *Pseudomonas fluorescens* from the aerials parts and roots of several plants and showed their antagonism to brown spot of pear caused by *Stemphillium vesicarium* (Wallr.) Simmons, which attacks leaves, fruits, and occasionally twigs of pear. They found that *P. fluorescnes* strain EPS 288 was the most effective and could reduce disease incidence by 57 percent.

Fire blight caused by *Erwinia amylovora* (Busill) Winslow is one of the most serious diseases limiting production of pear, apple, and other pome fruits in many parts of the world (Miller and Scroth, 1972; Adwinckle and Beer, 1979). Flowers are the most common site of infection by *Erwinia amylovora.* Establishment of large epiphytic populations on stigmatic surfaces precedes infection of flowers by the pathogen (Miller and Scroth, 1972). Epiphytotic populations of the pathogen are common even on flowers that subsequently do not become infected. The control of fire blight of pear has been commonly based on certain antibiotics and copper-containing compounds, but this approach had to be abandoned due to the emergence of antibiotic resistant strains of *E. amylovora* in most apple-pear-growing regions of the world. Resistant pome varieties have also not been identified. The high costs of chemical control, resistance to antibiotics/fungicides, and lack of effective control measures against fire blight of pear have been responsible for the biocontrol of this disease (Adwinckle and Beer, 1979; Lindow et al., 1996). Kearns and Hale (1995) isolated *Erwinia herbicola* and *Pseudomonas fleorescens* from blossoms in New Zealand apple orchards and demonstrated in vitro inhibition of *Erwinia amylovora* on pear and apple.

Peach

Peach twig blight caused by *Monilinia laxa* (Aderh. and Ruhl) Honey ex Whetzel; Mordue can be controlled by the application of spore and mycelial preparations of *Epicoccum nigrum* (Madrigal et al., 1994). *Penicillium frequentans* has also been shown to be effective against peach twig blight (De Cal et al., 1990).

Grape and Strawberries

Derckel et al. (1999) demonstrated differential behavior of *Botrytis cinerea* from grapevine, and reported that strain T4 was less aggressive and enhanced the accumulation of many secondary-defense metabolites that contributed to the increased resistance of grapevine against gray mold. Dubos and Bulit (1981), Bisiach et al. (1985), and Guillino and Garibaldi (1988) showed that *Botrytis cinerea* on grape could be controlled by certain isolates of *Trichoderma* and *Gliocladium* spp., *Cladosporium herbarum,* and *Penicillium* spp.

Falk et al. (1996) investigated downy mildew of grape caused by *Plasmopara viticola* (Berk. & Curtis ex de Bar) Berl. & de Toni, and reported its control by postinfection application of suspensions of *Fusarium proliferatum* G6 to grape.

Bhatt and Vaughan (1962) reported that blossom blight and green fruit rot of strawberry could be controlled by *C. herbarum*. Tronsmo and Dennis (1974) reported that fruit rots of strawberry caused by *B. cinerea* could also be controlled by *Trichoderma* spp. Peng and Sutton (1990) described various methods of biocontrol of gray mold of strawberry.

Citrus and Lemon

Fang and Tsao (1995) conducted greenhouse trials for the control of root rot of sweet orange [*Citrus sinensis* (L.) Osbeck], caused by *Phytophthora parasitica* Dastur, and found that *Penicillium funiculosum* could control the disease. Chalutz and Wilson (1989) showed that control of green and blue mold and sour rot of many citrus cultivars was possible by *Debaromyces hansenii*. Smilanick and Dennis-Arrue (1992) reported control of green mold of lemons with a species of *Pseudomonas*. Wilson and Chalutz (1989) also reported that certain antagonistic yeasts and bacteria could control *Penicillium* rots of citrus fruits.

Watermelon

Larkin et al. (1996) isolated nearly 400 bacteria, actinomycetes, and fungi from watermelon roots growing in soils suppressive and nonsuppressive to *Fusarium* wilt of watermelon. They found that specific isolates of nonpathogenic *F. oxysporum* from suppressive soil could reduce *Fusarium* wilt by 35 to 75 percent. Larkin et al. (1993) also showed the effect of successive watermelon plantings on *F. oxysporum* and other microbes in soils suppressive and conducive to wilt of watermelon.

Mango

Koomen and Jeffries (1993) isolated 648 microbes from mango blossoms, leaves, and fruits, and found that 121 isolates were antagonistic to the anthracnose pathogen of mango *Colletotrichum gleosporioides* (Penz.) Penz. & Sacc. However, out of these isolates only *Pseudomonas fluorescens* showed significant reduction of mango anthracnose.

Avocado

Avocado pear has become a favorite fruit during the past several decades. The fruit is prone to several postharvest diseases, and their biocontrol potential has been demonstrated by Korsten (1993) and Korsten and Jeffries (2000) by field sprays of *Bacillus subtilis*.

POSTHARVEST FRUIT AND VEGETABLE DISEASES

In the past decades postharvest spoilage of fruits and vegetables has become a challenging problem for plant pathologists. In apples alone, postharvest losses of 15 to 25 percent have been reported. The spoilage of fruits and vegetables is undoubtedly a pathological problem and occurs because they become infected by a handful of opportunistic organisms. This kind of spoilage is done chiefly by *Botrytis cinerea*, *Monilinia fruticola* (Winter) Honey; Mordue, *Penicillium digitatum*, *P. expansum*, *P. italicum*, and occasionally by *Rhizopus stolonifer* (Ehrenb.:Fr.) Lind. Application of synthetic fungicides before and after harvest has been common practice for the control of these diseases, but this method has become unpopular and is not favored by consumers due to concerns for the environment and human health. The use of benomyl for the postharvest treatment of fruit and vegetables has been banned in the United States (Sancez, 1990) for the reasons explained earlier. Pathogen resistance to several synthetic fungicides has also appeared in many pathogens (Rosenberger and Meyer, 1979; Viñas et al., 1991; Holmes and Eckert, 1999). The appearance of pathogen resistance to fungicides has also brought disrepute to them (Spotts and Cervantes, 1986; Viñas et al., 1993). A number of naturally occurring antagonistic fungi and bacteria have been recognized as potential biocontrol organisms to reduce incidence of postharvest diseases of fruits and vegetables (Janisiewicz, 1988; Droby and Chalutz, 1993; Wilson and El-Ghaouth, 1993). The biocontrol of many plant diseases is considered a promising alternative in comparison to the synthetic fungicides. The most recent works cover diseases of apple, pear, citrus fruits, stone fruits, avocado, chillies, bean, cucumber, pea, and tomato (Pusey and Wilson, 1984; Appel et al., 1988; Janisiewicz, 1988; Chalutz and Wilson, 1989; Droby et al., 1989, 1993; Wilson and Chalutz, 1989; Roberts, 1990; Droby and Chalutz, 1993; Janisiewicz and Bors, 1995; Sobiczewski and Bryk, 1996; Sobiczewski et al., 1996; Leibinger et al., 1997; Sutton et al., 1997; Lima et al., 1997, 1998; Zhou et al., 1999; Zahavi et al., 2000). Viñas et al. (1991) and Nunes et al. (2002) investigated the potential of biocontrol of Golden Delicious apples with the bacterium *Pantoea agglomerans* (CPA-2) and proved its efficacy as an antagonist against postharvest diseases of apple.

The efficacy of some yeasts in the biocontrol of apple and pea by *Candida sake* was shown by Chand-Goyal and Spotts (1997) and Usall et al. (2000). A virulent strain of *Geotrichum canidum* Link causes sour rot on grapefruit, oranges, lemons, and limes, but its avirulent strain was discovered and patented and could control green mold of citrus and also diseases of apples, pears, and strawberries. The biocontrol of *Botrytis cinerea* by *Paenibacillus polymyxa* isolated from mature strawberries has been shown (Helbig, 2001).

MECHANISMS OF BIOCONTROL

The role of arbuscular mycorrhizal fungi (AMF) in controlling pathogens of agricultural crops indicates that they can control diseases caused by *Aphanomyces, Phytophthora, Fusarium,* and *Sclerotium* (Hooker et al., 1994; Azcón-Aguilar and Barea, 1996; Mark and Cassells, 1996; Norman et al., 1996; Trotta et al., 1996; Cordier et al., 1998; Mukerji, 1999).

The proposed mechanisms underlying biocontrol of plant diseases by AMF include the following:

- changes in nutrient status of plants
- biochemical changes in plant tissues
- anatomical changes in host cells
- stress alleviation
- microbial changes in the rhizosphere
- changes to host root-system morphology

The beneficial effects of rhizobacteria in the control of soilborne diseases show that their activity is due to

- competition from substrate,
- niche exclusion,
- production of extracellular enzymes,
- induced plant resistance, and
- siderophore production by PGPR.

The evidence in support of the above mechanisms has been provided by many cases, for example, *Pythium* root rot of cucumber (Onega et al., 1999; Singh et al., 1999), *Fusarium* wilt of tomato (De Cal et al., 1999), *Phytophthora* root rot of citrus (Fang and Tsao, 1995), and soilborne diseases in general (Inbar and Chet, 1991). Madi et al. (1997) indicated that biocontrol of diseases is not mediated by a single but many mechanisms.

Rhizobacteria can produce antibiotics, enzymes, and phenols. Rhizobacteria control plant diseases and enhance plant growth (Kloepper and Scroth, 1978; Kloepper, 1991; Loper and Scroth, 1986). Chet (1998) and Meena et al. (2000) have shown production of phenylalanine ammnonialyase and phenols by rhizobacteria. Induction of systemic resistance against bacterial and fungal disease of cucumber has been demonstrated by several workers (Wei et al., 1991; Liu et al., 1992, 1995; Ryals et al., 1996; Zhang et al., 1996, 1998; M'Piga et al., 1997). Larkin et al. (1996) reported that suppressive soils could also induce systemic resistance in plants against many diseases (Kloepper et al., 1992; Hammerschmidt and Yang-Cashman, 1995; Hoffland et al., 1996).

Involvement of enzymes with potential for biocontrol of many diseases has been reported (Alfonso et al., 1992; Hidalgo et al., 1992; Larena and Melgarejo, 1993; Lorito et al., 1993; Liu et al., 1995; Derckel et al., 1999; Singh et al., 1999). The enzymes identified were lytic β-glucosidase and chinolytic.

Finstein and Alexander (1962) reported that competition for carbon and nitrogen by the pathogen *Fusarium* and rhizobacteria would also trigger a protective mechanism. The interactions between fluorescent pseudomonads and nonpathogenic *Fusarium* may also provide control of *Fusarium* diseases (Lemanceau and Alabouvette, 1993). In this kind of interaction the mechanisms of competition for nutrients and antibiosis seem to work (Weinhold and Bowian, 1968; Gurusiddaih et al., 1986).

Many examples illustrate that the biocontrol of plant diseases is possible by a nonpathogenic strain of its pathogenic type. Larkin et al. (1996) reported the control of *Fusarium* wilt of watermelon with a nonpathogenic species of *Fusarium*.

COMMERCIAL BIOCONTROL PRODUCTS

Several commercial preparations are now available that are reported to offer biocontrol of fruit and vegetables diseases. These products are based on fungal, yeast, or bacterial components. They may be used alone, in combination, or as part of an integrated control in order to minimize fungicide inputs. These products are Aspire (obtained from *Candida oleophila* strain 182) and Biosave 10 and 11 (obtained from *Pseudomonas syringae* strains ESC10 and ESC11). In South Africa, Avo-green and Yield Plus have been registered as biocontrol products for the control of fruit diseases. Plant-Shield is a fungal-based product of *Trichoderma harzianum,* and is manufactured by Bio-Works, Inc., Geneva, New York. Mycostop is another commercial product of *Streptomyces grisiovirdis* and is manufactured by Kemira Agro of Finland. Both Mycostop and Plant-Shield are recommended for the control of a number of vegetables diseases. Quantum 4000 (Weller, 1988) is also available as a commercial biocontrol product. In India, *Aspergillus niger* AN 27 has been developed as a biocontrol agent of many soilborne pathogens (Sen, 2000), and its formulations include Kalisena SL, Pusa Mrida, and Kala Sipahi, which could be used as a seed dresser.

CONCLUSIONS

Production of agricultural crops and their protection from diseases have been rapidly evolving during the past several decades.

Fruits and vegetables are now commodities of global commerce and are regularly airfreighted to world markets. However, this kind of transportation can result in postharvest decay and spoilage of fruits and vegetables. For the time being, this problem has been brilliantly contained by the application of biocontrol preparations. These products have been patented and are now commercially available. The biocontrol technology for soilborne diseases of fruits and vegetables using rhizobacteria, arbuscular mycorrhizal fungi, and antagonistic microbes is developing very rapidly and offers a satisfactory level of control of plant diseases. Although the biocontrol of plant diseases is a fashionable research area these days, it is hoped that it grows as a long-lasting method and does not fail us in the future.

REFERENCES

Abbatista, G.I., L. Ferrais, and A. Matta. (1988). Variations of phenoloxidase activities as a consequence of stresses that induce resistance to Fusarium wilt of tomato. *Journal of Phytopathology* 122: 45-53.

Adams, P.B. and W.A. Ayers. (1982). Biological control of Sclerotinia lettuce drop in the field by *Sporidesmium sclerotivorum*. *Phytopathology* 72: 485-488.

Adwinckle, H.S. and S.V. Beer. (1979). Fire blight and its control. *Horticultural Review* 1: 423-474.

Ahmad, J.S. and R. Baker. (1987). Competitive saprophytic ability and cellulolytic activity of rhizosphere-competent mutants of *Trichoderma harzianum*. *Phytopathology* 7: 358-362.

Ahmad, S.K. and J.S. Prasad. (1995). Efficacy of foliar extract against pre- and postharvest of sponge fruits. *Letters in Applied Microbiology* 21: 373-375.

Ahmed, S.R. and J.P. Agnihotri. (1977). Antifungal activity of some plant extracts. *Journal of Mycology and Plant Pathology* 7: 180-181.

Alfonso, C., F. Del Amo, O.M. Nuero, and F. Reyes. (1992). Physiological and biochemical studies on *Fusarium oxysporum* f.sp. *lycopersici* race 2 for its bicontrol by non-pathogenic fungi. *FEMS Microbiology Letters* 99: 169-174.

Appel, D.J., R. Gees, and M.D. Coffey. (1988). Biological control of the postharvest pathogen *Penicillium digitatum* on Eureka lemons. *Phytopathology* 78: 1593.

Azcón-Aguilar, C. and J.M. Barea. (1996). Arbuscular mycorrhiza and biological control of soil-borne plant pathogens: An overview of the mechanisms involved. *Mycorrhiza* 6: 457-464.

Baker, K.F. (1987). Evolving concepts of biological control of plant pathogens. *Annual Review of Phytopathology* 25: 67-85.

Baker, K.J. and R.J. Cook. (1974). *Biological Control of Plant Pathogens*. Freeman, San Francisco, CA.

Baker, R. (1985). Biological control of plant pathogens: Definitions. In *Biological Control in Agricultural IPM Systems*, eds. M.A. Hoy and D.C. Herzog. Academic Press, New York, pp. 25-39.

Bhatt, D.D. and E.K. Vaughan. (1962). Preliminary investigations on biological control of gray mould *(Botrytis cinerea)* of strawberries. *Plant Disease Reporter* 46: 342-345.

Bisiach, M., G. Minervini, A. Vercesi and F. Zerbetto. (1985). Six years of experimental trials in biological control against gray mould. *Proceedings Eighth Botrytis Symposium (Alba, Torino, Italy) Quademi di Viticultura ed Ecologia dell' Universitas di Torino* 9: 285-297.

Blakeman, J.P. and N.J. Fokkema. (1982). Potential for biological control of plant diseases on the phylloplane. *Annual Review of Phytopathology* 20: 167-192.

Brent, K.J. (1995). *Fungicide Resistance in Crop Pathogens: How Can it Be Managed?* FRAC Monograph No.1, GIFAP (Brussels).

Broadbent, P., K.F. Baker, N. Franks and J. Holland. (1971). Bacteria and actinomycetes antagonistic to fungal root pathogens in Australia soils. *Australian Journal of Biological Sciences* 24: 925-944.

Budge, S.P., M.P. McQuilken, J.S. Fenlon and J.M. Whipps. (1995). Use of *Coniothyrium minitans* and *Gliocladium virens* for biological control of Sclerotinia sclerotiorum in glasshouse lettuce. *Biological Control* 5: 513-522.

Budge, S.P. and J.M. Whipps. (1991). Glasshouse trials of *Coniothyrium minitans* and *Trichoderma* species for biological control of *Sclerotinia sclerotiorum* in celery and lettuce. *Plant Pathology* 40: 59-66.

Bunker, R.N. and K. Mathur. (2001). Antagonisms of a local biocontrol agent to *Rhizoctonia solani* inciting dry rot of chilli. *Journal of Mycology and Plant Pathology* 31: 50-53.

Burr, T.J. and A.J. Caesor. (1984). Beneficial Plant Bacteria. *CRC Critical Review of Plant Science* 2: 1-20.

Callan, N.W., D.E. Mathre and J.B. Miller. (1990). Bio-priming seed treatment for biological control of *Pythium ultimum* pre-emergence damping-off of Sh2 sweet corn. *Plant Disease* 74: 368-372.

Chalutz, E. and C.L. Wilson. (1989). Postharvest biological control of green and blue mold and sour rot on citrus fruit by *Debaromyces habsenii*. *Plant Disease* 74: 134-137.

Chand-Goyal, T. and R.A. Spotts. (1997). Biological control of postharvest diseases of apple and pear under semi-commercial and commercial conditions using three saprophytic yeasts. *Biological Control* 10: 199-206.

Chao, W.L., E.B. Nelson, G.E. Harman and H. Hach. (1986). Colonization of the rhizosphere of biological control agents applied to seeds. *Phytopathology* 76: 60-65.

Chet, I. (1998). Lytic enzymes and host recognition: main factors in mycoparasitism (Abstr.). Sixth International Mycological Congress, IMC6, Jerusalem, Israel, 1998.

Clark, C.A. and J.W. Lorbeer. (1977). The role of phyllosphere bacteria in pathogenesis by *Botrytis squamosa* and *B. cinerea* on onion leaves. *Phytopathology* 87: 96-100.

Condole, B.L. and C.S. Rothrock. (1997). Characterization of the suppressiveness of hairy Vetch-amended soils to *Thielaviopsis basicola*. *Phytopathology* 67: 197-202.

Cook, R.J. and A.D. Rovira. (1976). The role of bacteria in the biological control of *Gaeumanomyces graminis* by suppressive soil. *Soil Biology and Biochemistry* 8: 265-273.

Cordier, C., M.J. Pozo, J.M. Barea, S. Gianinazzi and V. Gianinazzi-Pearson. (1998). Cell defence response associated with locaised and systemic resistance to *Phytophthora parasitica* in tomato by an arbuscular mycorrhizal fungus. *Molecular Plant Microbe Interactions* 11: 1017-1028.

Cordoso, J.E. and E. Echandi. (1987). Biological control of *Rhizoctonia* root rot of snapbean with binucleate *Rhizoctonia*-like fungi. *Plant Disease* 71: 167-170.

Culleen, D., F.M. Berbee and J.H. Andrew. (1984). *Chaetomium globosum* antagonizes the apple scab pathogen, *Venturia inaequalis* under field conditions. *Canadian Journal of Botany* 62: 1814-1818.

Davis, D. (1968). Partial control of *Fusarium* wilt of tomato by formae of *Fusarium oxysporum*. *Phytopathology* 58: 121-122.

De Cal, A., R. Garcia-Lepe, S. Pascual and P. Melgarejo. (1999). Effects of timing and method of application of *Penicillium oxalicum* on efficacy and duration of control of Fusarium wilt of tomato. *Plant Pathology* 48: 260-266.

De Cal, A., S. Pascual, I. Larena and P. Melgarejo. (1995). Biological control of *Fusarium oxysporum* f.sp. *lycopersici*. *Plant Pathology* 44: 909-917.

De Cal, A., E. M-Sagasta and P. Melgarejo. (1990). Biological control of peach twig blight *(Monilinia laxa)* with *Penicillium frequentans*. *Plant Pathology* 39: 612-618.

Défago G., C.H. Berling, U. Burger, D. Hass, G. Kahr, C. Keel. C. Voisard, P. Wurthner and B. Wulthrich. (1990). Suppression of black root rot of tobacco and other root diseases by strains of *Pseudomonas fluorescens*. In: *Biological Control of Soil-Borne Plant Pathogens*. eds. Hornby, D. Cook, R.J., Henis Y., KO. W.H. Rovira, A.D., Shcippers B. and Scott, P.R. CAB International, Oxon, U.K. pp. 93-108.

Derckel, J.P., F. Bailliul, S. Manteau, J.C. Audran, B. Haye, B. Lambert and L. Langendre. (1999). Differential induction of grapevine defenses by two strains of *Botrytis cinerea*. *Phytopathology* 89: 197-203.

Droby, S. and E. Chalutz. (1993). Mode of action of biological control agents for postharvest diseases. In, *Biological Control of Postharvest Diseases of Fruits and Vegetables-Theory and Practice*. eds. C.L., Wilson and M.E. Wisniewski. CRC Press, Boca Raton, FL. pp. 63-75.

Droby, S., E. Chalutz, C. Wilson and M. Wisniewski. (1989). Characterization of the biocontrol activity of *Debaromyces hansenii* in the control of *Penicillium digitatum* on grapefruit. *Canadian Journal of Microbiology* 35: 794-800.

Droby, S., R. Hofstein. C.L. Wilson, M. Wisniewski, B. Fridlender, L. Cohen, B. Weiss, A. Daus, D. Timar and E. Chalutz. (1993). Pilot testing of *Pichia guilliermondii*: a biocontrol agent of postharvest diseases of citrus fruit. *Biological Control* 12: 3: 47-52.

Dube. H.C. (2001). Rhizobacteria in biological control and plant growth promotion. *Journal of Mycology and Plant pathology* 31: 9-21.

Dubos, B. and J. Bulit. (1981). Filamentous fungi as biocontrol agents on aerial plant surfaces. In: *Microbial Ecology of the Phyllosphere*, ed. J.P. Blakeman. Academic Press, London. pp. 353-367.

Elad, Y. and D. Shteinberg. (1994). Effect of compost water extracts on gray mould *(Botrytis cineria). Crop Protection* 13: 109-114.

Elad, Y. and G. Zimand. (1993). Use of *Trichoderma harzianum* in combination or alteration with fungicides to control cucumber gray mould *(Botrytis cinerea)* under commercial greenhouse conditions. *Plant Pathology* 42: 324-332.

El-Ghaouth. A. C.L. Wilson and M. Wisniewski. (1998). Ultrastructural and cytochemical aspects of the biological control of *Botrytis cineria* by *Candida saitoana* in apple fruit. *Phytopathology* 88: 282-291.

El-Tarabily, K.A., G.E. Hardy, J. St.. K. Sivasithamparam, A.M. Hussein and I.D. Kurtböke. (1997). The potential for the biological control of cavity spot disease of carrots caused by *Pythium coloratum* by streptomycete and a nonstreptomycete actinomyces in Western Australia. *New Phytologist* 137: 495-507.

El-Tarabily. K.A., M.H. Soliman, A.H.O. Nasser, H.A. Al-Hassani. K. Sivasitharam, F. McKenna and G.E. Hardy St. (2002). Biocontrol of *Sclerotinia minor* using a chitinolytic bacterium and actinomycetes. *Plant Pathology* 49: 573-583.

Falk, S.P., R.C. Pearson, D.M. Gadoury, R.C. Seem and A. Sztejnberg. (1996). *Fusarium proliferatum* as a biocontrol agent against grape downy mildew *Phytopathology* 86: 1010-1017.

Fang, J.G. and P.H. Tsao. (1995). Efficacy of *Penicillium funiculosum* as a biological control agent against *Phytophthora* root rots of azalea and citrus. *Phytopathology* 85: 871-878.

Finstein, M.S. and M. Alexander. (1962). Competition for carbon and nitrogen between *Fusarium* and bacteria. *Soil Science* 94: 334-339.

Flaishman, M.A.. Z. Eyal, A. Zilberstein, C. Voisard and D. Hass. (1996). Suppression of *Septoria tritici* blotch and leaf rust of wheat by recombinant cyanide-production strains of *Pseudomonas putida. Molecular Plant-Microbe Interactions* 9: 642-645.

Fuchs, J. and G. Défago. (1991). Protection of tomatoes against *Fusarium oxysporum* f. sp, *lycopersici* by combining non-pathogenic Fusarium with diffeent bacteria in untreated soil. In, *Plant Growth Promoting Rhizobacteria*, eds. C. Keel, B. Koller, G. Défago, Progress and Prospects, IOBC/WPRS pp. 51-56.

Fukeri, R.. H. Fukui and A.M. Alvarez. (1999). Comparisons of single versus multiple bacterial species on biological control of anthurium blight. *Phytopathology* 89: 366-373.

Guillino, M.L. and A. Garibaldi. (1988). Biological and integrated control of gray mould of grapevine: results in Italy. *Bulletin EPPO Bulletin* 18: 9-12.

Gurusiddaih, S., D.M. Weller, A. Sarkar and J.R. Cook. (1986). Characterization of an antibiotic produced by a strain of *Pseudomonas fluorescens* inhibitory to

Gaeumannomyces graminis var. *tritici* and *Pythium* spp. *Antimicrobial Agents Chemotherapy* 29: 488-495.

Halder, Y., G.E. Harman, A.G. Taylor and J.M. Norton. (1983). Effects of pregermination of pea and cucumber seeds and seed treatment with *Enterobacter cloacae* on roots caused by *Pythium* spp. *Phytopathology* 73: 1322-1325.

Hammerschmidt, R. and P. Yang-Cashman. (1995). Induced resistance in cucurbits. In, *Induced Resistance to Disease in Plants, Developments in Plant Pathology 4*. Kluwer Academic Publishers, Dordrecht, the Netherlands. pp. 63-86.

Harman, G.E., C.K. Hays, M. Lorito, R.M. Broadway, A. Di Pietro, C. Peterbauer and A. Tronsmo. (1993). Chinolytic enzymes of *Trichoderma harzianum*: Purification of chitobiosidase and endochitinase. *Phytopathology* 83: 313-318.

Hartman, G.E., I. Chet and R. Baker. (1980). *Trichoderma hamatum* effects on seed and seedling diseases induced in radish and peas by *Pythium* sp. or *Rhizoctonia solani*. *Phytopathology* 70: 1167-1172.

Helbig, J. (2001). Biological control of *Botrytis cinerea* Pers. Ex Fr. in strawberry by *Paenibacillus polymyxa* (Isolate 18191). *Journal of Phytopathology* 149: 265-273.

Herr, L.J. (1988). Biological control of *Rhizoctonia* crown root rot of sugarbeet by binucleate *Rhizoctonia* spp. and *Laetisaria arvalis*. *Annals of Applied Biology* 133: 107-118.

Hidalgo, M., J. Stein and J. Eyzaguirre. (1992). β-glucosidase from *Penicillium purpurogennum*: purification and properties. *Biotechnology and Applied Biochemistry* 15: 185-191.

Hoffland, E., J. Hakulinen and J.A. van Pelt. (1996). Comparison of systemic resistance induced by avirulent and non-pathogenic *Pseudomonas* sp. *Phytopathology* 86: 757-762.

Holmes, G.J. and J.W. Eckert. (1999). Sensitivity of *Penicillium digitatum* and *P. italicum* to postharvest citrus fungicides in California. *Phytopathology* 89: 716-721.

Holmes, K.A., S.D. Nayagam and G.D. Craig. (1998). Factors affecting the control of beet by *Pythium ultimum* damping off of sugar beet by *Pythium oligandrum*. *Plant Pathology* 47: 516-522.

Hooker, J.E., M. Jaime-Vega and D. Atkinson. (1994). Biocontrol of plant pathogens using arbuscular mycorrhizal fungi. In: *Impact of Arbuscular Mycorrhizas on Sustainable Agriculture and Natural Ecosystems*. eds. S. Gininazzi and H. Schuepp. Basel, Switzerland, Birkhäuser-Verlag. pp. 191-200.

Howell, C.R., R.C. Beier and R.D. Stipanovick. (1988). Production of ammonia by *Enterobacter cloacae* and its possible role in the biological control of *Pythium* pre-emergence damping off by the bacterium. *Phytopathology* 78: 1075-1078.

Inbar, J. and I. Chet. (1991). Evidence that chitinase produced by *Aeromonas caviae* is involved in the biological control of soil-borne plant pathogens by this bacterium. *Soil Biology and Biochemistry* 23: 973-978.

Janisiewicz, W.J. (1988). Biological control of diseases of fruit. In *Biological Control of Plant Disease*. eds. K.G. Mukerji and K.L. Garg. CRC Press, Boca Raton, Fl. pp. 153-165.

Janisiewicz, W.J. and B. Bors. (1995). Development of a microbial community of bacterial and yeast antagonists to control wound-invading postharvest pathogens of fruits. *Applied Environmental Microbiology* 61: 3261-3267.

Kay, S.J. and A. Stewart. (1994). Evaluation of fungal antagonists for control of onion white rot in soil boxes. *Plant Pathology* 43: 371-377. *Soil Biology and Biochemistry* 23: 973-978.

Kearns, L.P. and C.N. Hale. (1995). Incidence of bacteria inhibitory to Erwinia amylovora from blossoms in New Zealand apple orchards. *Plant Pathology* 44: 918-924.

Ketterer, N., B. Fisher and H.C. Eltzien. (1992). Biological control of *Botrytis cinerea* on grapevine by compost extracts and their microorganisms in pure culture. In: *Recent Advances in Botrytis Research. Proceedings 10th International Botrytis Symposium*. eds. K. Verhoeff, N.E. Malathrakis and B. Williamson. Pudoc Scientific Publishers, Wageningen, Netherlands. pp. 179-186.

Khan, M.N. and A.C. Verma. (2002). Biological control of root-knot nematode-*Meloidogyne incognita* on okra *(Abelmoschus esculentus)* by *Paecilomyces lilacinus* in microplot condition (Abstr.). *Journal of Mycology and Plant Pathology* 32: 144.

Kloepper, J.W. (1991). Development of in vitro assay for prescreening antagonists of *Rhizoctonia solani* on cotton. *Phytopathology* 81: 1006-1013.

Kloepper, J.W. and M.N. Schroth. (1978). Plant growth-promoting rhizobacteria on radish. In: *Proc. 4th International Conference. Plant Pathogenic Bacteria*. Angers, France. pp. 879-882.

Kloepper, J.W., S. Tuzun and J.A. Kuc. (1992). Proposed definitions related to induced disease resistance. *Biocontrol Science and Technology* 2: 349-351.

Koomen, I. and P. Jeffries. (1993). Effects of antagonistic microorganisms on thepostharvest development of *Colletotrichum gloeosporioides* on mango. *Plant Pathology* 42: 230-237.

Korsten, L. (1993). Biological control of avocado fruit diseases. Ph.D. thesis, University of Pretoria, South Africa.

Korsten, L. and P. Jeffries. (2000). Potential for biological control of diseases caused by *Colletotrichum*. In: *Colletotrichum Host Specificity, Pathology and Host-Pathogen Interactions*. eds. D. Prusky, S. Freeman and M.B. Dickman, St. Paul, Minnesota: APS Press. pp. 266-295.

Kun, D.G., D.G. Riggs and J.C. Correll. (1998). Isolation, characterization and distribution of a biocontrol fungus from cysts of *Heterodera glycines. Phytopathology* 88: 465-471.

Larena, I. and P. Melgarejo. (1993). The lytic enzymatic complex of *Penicillium purpurogenum* and its effects on *Monilinia laxa. Mycological Research* 97: 105-110.

Larkin, R.P., D.L. Hopkins and F.N. Martin. (1993). Effects of successive watermelon plantings on *Fusarium oxysporum* and other microorganims in soils suppressive and conducive to *Fusarium* wilt of watermelon. *Phytopathology* 83: 1097-1105.

Larkin, R.P., D.L. Hopkins, and F.N. Martin. (1996). Suppression of *Fusarium* wilt of watermelon by nonpathogenic *Fusarium oxysporum* and other microorganisms recovered from diseases suppressive soil. *Phytopathology* 86: 812-819.

Latunde-Dada, A.O. (1993). Biological control of southern blight disease of tomato caused by *Sclerotium rolfsii* with simplified mycelial formulations of *Trichoderma koningii*. *Plant Pathology* 42: 522-529.

Leibinger, W., B. Beuker, M. Hahn, and K. Mendgen. (1997). Control of postharvest pathogens and colonization of apple surface by antagonistic microorganisms in the field. *Phytopathology* 87: 1103-1110.

Leifert, C., D.C. Sigee, R. Stanley, C. Knight, and H.A.S. Epton. (1993). Biocontrol of *Botrytis cineria* and *Alternaria brassicola* on Dutch white cabbage by bacterial-antagonists at cold-store temperatures. *Plant Pathology* 42: 270-279.

Lemanceau, P. and C. Alabouvette. (1991). Biological control of *Fusarium* diseases by fluorescent *Pseudomonas* and nonpathogenic *Fusarium*. *Crop Protection* 10: 279-285.

Lemanceau, P. and C. Alabouvette. (1993). Suppression of Fusarium wilts by fluorescent Pseuodomonads: mechanisms and applications. *Biocontrol Science and Technology* 3: 219-234.

Lima, G., F. De Curtis, R. Castoria, and V. De Cicco. (1998). Activity of the yeasts *Cryptococcus laurentii* and *Rhodotorula glutinis* against post-harvest rots on different fruits. *Biocontrol Science and Technology* 8: 257-267.

Lima, G., A. Ippolito, F. Nigro, and M. Salerno. (1997). Effectiveness of *Aureobasidium pullulans* and *Candida oleophila* against postharvest strawberry rots. *Postharvest Biological Technology* 10: 169-178.

Lindow, S.E., G. McGourty, and R. Elkins. (1996). Interactions of antibiotics with *Pseudomonas fluorescens* strain A506 in the control if fire blight and frost injury to pear. *Phytopathology* 86: 841-848.

Liu, L., J.W. Kloepper, and S. Tuzun. (1992). Induction of systemic resistance against cucumber mosaic virus by seed inoculation with selected rhizobial strains. *Phytopathology* 82: 1108-1109.

Liu, L., J.W. Kloepper, and S. Tuzun. (1995). Induction of systemic resistance (ISR) in cucumber against bacterial angular leaf spot by plant growth-promoting rhizobacteria. *Phytopathology* 85: 843-847.

Loper, J.E. and M.N. Schroth. (1986). Influence of bacterial sources of IAA on root elongation of sugar beet. *Phytopathology* 76: 386-389.

Lorito, M., G.E. Harman, C.K. Hayers, R.M. Broadway, A. Tronsmo, S.L. Woo, and A. Di Pietro. (1993). Chinolytic enzymes produced by *Trichoderma harzianum*: Antifungal activity of purified endochitinase and chitobiosidase. *Phytopathology* 83: 302-307.

Madi, L., T. Katan, J. Katan, and Y. Henis. (1997). Biological control of *Sclerotium rofsii and Verticillium dahliae by Talaromyces flavus* is mediated by differential mechanisms. *Phytopathology* 87: 1054-1060.

Madrigal, C., S. Pascual, and P. Melgarejo. (1994). Biological control of peach twig blight *(Monilinia laxa)* with *Epicoccum nigrum*. *Plant Pathology* 43: 554-561.

Mandeed, Q. and R. Baker. (1991). Mechanisms involved in biological control of *Fusarium* wilt of cucumber with strains of nonpathogenic *Fusarium oxysporum*. *Phytopathology* 81: 462-469.

Manoranjitham, S.K.. V. Prakasam, K. Rajappan and G. Amutha. (2000). Control of chilli damping off using bioagents. *Journal of Mycology and Plant Pathology* 30: 225-228.

Mark, G.L. and A.G. Cassels. (1996). Genotype dependence in the interaction between *Glomus fasciculatum, Phytophthora fragariae* and the wild strawberry *(Fragaria vesca). Plant and Soil* 185: 233-238.

Marois, J.J., S.A. Johnston. M.T. Dunn and G.C. Papavizas. (1982). Biological control of Verticillium wilt of eggplant in the field. *Plant Disease* 66: 1166-1168.

Meena, B., V. Rammoorthy. T. Marimuthu and R. Velazahan. (2000). *Pseudomonas fluorescens* mediated systemic resistance against leaf spot of groundnut. *Journal of Mycology and Plant Pathology* 30: 151-158.

Merriman. P.R., R.D. Price, J.F. Kolmorgen. T. Peggot and E.H. Ridge. (1974). Effect of seed inoculation with *Bacillus subtilis* and *Streptomyces griseus* on the growth of cereals and carrots. *Australian Journal of Agricultural Research* 25: 219-226.

Miller, T.D. and M.N. Schroth. (1972). Monitoring the epiphytic population of *Erwinia amylovora* on pear with a selective medium. *Phytopathology* 62: 1175-1182.

Mistry, D., D.G. Vala and S.T. Patel. (2001). *In vitro* evaluation of phytoextracts against *Alternaria alternata* (Fr) Keissler (Abstr.). *Journal of Mycology and Plant Pathology* 31: 113.

Montesinos, E., A. Bonaterra, A. Ophir and S.V. Beer. (1996). Antagonism of selected bacterial strains to *Stemphillium vesicarium* and biological control of brown spot of pear under controlled environmental conditions. *Phytopathology* 86: 856-863.

M'Piga, P.. R.R. Belanger. T.C. Paulitz and N. Benhamou. (1997). Increased resistance to *Fusarium oxysporum* f. sp. *racidis-lycopersici* in tomato plants treated with the endophytic bacterium *Pseudomonas fluorescens* strain 63-28. *Physiological Molecular Plant Pathology* 50: 301-320.

Muehlchen. A.M.. R.E. Rand and J.L. Parke. (1990). Evaluation of crucifer green manuresfor controlling *Aphanomyces* root rot of peas. *Plant Disease* 74: 651-654.

Mukerji, K.G. (1999). Mycorrhiza in control of Plant Pathogens: Molecular Approaches. In: *Biotechnological Approaches in Biocontrol of Plant Pathogens*. eds. K.G. Mukerji, B.P. Chamola and R.K. Upadhyay. Dordrecht, London. Kluwer Academic Publishers. Netherlands. pp. 133-155.

Mukerji, K.G. (2002). Rhizosphere biology. In: *Techniques in Mycorrhizal Research*. eds. K.G. Mukerji, C. Manoharachary and B.P. Chamola. Dordrecht, London: Kluwer Academic Publishers. pp. 87-102.

Mukherjee, S. and H.S. Tripathi. (2000). Biological and chemical control of wilt complex of French bean. *Journal of Mycology and Plant Pathology* 30: 380-385.

Mukherjee, S., H.S. Tripathi and Y.P.S. Rathi. (2001). Integrated management of wilt complex of Frenchbean (*Phaseolus vulgaris* L.). *Journal of Mycology and Plant Pathology* 31: 213-215.

Narasiwa, K., S. Tokumasu and T. Hashiba. (1998). Suppression of clubroot formation in Chinese cabbage by the root endophyte fungus, *Heteroconium chaetospira*. *Plant Pathology* 47: 206-210.

Nelson, E.B. (1988). Biological control of Pythium seed rot and preemergence damping-off of cotton with *Enterobacter cloacae* and *Erwinia herbicola* applied as seed treatments. *Plant Disease* 72: 140-142.

Newhook, F.J. (1957). The relationship of saprophytic antagonism to control *Botrytis cinerea* Pers. on tomatoes. *New Zealand Journal of Science and Technology* 38: 473-481.

Nicholson, R.L. and J.E. Rahe. (2004). Apple scab and its management. In: *Disease Management of Fruits and Vegetables*. Vol. I. ed. K.G. Mukerji. Kluwer Academic Publishers, Netherlands. pp. 41-58.

Norman, J.R., D. Atkinson and J.E. Hooker. (1996). Arbuscular mycorrhizal fungal-induced alteration to root architecture in strawberry and induced resistance to the root pathogen *Phytophthora fragariae*. *Plant and Soil* 185: 191-198.

Nunes, C., J. Usall, N. Teixido, E. Fons and I. Viñas. (2002). Post-harvest biological control by *Pantoea agglomerans* (CPA-2) on golden delicious apples. *Journal of Applied Microbiology* 92: 247-255.

Onega, M., F. Daayf, S. Jacques, P. Thonart, N. Benhamou, T.C. Paulitz, P. Cornélis, N. Koedam and R.R. Bélanger. (1999). Protection of cucumber against *Pythium* root rot by fluorescent pseudomonads: predominant role of induced resistance over siderophores and antibiotics. *Plant Pathology* 48: 66-76.

Pandey, K.K. and J.P. Upadhyay. (2000). Microbial population from rhizosphere and non-rhizosphere soil of pigeon-pea: Screening for resident antagonist and mod of mycoparasitism. *Journal of Mycology and Plant Pathology* 30: 7-10.

Papavizas, G.C. and J.A. Lewis. (1989). Effect of *Gliocladium* and *Trichoderma* on damping-off and blight of snapbean caused by *Sclerotium rolfsii* in the greenhouse. *Plant Pathology* 38: 277-286.

Park, C.S., T.C. Paulitz and R. Baker. (1988). Biocontrol of *Fusarium* wilt of cucumber resulting from interactions between *Pseudomonas putida* and non-pathogenic isolates of Fusarium oxysporum. *Phytopathology* 78: 190-194.

Paulitz, T.C. and J.E. Loper. (1991). Lack of a role for fluorescent siderophore production in biological control of *Pythium* damping-off of cucumber by a stain of *Pseudomonas putida*. *Phytopathology* 81: 930-935.

Peng, G. and J.C. Sutton. (1990). Biological methods to control gray mould of strawberry. *Brighton Crop Protection Conference-Pests and Diseases* 3C: 233-240.

Pennock-Vos, M.G., E.J.A. Roebroek and C.Z. Skrzypezak. (1990). Preliminary results on biological control of *Botrytis cinerea* in forced tulips. *Acta Horticulturas* 226: 425-428.

Phillips, D.V., C. Leben and C.C. Allison. (1967). A mechanism for the reduction of Fusarium wilt by a *Cephalosporium* species. *Phytopathology* 57: 916-919.

Pusey, P.L., M.W. Hotchkiss, H.T. Dullmage, R.A. Baumgardner, E. Zehr, C.C. Reilly and C.L. Wilson. (1988). Pilot tests for commercial production and application of *Bacillus subtilis* (B-3) for postharvest control of peach brown rot. *Plant Disease* 72: 622-626.

Pusey, P.L. and C.L. Wilson. (1984). Postharvest biological control of stone fruits brown rot by *Bacillus subtilis*. *Plant Disease* 68: 753-756.

Raaijmakers, J.M., M. Leeman, M.M.P. van Oorschot, I. van der Sluis, B. Schippers and P.A.H.M. Bakker. (1995). Dose-response relationships in biological control of Fusarium wilt of radish by *Pseudomonas* spp. *Phytopathology* 85: 1075-1081.

Ramamoorthy, V. and R. Samiyappam. (2001). Induction of defense-related genes in *Pseudomonas fluorescens* treated chilli plants in response to infection by *Colletotrichum capsici*. *Journal of Mycology and Plant Pathology* 31: 146-155.

Raupach, G.S. and J.W. Kloepper. (1988). Mixtures of plant growth promoting Rhizobacteria enhance biological control of multiple cucumber pathogens. *Phytopathology* 88: 1158-1164.

Rawal, P. and B.B.L. Thakore. (2003). Investigations of *Fusarium* rot of sponge gourd fruits. *Journal of Mycology and Plant Pathology* 33: 15-20.

Roberts, R.G. (1990). Post harvest control of gray mould of apple by *Cryptococcus laurentii*. *Phytopathology* 80: 526-530 .

Roiger, D.J. and S.N. Jeffers. (1991). Evaluation of *Trcihoderma* spp. for biological control of *Phytophthora* crown and root rot of apple seedlings. *Phytopathology* 81: 910-917.

Rosenberger, D.A. and F.W. Meyer. (1979). Benomyl tolerant *Penicillium expansum* in apple packinghouses in eastern New York. *Plant Disease Reporter* 63: 37-40.

Ryals, J.A., U.H. Neuenschwander, M.G. Willits, A. Molina, H.Y. Steiner and M.D. Hunt. (1996). Systemic acquired resistance. *Plant Cells* 8: 1809-1819.

Sancez, D. (1990). Natural agents fight fruit spoilage. *Agricultural Research*. Washington DC 38: 15-17.

Sakthivel, N. and S.S. Gananamanickam. (1986). Toxicity of *Pseudomonas fluorescens* towards rice sheath rot pathogen *Sarocladium oryzae*. *Current Science* 55: 106-107.

Savitry, N. and S.S. Gananamanickam. (1987). Bacterization of peanut with *Pseudomonas fluorescens* for biological control of *Rhiozctonia solani* and for enhanced yield. *Plant and Soil* 102: 11-15.

Scher, F.M. and R. Baker. (1982). Effect of *Pseudomonas putida* and synthetic iron chelator on induction of soil suppressiveness to *Fusarium* wilt pathogens. *Phytopathology* 72: 1567-1573.

Schippers, B., A.W. Bakker and P.A.H.M. Bakker. (1987). Interactions of deleterious and beneficial rhizosphere microorganisms and the effect of cropping practices. *Annual Review of Phytopathology* 25: 339-358.

Sen, B. (2000). Biological control: A success story. Jeerasandhi Award Lecture. *Indian Phytopathology* 53: 809-812.

Sharma, P. and B.P. Mehta. (2001). Antibacterial activity of plants extracts to phyto-Pathogenic Xanthomonas campestris pv. campestris causing balck rot of cabbage (Abstr.). *Journal of Mycology and Plant Pathology* 31: 111-112.

Sid Ahmed, A., C. Pérez-Sanchez, C. Egea and M.E. Candela. (1999). Evaluation of *Trichoderma harzianum* for controlling root rot caused by *Phytophthora capsici* in pepper plants. *Plant Pathology* 48: 58-65.

Singh, J. and V.L. Majumdar. (2001). Efficacy of plant extracts against *Alternaria alternata*-the incitant of fruit rot of pomegranate (*Punica granatum* L.). *Journal of Mycology and Plant Pathology* 31: 346-349.

Singh, P.P., Y.C. Shin, C.S. Park and Y.R. Chung. (1999). Biological control of Fusarium wilt of cucumber by chitinolytic bacteria. *Phytopathology* 89: 92-99.

Singh, V., S.P. Pathak and S.B. Singh. (2002). Integrated disease management (IDM) for vine and fruit of Parwal (*Tichosanthes dioica* Roxb) (Abstr.) *Journal of Mycology and Plant Pathology* 32: 140.

Smilanick, J.L. and R. Denis-Arrue. (1992). Control of green mold of lemons with *Pseudomonas* sp. *Plant Disease* 76: 481-482.

Smith, G.E. (1957). Inhibition of *Fusarium oxysporum* f.sp. *lycopersici* by a species of Micromonospora isolated from tomato. *Phytopathology* 47: 429-432.

Smolinska, U., M.J. Morra, G.R. Kundsen and P.D. Brown. (1997). Toxicity of glucosinolate degradation products from *Brassica napus* seed meal toward *Aphanomyces eutiches* f. sp. *pisi*. *Phytopathology* 87: 77-82.

Sneh, B., M. Dupler, Y. Elad and R. Baker. (1984). Chlamydospore germination of *Fusarium oxysporum* f. sp. *cucumerinum* as affected by fluorescent and lytic bacteria from Fusarium-suppressive soil. *Phytopathology* 74: 1115-1124.

Sobiczewski, P. and H. Bryk. (1996). Biocontrol of *Botrytis cinerea* and *Penicillium expansum* on postharvest apples by antagonistic bacteria. *International Conference on Integrated Fruit Production*. eds. F. Polesny, W. Muller and R.W. Olszak 19: 344-345.

Sobiczewski, P., H. Bryk and S. Berezynski. (1996). Evaluation of epiphytic bacteria isolated from apple leaves in the control of postharvest apple diseases. *Journal of Fruit and Ornamental Research* 4: 35-45.

Spotts, R.A. and L.A. Cervantes. (1986). Population pathogenecity and benomyl resistance of *Botrytis* spp., *Penicillium* spp. and *Mucor piriformis* in packinghouses. *Plant Disease* 70: 106-108.

Suslow, T.V., J.W. Kloepper, M.N. Schroth and T.J. Burr. (1979). Beneficial bacteria enhance plant growth. *Californian Agriculture* 33: 15-17.

Sutton, J.C., D. Li, G. Peng, H. Yu, P. Zhang and R.M. Valdebenito-Sanhueza. (1997). *Gliocladium roseum*, a versatile adversary of *Botrytis cinerea* in crops. *Plant Disease* 81: 316-328.

Tamietti, G., L. Ferrris, A. Matta and I. Abbatista Gentile. (1993). Physiological response of plants grown in *Fusarium* suppressive soil. *Journal of Phytopathology* 138: 66-76.

Teixidó, N. I. Viñas, J. Usall and N. Magan. (1998). Control of blue mold of apples by preharvest application of *Candida sake* grown in media with different water activity. *Phytopathology* 88: 960-964.

Thomashow, L.S., D.W. Essar, D.K. Fujimoto, III Pierson, C. Thrane and D.M. Weller. (1993). Genetic and biochemical determinants of phenazine antibiotic production in fluorescent pseudomonads that suppress take-all disease of wheat. In. *Advances in Molecular Genetics of Plant-Microbe Interactions.* eds. E.W. Nester and D.P.S. Verma. Vol. 2 . Kluwer, Dordrecht. pp. 535-541.

Tronsmo, A. and C. Dennis. (1974). The use of *Trichoderma* species to control strawberry fruit rots. *Netherlands Journal of Plant Pathology* 83: 449-495.

Tronsmo. A. and J. Raa. (1977). Antagonistic action of *Trichoderma pseudokoningii* against the apple pathogen *Botrytis cinerea. Phytopathologische Zeitschrift* 89: 216-220.

Trotta, A., G.C. Varesee, E. Gnavi, A. Fusconi, S. Sampo, and G. Berta. (1996). Interactions between the soil-borne root pathogen *Phytophthora nicotinae* var. *parasitica* and the arbuscular mycorrhizal fungus *Glomus mossae* in tomato plants. *Plant and Soil* 185: 199-205.

Unnamalai, N. and S.S. Gananamanickam. (1984). *Pseudomonas fluorescens* is an antagonist to *Xanthomonas citri* (Hasse) Oxe. the incitant of citrus canker. *Current Science* 53: 703-704.

Usall, J., N. Teixidó, E. Fons, and I. Viñas. (2000). Biological control of blue mould on apple by a strain of *Candida sake* under several controlled atmosphere conditions. *International Journal of Food Microbiology* 58: 83-92.

Uthkhede, R.S. and J.F. Rahe. (1980). Biological control of onion white rot. *Soil Biology and Biochemistry* 12: 101-104.

Villajuan-Abgona, R., K. Kageyama, and M. Hayakumachi. (1996). Biocontrol of *Rhizoctonia* damping-off of cucumber by non-pathogenic binucleate *Rhizoctonia. European Journal of Plant Pathology* 102: 227-235.

Viñas, I., J. Usall, and V. Sanchis. (1991). Tolerance of *Penicillium expansum* to postharvest fungicide treatments in apple packinghouses in Lerida (Spain). *Mycopathologia* 113: 15-18.

Viñas, I., N. Vallverdú, S. Monllao. J. Usall, and V. Sanchis. (1993). Imazalid resistant *Penicillium* isolated from Spanish apple packing houses. *Mycopathologia* 123: 2733.

Voisard. C., C. Keel, D. Hass, and G. Défago. (1989). Cyanide production by *Pseudomonas fluorescens* helps suppress black root rot of tobacco under gnotobiotic conditions. *EMBO Journal* 8: 351-358.

Wei, G., J.W. Kloepper, and S. Tuzun. (1991). Induction of systemic resistance of cucumber to *Colletotrichum orbiculare* by selected strains of plant-growth promoting rhizobacteria. *Phytopathology* 81: 1508-1512.

Weinhold, A.R. and T. Bowian. (1968). Selective inhibition of the potato scab pathogen by antagonistic bacteria and substrate influence on antibiotic production. *Plant Soil* 27: 12-14.

Weller, D.M. (1988). Biological control of soil-borne plant pathogens in the rhizosphere with bacteria. *Annual Review of Phytopathology* 26: 379-407.

Wettzien, H.C. (1991). Biocontrol of foliar fungal diseases with compost extracts. In *Microbial Ecology of Leaves,* eds. J.H. Andrews and S.S. Hirano. Springer-Verlag, New York, pp. 430-450.

White, J.G., C.A. Lingfield. M.L. Lahdenpera, and J. Uloti. (1990). Mycostop-a novel biofungicide based on *Streptomyces griseovirdis. Brighton Crop Protection Conference-Pests and Diseases* 3C: 221-226.

Wilson, C.L. and E. Chalutz. (1989). Postharvest biocontrol of *Penicillium* rots of citrus with antagonistic yeasts and bacteria. *Scientia Horticulturae* 40: 105-112.

Wilson, C.L. and A. El-Ghaouth. (1993). Multifaceted biological control of post-harvest diseases of fruits and vegetables. In *Pest Management: Biologically Based Technologies,* eds. R.D. Lumsden and J.L. Vaughn. American Chemical Society Press, Wageneingen (DC), pp. 81-185.

Wilson, C.L., J.D. Franklin, and I. Pussey. (1987). Biological control of Rhizopus rot of peach with *Enterobacter cloacae. Phytopathology* 77: 303-305.

Wood, R.K.S. (1950). The control of diseases of lettuce by the use of antagonistic organisms: I. The control of *Botrytis cinerea* Pers. *Annals of Applied Biology* 38: 203-216.

Yohalem, D.S., E.V. Nordheim, and J.H. Andrews. (1996). The effect of water extracts of spent mushroom compost on apple scab in the field. *Phytopathology* 86: 914-922.

Yuen, G.Y, M.N. Schroth, and A.H. McCain. (1985). Reduction of *Fusarium* wilt of carnation with suppressive soils and antagonistic bacteria. *Plant Disease* 69: 1071-1075.

Zahavi, T., L. Cohen. B. Weiss, L. Schena, and A. Daus. (2000). Biological control of *Aspergillus* and *Rhizopus* rots of table and wine grapes in Israel. *Postharvest Biological Technology* 20: 115-124.

Zhang, W., W.A. Dick, and H.A.J. Hoitink. (1996). Compost-induced systemic acquired resistance in cucumber to Pythium root rot and anthracnose. *Phytopathology* 86: 1066-1070.

Zhang. W., D.Y. Han. W.A. Dick, K.R. Davis, and H.A.J. Hoitink. (1998). Compost and compost water extract-induced systemic acquired resistance in cucumber and arabidiopsis. *Phytopathology* 88: 450-455.

Zhou, T., J. Northover, and K.E. Schneider. (1999). Biological control of post-harvest diseases of peach with phyllosphere isolates of *Pseudomonas syringae. Canadian Journal of Plant Pathology* 21: 375-381.

Zhou, T. and R.D. Recleder. (1989). Application of *Epicoccum purpurascens* spores to control white mould of snapbean. *Plant Disease* 73: 639-642.

Chapter 9

Novel Biological Control Methods for Gray Mold Disease of Vegetables and Fruits Using *Bacillus subtilis* IK-1080

Yoshihiro Taguchi
Mitsuro Hyakumachi

INTRODUCTION

Bacillus subtilis is known as an antagonist that shows antibiotic activity on many bacteria and fungi (Sharon et al., 1954: Phae et al., 1990). There are many reports on the effectiveness of *B. subtilis* for the prevention of airborne and soilborne plant diseases, including gray mold and powdery mildew in various vegetables (Newhook, 1950), scab and gray mold in citrus fruits (Singh and Devarall, 1984), bacterial wilt and crown and root rot in tomatoes (Phae et al., 1992), and many *Rhizoctonia* and *Phytophthora* diseases (Broadbent et al., 1971; El-Goorani et al., 1976; Mew and Rosales, 1992). *B. subtilis* has also been used in postharvest treatment of peaches and cherries (Pusey et al., 1986).

Several studies have been carried out to identify antibiotic substances produced by *B. subtilis* that act on plant pathogenic fungi. In the 1950s, the antibiotic substances bulbiformin (Vasudeva et al., 1953), mycobacillin (Majumdar and Bose, 1958), and substilin (Dunleavy, 1955) were identified, and later antibiotic circular peptide substances, including iturin A (Gueldner et al., 1988; Loeffler et al., 1986), surfactin (Hiraoka et al., 1992), bacilysin, bacillomycin L, fengymycin, and subsporin (Loeffler et al., 1986), were identified. It has been shown that *B. subtilis* produces antibiotic substances at the vegetative cell stage (Phae et al., 1990; Shoda, 1998; Mckeen et al., 2000). Genes responsible for the synthesis of these antibiotic substances have also been identified (Hung et al., 1993).

Among the various antibiotic substances, iturin A showed a strong suppressive effect on the growth of rice pathogenic fungi such as blast fungus *(Pyricularia oryzae)* and brown spot fungus *(Cochliobolus miyabeanus)* (Phae et al., 1990).

It has also been shown that *B. subtilis* can induce resistance in plants, such as increase of peroxidase activity, accumulation of lignification and suberization in cell walls, and promotion of free radicals production (Mew and Rosales, 1992; Barry and Nina, 2000).

USE OF B. SUBTILIS IN JAPAN
AS A BIOCONTROL AGENT

Bacillus subtilis IK-1080 was registered in 1998 as a microbial pesticide as Botokira Wettable Powder (produced by Idemitsu Kosan Inc., Japan) (Kawane, 2000). This product (hereafter called IK-1080) is now widely used as a microbial pesticide for the prevention of gray mold and powdery mildew diseases in various vegetables and fruits, including tomatoes, eggplants, strawberries, and grapes. IK-1080 is a white-colored, wettable powder mixed with 1×10^{11} cfu/g bacterial endospores and fillers. Since it is an endospore formulation, the bacteria can survive even at temperatures as high as 60°C. The powder can, therefore, be stored at room temperature for up to three years. The product is recommended to be used at temperatures above 10°C, because it is less effective at temperatures lower than 10°C. IK-1080 has also been shown to have a strong preventive effect on blast disease, one of the major rice diseases in Japan (Taguchi, Hyakumachi, Horinouchi, and Kawane, 2003). Another *B. subtilis* QST-713 (named Impression Wettable Powder, SDS Biotech Inc., Japan; hereafter called SB-910) was also registered as a microbial pesticide in Japan in 2002 for the prevention of gray mold and powdery mildew diseases in tomatoes and grapes (Barry and Nina, 2000). There are two other recently registered microbial pesticides of *B. subtilis,* KUF-1401 (Kumiai Chemical Inc., Japan) and MBF-122 (Maruwa Biochemical Inc., Japan), for the prevention of gray mold and powdery mildew diseases in tomatoes, green peppers, eggplants, strawberries, cucumbers, melons, grapes, and citrus fruits.

EFFECTIVENESS OF B. SUBTILIS
FOR CONTROL OF CROP DISEASES

Table 9.1 shows the results of tests conducted in various locations throughout Japan in 2001 and 2002 on the effectiveness of *B. subtilis* formulations in preventing vegetable and fruit diseases (Japan Plant Protection Association, 2002). The protection values of the four formulations differed, depending on the crop. IK-1080, KUF-1401, SB-910, and MBF-122 formulations were applied to crops by spraying suspension solutions containing 1×10^8, 5×10^7, 1×10^7, and 1×10^6, or 5×10^6 cfu/ml of bacteria, respectively. The amounts of the formulations used in the tests ranged from 1500 to 3000 liters/ha. The effectiveness of the four formulations in preventing gray mold disease in tomatoes greatly differed, with protection values ranging from 31.8 to 88.3. IK-1080 and SB-910 were used in eggplants and showed protection values ranging from 46.5 to 82.3. IK-1080 and KUF-1401 were used in cucumbers and showed protection values ranging from 78 to 85.4. All four formulations were used in strawberries and showed lower protection values ranging from 27.9 to 64.3 in comparison to those for other crops. The effectiveness of the formulations applied by a compressed air-type fogging for preventing gray mold disease in tomatoes was similar to that of spraying with the suspension.

The KUF-1401, MBF-122, and SB-910 formulations have also been used against powdery mildew in sweet peppers, cucumbers, melons, watermelons, and strawberries. SB-910 formulation particularly showed a strong preventative effect on powdery mildew in melons and watermelons. IK-1080, KUF-1401, and SB-901 have also been used against gray mold disease in fruits, and they showed protection values of 16.5-93.8 in grapes and 34.3-71.9 in citrus. Many factors including the state of disease and environmental conditions, such as temperature and humidity, at the time of the formulation application, and the method of application are thought to contribute to such varied protection values in tests.

NEW METHODS OF APPLICATION
OF B. SUBTILIS FORMULATIONS

Various methods have been reported on the application of *B. subtilis* formulations to plants, including spraying, soil drenching (Broadbent et al., 1971), or seed coating (Dunleavy, 1955, Mew and Rosales, 1992) with a suspension of vegetative cells or endospores. In addition to these methods, new methods of application of *B. subtilis* formulation which are safe and have long-lasting effects have been tried. As a result, we could establish the

TABLE 9.1. The protection values of *B. subtilis* formulations to gray mold and powdery mildew diseases of vegetables and fruits in 2001 and 2002 in Japan.

Name of vegetable or fruit	Name of disease	Name of B. subtilis formulation in Japan	Concentration cfu/ml	Spray volume (L)	Part of plant	Protection value in 2002	Protection value in 2001
Tomato	Gray mold	IK-1080 KUF-1401 MBF-122 SB-910 IK-1080	1.0×10^8 5.0×10^7 5.0×10^6 1.0×10^7 1.0×10^9	200 200 300 490 2.1L fog	fruit	IS:31.8, ME:60.9, S:62.9, N:62.9 IS:60.5, ME:60.1 N:72.2	KN:75.9, KG:75.6 KG:66.7, N:88.3 MY:78.3, I:68.8, N:74.9, M:63KN:68.0, KG:72.6
Eggplant	Gray mold	SB-910 IK-1080	1.0×10^7 1.0×10^8	200 230	fruit		OO:82.3, N:48.2 N:46.5
Sweet pepper	Powdery mildew	IK-1080 MBF-122 SB-910	1.0×10^8 5.0×10^6 1.0×10^7	300 300 240-300	leaf	KY:81.6, N:38.8, M:46.3, N:57.1 KY:67.3, N:53.3, KG:82.4	MY:61.0 KY:52.1, N:55.8
Cucumber	Gray mold	IK-1080 KUF-1401	1.0×10^8 5.0×10^7	250 250	fruit	GF1:85.4, GF2:79.0, GF1:82.4, GF2:78	

Crop	Disease	Strain	Concentration	Dose	Part	Efficacy	Efficacy
	Powdery mildew	SB-910 MBF-122	1.0×10^7 5.0×10^6	300 300	leaf	SI:73.4, OO:28.7, HY:85.7	IW:58.2, OO:58.1, HY:87.9 IW:31.3, TO:26.5
Melon	Powdery mildew	SB-910	1.0×10^7	300	leaf	IB:97, SZ:84.1, AI:98.1, KO:100	IB:51, 89, SZ:100, AI:95.1
Watermelon	Powdery mildew	SB-910	1.0×10^7	119-250	leaf	TO:92.7	N:86.5, N:86.3, TT:84.5
Strawberry	Gray mold	IK-1080 MBF-122 KUF-1401 SB-910	1.0×10^8 1.0×10^6 5.0×10^7 1.0×10^7	200 200 200 200-300	fruit	TC:52.8, ME:60.9, S:62.9, N:62.9, TC:52.1, TC:64.3, IB:60.5, ME:60.1, TC:54.3, MY:35.1, FU:59.1	TO:43.3, N:87.5, NR:27.9
	Powdery mildew	MBF-122 SB-910	1.0×10^6 1.0×10^7	200-300 250	leaf	MY:76.2, TO:97.3, FK:67.8	MG:85.6, NG:57.1, N:86.8 MG:61.6, FK:89.4,
Grape	Gray mold	IK-1080 KUF-1401 SB-910	1.0×10^8 5.0×10^7 1.0×10^7	150-300 150- 300 150	bunch	AK:16.5, NG:51, TO:60.4	AM:75.1, FK:70.4, NG:60.0, FK:70.4, AM:93.8, AK:62.5, YN:69.0 NG:65.2, OO:58.8

227

TABLE 9.1. *(continued)*

Name of vegetable or fruit	Name of disease	Name of *B. subtilis* formulation in Japan	Concentration cfu/ml	Spray volume (L)	Part of plant	Protection value in 2002	in 2001
Citrus	Gray mold	IK-1080 KUF-1401	1.0×10^8 5.0×10^7	200 150-300	fruit	SZ:60.4, N:34.3, SZ:53.7, SZ:61.2, YG:71.9, N:44.1	

Source: These protection values were cited from "Report of biological control 2001 and 2002 in Japan" (Japan Plant Protective Association).

Note: The marks like IK or ME indicate the following prefectures in Japan that were tested: AI:Aichi, AK:Akitap, FK:Fukuoka, FU:Fukushima, GF:Gifu, HY:Hyougo, IB:Ibaragi, IS:Ishikawa, KG:Kagoshima, KN:Kanagawa, KO:Kouchi, KY:Kyoto, M:Miyagi, ME:Mie, MG:Miyagi, MY:Miyazaki, N:Nihon-shokubutu-bouekikyoukai, NG:Nagano, NR:Nara, SI:Saitama, SZ:Sizuoka, OO:Osaka, TC:Tochigi, TO:Tottori, YG:Yamaguchi, YN:Yamanashi.

following three methods: (1) duct dusting; (2) petal spraying with fruitage-enhancing hormone; and (3) carrying to petals using bumblebees. In this chapter, the results of these three methods using IK-1080 are presented, and the consistency and durability of their protection values are discussed.

Duct Dusting

Ducts are used to distribute heated air and to maintain a uniform temperature in a greenhouse. We have used the duct system in a greenhouse to dust plants with IK-1080 formulation in powder form to determine its effectiveness for the prevention of crop diseases (Taguchi, Hyakumachi, Tsueda, and Kawane, 2003). Daily duct dusting with IK-1080 during the cultivation period at 100 g/ha resulted in a significant long-term suppression of gray mold of tomatoes, eggplants, and cucumbers (Table 9.2). Protection values of IK-1080 were very high (98-100) and were equivalent to those of chemical pesticides (Japan Plant Protection Association, 2002; Taguchi, Watanabe, Baba, et al., 2003; Taguchi, Watanabe, Kawane, and Hyakumachi, 2003). Duct dusting of IK-1080 was also effective against gummy stem blight *(Mycosphaerella melonis)*, powdery mildew, and stem rot *(Sclerotinia sclerotiorum)* diseases of cucumber plants (Figure 9.1). Since water is not used in this method, there is no increase in humidity in greenhouses (Figure 9.2), and there is no adverse effect on the survival of natural enemy insects.

When duct dusting is used, a sufficient amount of bacteria must be applied evenly to all plants to evenly prevent the occurrence of a disease. To achieve this, branch ducts from the main ducts with air outlets of 5 cm in diameter facing upward should be set at intervals of 2 to 3 m along each ridge (Figure 9.3). In a recent greenhouse cultivation method, only main ducts set around the internal perimeter of the greenhouse were used, but such an ar-

TABLE 9.2. Protection values of gray mold disease in cucumbers, tomatoes, and eggplants using duct dusting of *B. subtilis* IK-1080.

Plants	Protection value[a]	
	IK-1080	Chemical fungicides[b]
Cucumber	98.2-100	86.4-100
Tomato	98.3-100	93.3-95.0
Eggplant	98.6	95.8

[a]Three hundred fruits were collected and examined for disease incidence (%).

[b]Fludioxonil, iprodione, and mepanipyrim were used cucumbers, tomatoes, and eggplants, respectively.

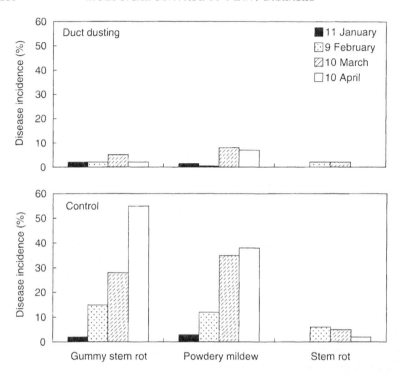

FIGURE 9.1. Protective effect of cucumber diseases by duct dusting of microbial fungicide "B. subtilis IK-1080 powder" in a greenhouse. Duct dusting of IK-1080 powder was done every day from November 1999 to April 2000.

rangement of ducts would not enable IK-1080 to reach the ridges in the center of the greenhouse. Thus, the arrangement of ducts and the direction in which the duct outlets face are very important for the effectiveness of duct dusting.

Gray mold disease occurs on all parts of the tomato plant, including petals, leaves, fruits, calyxes, and stems. Petals are affected first, with necrosis occurring after flowering, if humidity exceeds 93 percent (Tezuka et al., 1983). Suppression of occurrence of gray mold on petals is thought to be an effective means of reducing the pathogen density over an entire greenhouse and for delaying the development of the disease (Okada et al., 2001).

Population of *B. subtilis* on petals of cucumber plants 35 days after the introduction of duct dusting with IK-1080 (10^9 cfu/g of dried petal) was

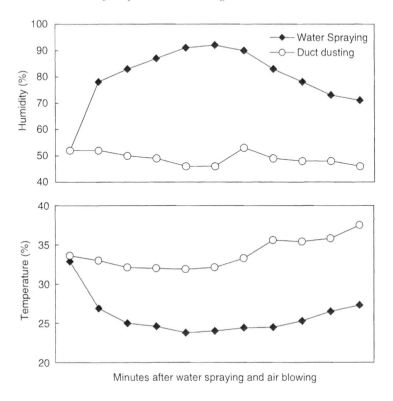

FIGURE 9.2. Influence of spraying water and blowing air through the heating ducts on the temperature and humidity in a greenhouse of cucumber. From 10:00 am to 10:30 am, the heating ducts were operated and air blown into the greenhouse. Water 2000 L/ha was sprayed. The temperature and humidity were measured by a thermometer and a hydrometer that were installed at 1 m high from the ground level in the greenhouse.

greater than the initial introduced population (10^6 cfu/g). During the period in which duct dusting was performed, the population of IK-1080 on leaves was maintained in the range of 10^4 to 10^5 cfu/cm^2. The bacterial population on petals of tomato plants 30 days after the start of duct dusting with IK-1080 were 1.3 to 4.3×10^7 cfu/g. The population was 1.5×10^5 cfu/g of dried petal on the day after the start of duct dusting.

Petals of cucumber that had been sprayed with IK-1080 using duct dusting were collected and stored in a humid room at 25°C for five days. The morphology of fungal conidia and conidiophores appearing on each petal was examined under a light microscope, and the species of fungi were determined. Only five genera including *Cladosporium* and *Aspergillus* were found on the treated petals, while more than 10 genera including *Cladosporium, Aspergillus, Fusarium, Ascochyta,* and *Botrytis* found on non-treated petals (Table 9.3).

On the treated petals of tomato, a similar phenomenon of simplification of fungal genera also occurred, and almost the same fungal genera seen on petals of cucumber appeared.

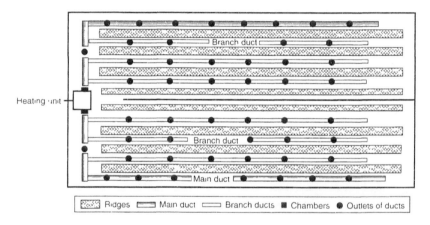

FIGURE 9.3. Outline of the greenhouses showing locations of the heating unit, and main and branch ducts. Treated and untreated greenhouses were 30 m × 20 m, respectively. A Nepon model NK-4020 was used as the heating unit.

TABLE 9.3. Rate (%) of fungal genera occurred on petals of cucumber plants sprayed with *B. subtilis* IK1080 using duct dusting[a].

	Fungal genera									
							Unidentified fungus			
Treatment	Botrytis	Alter-naria	Clado-sporium	Fusarium	Ascochyta	Asper-gillus	1	2	3	4
IK-1080	0	10	100	0	0	88	033	16	0	0
Control	31	100	100	9	73	0	91	45	18	9

[a]One hundred petals were collected and examined under the microscope after five days, incubated at 25°C in a humid chamber.

PETAL SPRAYING OF B. SUBTILIS
WITH A FRUITAGE-ENHANCING HORMONE

A solution containing 4-CPA (Tomato Tone, 150 ppm), which is a fruitage-enhancing hormone, and IK-1080 (1.0×10^8 cfu/ml) was sprayed using a hand sprayer onto petals of tomato plants at a dose of about 1.0 ml per flower cluster. At the time of spraying, the population of IK-1080 was 2 to 4×10^8 cfu/g of dried petals. A high protection effect (99-100) against gray mold disease of tomatoes was obtained as a result of this method. Petals were sampled 17 days after spraying, and the population of IK-1080 was investigated. IK-1080 was isolated at a concentration of 7.8×10^9 cfu/g of dried petals that had been treated with the solution containing both IK-1080 and 4-CPA. On the other hand, IK-1080 was not isolated at all from petals without the treatment or from petals that had been sprayed with a solution containing only the hormone (Taguchi, Hyakumachi, Tsueda, and Kawane, 2003). This high population isolated from the treated petals indicates that IK-1080 could multiply on the petals of tomato plants. In this method, simplification of fungal genera was also seen on the treated petals.

Bacillus subtilis *Inoculation Through Bumblebees*

Bumblebees have been used as pollinators of tomato plants since 1992 in Japan (Mastuura, 1993). Gray mold disease can occur easily in scarred areas on anthers of petals made by bumblebees. Spores of *B. cinerea* were found on petals just after flowering when bumblebees were used for pollination. It has, therefore, been pointed out that utilization of bumblebees as vectors can result in an increase in the occurrence of gray mold disease (Okada et al., 2001). However, it is believed that the gray mold disease on petals could be suppressed if IK-1080 carried by the bumblebees can multiply on the petals of tomato plants. In fact in our study, the application of IK-1080 to petals of tomato plants by using bumblebees resulted in a significant reduction of gray mold incidence (Table 9.4). The amount of IK-1080 carried by bumblebees was as low as 3.5×10^2 cfu/g of petals, but it increased to 10^5-10^6 cfu/g at 21 days after pollination (Table 9.5).

Botrytis cinerea was not isolated at all from the treated petals and the simplification of fungal genera that appeared on the petals was also seen. It is thought that gray mold disease is suppressed by such increased population of IK-1080 on the treated petals of tomato. Since bumblebees visit the same flowers many times during the period in which pollen is formed (Mastuura, 1993), there are many chances of adhereing IK-1080 to petals, and IK-1080 could easily proliferate.

TABLE 9.4. Rate (%) of gray mold disease occurrence on petals and fruits of tomato plants carried with B. subtilis IK1080 using bumblebees[a].

			Rate (%) and protection value of occurrence of gray mold disease[b]	
	Treatment	Month/Day	Petals	Fruits
Experiment 1	Control	7/10	1.0	0.0
		17	1.0	0.0
		24	4.0	0.0
	IK-1080	7/10	0.0	0.0
		17	0.0	0.0
		24	$0.0*^{c}(100)^{d}$	0.0
Experiment 2	Control	9/1	35.0	2.0
		7	58.0	3.5
		14	60.0	8.0
	IK-1080	9/1	4.0*(88.6)	1.0
		7	4.0*(93.1)	1.5*(57.1)
		14	8.0*(86.7)	1.0*(87.5)

[a]B. subtilis IK-1080 was carried to petals of tomato plants using bumblebees in the plastic house from June 16 to August 30, 1999.

[b]Two hundred petals were collected and examined under the microscope after five days, incubated at 25°C in a humid chamber.

[c]*indicates significant difference at $p = 0.05$ as determined by Student's t-test.

[d]Number in ( ) indicates a protection value.

TABLE 9.5. Changes in bacterial population of B. subtilis IK-1080 carried to petals usingbumblebees[a].

	Days after pollinated		
	7 days	14 days	21 days
Bacterial population (cfu/g of petals)	1.3×10^{3}	1.5×10^{5}	4.6×10^{5}

[a]Bacterial population of B. subtilis IK-1080 in 25 petals of tomato plants was investigated using nutrient agar medium.

It has been reported that antagonistic microorganisms carried to petals of raspberry and strawberry plants by pollinators, such as bumblebees and honeybees, could suppress B. cinerea and Erwinia amylovora (Peng et al., 1992; Thomson et al., 1992; Yu and Sutton, 1997). However, loading pollinators with large amounts of bacteria would result in a reduction of the homing rate that could affect the pollination activity. Since IK-1080 can proliferate easily on petals, a small amount of IK-1080 carried by bumblebees is enough to suppress plant pathogens, which brings no adverse effect on the functions of pollinators (Taguchi, Hyakumachi, Tsueda, and Kawane, 2003).

Consideration must be given to the fruiting rate and to the homing of bumblebees as vectors for IK-1080. The authors have constructed several prototype adaptors for attaching IK-1080 to bumblebees' hives. As a result, the use of the adaptor with separate entrance and exit (Figure 9.4) had no adverse effect on homing or pollination rate of bumblebees after allowing them two or three days to become used to the adaptor.

A dramatic decline in the rate of gray mold disease occurred on petals that had been treated with IK-1080, and there was low disease occurrence on fruits compared with that of untreated plants. It has been shown that removal of petals from tomato plants resulted in a significant decline in the rate of infection of fruits by *B. cinerea* (Kazumi and Ihori, 1994). However, the use of bumblebees as carriers of IK-1080 could eliminate such a laborious task of removing flowers from plants. It is expected that this pinpoint suppression method against *B. cinerea* on petals of tomatoes can also be applied for other fruits and vegetables.

FUTURE PROSPECTS AND CONCLUSIONS

The establishment of optimal treatment conditions working favorably for microbial pesticides rather than for pathogenic fungi are very important

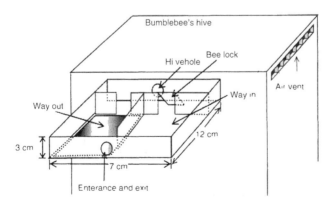

FIGURE 9.4. Device of a bumblebee's hive for carrying IK-1080 to tomato flowers by bumblebees. Entrance and exit were separated. IK 1080 was put in the way out.

for their utilization. One advantage of microbial pesticides is that there is no pesticide residue as in the case of chemical pesticides. This advantage enables microbial pesticides to be applied continuously to plants, which could enhance the effect of the microbial pesticides.

Although the suppressive effects of *B. subtilis* against gray mold disease have been the main focus of this report, *B. subtilis* could be applied to other vegetable and fruit diseases. Further studies are needed to clarify the population dynamics in greenhouse conditions to which *B. subtilis* have been applied, and to determine which range of pathogens *B. subtilis* could show high protection effects. Although the effects of *B. subtilis* on soilborne diseases have not been investigated in this study, there are several reports on the effectiveness of soil drenching with suspensions of *B. subtilis* for the prevention of bacterial wilt and crown and root rot in tomato plants (Phae et al., 1992; Kioka et al., 1999). It is therefore expected that the use of soil drenching with suspensions of *B. subtilis* against other soilborne diseases could also be used. To establish effective techniques for preventing plant diseases using *B. subtilis,* combinations with chemical pesticides or other microbial pesticides, should be also considered. Methods combining the use of IK-1080 and antagonistic microorganisms for the prevention of plant diseases will be valuable for the establishment of an environment-friendly agricultural system in the future.

Although IK-1080 has been shown to have no adverse effect on natural enemy insects, there has not been enough investigation of its possible effects on natural enemy microorganisms such as *Beauveria brongniartii, Paecilomyces fumosoroseus, Talaromyces flavus,* and *Verticillium lecanii.* Further studies on integrated pest management involving the use of IK-1080 are therefore required.

REFERENCES

Barry, J.J. and K.Z. Nina. (2000). Induction of disease resistance mechanisms in by *Bacillus subtilis* strain SB-910 and disease control of *Erwinia carotovora* subsp. *beatavasculorum* in sugar beet. Department of Plant Sciences, Montana State University, pp. 1-7.

Broadbent, P., K.F. Baker, and Y. Waterworth. (1971). Bacteria and actinomycetes antagonistic to fungal root pathgens in Australian soils. *Australian Journal of Biological Society* 24: 925-944.

Dunleavy, J. (1955). Control of damping off of sugar beet by *Bacillus subtilis. Phytopathology* 45: 252-258.

El-Goorani, M.A., S.A. Faraga, and M.R.A. Shehata. (1976). The effect of *Bacillus subtilis* and *Penicillium patulum* on in vitro growth and pathogenicity of *Rizoc-*

tonia solani and *Phytophthora cryptogea*. *Phytopathologische Zeitschrift* 85: 345-352.

Gueldner, R.C., C.C. Reilly, P.L. Pusey. C.E. Costello, R.F. Arrendale, R.H. Cox, D.S. Himmelshbach, F.G. Crumley, and H.G. Culter. (1988). Isolation and identification of iturins as antifungal peptides in biological control of peach brown rot with *Bacillus subtilis*. *Journal of Agricultural and Food Chemistry* 36: 366-370.

Hiraoka, H., T. Ano, and M. Shoda. (1992). Molecular cloning of a gene responsible for the biosynthesis of the lipopeptide antibiotics iturin and surfactin. *Journal of Fermentation and Bioengineering* 74: 323-326.

Hung, C.C., T. Ano, and M. Shoda. (1993). Nucleotide sequence and characteristics of the gene, lpa-14, responsible for biosynthesis of lipopeptide antibiotics iturinA and surfactin from *Bacillus subtilis* RB14. *Journal of Fermentation and Bioengineering* 76: 445-450.

Japan Plant Protection Association. 2002. *Report of Biological Control*. JPPA, Tokyo.

Kawane, F. (2000). Property of microbial pesticide. *Shokubutsu-boueki* 54: 342-345.

Kazumi, I. and S. Ihori. (1994). Control of tomato gray mold *(Botrytis cinerea)* by removing petal. *Proceedings of the Kansai Plant Protection Society* 36: 87-88.

Kioka, Y., T. Akazawa, H. Masumura, and H. Noguchi. (1999). Suppressive effect of antagonistic bacteria *Bacillus* sp. on soil-borne plant diseases. *Soil Microorganisms* 53: 103-109.

Loeffler, W., J.S.M. Tshchen, N. Vanittanakom, M. Kugler, E. Knorpp, T.F. Hsieh, and T.G. Wu. (1986). Antifungal effects of bacilysin and fengymycin from *Bacillus subtilis* F-29-3: A comparison with activities of other *Bacillus* antibiotics. *Journal of Phytopathology* 115: 204-213.

Majumdar, S.K. and S.K. Bose. (1958). Mycobacillin, new antifungal antibiotic produced by *Bacillus subtilis*. *Nature* 181: 134-135.

Mastuura, M. (1993). Pollination of tomato plants with introduced bumblebee, *Bombus terrestris*. *Shokubutsu-boueki* 47: 17-20.

Mckeen, C.D., C.C. Reilly, and P.L. Pusey. (2000). Production and partial characterization of antifungal substances antagonistic to *Monilia fructicola* from *Bacillus subtilis*. *Phytopathology* 76: 136-139.

Mew, T.W. and A.M. Rosales. (1992). Control of *Rhizoctonia* sheath blight and other disease of rice by seed bacterization. In *Biological Control of Plant Disease*, eds. E.S. Tjamos et al. Plenum Press, New York. pp.113-123.

Newhook, F.J. (1950). Microbiological control of *Botrytis cinera* Pers. *New Mycologist*, Plant Diseases Division, Auckland, New Zealand, pp. 169-202.

Okada, K., M. Kawaratani, S. Kusakari and W. Nakasone. (2001). Contaminated bumblebee *(Bumbus terrestris* L.) by *Botrytis cinerea* and its disease promotion. *Japanese Journal of Phytopathology* 67: 197.

Peng, G., J.C. Sutton, and P.G. Kevan. (1992). Effectiveness of honey bees for applying the biocontrol agent *Gliocladium roseum* to strawberry flowers to suppress *Botrytis cinerea*. *Canadian Journal of Plant Pathology* 14: 117-129.

Phae, C.G., M. Shoda, M. Kita, M. Nakano, and K. Ushiyama. (1992). Biological control of crown and root rot and bacterial wilt of tomato by *Bacillus subtilis* NB22. *Annals of the Phytopathological Society of Japan* 58: 329-339.

Phae, C.G., M. Shoda, and H. Kubota. (1990). Suppressive effect of *Bacillus subtilis* and its products on phytopathogenic microorganisms. *Journal of Fermentation and Bioengineering* 69: 1-7.

Pusey, P.L., C.L. Wilson, M.W. Hotchkiss, and L.D. Franklin. (1986). Compatibility of *Bacillus subtilis* for postharvest control of peach brown rot with commercial fruit waxes, dicloran, and cold storage condition. *Plant Disease* 70: 587-590.

Sharon, N., A. Pinsky, R. Turner-Graff, and J. Babad. (1954). Classification of the antifungal antibiotics from *Bacillus subtilis*. *Nature* 174: 1190-1191.

Shoda, M. (1998). Mechanism of biological control by *Bacillus subtilis*. *Shokubutsu-boueki* 49: 178-183.

Singh, V. and B.J. Devarall. (1984). *Bacillus subtilis* as control agent against fungal pathogens of citrus fruit. *Transaction of British Mycological Society* 83(3): 487-490.

Taguchi, Y., M. Hyakumachi, H. Horinouchi, and F. Kawane. (2003). Biological control of rice blast disease by *Bacillus subtilis* IK-1080. *Japan Journal of Phytopathology* 69: 85-93.

Taguchi, Y., M. Hyakumachi, H. Tsueda, and F. Kawane. (2003). Method for adhering *Bacillus subtilis* IK-1080 to bumblebees and control of gray mold disease of tomato. *Japan Journal of Phytopathology* 69: 94-101.

Taguchi, Y., H. Watanabe, M. Baba, S. Chida, and M. Hyakumachi. (2003). Biological control of gray mold disease of tomato in plastic house by *Bacillus subtilis* IK-1080 using duct dusting. *Japan Journal of Phytopathology* 69: 72-73.

Taguchi, Y., H. Watanabe, F. Kawane, and M. Hyakumachi. (2003). Development of a new dusting method for *Bacillus subtilis* (IK-1080) through hot-air heating ducts to control gray mold disease of cucumber. *Japan Journal of Phytopathology* 69: 107-116.

Tezuka, N., M. Ishii, and M. Watanabe. (1983). Effect of relative humidity on the development of gray mold of tomato in greenhouse cultivation. *Bulletin of the Yasai Agricultural Experimental Station* Japan, A-11: 105-111.

Thomson, S.V., D.R. Hansen, K.M. Flint, and J.D. Vandenberg. (1992). Dissemination of bacteria antagonistic to *Erwinia amylovora* by honeybees. *Plant Disease* 76: 1052-1056.

Vasudeva, R.S., T.V. Subbaiah, M.L.N. Sastry, G. Rangaswamy, and M.R.S. Iyengar. 1953. "Bulbiformin," an antibiotic produced by *Bacillus subtilis*. *Annals of Applied Biology* 46: 336-345.

Yu., H. and J.C. Sutton. (1997). Effectiveness of bumblebees and honeybees for delivering inoculum of *Gliocladium roseum* to raspberry flowers to control *Botrytis cinerea*. *Biological Control* 10: 113-122.

Chapter 10

Strategies for Biological Control of Fungal Diseases of Temperate Fruits

Ting Zhou
Hai Yu
Deena Errampalli

INTRODUCTION

In recent years, due to realization of healthy benefits of fresh fruits, many consumers are incorporating fruit as a part of their diet, seeking more in quantity and variety, and demanding higher standard in safety and quality of fruit. Findings of residues of pesticides and toxins produced by pathogens that cause diseases in fruit have been widely discussed in public forums and publications, and the public demand has grown for pesticide-residue-free and high-quality fruit. Also, consumer interest in how food is produced is rapidly increasing, particularly in the fresh fruit and vegetable sectors, evident from the 20 percent annual growth in organic food sales in North America and Europe (Granatstein and Dauer, 2000). The changes in the fruit markets are both opportunities and challenges to the fruit industry and researchers. One of the challenges is to improve fruit yield and quality and to maintain or extend fruit shelf life for long-distance transportation and off-season supplies with reduced use of synthetic pesticides.

An important limiting factor to fruit industry is fungal diseases, which could significantly reduce fruit yield and quality during production and result in complete loss of market value in storage. In temperate regions, berry fruit, pome fruit, and stone fruit are the major types of fruit, which mainly include grapes, strawberries, raspberries, apples, pears, nectarines, peaches, and plums. Although many pathogens can cause different diseases on these

Biological Control of Plant Diseases
doi:10.1300/5682_10

plants, fungal pathogens that attack fruit are relatively few and similar. Apple scab, caused by *Venturia inaequalis,* is the most economically important disease in most apple-growing areas; the pathogen infects different parts of apple trees, including the fruit. A similar pathogen, *Venturia pirina,* causes pear scab. *Podosphaera leucotricha,* which mainly infects apple leaves but can also infect fruit. Blue and gray mold, incited by *Penicillium expansum* and *Botrytis cinerea,* respectively, are the most important postharvest diseases of pome fruit. On stone fruit, brown rot is the most damaging fruit disease, both preharvest and postharvest; depending on geographic regions, it may be caused by one of the three closely related fungi namely, *Monilinia fructicola, M. laxa,* and *M. fructigena.* Rhizopus rot, a major disease on ripe fruit of stone fruit, is commonly caused by *Rhizopus stolonifer* (Ogawa et al., 1995). *Uncinula necator* and *Plasmopara viticola* cause powdery mildew and downy mildew, respectively, generally on leaves of grapes, but the pathogens also infect fruit clusters affecting yield and quality (Pearson and Goheen, 1988). The most common and damaging pathogen on berry fruit is *B. cinerea,* which causes bunch rots on grapes and gray mold on strawberries and raspberries (Maas, 1998).

Since the late 1940s, synthetic fungicides had been used as the sole control of fruit diseases and were considered economical. The development of modern pesticides and technologies have greatly enhanced the efficacy of fruit disease control in the orchards, particularly for the severe airborne diseases, such as apple scab, and for improved fruit quality and shelf life after the harvest. However, fruit losses due to pathogenic and spoilage microorganisms are still considerably high. It has been estimated that postharvest losses of fruit is from 5 percent to more than 20 percent in the developed countries, depending on the commodity, and can be as high as 50 percent in developing countries (Janisiewicz and Korsten, 2002). The exclusive use of chemical fungicides and/or the use of fungicides with similar modes of action led to the development of fungicide resistance in pathogens. Concerns over the impact of synthetic fungicides used in the agricultural production on the environment and human health (NRC, 1987) resulted in the withdrawal of key fungicides used for controlling fruit diseases (Hileman, 1993). The negative perception to fungicide has also promoted governmental policies restricting use of fungicides. All these have contributed to the increasing attempts in actively searching for alternatives to synthetic pesticides by the scientific societies, fruit industry, and even chemical companies (Froyd, 1997). Biological control as an effective alternative has emerged and gained popularity in fruit disease management.

There are many factors determining the success of a biocontrol system. A primary one is to have an effective biocontrol agent (BCA), which relys on an appropriate screening strategy with consideration of the targeting pathosystem, selecting sources, types of microorganisms, possible modes of action, and evaluation methodologies. It is not difficult to find a microorganism with antagonistic activities; however, it is a great challenge to obtain a BCA that can ultimately be used in the commercial biocontrol. Also, application strategies should be developed specifically for each biocontrol system. Since a biocontrol system is different from the chemical control, a BCA should not be used as a simple replacement of a synthetic chemical; BCAs with different modes of action may be applied with different method and time schemes. Unless innovative application strategies are developed to individual biocontrol systems, the success of biocontrol may be limited and shadowed by chemicals. In addition, efficacy and consistency are key issues for acceptance of BCAs by the fruit industry. Various strategies have been used to enhance biocontrol systems, including manipulation of BCAs themselves, modification of chemical and physical conditions affecting either the BCA or pathogen, either singly or combined, and integration of BCAs with other control means. Biocontrol is only a component of an integrated pest management system; and appropriate integration may result in enhanced control efficacy to a level satisfactory to fruit producers, thus providing an opportunity for a BCA being accepted by the industry.

Biocontrol has been successful in certain disease systems, but not in the others. Compared with annual crop production systems, very few studies have been reported with the use of biological control as a disease management method in fruit orchards. This, in part, may be due to the perennial nature of the crop and considerably long time required to test the efficacy of biological control on perennial crops. However, over the past two decades, a significant advance has been made in the biological control of postharvest diseases (Wilson and Wisniewski, 1989; Chalutz and Droby, 1997; Janisiewicz and Korsten, 2002), which may be due to the uniqueness of the postharvest system (Janisiewicz and Korsten, 2002). In this chapter, strategies used for developing effective biocontrol systems for controlling major fungal diseases of fruits in temperate regions will be summarized and discussed, with a focus on the practical applications of these strategies.

DEVELOPMENT OF BIOLOGICAL CONTROL AGENTS

Development of a BCA is a very long process and can be costly and very time-consuming; therefore, intense consideration and utmost care must be taken before and during its development. Because of the improperly conceived selection strategies, many BCAs do not reach the stage of commercialization (Schisler and Slininger, 1997). Although many basic plant pathological methodologies can be applied for the selection of BCAs, strategies used for the development of BCAs should be specific to the targeted patho-systems. Different pathosystems may require searches for different types of BCAs from different sources. The modes of action and future applications of the potential BCAs should also be considered in determining isolation methodologies, selection criteria, and in vitro bioassays.

Sources and Types of BCAs

In 1974, Baker and Cook in their classic book on the biological control of plant pathogens suggested that antagonists must be searched where the disease does occur but declines or cannot develop despite the presence of a susceptible host. Subsequently, many researchers have isolated ecologi-cally fit organisms that possess antagonistic activity against the pathogens of fruit from healthy fruits in the orchard and storage (Table 10.1). Natu-rally occurring microbial communities present in the phyllosphere and carposphere have been the primary source of BCAs for the control of fungal diseases of temperate fruits (Table 10.1), which have been discussed in pre-vious reviews (Chalutz and Droby, 1997; Janisiewicz, 1997). Soil and plant debris are another popular source for antagonists (Table 10.1); for example, *Bacillus subtilis* B3, a contaminatant isolated from peach roots, was able to inhibit *M. fructicola* on peach fruit (Pusey and Wilson, 1984); isolate S10B of *Trichoderma harzianum*, a BCA against *B. cinerea* on grape, was an iso-late from soil (Latorre el al., 1997).

Microorganisms, including bacteria, yeasts, and filamentous fungi, have historically been used in natural preservation systems for fermenting fruit, meat, vegetables, and dairy products. Many of these beneficial "food mi-crobes" have antimicrobial activity and produce bactericins and antifungal substance, and thus may have potential use as BCAs to reduce fresh-fruit decay (Wilson and Chalutz, 1991; Zhou, 2001). After screening of over 100 food microbes, several bacterial isolates with strong antifungal activities against *M. fructicola* were discovered, and the application of the isolates significantly controlled peach brown rot (Zhou, 2001). Food microbes may not survive in field conditions. However, the use of food microbes for the control of postharvest diseases may have additional advantages: they have proven to be generally safe, and thus may be easier in going through the reg-istration processes; many of them are already in commercial production, and thus should be ready to be developed into BCA products.

The biological agents selected for the control of fruit diseases include bacteria, yeast, filamentous, and yeastlike fungi (Table 10.1), and poten-tially, mycoviruses.

The most popularly known example of bacteria for the control of fungal diseases in storage may be the *Pseudomonas syringae* strains ESC-10 and

TABLE 10.1. Biocontrol agents and their applications in control of fungal diseases on temperate fruit.

Crops	BCA / Pathogen	BCAs' Origin	Application				Reference
			Method	Target	Timing	Concentration	
Apple	Aureobasidium pullulans, Rhodotorula glutinis, Bacillus subtilis / Penicillium expansum, Botrytis cinerea, Pezicula malicorticis	Apple surface	Spray	Fruit	Preharvest	10^7 cells/ml for yeast and 10^8 for bacteria	Leibinger et al., 1997
	Candida sake CPA-1 / P. expansum	Apple surface	Spray / dipping	Fruit	Preharvest postharvest	3×10^6 cfu/ml	Teixidó et al., 1999
	C. sake CPA-1 / P. expansum	Apple surface	Submerging / drenching	Fruit	Postharvest	1.6×10^4 or $1.6 \times 10^6 / 10^{5-7}$	Usall et al., 2001
	Chaetomium globosum, A. pullulans / Venturia / inaequalis	Apple leaves	Spray	Foliage	Starting in April at 1-2 wk intervals	10^6 spores/ml	Cullen et al., 1984
	Cryptococcus infirmo-miniatus YY6, C. laurentii HRA5, R. glutinis HRB6 / P. expansum, B. cinerea	Pear surface	Drenching / line spray	Fruit	Postharvest	4×10^9 cells/ml	Chand-Goyal & Spotts, 1997
	Metschnikowia pulcherrima / B. cinerea, P. expansum, Alternaria sp., Monilia sp.	Apple carposphere	Dipping	Fruit	Postharvest	10^7 cells/ml	Spadaro et al., 2002
	Microsphaeropsis sp. / V. inaequalis	Apple leaf litter	Spray	Foliage, fallen leaves	At leaf fall	4.5×10^5 spores/ml	Carisse et al., 2000
	Muscodor albus / B. cinerea. P. expansum	Cinnamon tree	Biofumigation	Fruit	Postharvest		Mercier & Jimenez, 2004
	Psuedomonas syringae (ESC-10 WP or ESC-11 WP) / P. expansum, B. cinerea		Line spray	Fruit	Postharvest	10g/L $_{10}$ (1×10^{10} cfu/g)	Janisiewicz & Jeffers, 1997
	R. glutinis LS11, C. laurentii LS28, A. pullulans LS30 / B. cinerea, P. expansum	Olives, Apple	Spray	Fruit	Postharvest	10^7 cfu/ml	Lima et al., 2003

(continued)

243

TABLE 10.1 (continued)

Crops	BCA / Pathogen	BCAs' Origin	Application				
			Method	Target	Timing	Concentration	Reference
Pear	C. infirmo-miniatus YY6, C. laurentii RR87-108, R. glutinis HRB6 / P. expansum, B. cinerea	Apple or pear surface	spray	Fruit	3 wk or 1 day prior to harvest	10^8 cfu/ml	Benbow & Sugar, 1999
	Pantoea agglomerans CPA-2 / P. expansum, B. cinerea, Rhizopus stolonifer	Apple & pear fruit & leaf surface	Dipping	Fruit	Postharvest	8×10^8 cfu/ml	Nunes et al., 2001
Peach	B. subtilis (B-3) / M. fructicola	Apple rhizosphere	Line spray	Fruit	Postharvest	5×10^8 cfu/ml	Pusey et al., 1988
	P. syringae / M. fructicola, R. stolonifer	apple leaf	Soaking	Fruit	Postharvest	10^7 cfu/ml	Zhou et al., 1999
	Penicillium frequentans / M. laxa	Peach twigs	Spray	Shoots	Early growing season	10^{8-9} spores/ml	De Cal et al., 1990
Apricot, peach, nectarine	P. agglomerans EPS125 / M. laxa, R. stolonifer	Pear surface	Immersion	Fruit	Postharvest	10^8 cfu/ml	Bonaterra et al., 2003
Cherry	A. pullulans. Epicoccum purpurascens / M. fructicola	Peach blossoms	Spray	Flowers	50% full bloom & full bloom	10^5, 10^6 spores/ml	Wittig et al., 1997
	A. pullulans / B. cinerea, M. laxa	Cherry flesh	Spray / dipping	Fruit	Preharvest & postharvest	10^7 cells/ml	Schena et al., 2003
Grape	Ampelomyces quisqualis / Uncinula necator	Powdery nildew colonies on grape leaf	Dispersal from wick cultures	Vines	At 15 cm of shoot growth & bloom		Falk et al., 1995
	C. guilliermondii A42, Acremonium cephalosporium B11 / B. cinerea, Aspergillus niger, R. stolonifer	Grape berry surface	Spray	Fruit	Veraison to har-vest weekly	10^{7-8} cells/ml	Zahavi et al., 2000

244

Crop	Antagonist / Pathogen	Substrate	Application	Plant part	Timing	Concentration	Reference
	Fusarium proliferatum / Plasmopara viticola	Grape downy mildew lesions	Spray	Vines	Weekly spray start at 15 cm of shoot growth	10^6 spores/ml	Falk et al., 1996
	Trichoderma harzianum T39 (Trichodex 25 WP or 25 P) / *B. cinerea*	Cucumber fruit	Spray	Flowers & fruit	Late flowering to 3 wk before harvest	2-4 kg of trichodex / ha	Elad 1994; O'Neill et al., 1996
	Trichoderma spp. / *B. cinerea*	Soil, wood shavings	Spray	Flowers & fruit	Late flowering to 3 wk prior to harvest	10^{7-8} spores/ml	Dubos 1984, 1987; Gullino 1992; Harman et al., 1996
Strawberry	*A. pullulans, Candida oleophila / B. cinerea, R. stolonifer*	Fruit & vegetables	Spray	Flowers & fruit	Flowering, just before or after harvest	10^{7-8} cfu/ml	Lima et al., 1997
	Bacillus pumilus, P. fluorescens / B. cinerea	Strawberry flowers	Spray	Flowers & fruit	White flower-bud to pink fruit		Swadling & Jeffries, 1996
	C. albidus / B. cinerea	Strawberry fruits	Spray	Flowers	Flowering	10^7 cells/ml	Helbig, 2002
	Gliocladium roseum / B. cinerea	Strawberry fruit	Bee vectoring	Flowers	Flowering	10^9 cfu/g of powder	Peng et al., 1992
	G. roseum / B. cinerea	Strawberry fruit	Spray	Flowers & fruit	Green flower-bud to pink fruit	10^6 spores/ml	Peng & Sutton, 1991
	T. harzianum and *T. viride / B. cinerea*		Spray	Flowers & fruit	Early flowering to 2 wk before harvest		Tronsmo & Dennis, 1977
	T. harzianum 1295-22 (Root-shield T22) / *B. cinerea*	Protoplast fusion between two isolates	Bee vectoring	Flowers	Flowering	5×10^8 cfu/g of powder formulation	Kovach et al., 2000
	Ulocladium atrum / B. cinerea	Onion necrotic leaves	Spray	Flowers & fruit	First open flower to first fruit turned red	2×10^6 spores/ml	Boff, Köhl, Gerlagh, and DeKraker, 2002a
Raspberry	*G. roseum / B. cinerea*	Strawberry fruit	Bee vectoring	Flowers	Flowering	10^9 cfu/g of grain powder	Yu & Sutton. 1997

ESC-11 (Janisiewicz and Jeffers, 1997), which are the bases of the bio-fungicide Bio-Save (EcoScience Corp., USA) targeting postharvest decay of fruit, including pome and stone fruits; *P. syringae* MA-4 isolated from leaves of apple trees was effective against postharvest pathogens of pome and stone fruit (Zhou et al., 1999, 2002); and different strains of *P. syringae* were also reported as potential BCAs (Janisiewicz and Marchi, 1992; Jeffers and Wright, 1994). Other species of *Pseudomonas* investigated as BCAs, include *P. cepacia, P. corrugata, P. fluorescens, P. gladioli,* and *P. putida* (Janisiewicz and Roitman, 1988; Smilanick et al., 1993; Guetsky et al., 2001). Another group of BCAs belongs to *Bacillus,* such as *B. subtilis* against peach brown rot (Pusey and Wilson, 1984) and apple postharvest decay (Leibinger et al., 1997), *B. amyloliquefaciens* against *B. cinerea* on pear (Mari et al., 1996), and *B. pumilus* and *B. mycoides* for gray mold on strawberry (Guetsky et al., 2001). Isolates of *Enterobacteriacae* from fruit were effective against *B. cinerea* on strawberry fruits (Guinebretiere et al., 2000), and *Enterobacter cloacae* was effective in controlling rhizopus rot of peach (Wilson et al., 1987). *Pantoea agglomerans* (CPA-2), an isolate from apple surface, showed high efficacy in controlling apple blue mold (Nunes et al., 2002). In addition, endophytic bacteria from fruit (Sholberg, 1995) and epiphytic bacteria (Sobiczewski et al., 1996) were reported with potential in controlling fungal diseases on fruit.

Yeasts that occur naturally on fruit surfaces have proven to be good bio-logical control agents (Table 10.1). Several yeast BCAs are species of *Candida: C. oleophila* I-182 has been developed into a biofungicide Aspire (Ecogen, Inc., USA and Israel) targeting gray and green mold on citrus and pome fruit; *C. guilliermondii* and *C. saitoana* showed effectiveness in con-trolling postharvest diseases (McLaughlin et al., 1990, 1992; Wisniewski et al., 1991). *C. sake* has also been intensively studied for the control of major postharvest pathogens on pome fruit (Teixidó, Viñas, Usall, and Magan, 1998; Usall et al., 2001). Species of *Cryptococcus* have been inves-tigated as BCAs. The biofungicide YieldPlus, based on *C. albidus,* that con-trols postharvest decay on pome fruit, has been registered in South Africa. *Cryptococcus laurentii* and *C. infirmo-miniatus* were effective for control of postharvest diseases on apple, pear, strawberry, and sweet cherry (Chand-Goyal and Spotts, 1996, 1997; Lima et al., 1998; Qin et al., 2003). Different strains of *Rhodotorula glutinis* were studied for the control of various pathogens on apple, pear, and sweet cherry (Chand-Goyal and Spotts, 1997; Leibinger et al., 1997; Lima et al., 1998; Calvente et al., 1999; Qin et al., 2003). Also, *Kloeckera apiculata, Metchnikowia pulcherrima, Pichia mem-branefaciens,* and *Sporobolomyces roseus* were reported to be effective in controlling pathogens on a variety of temperate fruit (Janisiewicz et al., 1994; Spadaro et al., 2002; Qin et al., 2004).

Aurebasidium pullulans, very common in phyllosphere, is the only yeast-like fungus that was reported for BCA on temperate fruit. Several strains of the yeastlike fungus were found effective on apple, sweet cherry, strawberry, and table grapes for the control of various fungal pathogens, including *B. cinerea, P. expansum,* and *V. inaequalis* (Cullen et al., 1984; Lima et al., 1997; Ippolito et al., 2000; Schena et al., 2003).

Filamentous fungi as BCAs are often applied in the field to target preharvest diseases of temperate fruit, unlike bacteria and yeasts that are mainly used to control postharvest diseases. *Ampelomyces quisqualis, Sporothix fliocculosa,* and *Tilletiopsis* spp. were reported to be effective against powdery mildew (Hoch and Provvidenti, 1979; Grove and Boal, 1997). *A. quisqualis* and *S. fliocculosa* have been developed into biofungicides AQ10 (Ecogen Inc., USA) and Sporodex (Plant Products Co. Ltd., Canada), respectively, for controlling powdery mildew on apple, grape, and strawberry. When applied during the growing season, *Chaetomium globosum* reduced apple scab (Andrews et al., 1983). On the other hand, *Athelia bombacina* and *Microshaeropsis ochracea* were effective in reducing primary inoculum of *V. inaequalis,* and the fall application of the BCAs resulted in less apple scab in the field (Heye and Andrews, 1983; Carisse and Dewdney, 2002). *Epicoccum nigrum* showed strong antifungal activity against *M. laxa* and gave good control of peach fruit rot (Foschi et al., 1995), although it has been mostly applied to control peach twig blight (Madrigal and Melgarejo, 1995). *Muscodor albus* was effective against *M. fructicola* on peach (Mercier and Jiménez, 2004). In addition, application of *Fusarium proliferatum* microconidia in grape vinery significantly reduced downy mildew on both leaves and fruit clusters (Falk et al., 1996). *Trichoderma harzianum* was proven effective in controlling grape bunch rot and strawberry gray mold (Elad, 1994; Elad and Shtienberg, 1994; O'Neill et al., 1996; Harman, 2000). *Gliocladium roseum* has been studied intensively as a BCA to control gray mold on berry fruit (Sutton et al., 1997), and *Ulocladium atrum* is another BCA for gray mold on strawberry (Boff, Kohl, Jansen, et al., 2002).

Mycoviruses were reported in hypovirulent isolates of *V. inaequalis* (Zhou and DeYoung, 1997), and *B. cinerea* and may be studied for their potential as BCAs to control apple scab and gray mold on different fruits (Castro et al., 2003).

Selection of BCAs

General steps and methodologies for the isolation and characterization of potential BCAs for the control of plant diseases can be adopted from basic plant pathology, specific reviews (Wilson et al., 1993; Smilanick, 1994;

Chalutz and Droby, 1997; Janisiewicz, 1997; Schisler and Slininger, 1997), and relevant individual scientific papers (Table 10.1). However, it is critical to set up appropriate criteria before the selection. Depending on the target pathosystem, the following criteria have been considered: capability to multiply in high numbers, without causing any damage to the host; antagonistic activity against the targeting pathogen; ability to survive and multiply in the similar conditions that favor the pathogen growth and multiplication; suitability in current practices, or as a part of the integrated disease management; tolerance to environmental stresses; amenability to commercial development; and most importantly, safety to humans. These criteria have been discussed in several reviews (Chalutz and Droby, 1997; Schisler and Slininger, 1997; Janisiewicz and Korsten, 2002).

Specific strategies could be used to obtain BCAs with certain mode(s) of action. Mechanisms that contribute to the biocontrol of fruit diseases can be broadly categorized under competition, antibiosis, parasitism, and induced resistance. Although in most cases more than one of these mechanisms are implicated in a biocontrol system, one mechanism may play more important role than the others. Generally, the mode of action of a BCA is studied after it has been identified. However, more BCAs may reach the commercialization stage if mode of action is well considered before the selection processes are started.

Competition is one of the most often cited mechanisms in biocontrol of fruit diseases, especially between yeasts and postharvest pathogens. Nutrient competition was cited between *S. roseus* and *P. expansum* and *B. cinerea* on apple (Janisiewicz et al., 1994), *E. cloacae* and *R. stolonifer* on peach (Wisniewski et al., 1989.), and *C. laurentii* and *B. cinerea* in apple wound (Roberts, 1990). The biocontrol mechanism of *P. syringae* ESC-10 and ESC-11 has not been clearly identified, but one of the possible mechanisms implicated is competition for nutrients. Production of antimicrobial compounds is the mechanism involved in many biocontrol systems, and a BCA producing antibiotics may have a greater challenge to accept. Antifungal peptides, iturins, were identified from *B. subtilis* B-3 and were proven to play important role in the biocontrol of brown rot of stone fruit (Gueldner et al., 1988). An antibiotic, pyrrolnitrin from *P. cepacia* LT-4-12W, reduced growth and conidia germination of *B. cinerea* and *P. expansum* in apples (Janisiwicz et al., 1991). The pyrronitril production by *P. cepacia* B37 was also reported (Burkhead et al., 1994).

Biocontrol can also be achieved through direct interactions between the BCA and the pathogen, or parasitism. *Ampelomyces quisqualis, T. harzianum*, and *M. ochracea* are all mycoparasites to their target pathogens (Falk et al., 1995; Carisse and Dewdney, 2002; Kiss, 2003). The attachment of yeast cells to the mycelium of pathogens was associated with degradation of the pathogen cell walls. Cell wall–degrading enzymes, particularly, β-1-3 glucanase, were implicated as biocontrol mechanisms of different yeasts. Lysis of cell walls of *B. cinerea* by *P. guelliermondii* on apple (Wisniewski et al., 1991), by *Pichia anomala* strain K on apple (Jijakli and Lepoivre, 1998), and by *A. pullulans* on apples, table grape, and other fruits (Ippolito et al., 2000) were reported.

The induction of fruit resistance has been reported in the biocontrol of fungal pathogens on fruit. It was found that *C. saitoana* induced systemic resistance in apple fruit against *B. cinerea,* evident from the reduction of lesion diameters. *C. saitoana* increased chitinase and β-1,3-glucanase activity with a higher induction in fresh apples than in stored apples (El-Ghaouth et al., 2001).

Following the identification of the pathosystem, a rigorous screening program can be determined with consideration of mode of action of the potential BCAs. A dual culture system on agar plates can detect antibiosis, competition, and parasitism between the potential antagonist and the pathogen (Singh and Faull, 1988). However, since in vitro antagonistic tests do not correlate with in situ biocontrol results in many cases, a realistic in vivo bioassay should be developed (Schisler and Slininger, 1997). Antagonism in the phyllosphere or fructoplane and within the plant tissue can all be identified by in vivo methods (Blakeman and Brodie, 1977). Although competition for nutrients in most cases was difficult to separate from the other mechanisms, a unique technique was developed to study the competitions (Janisiewicz et al., 2000). Also, biosensors or molecular markers, such as green fluorescent protein (GFP) and ice nucleation protein, will aid in determining the gene expression on and in plant tissues.

Strategies for the Application of BCAs

Diseases on temperate fruit are incited by a variety of fungal pathogens, and these pathogens generally have different disease cycles. Specific biocontrol strategies have been developed for different fungal diseases on temperate fruit as summarized in Table 10.1, and these strategies applied to the biocontrol of major diseases of temperate fruit are discussed, mainly using research conducted under field and/or commercial storage conditions.

POSTHARVEST DISEASES OF POME FRUITS

The postharvest pathogens, including *P. expansum, B. cinerea, Mucor pyroformis,* and *R. stolonifer,* initiate infections at fruit wounds that are generated through the harvest processing and postharvest handling. Therefore, wound protection is a key strategy for the control of postharvest diseases.

Candida sake (strain CPA-1) was tested under commercial conditions over a period of three seasons. Apple fruits were drenched with the yeast and then stored under controlled atmosphere conditions for eight months. The yeast significantly suppressed fruit rot, even more effectively than the standard commercial treatment (thiabendazole + folpet) when it was applied at 10^7 cfu/ml (Usall et al., 2001). In another commercial trial, the yeasts *C. laurentii* (HRA5), *C. infirmo-miniatus* (YY6), and *R. glutinis* (HRB6) were sprayed on apples in a processing line, and the BCAs alone significantly reduced the incidence of blue mold. A combination of *C. laurentii* and *C. infirmo-miniatus* controlled pear blue mold by 91 percent, compared with TBZ that controlled it by 88 percent; the same trend was found on apples (Chand-Goyal and Spotts, 1997).

Pseudomonas syringae ESC 10 and ESC11 are the base of Bio-Save 10 and 11. When applied using line-spray method, commercial formulation ESC-11 WP was equally effective as the fresh cells in controlling *B. cinerea* and *P. expansum* on apples and comparable to fungicide TBZ; however, ESC-10 WP was often less effective than ESC-11 WP or the fresh cell of the bacterium (Janisiewicz and Jeffers, 1997). Under semicommercial conditions, *P. agglomerans* CPA-2 consistently reduced blue mold and gray mold on pear fruits in three years, and often performed as well as the fungicide imazalil (Nunes et al., 2001). Similar results were obtained on apple as well (Nunes et al., 2002).

Although most research efforts have been focused on biocontrol in postharvest treatments, preharvest application of BCAs to control postharvest fruit decay has drawn an increasing attention (Ippolito and Nigro, 2000; Janisiewicz and Korsten, 2002). Because competition for nutrients and space is the main mode of action for most of BCAs for control of wound pathogens, preharvest application would be advantageous for BCAs to colonize the wound immediately after it has occurred during harvest or subsequent postharvest handling. Mercier and Wilson (1995) found that *C. oleophila* controlled apple gray mold more effectively when applied to fresh wounds than to one-day-old wounds. Combinations of *A. pullulans* with *R. glutinis*, or *A. pullulans* with *B. subtilis* were sprayed on apple trees about six weeks prior to harvest in two consecutive years. The BCAs, in particular of *A. pullulans*, maintained stable populations on the fruit surface in the field and cold storage, and reduced fruit decay to the same extent as the fungicide euparen (Leibinger et al., 1997). However, preharvest application of *C. sake* was less effective in control of apple blue mold than postharvest treatment, and combination of pre- and post-harvest application did not improve biocontrol efficacy. The poor survival of the yeast, applied two days prior to the harvest, during cold storage may account for the results (Teixidó et al., 1999). In aiming for preharvest application, antagonistic strains capable of adapting to both the field and cold storage conditions should be selected, such as isolates of *A. pullulans* used by Leibinger et al. (1997).

Preharvest application is a vital strategy to control latent infection that cannot be satisfactorily controlled by postharvest treatments (Roberts, 1994). Dry eye rot of apple fruits is caused by *B. cinerea* that infects the sepals of the flower and remains latent inside the tissue. When fruits mature, the pathogen invades the fruit and causes fruit rots in the field or during storage (Tronsmo, 1991). A cold-tolerant strain of *T. harzianum* sprayed on apple trees during flowering period provided significant reduction of fruit rot (Tronsmo, 1991).

Apple Scab Caused by Venturia inaequalis

Biocontrol has been of interest in reducing fungicide use for control of apple scab, and it has been recently reviewed by MacHardy et al. (2001) and by Carisse and Dewdney (2002). Andrews et al. (1983) tested phylloplane microbes antagonistic to *V. inaequalis,* and subsequently selected *C. globosum* as a potential BCA. When sprayed to apple foliage in the growing season, *C. globosum* provided only partial control to scab in the two-year field tests (Cullen et al., 1984). Recently, the same strategy against leaf infection was also used by other researchers (Burr et al., 1996, Ouimet et al., 1997). This approach was fraught with difficulties because of prolonged development of pathogens on foliage and fruit and unfavorable conditions for the applied antagonists to maintain active populations. On the other hand, the use of antagonists to reduce initial inoculum on apple leaf residues has been more promising. The basidiomycete *A. bombacina* completely suppressed ascospore production on intact leaves incubated in the orchard (Heye and Andrews, 1983). However, the amount of inoculum required, and the cost of inoculum production in large scales, may limit the commercialization of the antagonist (Carisse and Dewdney, 2002). More promisingly, another filamentous fungus, *Microsphaeropsis* sp., suppressed ascospore production by 70 to 79 percent and 60 to 62 percent, respectively, as applied to foliage at about 10 percent leaf fall or to leaves on the ground at 90 percent leaf fall in a 0.41 ha orchard over two years (Carisse et al., 2000). This first large plot test shows that biocontrol has a potential to be a part of the integrated scab management.

Diseases Caused by Monilinia spp. on Stone Fruit

Biocontrol of twig and blossom blight as well as fruit rots has been investigated on stone fruits in field trials. Various preparations of *Penicillium frequentans* were applied to the peach shoots inoculated with *M. laxa*. The preparations with nutrients resulted in significant reductions in severity of twig blight, comparable to that given by captan (De Cal et al., 1990). *A. pullulans* and *E. nigrum* were sprayed onto sweet cherry blossoms and highly suppressed blossom blight and fruit latent infections by *M. fructicola,* though they were slightly less effective than fungicides (Wittig et al., 1997). Sprayed on cheery fruits one week before harvest, *A. pullulans* 547 controlled fruit rots as effectively as fungicide folicur (Schena et al., 2003).

Many antagonists have been evaluated on naturally infected fruits or on nonartificially wounded fruits inoculated with pathogens. *Rhodotorula* sp. completely suppressed brown rot of peaches inoculated with *M. fructicola,* and *Trichoderma* isolates reduced the disease by 63 to 98 percent (Hong et al., 1998). *C. oleophila* significantly reduced natural infections of nectarine fruits (Lurie et al., 1995). In pilot tests, *B. subtilis* B-3 often performed as well as fungicides in control of brown rot and rhizopus rot of peaches (Pusey et al., 1988). *P. agglomerans* controlled stone fruit decay as effectively as did the fungicides (Bonaterra et al., 2003). Although *Pseudomonas* species were very effective in the artificially inoculated wounds, they failed to control decay on peaches with natural infections. Preharvest applications were suggested for the effective use of the antagonists (Smilanick et al., 1993; Zhou et al., 1999).

Gray Mold of Grape

Biocontrol of gray mold has been developed in vineyards since the 1970s. Sprayed spore suspensions on vines at late flowering, before bunch closure, the beginning of berry ripening (véraison), and three weeks before harvest, *T. harzianum* suppressed the incidence of berry rot by 56 to 84 percent in seven-year Bordeaux vineyard trials, compared with 80 to 95 percent in fungicide treatments (Dubos, 1984). Similar results were achieved in Italian vineyards (Dubos, 1987; Gullino, 1992). Commercial products Trichodex 25 WP and 25 P (Makhteshim Ltd., Israel) have been extensively examined in 19 countries under diverse conditions for control of grape gray mold. The results showed that Trichodex was generally less effective than fungicides, but alternating Trichodex with fungicides provided the same level control as that given by fungicides. The alternation reduced fungicide use by 50 percent (Elad, 1994; O'Neill et al., 1996).

To control postharvest fruit rot, grape clusters were dipped in yeast suspensions. Although the yeasts were effective, the dipping process removed some of the bloom, an important quality characteristic. Therefore, the attention was turned to preharvest applications (Ben Arie et al., 1991). More recently, efforts have been made to select antagonists capable of adapting both the field and cold storage conditions, so that they may be used for controlling postharvest fruit rots by preharvest applications (Zahavi et al., 2000; Schena et al., 2003).

Grape Diseases Caused by Biotrophic Fungal Pathogens

Biocontrol of biotrophic fungal pathogens has been achieved primarily by using mycoparasites to suppress pathogen sporulation and dissemination (Wilson, 1997). To date for biocontrol of powdery mildew, two biofungicides, AQ-10 and Sporodex, and many other BCAs have been tested in different crop systems (Kiss, 2003). In New York vineyards, *A. quisqualis* G273 was applied by means of wrapping cotton wick cultures of G273 around a wire suspended 30 cm above the top wire of the trellis. Conidia of G273 were naturally released and dispersed by rain splash. *A. quisqualis* significantly reduced powdery mildew on leaves and fruit clusters in two out of three trials (Falk et al., 1995). By the same research group, a mycoparasite *F. proliferatum* G6 has been identified and evaluated in vineyards for control of grape downy mildew by *P. viticola*. Microconidial suspensions of the antagonist were weekly sprayed on vines, resulting in reduction of downy mildew on clusters by 53 to 99 percent over four years at two locations. The antagonist does not produce mycotoxins and resists most of the fungicides used for control of the disease. Thus, biocontrol with *F. proliferatum* might be practical as a part of the integrated disease management program (Falk et al., 1996).

Gray Mold Fruit Rot on Strawberry and Raspberry

Gray mold fruit rot is a serious problem in berry fruit production worldwide. The pathogen, *B. cinerea*, establishes latent infections in floral parts and invades fruits from the infected tissues. The pathogen becomes aggressive in ripe fruit inciting gray mold fruit rot (Powelson, 1960; Jarvis, 1962). Therefore, flower protection is crucial in controlling this disease.

By protecting strawberry flowers, *Trichoderma* spp. effectively reduced gray mold to the same level as fungicide dichlofluanid in an early field trial (Tronsmo and Dennis, 1977). However variable results were reported in other field trials (Tronsmo, 1992). *T. harziannum* T39 integrated with fungicides significantly controlled gray mold on strawberries (Elad and Shtienberg, 1994). More recently, very promising results in the control of gray mold were achieved, using a commercial formulation of *T. harzianum* 1295-22 disseminated by bees during flowering periods in a commercial field. The antagonist controlled the disease remarkably to the same extent as the standard fungicide treatments (Kovach et al., 2000). In Norway, three commercial *Trichoderma* products (Trichodex 25 WP, Rootshield Drench, and Binab TF WP) and *T. harzianum* P1 were weekly sprayed on strawberries grown in greenhouse during flowering period but failed to control berry infections by *B. cinerea* and *M. piriformis*. The reason was that the conidia of the antagonists took a much longer time to germinate than the conidia of the pathogens under relative low temperatures and low level of nutrients (Hjeljord et al., 2001). In the further study, individual strawberry flowers at various stages were inoculated with *B. cinerea* conidia before being sprayed with suspension of quiescent or nutrient-activated *T. harzianum* P1 conidia in the field, and it was found that the nutrient-activated *T. harzianum* P1 conidia that were applied to the newly opened flowers provided a significant reduction of gray mold (Hjeljord et al., 2001).

When sprayed weekly in the period between the green flower bud to the white pink fruit stages, *Gliocladium roseum*, reclassified as *Clonostachys rosea* (Schroers et al., 1999), provided effective and consistent control of gray mold in the field (Peng and Sutton, 1990, 1991). Propagules of *G. roseum* transferred to strawberry and raspberry flowers by bees suppressed *B. cinerea* on flowers and fruits (Peng et al., 1992; Yu and Sutton, 1997). Conidial suspensions of *U. atrum* were sprayed on strawberry plants once or twice weekly starting at transplanting or at the beginning of flowering to the first fruit turned red in annual strawberry crops, and the results showed that the flowering period was the best time for *U. atrum* application (Boff, Kohl, Jansen, et al. 2002). In another field study, Boff, Kohl, Gerlagh, and De Kraker (2002) further defined that the late flowering or early fruit stages were the best time to apply *U. atrum*.

There is more evidence showing the importance of application timing in the control of fruit gray mold. Tested on strawberries grown under plastic tunnels, *C. oleophila* L66 and *A. pullulans* L47 were sprayed on flowers at full bloom and at late petal fall, or on berries just before or just after harvest (Lima et al., 1997). Both the BCAs applied at the flower stage maintained high and stable populations, and significantly suppressed botrytis fruit rot originated from flower infections. In addition, only L47 significantly reduced botrytis fruit rot originated from the surface lesions on the fruits and suppressed rhizopus rot by 78 percent, while isolate L66 and the fungicide failed to control rhizopus rot. On fruits treated just before harvest, L47 reduced rhizopus rot by 72 percent but did not control botrytis fruit rot. This indicates that application to fruits before harvest cannot control latent infection but is effective to control superficially grown pathogens, like *R. stolonifer*. On fruits treated just after harvest, L47 and L66 maintained high population on fruits over seven-day cold storage followed by five days of shelf life at 20°C and suppressed botrytis fruit rot.

Suppression of *B. cinerea* at the source is an alternative strategy for managing strawberry fruit rot (Sutton and Peng, 1993b). The pathogen establishes latent infection in emerging leaves and subsequently colonizes the tissues and sporulates when the tissues senesce and die (Braun and Sutton, 1988). Mycelium of the pathogen in dead leaves is a chief source of initial inoculum in epidemics of gray mold fruit rot in perennial strawberry crops (Braun and Sutton, 1987). In field tests, *G. roseum* suppressed sporulation potential of *B. cinerea* by 90 to 100 percent on green leaves inoculated with the pathogen and about 60 percent on semisenescent overwintered leaves (Sutton and Peng, 1993a). Ability of suppressing *B. cinerea* on strawberry leaves by *U. atrum* was also examined in field trials (Boff et al., 2001). In addition, combination of *Pichia guilermondii* with *B. mycoides* suppressed *B. cinerea* on strawberry leaves by 80 to 99 percent under various temperature and relative humidity conditions (Guetsky et al., 2001). The contribution to reduction in initial inoculum and fruit rot by antagonists needs to be examined in strawberry fields, and the application of timing and frequency should be optimized.

Delivery Systems Specific for BCAs

Adoption of existing methods used for fungicides is a general approach for application of BCAs. Spray is a chief delivery system for applications of biocontrol agents to foliage, flowers, and fruits. Dipping, drenching, or in-line spray are mainly used for applications of BCAs to fruit in commercial packinghouse. Hofstein and Chapple (1999) discussed the quality of spray applications. They found that the distribution of spores of *A. quisqualis* in a spray tank changed markedly after passage through a pump, and the clumping spores of the BCAs could decrease the efficiency of application significantly. They also noted that it was difficult to find nonfungitoxic surfactant for suspending spores. Although the adopted application methods need to be refined for the best fit to the biocontrol system, there are very few publications on this area (Gan-Mor and Mathews, 2003; Hidalgo et al., 2003). In contrast, significant development of delivery systems for the purpose of biocontrol has been reported, such as production and dispersal of pycnidiospores of *A. quisqualis* in vineyards, bee vectoring BCAs to flowers, both reviewed by Sutton and Peng (1993b), and recently published biofumigation (Strobel et al., 2001; Mercier and Jiménez, 2004). The biofumigation and bee vectoring systems will be discussed in detail.

BIOLOGICAL FUMIGATION OR BIOFUMIGATION

Biofumigation is to use biofumigant, that is, microorganisms capable of producing antimicrobial volatiles, for controlling plant disease (Strobel

et al., 2001; Mercier and Jiménez, 2004). This new concept was first demonstrated by a greenhouse test. Barley seeds, naturally contaminated with
the covered smut fungus *Ustilago bordei*, were exposed to the gases of
Muscodor albus in an agar plate for four days and then planted into pots and
grown in a greenhouse. The fumigation completely controlled smut disease
but did not appear to have any inhibition or damage to the plants (Strobel
et al., 2001). Further investigation of biofumigation with *M. albus* was
conducted for control of apple and peach fruit rots. Artificially wounded apples were inoculated with *P. expansum* or *B. cinerea*, and peaches with
M. fructicola. The fruits were placed in lid-closed boxes, in which *M. albus*
grown on rye grains was added and incubated for various periods. Regardless of the fumigation dose, application immediately or 24 h after inoculation of the fruits, the treatment provided excellent control of blue mold and
gray mold of apples, and brown rot of peaches. In most cases, fumigation
killed the pathogens in wounds (Mercier and Jiménez, 2004). The successes
with biofumigation have opened up a new opportunity for control of
postharvest fruit decay. The biofumigation is particularly important to fruits
that are easily damaged or less appealing with postharvest treatments, such
as strawberry, raspberry, and grapes.

Muscodor albus is an endophytic fungus isolated from *Cinnamomum
zeylanicum* (cinnamon tree). The fungus produces various kinds of volatiles; among them, twenty-eight compounds were identified to be alcohols,
esters, ketones, acids, and lipids; and ester derivatives are the most biologically effective. Many major plant and human fungal, yeast, and bacterial
pathogens were sensitive to the gases of *M. albus*; however, a few microbes
for example, *Xylaria* sp., a close relative of *M. albus*, and *Fusarium salani*
were only partially inhibited. This indicates that volatiles of *M. albus* are
biologically selective. The results from the further tests implicated that the
M. albus volatiles operate additively or synergistically (Strobel et al.,
2001). Thus, it would be very difficult for treated plant pathogens to develop resistance to the set of the fungal volatiles. To use biofumigation
commercially, large-scale fermentation needs to be worked out, and fermentation substrates and conditions should be examined and optimized.
Instead of applying *M. albus* cultures, direct use of the fungal gases may
have advantage for operating fumigation under various environmental conditions such as during cold storage and transportation. Besides *M. albus*,
two other fungal endophytes were isolated from rainforests. They have fumigation functions but produce a different set of volatile compounds from
those of *M. albus* (Strobel et al., 2001).

Bee Vectoring

Treatment of flowers with BCAs applied as sprays is inefficient, because a large portion of the microbial inoculum fails to meet the target and is deposited on the foliage or ground. Targeting flowers with sprays is often difficult on account of the small size of the target, poor exposure of flowers to directed sprays, and continual opening of new flowers after sprays have been applied (Sutton and Peng, 1993b). Better targeting of BCAs can be achieved in flowers of some kinds of plants using bees vectoring the BCA (Peng et al., 1992; Yu and Sutton, 1997; Kovach et al., 2000). John Sutton and his colleagues pioneered a concept of using honeybees *(Apis mellifera)* as vehicles to deliver a fungal BCA to strawberry flowers for control of gray mold (Peng et al., 1992). An inoculum dispenser was employed for dusting bees with a powder formulation of the BCA before the bees left the hive. In small-scale trials with strawberry plants in a greenhouse and in field plots, propagule density of *G. roseum* on flowers forged by bees was more stable and often higher than on flowers sprayed weekly with conidial suspensions of the agent. The two treatments were about equally effective in suppressing sporulation potential of *B. cinerea* on stamens and petals, and in controlling gray mold fruit rot.

Bumblebees *(Bombus impatiens)* have also proved to be good vectors of *G. roseum* (Yu and Sutton, 1997). A special dispenser for contaminating bumblebees with inoculum was designed to fit on the front of the bumblebee colony box. *G. roseum* was recovered from more than 85 percent of flowers in plots with bumblebees or honeybees compared to 45 percent in plots sprayed with spore suspensions. Bumblebees foraged more actively and vectored more inoculum to raspberry flowers than did honeybees under conditions of cool temperatures, persistent high humidity, and periods of rain. On the other hand, honeybees were more active on warm days with bright sun. This suggested that combined use of bumblebees and honeybees might help to optimize inoculum delivery when weather conditions vary markedly during the flowering periods.

Large-scale tests of bee vectoring were conducted in strawberry fields in New York State (Kovack et al., 2000). Honeybees and bumblebees were employed to deliver *T. harzianum* 1295-22. Treatment applied by bees required only 6 percent as much inoculum as the spray applications. However the bee-vectored inoculum controlled fruit rot more effectively than did inoculum applied as sprays and was as or more effective than were the fungicide treatments.

Two principal questions arise when the efficiency of bee-vectoring systems is considered in relation to disease control: (1) How efficient are bees in delivering inoculum to the flowers? and (2) How effective is the delivered inoculum in controlling diseases? The propagule density of BCAs (cfu per bee and per flower), the percentage of flowers with detectable levels of the agents, inoculum distribution on floral parts, and effective delivery radius from hives and colony boxes are primary parameters to be considered for assessing bee vectoring. In experiments with bee delivery of *G. roseum* and *T. harzianum* to strawberry flowers, and of *G. roseum* to raspberry flowers, the propagule density of the fungal agents was commonly in the range of 5×10^2 to 4×10^4 cfu per flower. This density was sufficient to effectively suppress *B. cinerea* on flowers and to control gray mold fruit rot, except when disease pressure was high (Yu and Sutton, 1997). An ability for bees to deliver greater quantities of inoculum to flowers to provide adequate control, even when conditions are extremely favorable for the pathogen and disease, may be possible through modification of the inoculum formulation.

Distribution of bee-vectored inoculum of BCAs on flowers may influence the effectiveness of control, but little is known of these relationships. In the large-scale studies (Kovach et al., 2002), *T. harzianum* recovered from fewer flowers, and propagule density per flower was lower, when the BCA was applied by bees than as sprays. However, the incidence of gray mold was 32 percent lower when the BCA was delivered by bees than as sprays. It was suggested that the bees delivered the agent more effectively to portions of the flowers, such as the stamens that are important sites of infection of the flowers and subsequently fruits (Powelson, 1960). GFP and GUS transformants of BCAs would be useful tools for study of inoculum distribution on floral parts (Lübeck et al., 2002; Rav-David and Elad, 2002). Use of transformants and other molecular approaches to allow the bee-vectored BCA to be distinguished from naturally occurring microbes in the field would greatly facilitate tracking of agents on flowers and elsewhere in crops (Schena et al., 2000). Bulat et al. (2000) developed a strain-specific PCR detection marker for *C. rosea* which has application, for example, in studying the radius and patterns of delivery of the BCA by bees in relation to hives or colony boxes, which is a base for determining how many bees should be employed in an unit area of the crop fields.

Effective delivery of a microbial agent to flowers may not result in good protection against a flower-infecting pathogen, should conditions in the field be unfavorable for the agent to survive and associate effectively with the flower tissues. The bee-delivered *T. harzianum* failed to control gray mold, even though a high density of the BCA was detected on bee-visited flowers (Maccagnani et al., 1999). Survival of BCAs, either in the inoculum dispenser or on bee-vectored flowers, has not been reported. This information is important to determine how often inoculum should be replaced and how the BCA interacts with pathogens. Bee-vectored BCAs protect only flowers from infection. Direct infection of fruits, such as grape and raspberry fruits infected by *B. cinerea*, may result in poor control of fruit rot (Yu and Sutton, 1997).

Bee delivery is affected by numerous interacting factors, including inoculum formulation, loading of bees in inoculum dispensers, foraging activities, and environmental variables (Sutton, 1995). Inoculum formulation and the architecture of the dispenser can each markedly influence inoculum acquisition by bees and require critical consideration for optimizing the efficiency of bee loading. It is also important to minimize disturbance of bees when developing the dispensers and formulations. When a bee-delivery system is employed in the field, potential competitive sources of pollen and nectar from surrounding vegetation must be considered. Although strawberries are not one of the most favorable nectar sources for bees, bee delivery was successful in the growers' fields when more attractive nectar sources were not available outside of strawberry fields (Kovach et al., 2000). Temperature, rain, overcast skies, and food availability are major factors affecting bee-foraging activity.

In other biocontrol studies on fruit, honeybees successfully vectored *Pseudomonas fluorescens* to apple and pear flowers (Thomson et al., 1992; Johnson et al., 1993). Bee vectoring may also have application for biocontrol of flower-pathogens in other fruit crops, such as peaches, cherries, blackberry, and blueberry (Sutton and Peng, 1993b). Combination of inoculum delivery and host pollination by bees has potential for controlling diseases and improving yield. Bee vectoring is a flower-orientated system, and the concept can be extended to explore wound-orientated insects for delivering BCAs to wounds. Successes in bee vectoring BCAs on strawberry, raspberry, and other crops provide a viable option for growers to manage diseases with environment-friendly means.

STRATEGIES FOR ENHANCING BIOCONTROL EFFICACY

Many research projects have been carried out in the area of search for effective BCAs; however, there are very few biocontrol products commercially available, and their market share is still marginal, compared with synthetic fungicides. Although certain biocontrol systems have already achieved great success (Harman, 2000), improvement in efficiency, consistency, economical feasibility, and so forth are still needed for many others. Therefore, while searching for more effective BCAs, extensive attempts have been made to improve the achieved biocontrol systems. The improvement strategies used in biocontrol systems related to temperate fruit are summarized in Table 10.2 and will be explained with practical examples.

TABLE 10.2. Strategies used to enhance biocontrol efficacy.

Strategies	BCA	Approaches / additions	Pathogen / host	Achievements	References
BCA manipulation					
Physiological	Candida sake	Improvements in water stress tolerance	Penicillium expansum/ apple	Improved biocontrol	Texidó et al., 1998
Genetic	Candida oleophila	Genetic markers	Postharvest diseases	Gene transformation	Chand-Goyal et al., 1998, 1999
Manipulating environments					
Chemical	C. sake	Ammonium molybdate	P. expansum & pear	Enhanced efficacy in postharvest but not in preharvest experiment	Nunes et al., 2002
	C. saitoana	2-deoxy-D-glucose	B. cinerea & P. expansum / apple	Increased efficacy	El-Ghaouth et al., 2000a
	Chaetomium globosum	Cellulose	Zygophiala jamaicensis & Gleodes pomigena / apple	Improved colonization and control efficacy	Davis et al., 1992
	Rhodotorula glutinis	Iron / Siderophore	P. expansum	Increased efficacy	Calvente et al., 2000
	Pseudomonas syringae	L-Asparagine & L-proline	P. expansum	Both amino acids enhanced the population of the BCA	Janisiewicz, 1992
Physical	P. syringae, yeasts	Heat / 38 C- 4days	P. expansum / apple	Additive and complementary effect	Leverentz et al., 2000, 2003
Integrations with other control means					
BCA mixtures	Acremonium breve	Pseudomonas sp.	P. expansum and B. cinerea / apple	Controlled two major diseases	Janisiewicz, 1987, 1996
	Rhodotorula glutinis	Cryptococcus albidus	P. expansum & B. cinerea / apple	Synergistic effect	Calvo et al., 2003
	Pichia guilermondii	Bacillus mycoides	B. cinerea / strawberry	Reduce the variability of biocontrol	Guestky et al., 2001

Category	BCA	Additive treatment	Pathogen / host	Effect	Reference
Antibiotics	C. oleophila	Nisin	B. cinerea & P. expansum / apple	Inhibited pathogen spore germination thus enhanced biocontrol	El-Neshawy & Wilson, 1995
Fungicides	Bacillus subtilis	dicloran	Monilinia fructicola / peach	Dicloran provided additive effect	Pusey et al., 1986
	P. syringae	Cyprodinil	P. expansum / apple	Additive effect	Zhou et al., 2002
	Cryptococcus infirmo-miniatus	Single preharvest applicaton of iprodione	M. Fructicola & P. expansum / sweet cherry	Effectively controlled diseases	Spotts et al., 1998
	C. laurentii HRA5 & R. glutinis HRB6	Thiabendazole (TBZ) at low dosage	P. expansum / pear	Control activity same as using TBZ alone at full dosage	Chand-Goyal & Spotts, 1996
	C. Laurentii LS28, or Aureobasidium pullulans LS30	Benomyl at low dosage	B. cinerea & P. expansum / apple	Control activity same as using benomyl alone at full dosage	Lima et al., 2003
Inorganic or organic salts	C. oleophila (Aspire)	Sodium bicarbonate	B. cinerea & P. Expansum / apple; Monilinia & Rhizopus peach	Improved BCA efficacy	Droby et al., 2003
	C. guilliermondii & Pichia membranefaciens	Calcium chloride	Rhizopus stolonifer / peach & nectarine	Increased efficacy	Tian et al., 2002
	P. syringae	Calcium chloride	P. expansum / apple .	More effective control	Janisiewicz, 1995
	P. syringae	Calcium chloride	M. fructicola / peach	Increased efficacy	Zhou et al., 1999
Elicitic chemical	C. saitoana	Chitosan / Glycolchitosan	B. cinerea & P. expansum / apple	Increased efficacy	El-Ghaouth et al., 2000
	R. glutinis & Cryptococcus laurentii	Salicylic acid	P. expansum & Alternaria alternata / sweet cherry	Induced resistance, thus enhanced disease control	Qin et al., 2003
Other	Saccharomyces cerevisae	Ethanol	B. cinerea / apple	Synergistic effect	Mari & Carati, 1998

MANIPULATION OF BIOCONTROL AGENTS

Certain physiological characteristics of a BCA can be modified through selections under specifically designed procedures and conditions, and several reports have demonstrated that the physiological manipulation can be a means for developing a BCA with improved ecological fitness, resulting in enhanced biocontrol potential. One of the examples is the manipulation of temperature tolerance of *Trichoderma* isolates (Tronsmo and Ystaas, 1980). Through series selections, *Trichoderma* isolates capable of growing at relatively low temperatures were obtained; these isolates were able to tolerate low temperatures occurring during blossom period when *B. cinerea* infection initiated and provided more effective control of *B. cinerea* on apple. BCAs have also been selected for heat resistance to allow being used in prestorage heating systems (Leverentz et al., 2000). In addition, a research group in Spain (Teixidó et al., 1998) has manipulated water stress tolerance of *C. sake*, which was effective against major pathogens of apple and pear, particularly at high humidity, but had poor establishment on fruit, and less disease control efficacy at reduced water availability, such as in the field. To improve water stress tolerance of the BCA, the yeast was cultured on media modified by the addition of ether glycerol glucose or trehalose, aiming to reduce water activity, which resulted in a modification of intracellular sugar alcohol and sugar content of the yeast cells. The viability of the cells with modified polyols/sugars was significantly improved, compared with unmodified yeast cells; the modified cells provided much greater control toward apple blue mold. Improved tolerance to water stress can also be a useful trait in the development of formulation for this BCA (Abadias et al., 2001).

Enhancement of biocontrol efficacy may also be achieved by genetically manipulating a BCA by addition or elevation of desirable gene-mediated characteristics. The benefits and feasibility of genetic manipulation of BCAs were realized decades ago and were discussed in several reviews (Gutterson et al., 1989; Pusey, 1994; Sumeet and Mukerji, 2000; Spadaro and Gullino, 2003). There are mainly two approaches to genetically manipulate BCAs: the mutagenicity and recombinant methods, including protoplast fusion and gene transformation. To date, many reports on genetic manipulation were in the area of using mutagens, such as ionizing radiation and ultraviolet light, mutagenic chemicals and fungicide or antibiotics, to induce mutants for the selection of improved BCA. The selected characteristics often include increased fungicide tolerance/resistance, capability of utilizing certain compounds, elevated capability in producing antagonism-related compounds, and improved tolerance to environmental stresses (Pusey, 1994). Recently, a cold-tolerant strain of *F. proliferatum* has been selected following UV mutagenesis (Bakshi et al., 2001).

Protoplast fusion techniques have been used for the improvement of fungal BCAs (Sumeet and Mukerji, 2000). In fact, the strain *T. harzianum* T-22, the base of the commercial biocontrol products by BioWorks, Inc, was obtained by protoplast fusion between two *T. harzianum* strains of T-95 and T12 (Harman, 2000). Protoplast fusion may also be interspecies. The fusion between *T. harzianum* and *T. longibrachiatum* produced fusants with enhanced antagonistic potential along with tolerance of copper sulphate and carbendazin (Lalithakumari et al., 1996). The use of protoplast fusion to overcome incompatibility barriers is an effective technique to induce genetic recombination in whole genomes, even between incompatible strains/ species.

As more and more functional genes are being identified and advanced genetic engineering techniques being are developed, gene transformation should become an important and unique tool for the purpose of enhancing biocontrol efficacy, even of producing new BCAs. Through transformation, a gene or genes with traits desired for the biocontrol of fruit diseases, such as those suggested in previous reviews (Pusey, 1994; Spadaro and Gullino, 2003), may be expressed in different microorganisms, aiming for various objectives. Although disease-resistant cultivars are ideal for managing plant diseases, an "ideal" BCA from gene transformation may have the advantage over the transgenic plant in being allowed to apply to a wide variety of fruits and vegetables. Moreover, compared with plant, yeast and bacteria, which most BCAs on fruit belong to, generally are readily transformed and various biocontrol-related compounds, such as antimicrobial peptides, may be rapidly introduced and expressed in the microorganisms (Klinner and Schäfer, 2003). There are already successful transformations relative to BCAs. *C. oleophila* was transformed with the β-glucuronnidase gene (Chand-Goyal et al., 1998), and histidine auxotrophs of *C. oleophila* were transformed with HIS3, HIS4, and HIS5 genes (Chand-Goyal et al., 1999). Even though these transformations were accomplished for obtaining variants of the BCA with a genetically stable marker in order to study the ecology of the BCA, the efficient transformation procedures developed from these researches will be very useful in transforming genes with desired biocontrol traits. The successful expression of a DNA sequence encoding a cecropin A-based peptide in *Saccharomyces cerevisiae* provided more positive evidence in the feasibility and benefits of using gene manipulation to achieve biocontrol improvement. The yeast transformants secrete the antifungal peptide, which inhibited the growth of germinated *Colletotrichum coccodes* spores, and controlled fruit decay on tomato inoculated with the pathogen (Jones and Prusky, 2002). The use of *S. cerevisiae* as a delivery system suggests that this method could be a safe alternative for the control of fruit diseases, and its product may be easier to be accepted by consumers.

Several yeasts have been reported as potential BCAs for fruit diseases. As mentioned in the first part of the review, manipulation of the antagonistic yeasts or expressing their biocontrol traits in other safe and ecologically sound microorganisms by gene transformation may result in effective biocontrol products in the future. However, more research is needed to discover genes with biocontrol functions, to select suitable microorganisms for their delivery, and even more importantly, to understand the safety and ecological fate of the transformed BCAs.

MANIPULATION OF CHEMICAL
AND PHYSICAL ENVIRONMENTS

Chemically and/or physically manipulating the environmental conditions of a biocontrol system is another approach that has been used for enhancing biocontrol efficacy. For a satisfactory control, an essential population of BCAs is often needed, and in many cases, poor performance of a BCA often appeared under high pressure of pathogen populations (Yu and Sutton, 1997). Thus, improved biocontrol can be achieved by increasing BCA population while suppressing pathogen level through manipulating environmental conditions.

Enhancement of biocontrol by adding nutrients that are favorable to BCAs but not so to pathogens has been reported for several BCA-pathogen interactions (Janisiewicz and Korsten, 2002). For example, on pome fruit, nitrogen can be a limiting factor for BCAs, because they, though rich in carbohydrates, are poor in nitrogen. The addition of L-asparagine and L-proline showed great enhancement in the control of apple blue mold by *P. syringae* (Janisiewicz et al., 1992). Also, L-serine and L-aspartic acid significantly improved the efficacy of *C. sake* against *P. expansum* on apple, and the addition of ammonium molybdate allowed *C. sake* to be used at ten times less without diminishing the control efficacy (Nunes et al., 2001). The explanation of this enhancement is that those nitrogenous compounds improved the colonization of the BCAs. Furthermore, compounds that are fungicidal to the targeting pathogens but compatible with the BCAs can also be useful additives, such as 2-deoxy-D-glucose, a sugar analogue, which has antifungal activity toward major postharvest pathogens of apple and peach but is compatible with the BCAs *P. syringae*, *C. saitoana*, and *C. sake*. The addition of this sugar analogue to BCAs has resulted in the increased control against *B. cinerea* and *P. expansum* on apples and pears (El-Ghaouth, Smilanick, Wisniewski, and Wilson, 2000; Nunes et al., 2001). Siderophores, low-molecular-weight ferric chelating agents, which inhibit pathogen growth or metabolic activity by sequestering iron (Riquelme, 1996), may also be a useful additive for certain biocontrol systems in which iron is a limiting factor. It was found that apple blue mold was controlled more effectively by *R. glutinis* plus siderophore than by the BCA alone (Calvente et al., 1999).

Prestorage heating treatment can be beneficial to biocontrol when heat-resistant BCAs are used (Leverentz et al., 2000, 2003). Because the heat treatment (38°C) reduces the pathogen population pressure and competition of other microorganisms in the fruit wound, BCAs likely have an advantage in colonizing the fruit surface. It was noted that the populations of two heat-resistant yeasts in apple wounds increased during the heat treatment, and the combination of heat and BCA resulted in excellent decay control against *P. expansum* on apple (Leverentz et al., 2000, 2003).

INTEGRATION FOR ENHANCED DISEASE CONTROL

Improved biocontrol may be achieved by integrating a BCA with other control measures, resulting in additive or synergistic or complementary effect. The integrations may have different aims, such as to widen the control spectrum in order to cover multiple diseases, to increase control efficacy to an acceptable level, to improve biocontrol consistency to meet producers' satisfaction, and/or to reduce the application dosage for economical feasibility.

A mixture containing BCAs, that are not only complementary but also compatible, can be carefully developed to improve biocontrol, as discussed in other reviews (Janisiewicz, 1997; Janisiewicz and Korsten, 2002). Some BCAs for diseases on fruit are not equally effective for all fungal pathogens that should be controlled, and an appropriate combination of BCAs may provide a solution. A mixture of *Acremonium brevie* and *Pseudomonas* sp. protected apples from infection by both *B. cinerea* and *P. expansum* (Janisiewicz, 1988). BCA mixtures may also have additive or synergistic effects on biocontrol. A mixture of *P. syringae* and *S. roseus* selected from various combinations was more effective than both BCAs applied alone in control of apple blue mold (Janisiewicz and Bors, 1995). Mixtures may be specific to different pathogens. When evaluated on apple fruits, mixtures of *R. glutinis* SL 1 with *C. albidus* SL 43 and *R. glutinis* SL 30 with *C. albidus* SL 43 showed synergistic effect against *P. expansum*, but not against *B. cinerea* (Calvo et al., 2003). The research also indicated that it was possible to improve the efficacy of biocontrol without increasing the total amount of BCA applied once an appropriate BCA mixture is developed. However, BCA mixtures have not been practically implemented in the biocontrol of fruit diseases because of the concern of potential high cost in commercializing the mixtures, compared with a single BCA (Harman, 2000; Janisiewicz and Korsten, 2002).

The most commonly used integration is combining a BCA with a fungicide or fungicides that are being used commercially, at a fraction of the recommended concentration, since fungicides are still the primary means to control fruit diseases when permitted. However, this approach may only be suitable for BCAs (in most cases, bacterial and certain yeast BCAs) insusceptible to the fungicide. For fungal BCAs, fungicide-resistant mutants may be required, which may be achieved through genetic manipulations. This integration has been investigated particularly for the control of postharvest diseases of fruit (Pusey et al., 1986; Chand-Goyal and Spotts, 1996, 1997; Sugar and Spotts, 1999). In a semicommercial trial, Chand-Goyal and Spotts (1997) showed that the control of apple blue mold by *C. laurentii* HRA5 was increased by combining it with TBZ. This fungicide was also used in the integrations with two commercial biofungicides: BioSave 110 and Aspire. In a packinghouse trial, combinations of Bio-Save 110 or Aspire with 100 µg/mL TBZ (about 17.6 percent of the label rate) provided control of blue mold and gray mold of pears, similar to that of TBZ alone used at the label rate (569 µg/mL) (Sugar and Spotts, 1999). In the tests on grapes conducted in New York, *Trichoderma* spp. significantly suppressed botrytis bunch rots in three out of four years, and were often inferior to fungicides. However, application of *T. harzianum* 1293-22 at bloom and early fruit development, followed by a tank-mix application of the antagonist and half rates of iprodione, suppressed the disease by 98 percent (Harman et al., 1996). Moreover, integration with new "reduced-risk fungicide," such as cyprodinil, may have additional merits in toxicity and resistant management. Combination of *P. syringae* MA-4 at $1\text{-}3 \times 10^7$ cfu/mL with cyprodinil at 5 to 10 µg/mL controlled both blue and gray mold by >90 percent on apple, demonstrating that the integration could not only improve disease control efficacy but also extend the degree of control to more than one important disease (Zhou et al., 2002).

Combination of a BCA with a fungicide, both at significantly reduced concentrations from what would be required if used alone, may greatly reduce their residues on fruit. Studies have shown that residue levels of chemicals were proportional to the concentrations used in many cases (Papadopoulou-Mourkidou, 1991), and the amount of BCA recovered from the surface of fruits was correlated with the concentration applied (Usall et al., 2001). Considerably high concentrations of BCAs, such as 1×10^8 cfu/mL for yeast and 1×10^9 cfu/mL for bacteria, have been used to obtain adequate postharvest disease control (Chand-Goyal and Spotts, 1997; Table 10.2). Lowering concentration by integration would significantly reduce not only residues but also the cost of the treatment. The latter can be an important factor in commercializing a BCA (Froyd, 1997). Until more effective alternatives of fungicides are available, integrating limited biofungicides with fungicides may be a practical and complementary strategy in the control of certain fruit diseases. This combination may also provide opportunity for reducing the pressure toward development of fungicide-resistant populations of pathogens (O'Neill et al., 1996).

Another approach often used is addition of calcium to a BCA. Application of calcium chloride in the orchard increased apple fruit resistance to infection by *B. cinerea* and was effective against blue mold and side rot of pears (Sugar et al., 2003). Postharvest applications of calcium, using various techniques, such as heating, dipping, vacuum filtration, pressure filtration, and coating (Chardonnet et al., 2003), were also reported to be effective in delaying fruit senescence, increasing fruit firmness and quality (Sams et al., 1993), and reducing diseases during storage (Conway et al., 1991). Several studies have reported the effect of calcium on the biocontrol of fruit diseases. McLaughlin et al. (1990) showed that addition of 2 percent $CaCl_2$ to a cell suspension of yeast *C. guilliermondii* significantly increased its efficacy in control of apple postharvest diseases, compared with either yeast or calcium used alone. In another report (Wisniewski et al., 1995), addition of 90 or 180 mM $CaCl_2$ enhanced biocontrol activity of *C. oleophila* isolate 182 against *B. cinerea* and *P. expansum*, but had no effect on its activity of isolate 247. It was postulated that the enhancement of the biocontrol activity in this case was directly due to the inhibitory effects of calcium ions on pathogen spore germination and metabolism, and indirectly due to the tolerance of isolate 182 to the calcium "toxicity." The addition of calcium affected the ability of isolate 247 to proliferate in apple wound, thus resulting in less control. However, calcium may enhance biocontrol by multiple actions. Zhou et al. (1999) reported that addition of 0.5 percent $CaCl_2$ to cell suspension of *P. syringae* MA-4 resulted in a greater reduction of peach brown rot incidence when sprayed on peaches naturally infected by *M. fructicola*, but the improvement by calcium addition did not appear on peaches artificially wound-inoculated with the pathogen. Cells of the BCA suspended in 0.5 percent $CaCl_2$ had higher recovery rate than the ones in water in serial dilution test, and had no effect on the germination of *M. fructicola* spores, although 2 percent $CaCl_2$ was inhibitory to both the BCA and pathogen. Preharvest application of *P. syringae* MA-4 with a foliar calcium fertilizer also significantly increased biocontrol efficacy against peach brown rot (Zhou and Schneider, 1998).

Numerous organic and inorganic salts are widely used as additives in the food industry, and some of them possess antimicrobial activities (Karabulut et al., 2001). These additives are generally safe for human consumption and less expensive; and thus, they should be a potential source for integration with BCAs for controlling fruit disease. Several food additives have recently been evaluated for their effects on Aspire (Droby et al., 2003). When combined with 2 percent sodium bicarbonate, Aspire consistently showed enhanced efficacy in the control of *Botrytis* and *Penicillium* rot in apple and *Monilinia* and *Rhizopus* rot in peach, compared with that used alone. The substance, which is not only antimicrobial but also possesses eliciting property to induce plant resistance, has additional advantage when integrated with BCAs. Chitosan and its derivatives, for example, are inhibitory to a number of pathogens (Romanazzi et al., 2001) and can also induce host-defense responses (Fajardo et al., 1998; Zhang and Quantick, 1998; Romanazzi et al., 2002). Chitosan, applied either preharvest or postharvest, significantly reduced storage decay of table grapes and sweet cherries (Romanazzi et al., 2002, 2003). Under semicommercial conditions, combination of BCA, *C. saitoana*, with 0.2 percent glycolchitosan gave a greater control of fruit decay on certain apple cultivars than either the BCA or glycochitosan alone (El-Ghaouth, Smilanick, Brown, et al., 2000). Another substance with eliciting property is salicylic acid (SA), a natural compound present in many plants, and an important component in the signal transduction pathway, is involved in local and systemic resistance to pathogens. When cherry fruits were pressure-infiltrated with SA solution before they were wounded and treated with two yeast BCAs separately and then inoculated with two pathogens, *P. expansum* and *Alternaria alternate*, the biocontrol efficacy of *R. glutinis* was significantly enhanced, although SA did not affect control efficacy by *C. laurentii*. It was found that SA at low concentration, such as 0.5 mM, had little effect on the growth of both the BCAs and pathogens. However, SA treatment produced significant increase in activity of several resistance-related enzymes in cherry fruit, indicating that the enhancement of disease control might be due to the ability of SA to induce a biochemical defense response (Qin et al., 2003).

CONCLUSIONS

Considerable progress has been made in the area of biocontrol of temperate fruits, as evident from the development of effective biocontrol strategies, and from the registration and commercial application of several biofungicides, particularly for the control of postharvest diseases. Compared with hundreds of research publications, biofungicides used for disease control in commercial agriculture are still very few; however, to both the biocontrol developers and users, the accumulated knowledge in understanding biocontrol systems is an essential asset and can certainly speed up the development of biocontrol products in the future.

Naturally occurring microbial communities present in the phyllosphere and carposphere have been the primary source of BCAs for the control of fungal diseases of temperate fruits; soil and plant debris are also popular sources for antagonists. In addition, food microbes, as BCAs for postharvest treatment, may have unique advantages. Bacteria and yeasts have been mainly used in biocontrol by postharvest treatments; filamentous fungi, on the other hand, are often applied in the field conditions; and the detection of mycoviruses in fruit pathogens provides a new resource for potential BCAs. Also, the knowledge of biocontrol systems achieved so far has allowed selection of BCAs with clear objectives, such as designating certain mode of action and/or specific applications.

Specific application strategies should be developed for each biocontrol system. BCAs are not simple replacements of synthetic chemicals; even BCAs with different modes of action may need to be applied with different methods and time schemes. Bee vectoring and biofumigation are innovative application methods specifically developed for BCAs and have showed great successes. Unique BCA application strategies may be developed to fit cultivation systems where biocontrol may have a great potential such as organic growing systems.

Biocontrol should be considered as one component of an integrated disease control system. Integrating BCAs with other control measures, such as fungicides, calcium, and plant resistance elicitors, have achieved successes in widening the disease control spectrum, increasing efficacy, and improving the consistency of BCAs.

Biocontrol has not been used in the control of fungal diseases of temperate fruits as one expected, and there still is a long journey before it can play a significant role in disease management. However, the newly developed biotechnologies may accelerate the biocontrol development by providing effective tools; for instance, the use of molecular tools, such as biosensors and other genetic markers, can help elucidate the modes of antagonism by the BCAs; and the use of PCR markers, strain-specific probes, GFP or GUS transformants, and so forth can provide insight into population interactions between BCAs and pathogens in the field. BCAs may also be manipulated effectively and acutely with gene transformations when necessary.

REFERENCES

Abadias, M., N. Teixido, I. Vinas, J. Usall, and N. Magan. (2001). Improving water stress tolerance of the biocontrol yeast *Candida sake* grown in molasses-based media by physiological manipulation. *Canadian Journal of Microbiology* 47: 123-129.

Andrews, J.H., F.M. Berbee, and E.V. Nordheim. (1983). Microbial antagonism to the imperfect stage of the apple scab pathogen, *Venturia inaequalis. Phytopathology* 73: 228-234.

Baker, K.F. and R.J. Cook. (eds.). (1974). *Biological Control of Plant Pathogens.* WH Freeman and Company, San Francisco, CA.

Bakshi, S., A. Sztejnberg, and O. Yarden. (2001). Isolation and characterization of a cold-tolerant strain of *Fusarium proliferatum,* a biocontrol agent of grape downy mildew. *Phytopathology* 91: 1062-1068.

Ben-Arie, R., S. Droby, J. Zuthhi, L. Cohen, B. Weiss, P. Sarig, M. Zeidman, A. Daus, and E. Chalutz. (1991). Preharvest and postharvest biological control of Rhizopus and Botrytis bunch rots of table grapes with antagonistic yeasts. In *Biological Control of Postharvest Diseases of Fruits and Vegetables, Workshop Proceeding,* eds. C.L. Wilson and E. Chalutz. ARS Publication 92, USDA-ARS, Washington, DC, pp. 100-113.

Benbow, J.M. and D. Sugar. (1999). Fruit surface colonization and biological control postharvest diseases of pear by preharvest yeast application. *Plant Disease* 83: 839-844.

Blakeman, J.P. and I.D.S. Brodie. (1977). Competition for nutrients between epiphytic microorganisms and germination of spores of plant pathogens on beetroot leaves. *Physiological Plant Pathology* 10: 227-239.

Boff, P., J. De Kraker, A.H. Van Bruggen, M. Gerlagh, and J. Köhl. (2001). Conidial persistence and competitive ability of the antagonist *Ulocladium atrum* on strawberry leaves. *Biocontrol Science and Technology* 11: 623-636.

Boff, P., J. Köhl, M. Gerlagh, and J. De Kraker. (2002). Biocontrol of grey mould by *Ulocladium atrum* applied at different flower and fruit stages of strawberry. *Biocontrol* 47: 193-206.

Boff, P., J. Köhl, M. Jansen, P.J.F.M. Horsten, C. Lombaers-van der Plas and M. Gerlagh. (2002). Biological control of gray mold with *Ulocladium atrum* in annual strawberry crops. *Plant Disease* 86: 220-224.

Bonaterra, A., M. Mari, L. Casalini, and E. Montesinos. (2003). Biological control of *Monilinia laxa* and *Rhizopus stolonifer* in postharvest of stone fruit by *Pantoea agglomerans* EPS125 and putative mechanisms of antagonism. *International Journal of Food Microbiology* 84: 93-104.

Braun, P.G. and J.C. Sutton. (1987). Inoculum sources of *Botrytis cinerea* in fruit rot of strawberries in Ontario. *Canadian Journal of Plant Pathology* 9: 1-5.

Braun, P.G. and J.C. Sutton. (1988). Infection cycles and population dynamics of *Botrytis cinerea* in strawberry leaves. *Canadian Journal of Plant Pathology* 10: 133-141.

Bulat, S.A., M. Lübeck, I.A. Alekhina, D.F. Jensen, I.M.B. Knudsen, and P.S. Lübeck. (2000). Identification of a universally primed-PCR-derived sequence-characterized amplified region marker for an antagonistic stain of *Clonostachys rosea* and development of a strain-specific PCR detection assay. *Applied and Environmental Microbiology* 66: 4758-4763.

Burkhead, K.D., D.A. Schisler, and P.J. Slininger. (1994). Pyrrolnitrin production by biological control agent *Pseudomonas cepacia* B37w in culture and in colonized wounds of potato. *Applied and Environmental Microbiology* 60: 2031-2039.

Burr, T.J., M.C. Matteson, C.A. Smith, M.R. Corral-Garcia, and T.C. Huang. (1996). Effectiveness of bacteria and yeasts from apple orchards as biological control agents of apple scab. *Biological Control* 6: 151-157.

Calvente, V., D. Benuzzi, and M.I.S. de Tosetti. (1999). Antogonistic action of siderophore from *Rhodotorula glutinis* upon the postharvest pathogen *Penicillium expansum*. *International Biodeterioration* and *Biodegradation* 43: 167-172.

Calvo, J., V. Calvente, M.E. De Orellano, D. Benuzzi, and M.I.S. De Tosetti. (2003). Improvement in the biocontrol of postharvest diseases of apples with the use of yeast mixtures. *BioControl* 48: 579-593.

Carisse, O. and M. Dewdney. (2002). A review of non-fungicidal approaches for the control of apple scab. *Phytoprotection* 83: 1-29.

Carisse, O., V. Philion, D. Rolland, and J. Bernier. (2000). Effect of fall application of fungal antagonists on spring ascospore production of the apple scab pathogen, *Venturia inaequalis*. *Phytopathlogy* 90: 31-37.

Castro, M., K. Kramer, L. Valdivia, S. Ortiz, and A. Castillo. (2003). A double-stranded RNA mycovirus confers hypovirulence-associated traits to *Botrytis cinerea*. *FEMS Microbiology Letters* 228: 87-91.

Chalutz, E. and S. Droby. (1997). Biological control of postharvest disease. In *Plant-Microbe Interactions and Biological Control*. eds. G.J. Boland and L.D. Kuykendall. Marcel Dekker, Inc., New York, pp. 157-170.

Chand-Goyal, T., J.W. Eckert, S. Droby, and K. Atkinson. (1998). A method for studying the population dynamics of *Candida oleophila* on oranges in the grove, using a selective isolation medium and PCR technique. *Journal of Microbiological Research* 153: 265-270.

Chand-Goyal, T., J.W. Eckert, S. Droby, E. Glickmann, and K. Atkinson. (1999). Transformation of *Candida oleophila* and survival of a transformant on orange fruit under field conditions. *Current Genetics* 35: 51-57.

Chand-Goyal, T. and R.A. Spotts. (1996). Postharvest biological control of blue mold of apple and brown rot of sweet cherry by natural saprophytic yeasts alone or in combination with low doses of fungicides. *Biological Control* 6: 252-259.

Chand-Goyal, T. and R.A. Spotts. (1997). Biological control of postharvest diseases of apple and pear under semi-commercial and commercial conditions using three saprophytic yeasts. *Biological Control* 10: 199-206.

Chardonnet, C.O., C.S. Charron, C.E. Sams, and W.S. Conway. (2003). Chemical changes in the cortical tissue and cell walls of calcium-infiltrated "Golden Delicious" apples during storage. *Postharvest Biology and Technology* 28: 97-111.

Conway, W.S., J.A. Abbott, and B.D. Bruton. (1991). Postharvest calcium treatment of apple fruit to provide broad-spectrum protection against postharvest pathogens. *Plant Disease* 75: 620-622.

Cullen, D., F.M. Barbee, and J.H. Andrews. (1984). *Chaetomium globosum* antagonizes the apple scab pathogen, *Venturia inaequalis*, under field conditions. *Canadian Journal of Botany* 62: 1814-1818.

Davis, R.F., P.A. Backman, R. Rodriguez-Kabana, and N. Kokalis-Burelle. (1992). Biological control of apple fruit diseases by *Chaetomium globosum* formulations containing cellulose. *Biological Control* 2: 118.

De Cal, A., E.M. Sagasta, and P. Melgarejo. (1990). Biological control of peach twig blight *(Monilinia laxa)* with *Penicillium frequentans. Plant Pathology* 39: 612-618.

Droby, S.. M. Wisniewski. A. El Ghaouth, and C. Wilson. (2003). Influence of food additives on the control of postharvest rots of apple and peach and efficacy of the yeast-based biocontrol product Aspire. *Postharvest Biology and Technology* 27: 127-135.

Dubos, B. (1984). Biocontrol of *Botrytis cinerea* on grapevines by an antagonistic strain of *Trichoderma harzianum*. In *Current Perspectives in Microbial Ecology,* eds. M.J. Klub and C.A. Reedy. American Society of Microbiology. Washington, DC. pp. 370-373.

Dubos, B. (1987). Fungal antagonism in aerial agrobiocenoses. In *Innovative Approaches to Plant Disease Control,* ed. Chen, I. John Wiley & Sons Inc., New York, pp. 107-135.

Elad, Y. (1994). Biological control of grape grey mould by *Trichoderma harzianum. Crop Protection* 13: 35-38.

Elad, Y. and D. Shtienberg. (1994). *Trichoderma harzianum* T39 integrated with fungicides: improved biocontrol of grey mold. *Brighton Crop Protection Conference—Pests and Diseases,* pp. 1109-1113.

El-Ghaouth, A., J. Smilanick, G.E. Brown, A. Ippolito, and C.L. Wilson. (2001). Control of decay of apple and citrus fruits in semicommercial tests with *Candida saitoana* and 2-Deoxy-D-glucose. *Biological Control* 20: 96-101.

El-Ghaouth, A.. J.L. Smilanick, G.E. Brown. A. Ippolito, M., Wisniewski, and C.L. Wilson. (2000). Applications of *Candida saitoana* and glycolchitosan for the control of postharvest diseases of apple and citrus fruit under semi-commercial conditions. *Plant Disease* 84: 243-248.

El-Ghaouth, A., J.L. Smilanick, and C.L. Wilson. (2000). Enhancement of the performance of *Candida saitoana* by the addition of glycolchitosan for the control of postharvest decay of apple and citrus fruit. *Postharvest Biology and Technology* 19: 103-110.

El-Ghaouth, A.. J.L. Smilanick. M. Wisniewski, and C.L. Wilson. (2000). Improved control of apple and citrus fruit decay with a combination of *Candida saitoana* and 2-deoxy-D-glucose. *Plant Disease* 84: 249–253.

El-Neshawy, S. and C.L. Wilson. (1995). Enhancement of *Candida oleophila* for the biocontrol of postharvest diseases of apple with Nisin. *Postharvest physiology, pathology and technologies for horticultural commodities: recent advances. Proceedings of an international symposium Agadir, Morocco* 16-21 January 1994, 419-425.

Fajardo, J.E.. T.G. McCollum, R.E. McDonald, and R.T. Mayer. (1998). Differential induction of proteins in orange flavedo by biologically based elicitors and challenged by Penicillium digitatum Sacc. *Biological Control* 13: 143-151.

Falk, S.P.. D. Gadoury, R.C. Pearson, and R.C. Seem. (1995). Partial control of grape powdery mildew by the mycoparasite *Ampelomyces quisqualis. Plant Disease* 79: 483-490.

Falk, S.P., R.C. Pearson, D.M. Gadoury, R.C. Seem, and A. Sztejnberg. (1996). *Fusarium proliferatum* as a biocontrol agent against grape downy mildew. *Phytopathology* 86: 1010-1017.

Filonow, A.B., H. Vishniac, J. Anderson, and W.J. Janisiewicz. (1996). Biological control of *Botrytis cinerea* in apple by yeasts from various habitats and their putative mechanism of antagonism. *Biological Control* 7: 212-220.

Foschi, S., R. Roberti, and A. Brunelli. (1995). Application of antagonistic fungi against *Monilinia laxa* agent of fruit rot of peach. *Bulletin OILB/SROP* 18: 79-82.

Froyd, J.D. (1997). Can synthetic pesticides be replaced with biological-based alternative?—An industry perspective. *Journal of Industrial Microbiology and Biotechnology* 19: 192-195.

Gan-Mor, S. and G.A. Matthews. (2003). Recent development in sprayers for application of biopesticides—An overview. *Biosystems Engineering* 84: 119-125.

Granatstein, D. and P. Dauer. (2000). *Trends in organic tree fruit production in Washington State.* CSANR Report No.1, Washington State University, Wenatchee, WA.

Grove, G.G. and R.J. Boal. (1997). Apple powdery mildew control trials using the mycoparasite *Ampelomyces quisqualis* (AQ10). *Fungicide and Nematicide Tests* 52: 7.

Gueldner, R.C., C.C. Riley, P.L. Pusy, C.E. Costello, and R.F. Arrendale. (1988). Isolation and identification of iturins as antifungal peptides in biological control of peach brown rot with *Bacillus subtilis. Journal of Agricultural and Food Chemistry* 36: 366-370.

Guetsky, R., D. Shtienberg, Y. Elad, and A. Dinoor. (2001). Combining biocontrol agents to reduce the variability of biological control. *Phytopathology* 91: 621-627.

Guinebretiere, M.H., C. Nguyen-The, N. Morrison, M. Reich, and P. Nictot. (2000). Isolation and characterization of antagonists for the biocontrol of postharvest wound pathogen *Botrytis cinerea* on strawberry fruit. *Journal of Food Protection* 63: 386-34.

Gullino, L.M. (1992). Control of Botrytis rot of grapes and vegetables with *Trichoderma* spp. In *Biological Control of Plant Diseases*, eds. E.C. Tjamos, G.C. Papavizas, and R.J. Cool. Plenum Press, New York, pp. 125-132.

Harman, G.E. (2000). Myths and dogmas of biocontrol—changes in perceptions derived from research on *Trichoderma harzianum* T-22. *Plant Disease* 84: 377-393.

Harman, G., B. Latorre, E. Agosin, R. San Martin, D.G. Riegel, P.A. Nielsen, A. Tronsmo, and R.C. Pearson. (1996). Biological and integrated control of Botrytis bunch rot of grape using *Trichoderma* spp. *Biological Control* 7: 259-266.

Helbig, J. (2002). Ability of the antagonistic yeast *Cryptococcus albidus* to control *Botrytis cinerea* in strawberry. *BioControl* 47: 85-99.

Heye, C.C. and J.H. Andrews. (1983). Antagonism of *Athelia bombacina* and *Chaetomium globosum* to the apple scab pathogen, *Venturia inaequalis. Phytopathology* 73: 650-654.

Hidalgo, E., R. Bateman, U. Krauss, M. Hoopen, and A. Martínez. (2003). A field investigation into delivery systems for agents to control *Moniliophthora roreri*. *European Journal of Plant Pathology* 109: 953-961.

Hileman, B. (1993). Food supply safety: U.S. seeks to cut pesticide use. *Chem Eng News* 71: 3-4.

Hjeljoid, L.G., A. Stensvand, and A. Tronsmo. (2001). Antagonism of nutrient-activated conidia of *Trichoderma harzianum (atroviride)* P1 against *Botrytis cinerea*. *Phytopathology* 91: 1172-1180.

Hoch, H.C. and R. Provvidenti. (1979). Mycoparasitic relationships: Cytology of *Spearotheca fuligiana* and *Tilletiopsis* sp. interaction. *Phytopathology* 69: 359-362.

Hofstein, R. and A. Chapple. (1999). Commercial development of biofungicides. In *Biopesticides Use and Delivery*, eds. F.R. Hall and J.J. Menn. Humana Press Inc., Totowa, NJ, pp. 77-102.

Hong, C.X., T.J. Michailides, and B. Holtz. (1998). Effects of wounding, inoculum density, and biological control agents on postharvest brown rot of stone fruits. *Plant Disease* 82: 1210-1216.

Ippolito, A., A. El-Ghaouth, A, C.L. Wilson, and M. Wisniewski. (2000). Control of postharvest decay of apple fruit by *Aureobasidium pullulans* and induction of defense responses. *Postharvest Biology and Technology* 19: 265-272.

Ippolito, A. and F. Nigro. (2000). Impact of preharvest application of biological control agents on postharvest diseases of fresh fruits and vegetables. *Crop Protection* 19: 715-723.

Janisiewicz, W.J. (1997). Biocontrol of postharvest diseases of temperate fruits: Challenges and opportunities. In *Plant-Microbe Interactions and Biological Control*, eds. G.J. Boland and L.D. Kuykendall. Marcel Dekker, Inc., New York, pp. 171-198.

Janisiewicz, W.J. (1988). Biocontrol of postharvest diseases of apples with antagonist mixtures. *Phytopathology* 78: 194-198.

Janisiewicz, W.J. and B. Bors. (1995). Development of microbial community of bacterial and yeast antagonists to control wound invading postharvest pathogens of fruit. *Applied and Environmental Microbiology* 61: 3261-3267.

Janisiewicz, W.J. and S.N. Jeffers. (1997). Efficacy of commercial formulation of two biofungicides for control of blue mold and gray mold of apples in cold storage. *Crop Protection* 16: 629-633.

Janisiewicz, W.J. and L. Korsten. (2002). Biological control of postharvest diseases of fruits. *Annual Review of Phytopathology* 40: 411-441.

Janisiewicz, W.J. and A. Marchi. (1992). Control of storage rots on various pear cultivars with a saprophytic strain of *Pseudomonas syringae*. *Plant Disease* 76: 555-560.

Janisiewicz, W.J., D.L. Peterson, and R. Bors. (1994). Control of storage decay of apples with *Sporobolomyces roseus*. *Plant Disease* 78: 446-470.

Janisiewicz, W.J. and J. Roitman. (1988). Biological control of blue mold and gray mold on apple and pear with *Pseudomonas cepacia*. *Phytopathology* 78: 1697-1700.

Janisiewicz, W.J., T.J. Tworkoski, and C.P. Kurtzman. (2000). Characterizing the mechanism of biocontrol of postharvest diseases on fruits with a simple method to study competition for nutrients. *Phytopathology* 90: 1196-1200.

Janisiewicz, W.J., T.J. Tworkoski, and C.P. Kurtzman. (2001). Biocontrol potential of *Mechnikowia pulcherrima* strains against blue mold of apple. *Phytopathology* 91: 1196-1200.

Janisiewicz, W.J., J. Usall, and B. Bors. (1992). Nutritional enhancement of biocontrol of blue mold on apples. *Phytopathology* 82: 1364-1370.

Janisiewicz, W.J., L. Yourman, J. Roitman, and N. Mahoney. (1991). Postharvest control of blue mold and gray mold of apples and pears by dip treatment with pyrrolnitrin, a metabolite of *Pseudomonas cepacia. Plant Disease* 75: 490-494.

Jarvis, W.R. (1962). The infection of strawberry and raspberry fruits by *Botrytis cinerea* Pers. *Annual Applied Biology* 50: 569-575.

Jeffers, S.N. and T.S. Wright. (1994). Biological control of postharvest diseases of apples: progress and commercial potential. *N Engl Fruit Meet* 100: 100-106.

Jijakli, M.H. and P. Lepoivre. (1998). Characterization of an exo-B-1,3-glucanase produced by *Pichia anomala* strain K, antagonist of *Botrytis cinerea* on apples. *Phytopathology* 88: 335-343.

Johnson, K.B., V.O. Stockwell, D.M. Burgett, D. Sugar, and J.E. Loper. (1993). Dispersal of *Erwinia amylovora* and *Pseudomonas fluoresces* by honey bees from hives to apple and pear blossoms. *Phytopathology* 83: 478-484.

Jones, A.L. and H.S. Aldwinckle. (1990). *Compendium of Apple and Pear Diseases.* APS Press, St. Paul, MN.

Jones, R.W. and D. Prusky. (2002). Expression of an antifungal peptide in *Saccharomyces*: A new approach for biological control of the postharvest diseases caused by *Colletotrichum coccodes. Phytopathology* 92: 33-37.

Karabulut, O.A., S. Lurie, and S. Droby. (2001). Evaluation of the use sodium bicarbonate, potassium sorbate and yeast antagonists for decreasing postharvest decay of sweet cherries. *Postharvest Biology and Technology* 23: 233-236.

Kiss, L. (2003). A review of fungal antagonists of powdery mildews and their potential as biocontrol agents. *Pest Management Science* 59: 475-483.

Klinner, U. and Schäfer, B. (2003). Genetic aspects of targeted insertion mutagenesis in yeasts. *FEMS Microbiology Reviews* in press.

Kovach, J., R. Petzoldt, and G.E. Harman. (2000). Use of honey bees and bumble bees to disseminate *Trichoderma harzianum* 1295-22 to strawberries for Botrytis control. *Biological Control* 18: 235-242.

Lalithakumari, D., C. Marinalini, A.B. Chandra, and P. Annamalai. (1996). Strain improvement by protoplast fusion for enhancement of biocontrol potential integrated with fungicide tolerance in *Trichoderma spp. Zeitschrift Furpflanzenkrankeiten und Pflanzenschutz* 103: 206-212.

Latorre, B.A., E. Agosin, R. San Martin, and G.S. Vásquez. (1997). Effectiveness of conidia of *Trichoderma harzianum* produced by liquid fermentation against Botrytis bunch rot of table grape in Chile. *Crop Protection* 16: 209-214.

Leibinger, W., B. Breuker, M. Hahn, and K. Mendgen. (1997). Control of postharvest pathogens and colonization of the apple surface by antagonistic microorganisms in the field. *Phytopathology* 87: 1103-1110.

Leverentz, B., W.S. Conway, W.J. Janisiewicz, R.A. Saftner, and M.J. Camp. (2003). Effect of combination MCP treatment, heat treatment, and biocontrol on the reduction of postharvest decay of "Golden Delicious" apples. *Postharvest Biology and Technology* 27: 221-233.

Leverentz, B., W.J. Janisiewicz, W.S. Conway, R.A. Saftner, and Y. Fuchs. (2000). Combining yeasts or a bacterial biocontrol agent and heat treatment to reduce postharvest decay of "Gala" apples. *Postharvest Biology and Technology* 21: 87-94.

Lima, G., F. De Curtis, R. Castoria, and V. De Cicco. (1998). Activity of the yeasts *Cryptococcus laurentii* and *Rhodotorula glutinis* against post-harvest rots on different fruits. *Biocontrol Science and Technology* 8: 257-267.

Lima, G., F. De Curtis, R. Castoria, and V. De Cicco. (2003). Integrated control of apple postharvest pathogens and survival of biocontrol yeasts in semi-commercial conditions. *European Journal of Plant Pathology* 109: 341-349.

Lima, G. A. Ippolito, F. Nigro, and M. Salerno. (1997). Effectiveness of *Aureobasidium pullulans* and *Candida oleophila* against postharvest strawberry rots. *Postharvest Biology and Technology* 10: 169-179.

Lübeck, M., I.M.B. Knudsen, B. Jensen, U. Thrane, C. Janvier, and D.F. Jensen. (2002). GUS and GFP transformation of the biocontrol atrain *Clonostachys rosea* IK726 and the use of these marker genes in ecological studies. *Mycological Research* 106: 815-826.

Lurie, S., S. Droby, L. Chalupowicz, and E. Chalutz. (1995). Efficacy of *Candida oleophila* strain 182 in preventing *Penicillium expansum* infection of nectarine fruits. *Phytoparasitica* 23: 231-234.

Maas, J.L. (eds.) (1998). *Compendium of Strawberry Diseases*, 2nd Edition. APS Press, St. Paul, MN.

Maccagnani, B., M. Mocioni, M.L. Gullino, and E. Ladurner. (1999). Application of *Trichoderma harzianum* by using *Apis mellifera* as a vector for the control of gray mould of strawberry: First results. *IOBC Bulletin* 22: 161-164.

MacHardy, W.E, D.M. Gadoury, and C. Gessler. (2001). Parasitic and biological fitness of *Venturia inaequalis:* Relationship to disease management strategies. *Plant Diseases* 85: 1036-1051.

Madrigal, C. and P. Melgarejo. (1995). Morphological effects of *Epicoccum nigrum* and its antibiotic flavipin on *Monilinia laxa*. *Canadian Journal of Botany* 73: 425-431.

Mari, M., M. Guizzardi, and G.C. Pratella. (1996). Biological control of gray gold in pears by antagonistic bacteria. *Biological Control* 7: 30-37.

McLaughlin, R.J., C.L. Wilson, S. Droby, R. Ben-Arie, and E. Chalutz. (1992). Biological control of postharvest diseases of grape, peach, and apple with the yeast *Kloeckera apiculata* and *Candida guilliermondii*. *Plant Disease* 76: 470-473.

McLaughlin, R.J., M.E. Wisniewski, C.L. Wilson, and E. Chalutz. (1990). Effect of inoculum concentration and salt solutions on the biological control of postharvest diseases of apples with *Candida* sp. *Phytopathology* 80: 456-461.

Mercier, J. and J.I. Jiménez. (2004). Control of fungal decay of apples and peaches by the biofumigant fungus *Muscodor albus*. *Postharvest Biology and Technology* 31: 1-8.

Mercier, J. and C.L. Wilson. (1995). Effect of wound moisture on the biocontrol by *Candida oleophila* of gray mold rot *(Botrytis cinerea)* of apple. *Postharvest Biology and Technology* 6: 9-15.

NRC (National Research Council). (1987). *Regulating Pesticides in Foods—The Delaney Paradox.* Board of Agriculture, National Research Council, National Academy Press, Washington, DC.

Nunes, C., J. Usall, N. Teixidó, E. Fons, and I. Viñas. (2002). Post-harvest biological control by *Pantoea agglomerans* CPA-2 on golden delicious apples. *Journal of Applied Microbiology* 92: 247-255.

Nunes. C., J. Usall, N. Teixidó, and I. Viñas. (2001). Biological control of postharvest pear diseases using a bacterium, *Pantoea agglomerans* CPA-2. *International Journal of Food Microbiology* 70: 53-61.

Ogawa, J.M. E.I. Zehr, G.W. Bird, D.R. Ritchie, K. Uriu, and J.K. Uyemoto. (eds). (1995). *Compendium of Stone Fruit Diseases.* APS Press, St. Paul, MN.

O'Neill, T.M, Y. Elad. D. Shtienberg, and A. Cohen. (1996). Control of grapevine grey mould with *Trichoderma harzianum* T39. *Biocontrol Science and Technology* 6: 139-146.

Ouimet, A., O. Carisse, and P. Neumann. (1997). Evaluation of fungal isolates for the inhibition of vegetative growth of *Venturia inaequalis. Canadian Journal of Botany* 75: 626-631.

Papadopoulou-Mourkidou, E. (1991). Postharvest–applied agrochemicals and their residues in fresh fruits and vegetables. *Journal Association of Annual Chemistry* 74: 743-765.

Pearson, R.C. and A.C. Goheen. (1988). *Compendium of Grape Diseases.* APS Press, St. Paul, MN.

Peng. G. and J.C. Sutton. (1990). Biological methods to control grey mould of strawberry. *Brighton Crop Protection Conference—Pests and Diseases*, pp. 233-240.

Peng. G. and J.C. Sutton. (1991). Evaluation of microorganisms for biocontrol of *Botrytis cinerea* in strawberry. *Canadian Journal of Plant Pathology* 13: 247-257.

Peng, G., J.C. Sutton, and P.G. Kevan. (1992). Effectiveness of honey bees for applying the biocontrol agent *Gliocladium roseum* to strawberry flowers to suppress *Botrytis cinerea. Canadian Journal of Plant Pathology* 14: 117-129.

Powelson, R.L. (1960). Initiation of strawberry fruit rot caused by *Botrytis cinerea. Phytopathology* 50: 491-494.

Pusey, P.L. (1994). Enhancement of biocontrol agents for postharvest diseases and their integration with other control strategies. In *Biological Control of Postharvest Diseases—Theory and Practice,* eds. C.L. Wilson and M.E. Wisniewski. CRC Press, Boca Raton, FL, pp. 77-87.

Pusey, P.L., M.W. Hotchkiss, H.T. Dulmage, R.A. Baumgardner, E.I. Zehr, C.C. Reilly, and C.L. Wilson. (1988). Pilot tests for commercial production and application of *Bacillus subtillis* (B-3) for postharvest control of peach brown rot. *Plant Disease* 72: 622-626.

Pusey, P.L. and C.L. Wilson. (1984). Postharvest biological control of stone fruit brown rot by *Bacillus subtilis. Plant Disease* 68: 753-756.

Pusey, P.L., C.L. Wilson, M.W. Hotchkiss, and J.D. Franklin. (1986). Compatibility of *Bacillus subtilis* for postharvest control of peach brown rot with commercial fruit waxes, dicloran and cold storage conditions. *Plant Disease* 70: 587-590.

Qin, G.Z., S.P. Tian, and Y. Xu. (2004). Biocontrol of postharvest diseases on sweet cherries by four antagonistic yeasts in different storage conditions. *Postharvest Biology and Technology* 31: 51-58.

Qin, G.Z., S.P. Tian, Y. Xu, and Y.K. Wan. (2003). Enhancement of biocontrol efficacy of antagonistic yeast by salicylic acid in sweet cherry fruit. *Physiological and Molecular Plant Pathology* 62: 147-154.

Rav-David, D. and Y. Elad. (2002). Use of GUS transformants of *Trichoderma harnianum* isolate T39 (Trichodex) for studying interactions on leaf surfaces. *Biocontrol Science and Technology* 12: 401-407.

Riquelme, M. (1996). Fungal siderophores in plant-microbe interactions. *Microbiologia Sem* 12: 537-546.

Roberts, R.G. (1990). Postharvest biological control of gray mold of apple by *Cryptococcus laurentii*. *Phytopaphology* 80: 526-530.

Roberts, R.G. (1994). Integrating biological control into postharvest disease management strategies. *Horticultural Science* 29: 758-762.

Romanazzi, G., F. Nigro, and A. Ippolito. (2001). Chitosan in the control of postharvest decay of some Mediterranean fruits. In *Chitin Enzymology*, ed. R.A.A. Muzzarelli. Atec, Italy, pp. 141-146.

Romanazzi, G., F. Nigro, and A. Ippolito. (2003). Short hypobaric treatments potentiate the effect of chitosan in reducing storage decay of sweet cherries. *Postharvest Biology and Technology* 29: 73-80.

Romanazzi, G., F. Nigro, A. Ippolito, D. Di Venere, and M. Salerno. (2002). Effects of pre and postharvest chitosan treatments to control storage grey mould of table grapes. *Journal of Food Science* 67: 1862-1867.

Sams, C.E., W.S. Conway, J.A. Abbott, R.J. Lewis, and N. Ben-Shalom. (1993). Firmness and decay of apples following postharvest pressure infiltration of calcium and heat treatment. *Journal American Society of Horticulture Science* 118: 623-627.

Schena, L., A. Ippolito, T. Zahavi, L. Cohen, and S. Droby. (2000). Molecular approaches to assist the screening and monitoring of postharvest biocontrol yeasts. *European Journal of Plant Pathology* 106: 681-691.

Schena, L., F. Nigro, I. Pentimone, A. Ligorio, and A. Ippolito. (2003). Control of postharvest rots of sweet cherries and table grapes with endophytic isolates of *Aureobasidium pullulans*. *Postharvest Biology and Technology* 30: 209-220.

Schisler, D.A. and P.J. Slininger. (1997). Microbial selection strategies that enhance the likehood of developing commercial biological control products. *Journal of Industrial Microbiology and Biotechnology* 19: 172-179.

Schroers, H., G.J. Samuels, K.A. Seifert, and W. Game. (1999). Classification of the mycoparasite *Gliocladium roseum* in *Clonostachys* as *C. rosea*, its relationship to *Bionectria ochroleuca*, and notes on other *Gliocladium*-like species. *Mycologia* 91: 365-385.

Sholberg. P.L.. A. Marchi, and J. Bechard. (1995). Biocontrol of postharvest diseases of apple using *Bacillus* spp. isolated from stored apples. *Canadian Journal of Microbiology* 41: 247-252.

Singh, J. and J.L. Faull. (1988). Antagonism and biological control. In *Biocontrol of Plant Diseases,* eds. K.G. Mukerji and K.L. Garg. CRC Press, Boca Raton. FL, Vol. 2. pp. 167-177.

Smilanick, J.L. (1994). Strategies for the isolation and testing of biocontrol agents. In *Biological Control of Postharvest Diseases: Theory and Practice,* eds. C.L. Wilson and M.E. Wisniewski. CRC. Press. Boca Raton, FL, pp. 63-75.

Smilanick, J.L., R. Denis-Arrue, J.R. Bosch, A.R. Gonzalez. D. Henson, and W.J. Janisiewicz. (1993). Control of postharvest brown rot of nectarines and peaches by *Pseudomonas* species. *Crop Protection* 12: 513-520.

Sobiczewski P., H. Bryk, and S. Berczynski. (1996). Evaluation of epiphytic bacteria isolated from apple leaves in the control of postharvest apple diseases. *Journal of Fruit Ornamental Research* 4: 35-45.

Spadaro. D. and M.L. Gullino. (2003). Sate of the art and future prospects of the biological control of postharvest fruit disease. *International Journal of Food Microbiology,* in press.

Spadaro, D.. R. Vola, S. Piano, and M.L. Gullino. (2002). Mechanisms of action and efficacy of four isolates of the yeast *Metschnikowia pulcherrima* active against postharvest pathogens on apples. *Postharvest Biology and Technology* 24: 123-134.

Strobel, G.A., E. Dirkse. J. Sears, and C. Markworth. (2001). Volatile antimicrobials from *Muscodor albus,* a novel endophytic fungus. *Microbiology* 147: 2943-2950.

Sugar, D.. J.M. Benbow, K.A. Powers, and S.R. Basile. (2003). Effects of sequential calcium chloride. ziram, and yeast orchard sprays on postharvest decay of pear. *Plant Disease* 87: 1260-1262.

Sugar, D. and R.A. Spotts. (1999). Control of postharvest decay in pear by four laboratory-grown yeasts and two registered biocontrol products. *Plant Disease* 83: 155-158.

Sumeet and K.G. Mukerji. (2000). Exploitation of protoplast fusion technology in improving biocontrol potential. In *Biocontrol Potential and its Exploitation in Sustainable Agriculture,* Vol. 1. *Crop Diseases, Weeds, and Nematodes.* eds. Kluwer Acadamic/Plenum Publishers, New York, pp. 39-48.

Sutton. J.C. (1995). Evaluation of micro-organisms for biocontrol: *Botrytis cinerea* and strawberry, a case study. In *Advances in Plant Pathology,* Vol. II, *Sustainability and Plant Pathology,* eds. J.H. Andrews and I.C. Tommerup. Academic Press, London, pp. 173-190.

Sutton, J.C., D.W. Li, G. Peng, H. Yu. P.G. Zhang. and R.M. Valdebenito-Sanhueza. (1997). *Gliocladium roseum:* A versatile adversary of *Botrytis cinerea* in crops. *Plant Disease* 81: 316-328.

Sutton, J.C. and G. Peng. (1993a). Biocontrol of *Botrytis cinerea* in strawberry leaves. *Phytopathology* 83: 615-621.

Sutton, J.C. and G. Peng. (1993b). Manipulation and vectoring of biocontrol organisms to manage foliage and fruit diseases in cropping systems. *Annual Review of Phytopathology* 31: 473-493.

Swaddling, I.R. and P. Jeffries. (1996). Isolation of microbial antagonists for biocontrol of grey mould disease of strawberries. *Biocontrol Science and Technology* 6: 125-136.

Teixidó, N., J. Usall, and I. Viñas. (1999). Efficacy of preharvest and postharvest *Candida sake* biocontrol treatments to prevent blue mould on apples during could storage. *International Journal of Food Microbiology* 50: 203-210.

Teixidó, N. I. Viñas, J. Usall, and N. Magan. (1998). Control of blue mold of apples by preharvest application of *Candida sake* grown in media with different water activity. *Phytopathology* 88: 960-964.

Teixidó. N. I. Viñas, J. Usall, V. Sanchis, and N. Magan. (1998). Ecophysiological responses of the biocontrol yeast *Candida sake* to water, temperature and pH stress. *Journal of Applied Microbiology* 84: 192-200.

Thomson, S.V., D.R. Hansen, K.M. Flint, and J.D. Vandenberg. (1992). Dissemination of bacteria antagonistic to *Erwinia amylovora* by honey bees. *Plant Disease* 76: 1052-1056.

Tronsmo, A. (1991). Biological and integrated control of *Botrytis cinerea* on apple with *Trichoderma harzianum*. *Biological Control* 1: 59-62.

Tronsmo, A. and C. Dennis. (1977). The use of *Trichoderma* species to control strawberry fruit rots, *Netherlands Journal of Plant Pathology* 83 (supplement 1): 449-455.

Tronsmo, A. and J. Ystaas. (1980). Biological control of *Botrytis cinerea* on apple. *Plant Disease* 64: 1009.

Usall, J., N. Teixidó, R. Torres, X. Ochoa de Eribe, and I. Viñas. (2001). Pilot tests of Candida sake (CPA-1) applications to control postharvest blue mold on apple fruit. *Postharvest Biology and Technology* 21: 147-156.

Wilson, M. (1997). Biocontrol of aerial plant diseases in agriculture and horticulture: current approaches and future prospects. *Journal of Industrial Microbiology and Biotechnology* 19: 188-191.

Wilson, C.L. and E. Chalutz. (eds.) (1991). *Biological Control of Postharvest Diseases of Fruits and Vegetables.* Workshop Proceedings, Washington, DC, U.S. GPO.

Wilson. C.L., J.D. Franklin, and P.L. Pusey. (1987). Biological control of Rhizopus rot of peach with *Enterobacter cloacae. Phytopathology* 77: 303-305.

Wilson, C.L. and M.E. Wisniewski. (1989). Biological control of postharvest diseases. *Annual Review of Phytopathology* 27: 425-441.

Wilson, C.L., M.E. Wisniewski, S. Droby, and E. Chalutz. (1993). A selection strategy for microbial antagonists to control postharvest diseases of fruits and vegetables. *Science Horticulture* 53: 183-189.

Wisniewski, M., C. Biles, S. Droby, R. McLaughlin, C. Wilson, and E. Chalutz. (1991). Mode of action of the postharvest biocontrol yeast, *Pichia guilliermondi*. 1. Characterization of attachment to *Botrytis cinerea. Physiological and Molecular Plant Pathology* 39: 245-258.

Wisniewski, M., S. Droby. E. Chalutz. and Y. Eilam. (1995). Effects of Ca^{2+} and Mg^{2+} on *Botrytis cinerea* and *Penicillium expansum* in vitro and on the biocontrol activity of *Candida oleophila. Plant Pathology* 44: 1016-1024.

Wisniewski, M.E., C.L. Wilson, and W. Hershberger. (1989). Characterization of inhibition of *Rhizopus stolonifer* germination and growth by *Enterobacter cloacae*. *Canadian Journal of Botany* 67: 2317-2323.

Wittig, H.P.P., K.B. Johnson, and J.W. Pscheidt. (1997). Effect of epiphytic fungi on brown rot blossom blight and latent infection in sweet cherry. *Plant Disease* 81: 383-387.

Yu, H. and J.C. Sutton. (1997). Effectiveness of bumblebees and honeybees for delivering inoculum of *Gliocladium roseum* to raspberry flowers to control *Botrytis cinerea*. *Biological Control* 10: 133-122.

Zahavi, T., L. Cohen, B. Weiss, L. Schena, A. Daus, T. Kaplunov, J. Zutkhi, R. Ben-Arie, and S. Droby. (2000). Biological control of *Botrytis, Aspergillus,* and *Rhizopus* rots on table and wine grapes in Israel. *Postharvest Biology and technology* 20: 115-124.

Zhang, D. and P.C. Quantick. (1998). Antifungal effects of chitosan coating on fresh strawberries and raspberries during storage. *Journal Horticultural Science Biotechnology* 73: 763-767.

Zhou, T. (2001). *Anti-fungal Activity of Food Microbes against Monilinia fructicola and their Potential Application in Controlling Peach Decay. 9th International Symposium on Microbial Ecology.* Amsterdam, the Netherlands, pp. 19-31.

Zhou, T., C.L. Chu, W.T. Liu, and K.E. Schneider. (2001). Postharvest control of blue mold and gray mold on apples using isolates of *Pseudomonas syringae*. *Canadian Journal of Plant Pathology* 23: 246-252.

Zhou, T. and R. DeYoung. (1997). Evaluation of hypovirulent isolates of *Venturia inaequalis* for the suppression of apple scab disease. *Phytopathology* 87: s109.

Zhou, T., J. Northover, and K. Schneider. (1999). Biological control of postharvest diseases of peach with phyllosphere isolates of *Pseudomonas syringae*. *Canadian Journal of Plant Pathology* 21: 375-381.

Zhou, T., J. Northover, K.E. Schneider, and X. Lu. (2002). Interaction between *Pseudomonas syringae* MA-4 and cyprodinil in the control of blue mold and gray mold of apple. *Canadian Journal of Plant Pathology* 24: 154-161.

Zhou, T. and K. Schneider. (1998). Control of peach brown rot by preharvest applications of an isolate of *Pseudomonas syringae*. *Abstracts of 7th International Congress of Plant Pathology* 3: 20.

Chapter 11

Biological Control of Various Diseases of Major Vegetables in Korea

Youn Su Lee
Min Woong Lee

INTRODUCTION

There has been a great interest in biological disease control recently, showing ever-increasing environmental concern over pesticide use. This great interest has been further stimulated by the occurrence of pesticide resistance in some plant pathogens and for some soilborne plant diseases, the lack of reliable chemical controls, or resistant plant varieties (Whipps, 1997). Over the past 20 years, the amount of research in this biological control area has increased dramatically. Within the past 10 years, over 40 biological control products have appeared on the commercial market; however, these are still a small fraction of the total number and sales of chemical fungicides in field crops and trees (Paulitz and Belanger, 2001). In a 1993 report, sales of biofungicides represented less than $1 million, whereas total fungicide sales were in excess of $5.5 billion. Optimistic estimates projected that sales of biocontrol products could reach $15 million before the year 2000 (Powell and Jutsum, 1993). Laws for the registration of microbial pesticides were formulated in 2000 by RDA, Korea. There are 14 microbial pesticides registered in Korea as of May 2004.

In Korea, foliar and stem blight of pepper (*Capsicum annum* L.) caused by *Phytophthora capsici* L. is one of the most widespread and destructive

The authors would like to express sincere thanks to Ms. Hee Sun Chung for her help in the preparation of the manuscript.

doi:10.1300/5682_11

soilborne diseases in many pepper-growing areas in Korea. In order to minimize the yield losses of pepper caused by *Phytophthora* blight, fungicides, crop rotation, and resistant cultivars have been widely employed during the cultivation of pepper plants. However, *Phytophthora* blight is not readily controlled by soil or foliar fungicides, and new resistance sources for effective control of the disease should be sought. Therefore, in Korea, many researchers tried to select and develop biological antagonists against *Phytophthora capsici*. The biological control of pepper diseases, especially soilborne fungal diseases including *Phytophthora* blight diseases, were mainly performed with soil- or rhizosphere-isolated fungi or bacteria. Therefore, isolation of antagonists was done mainly from the soil or rhizosphere. However, Paik and Kim (1995) screened phyllospheral-antagonistic microorganisms for the control of red-pepper anthracnose *(Colletotrichum gloeosporioides).*

Cucumber *(Cucumis sativus* L.) is also a crop of importance in Korea. However, *Fusarium* wilt of cucumber, caused by *Fusarium oxysporum* f. sp. *cucumerinum*, is a serious vascular disease in places where the cucumber is cultivated. The biological control of *Fusarium* wilt of several crops has been accomplished by introduction of nonpathogenic *Fusarium* spp. in the soil or in infection courts (Lee, 1994). There are many different mechanisms involved (Jee and Kim, 1987; Kim and Jee, 1988). Fluorescent pseudomonads have also been applied to seeds and rhizosphere soils or other infection courts for biological control (Weller, 1988; Alabouvette et al., 1996). In cucumber, powdery mildew also has been a major hindrance for production, especially in the greenhouse system. Even though various chemical fungicides have been used quite successfully for the control of powdery mildews, there was need for the selection of antagonists or selection of bioactive substances extracted from various beneficial plant species.

Among the various diseases that affect cucumber, damping-off diseases caused by *Pythium* spp. and *Rhizoctonia solani,* and *Fusarium* wilts are the major ones. Damping-off, especially, occurs during and/or at germination of seeds, and right after the germination of seeds by *Pythium* spp. and *Rhizoctonia* spp. in the seedbed. In some cases, dual infection of *Pythium* spp. and *Rhizoctonia* spp. makes the matter worse. Baker (1987) and Hornby (1983) worked on biological control of damping-off diseases.

Sugar beet damping-off diseases are caused mainly by *Aphanomyces cochlioides, Pythium* sp., and *Rhizoctonia solani* in seedbeds (Lee and Kobayashi, 1988). They found that *Pseudomonas* sp. B-1218 showed antagonistic effect on both *A. cochlioides* and *Pythium* sp. In another study, bacterization of sugarbeet seed was found to be effective for the production increase (Suslow and Schroth, 1982).

Fusarium wilt diseases are responsible for yield losses on numerous crops. Some soils are known for their natural suppressiveness to *Fusarium* wilts. The soil microflora is responsible for the natural suppressiveness of these soils (Scher and Baker, 1980; Alabouvette, 1986). When the biological control organisms were introduced to the disease-conducive soil, *Fusarium* wilt was significantly suppressed (Kolepper et al., 1980; Alabouvette, 1986; Paulitz et al., 1987; Leeman, 1995). However, the inoculation of disease-suppressing microorganisms to the disease-conducive soil never reach the level of suppression observed in the natural suppressive soils, and the effects are often inconsistent (Weller, 1988). A multitude of factors could account for the inconsistent performance of biocontrol agents due to the complex interactions among the antagonists, the pathogen, the host, and the environment (Weller, 1988; Schippers, 1992).

Fusarium wilt of strawberry in Korea, especially in southern regions, is also a major problem in strawberry production, as in other parts of the world (Moon et al., 1988, 1990). As in other crops, the control of *Fusarium* wilt diseases of strawberry mainly depended on the use of fungicides. In recent years, however, there was a great interest in biological control of *Fusarium* wilt diseases in Korea.

Bud rot of strawberry caused by *R. solani* is also a major hindrance to the production of strawberry, especially under low-temperature conditions in a greenhouse system (Shin et al., 1994). The biological control of damping-off and root rot diseases of strawberry was reported to be possible by using of *Trichoderma* spp., nonpathogenic *Rhizoctonia* sp., and binucleate *Rhizoctonia* sp. (Cook and Baker, 1983; Chet, 1987; Murkerji and Garg, 1988; Hornby, 1990). However, bud rot diseases of strawberry occurs under different environmental conditions. The causal agent of bud rot of strawberry, *R. solani* AG2-1 type, usually causes the disease under low-temperature conditions in a greenhouse system.

Watermelon [*Citrullus lanatus* (Thumb.) Matsum. & Nakai] is one of the most extensively planted fruit crops in greenhouses and fields of Korea. The gummy stem rot caused by the soil-borne fungus *Didymella bryoniae* is one of the most serious diseases of watermelon, and is a major factor that limits production in subsequent years. In response to environmental and health concerns about extended use of chemicals, there is a considerable interest in finding alternative control methods for use in integrated disease management strategies. One of the alternative control methods is the use of beneficial microorganisms for biological control. Among the beneficial microorganisms, fluorescent pseudomonads, which are characterized by the production of yellow-green pigments that fluoresce under UV light and function as siderophores, are one of the most widely investigated bacteria. There have been many reports in recent years that plant growth-promoting

Pseudomonas spp. induce systemic resistance to bacterial (Alstrom, 1991 and 1995), fungal (Van Peer et al., 1991; Van Peer and Schippers, 1992; Zhou et al., 1992; Zhou and Paulitz, 1994), and viral pathogens (Maurhofer et al., 1994; Raupach et al., 1996).

Systemic resistance is the phenomenon in which a plant exhibits an increased level of resistance to pathogen infection after appropriate stimulation. Systemic resistance comprises systemic acquired resistance and induced systemic resistance. Induced systemic resistance-mediated plant protection is induced by colonization of the rhizosphere with plant growth- promoting rhizobacteria (PGPR). Seed-coating or soil-drenching with PGPR protected plants against various pathogens, including fungi, bacteria, and viruses (Alstrom, 1991; Hoffland et al., 1996; Wei et al., 1996; Van Wee et al., 1997), in greenhouses and field conditions.

Various plant species, including various medicinal plant species, were tested for their antagonistic effect against plant pathogens. Paik (1989) tested 100 species of plants in 54 families of plants for their antifungal activities. Leaf extracts from various plant species were inhibitory on mycelium growth of *Phytophthora* spp. His research results indicated the possibility of finding antagonistic plants in the nature for the control of certain pathogens in soil. In a subsequent study, Paik and Oh (1990) screened various medicinal plants for antifungal activities against the soilborne pathogen, *Pythium ultimum,* and selected plant extracts from nine species were strongly inhibitory to zoosporangium germination of *P. ultimum*. Plant extracts from three species were strongly inhibitory to mycelium growth of *P. ultimum*.

Culture filtrates or extracts of various bacteria were used for biological control of plant pathogens. Kim et al. (1990) selected a strain of *Erwinia* spp. from the soil for the production of bacteriocin effective against the root rot plant pathogen. In another study, Kim et al. (1992) selected excellent strains (S4, S14, S65) of *Pseudomonas* sp. from 1,196 strains of bacteria, which were isolated from the rhizosphere in vegetable root rot-suppressive soil, for the biological control of soilborne *Erwinia carotovora* subsp. *carotora* which causes rot of vegetables. In a subsequent study, Kim et al. (1994) further selected excellent strains (S43, S62) of *Pseudomonas* sp. for biological control of soilborne *Erwinia rhapontici* which causes rot of vegetables and fruits. In other studies, Yoon et al. (1994) purified an antifungal substance from *Pseudomonas maltophilia* sp. B14 which inhibits the growth of *Rhizoctonia solani*. The causal active substance, PM3F, inhibited the growth of *Rhizoctonia solani*. Lee (2000) selected an isolate of *P. fluorescence* for the control of several vegetable diseases and for the promotion of plant growth at the same time. He tested with *Lactuca sativa* L., *Daucus carota* var. *sativa* DC., *Allium cepa* L., *Allium fistulosum* L, *Brassica campestris, Brassica oleracea* var. *capitata* L., and *Raphanus sativus* L.

In another case, entomopathogenic fungus was used for the control of phytopathogenic fungi, *Fusarium oxysporum, Botrytis cinerea,* and *Alternaria solani.* Kang et al. (1996) conducted a study to find out antifungal activities of entomopathogenic fungus, *Metarhizium anisopliae,* against phytopathogenic fungi, *Fusarium oxysporum, Botrytis cinerea,* and *Alternaria solani.*

Due to the recent explosion in interest in biological control of plant pathogens, it is impossible to cite all the relevant publications that have published. Therefore, by necessity, this review will limit the consideration of biological control in Korea of major vegetable diseases.

BIOLOGICAL CONTROL OF MAJOR VEGETABLE DISEASES IN KOREA

Pepper

Phytopthora *Blight*

Foliar and stem blight of pepper (*Capsicum annum* L.), caused by *Phytophthora capsici* L., is one of the most widespread and destructive soil-borne diseases in many pepper-growing areas in Korea. In order to minimize the yield losses of pepper caused by *Phytopthora* blight, fungicides, crop rotation, and resistant cultivars have been widely employed during the cultivation of pepper plants. However, the *Phytopthora* blight is not readily controlled by soil or foliar fungicides, and new resistance sources for the effective control of the disease should be sought. Therefore, in Korea, many researchers tried to select and develop biological antagonists against *Phytophthora capsici.*

Jee et al. (1988) isolated a total of 846 microorganisms from the rhizopshere or nonrhizosphere soils collected from red-pepper (*C. annum* L.)-growing areas. Among the selected microorganisms, *T. harzianum, P. cepacia,* and *B. polymyxa* showed antagonistic ability against *Phytophthora capsici* in vitro. In their studies, the selected antagonists inhibited the mycelial growth and zoosporangia germination of the pathogen in solid or dual- liquid media. Antibiosis was also observed between any two antagonist combinations. None of the three antagonists showed harmful effects on seed germination or plant growth of red-pepper in either treatment with seed-dipping or soil-drenching of antagonistic suspension. In some cases, *T. harzianum* stimulated plant growth. Application of the antagonistic suspensions in naturally or artificially infested pot soil reduced disease incidence of red-pepper plants, but the extent of the suppression varied with the application concentration, or the application method. Disease suppression by the antagonists tended to decrease gradually as time after application prolonged. Increasing concentration of antagonist application in soil resulted in reduced disease incidence, but this tendency was not always consistent being affected by the inoculation method and the antagonist species used. Combined applications with *T. harzianum* and *P. cepacia* increased disease-suppression effect, greatly compared with that with *T. harzianum* only.

Park and Kim (1989) selected *Trichoderma* spp., *T. harzianum, T. hamatum, B. lentus, Enterobacter agglomerans,* and *P. mendocina* as antagonists against the pepper crown and root rot pathogen *P. capsici*. Among the selected antagonists, *T. harzianum* T873, T77, and *E. agglomerans* reduced the disease incidence when the soil electrical conductivity (EC) was increased from three to five.

Trichoderma harzianum and *P. cepacia,* antagonistic to *P. capsici,* were formulated into granule- and/or powder-using sodium alginate, zeolite, or diatomaceous earth with the intention to develop biofungicides for commercial purposes by Park et al. (1989). In their studies, the population of *T. harzianum* formulated in granules increased 180 times for 16wk preservation at 25°C. However, the population decreased by 30 percent at 5°C. Viability of *P. cepacia* formulated in wet granules remained better at 25°C that at 5°C. When the granules were kept dry, the density of *P. cepacia* was reduced to 1.4 percent and 0.4 percent of the initial density during eight weeks of preservation at 5°C and 25°C, respectively. Viability of *T. harzianum* in the powder formulation was much poorer compared with that of granule formulation, resulting in significant decrease in antagonistic populations regardless of preservation temperature. Amendment of rice bran as a bulking agent in alginate formulation markedly increased the viability of *P. cepacia*. Application of *P. cepacia* granules into soil provided better suppression of *Phytopthora* blight on red-pepper seedlings compared with that of direct drenching of *P. cepacia* suspension. whereas the case was not true for *T. harzianum*.

Lee, Jee, et al. (1990) tested two types of formulation of antagonists ASI87LH3 *(Psedomonas cepacia)* and ASI87TH1 *(Trichoderma hazianum)* for their effects on *Phytophthora* blight incidence on red-pepper in two fields under polyethylene filmhouse. When the seedlings were raised in pot containing seedbed soil formulated with ASI87LH3 and transplanted into fields with pot soil, the initial incidence of the disease was delayed for 30-79 days. Disease development after initial incidence, however, was not reduced. When granules formulated with the same antagonist were applied into the rhizosphere at field transplanting, less delay of initial incidence was observed. The other antagonist failed to suppress disease development.

Kim, Jee, et al. (1990) prepared two types of formulations of *T. harzianum* (TH1) and *Pseudomonas cepacia* (LH3) antagonistic to *P. cepacia,* and examined them for their effects on the suppression of *Phytophthora* blight of red-pepper in three commercial fields. The efficacy of antagonists in the suppression of red-pepper blight was better in the treatment of direct incorporation of the antagonists into seedbed soil than that of surface application of the antagonist's granules at all tested fields. Of the two antagonists, LH3 performed better in disease suppression. Final disease severity in plots treated with the antagonist-incorporated seedbed soil was 29.8 percent and 40 percent at two different locations, respectively. This result was significantly lower than those of the plots which received regular fungicide applications with 42.1 percent and 56 percent. Suppression effects of TH1, however, varied greatly with fields. The population density of TH1 in sol decreased rapidly after application, regardless of type of formulations, but the rate of reduction was a little slower with the antagonist's granules. Red-pepper yields tended to increase in all plots treated with antagonists, particularly with TH1, although the differences were not significant.

Lee, Lim, et al. (1990) selected a strain of *Streptomyces parvullus*, which produced an antifungal antibiotic compound against *P. capsici* and *P. parasitica*. The active compound was purified from culture broth by ion-exchange, adsorption, gel permeation, and partition column chromatography, and identified as polyoxin by the physicochemical and biological properties. Kim et al. (2001) also selected *Chryseomonas luteola* 5042, which is antagonistic to *P. capsici*.

Kim et al. (1991) tested the effects of combined application of antagonists *T. harzianum* and *Pseudomonas cepacia* and fungicide for the control of red-pepper *Phytophthora* blight in vitro and greenhouse. Their studies indicated the use of fungicide metalaxyl together with either of two antagonists can greatly enhance the control efficacy of *Phytophthora* blight of red-pepper where single treatment of antagonist or fungicide was not sufficiently effective.

Ahn and Hwang (1992) selected fifty-three actinomycetes antagonistic to *P. capsici* from rhizosphere soils in six pepper-growing areas and shore soils. The antagonistic activity against *P. capsici* varied greatly. In their study, butanol extracts of culture filtrates from antagonistic actinomycetes inhibited mycelial growth of *P. capsici* and symptom development.

Hwang and Kim (1992) studied the resistance induced by an avirulent isolate of *P. capsici*. Avirlent isolate retarded the lesion development caused by virulent isolate in the pepper cultivars Hanbyul and Kingkun. Prior inoculation of pepper plants with the avirulent isolate, and the simultaneous inoculation with both avirulent and virulent isolates effectively inhibited the development of *Phtyophthora* blight in the two pepper cultivars, as compared with the inoculation with only the virulent isolate. The subsequent inoculation of pepper stems with the avirulent isolate did not protect pepper plants from the prior infection by the virulent isolate. Development of the *Phytophthora* blight in pepper plants inoculated with a mixture of the avirulent and virulent isolates was proportional to the amounts of the virulent inoculum.

Chang, Chang, Lee, and Choi (1996) isolated three strains, A-35, A-67, and A-183, of *Pseudomonas* sp. from the rhizosphere in soil where red-pepper had been cultivated continuously for a long period. They tested the isolates against the pathogen *P. capsici*, and found maximum antifungal activity when the antagonist was cultured at 30°C for five days in potato-extract medium (pH 6.5) with 2 percent mannitol and 0.3 percent peptone. In a subsequent study, Chang, Chang, Lee, Choi, and Lee (1996a) fractionated three different antifungal substances out of the culture media. Substances A and B were known to be effective against *P. capsici* at the level of 5mg/kg, and substance C was found to be effective above the level of 1mg/kg. In pot experiments, the three substances effectively controlled the Phytophthora blight of the red-pepper plant grown in the soil inoculated with *P. capsici*. Chang, Chang, Lee, Choi, and Lee (1996b) identified the three antifungal substances isolated from the culture medium of *Pseudomonas* sp. A-183 which is antagonistic against *P. capsici*. Substances A and B showed positive reactions in the Molish test and Anthrone test, but a negative one in the Fehling test. Substance C only exhibited the phenomenon of UV-induced fluorescence. Based on the qualitative analysis with the spectroscopic techniques including UV, Mass, IR and NMR, substances A and B were known to be composed of sugar and fatty acid, and showed a base peak of 171(m/e). As a result, substance A was 2-O-L-rhamnosyl-α-L-rhamnosyl-β-hydroxydecanoyl-β-hydroxy decanoic acid, and substance B was α-L-rhamnosyl-β-hydroxydecanoyl-β-hydroxydecanoic acid. Substance C was identified as a phenazine from the results of qualitative analysis with the spectroscopic techniques such as UV, Mass, IR, and NMR.

Stanghellini et al. (1996) controlled the root rot of peppers caused by *Phytophthora capsici* with a nonionic surfactant in a recirculating rock wool cultural system, and suggested the potential significance of surfactants for the control of polycyclic soilborne diseases attributed to *Phytophthora* spp. and other zoosporic pathogens. In their experiment, amending the nutrient solution with a nonionic surfactant resulted in the elimination of zoospores and 100 percent control of the spread of the root pathogen from a point source. In the absence of the surfactant, all of the pepper plants within the cultural system, irrespective of plant age, died within two weeks following hypocotyl-inoculation of a single plant, which served as the source of secondary inoculum.

Kim et al. (2000) conducted two field tests to examine the effects of composts and soil amendments on physicochemical properties of soil in relation to *Phytohphthora* root and crown rot of bell pepper. Chitosan, crab shell waste, humate, sewage sludge-yard trimmings, and wood chips were applied to test plots. In their studies, they found physicochemical properties were not correlated with disease incidence, but percent organic matter, estimated nitrogen release, K, and Mg were correlated with the total microbial activity.

Lee and Kim (2000) selected antagonistic bacterium *Pseudomonas* sp. 2112 against red-pepper rotting *Phytophthora capsici*. The selected strain was identified as *Pseudomonas fluorescens* biotype F. The antibiotic produced from *P. fluorescens* 2112 inhibited hyphal growth and the zoospore germination of *Phytophthora capsici*. The favorable carbon, nitrogen source, and salts for the production of antagonistic from *P. fluorescens* 2112 were glycerol, beef extract, and LiCl at 1.0 percent, 0.5 percent and 5mM, respectively. And the antagonistic activity of *P. fluorescens* 2112 was confirmed against *P. capsici* in vivo.

Chang et al. (2000) selected antifungal bacteria for the biological control of plant disease or production of novel antibiotics to plant pathogens. They isolated, in 1997, from various soils of Ansung, Chunan, Koyang, and Paju in Korea, sixty-four bacterial strains, and tested them on V-8 juice agar against eight plant pathogenic fungi, using in vitro bio-assay technique for inhibition of mycelial growth. Among the test pathogens, *Phytophthora capsici,* a root rot pathogen of pepper, was included. A wide range of antifungal activity of bacterial strains was found against the pathogenic fungi tested, and strain RC-B77 showed the best antifungal activity. Correlation analysis between inhibition of each fungus and mean inhibition of all eight fungi by 64 bacterial strains revealed that *C. gloeosporiodes* would be best appropriate for detecting bacterial strains, producing antibiotics with potential as biocontrol agents for plant pathogens.

Shen et al. (2002) used *Serratia plymuthica* strain A21-4 to evaluate for the control of Phytophthora blight of pepper. They found that *Serratia plymuthica* strain A21-4 inhibited mycelial growth, germination of zoosporangia and cystospores, and formation of zoospore and zoosporangia of *P. capsici* in vitro. In the pot experiment, no disease was observed in the pots treated with *Serratia plymuthica* strain A21-4. In the greenhouse test, the same result was obtained. In their subsequent studies, Shen et al. (2005) tested *Serratia plymuthica* A21-4 for the control of *Phytophthora* blight of pepper in the field. In their experiments, they found the antagonist colonized on the root of pepper readily and moved to newly emerging roots continuously, thus providing protection. In the field trials, the blight incidence treated with the antagonist was 12.6 percent, while in the nontreated case, it was 74.5 percent. Thus, they established the possibility of using *Serratia plymuthica* as a biocontrol agent against *P. capsici*. Shen, Park, and Park

(2005) also studied enhancement of the efficacy of the antagonist by improving the inoculation buffer solution.

Anthracnose (Colletotrichum gloeosporioides)

Biological control of pepper diseases, especially soilborne fungal diseases including *Phytophthora* blight diseases, was mainly performed with the soil- or rhizosphere-isolated fungi or bacteria. Therefore, the isolation of antagonists was performed mainly from the soil or the rhizosphere. However, Paik and Kim (1995) screened phyllospheral-antagonistic microorganisms for the control of red-pepper anthracnose *(Colletotrichum gloeosporioides)* (Figure 11.1). In their dual-culture tests on PDA, four isolates of *B. subtilis,* KB6, KB12, KB13, and KB14 were shown to be highly antagonistic to *C. gloeosporioides* compared with others. Among the four bacterial isolates, the culture filtrate of the isolate KB12 showed the highest inhibition of *C. gloeosporioides* on PDA. Culture filtrates of four isolates controlled anthracnose on the red fruits, but not on the green fruits. In the living bacterial cell test, high control effect was observed on the red and green fruits. In another country, Jetiyanon and Kloepperb (2002) selected mixtures of compatible PGPR strains with the capacity to elicit induced systemic resistance in long cayenne pepper *(Capsicum annuum* var. *acuminatum).* The specific diseases and hosts tested in their study included anthracnose of long cayenne pepper caused by *C. gloeosporioides,* among others. In order to examine compatibility, seven selected PGPR strains were individually tested for

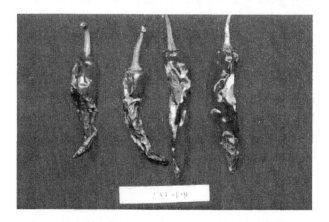

FIGURE 11.1. Anthracnose on red hot pepper *(Capsicum annuum)* caused by *Colletotrichum gloeosporioides.*

in vitro antibiosis against all other PGPR strains, and against the pathogen, *C. gloeosporioides*. No in vitro antibiosis was observed among PGPR strains or against the pathogen. Twenty-one combinations of PGPR and seven individual PGPR were tested in greenhouse for induced resistance activity. Results indicated that four mixtures of PGPR and one individual strain treatment significantly reduced the severity of the diseases compared with the nonbacterized control: 18 mixtures reduced anthracnose of long cayenne pepper. Most mixtures of PGPR provided a greater disease suppression than individual PGPR strains. These results suggest that mixtures of PGPR can elicit induced systemic resistance to fungal disease.

Rhizoctonia solani

Lewis, Fravel, Lumsden, and Shasha (1995) contrived a method to apply biocontrol agents. They found that pregelatinized starch-flour granules containing fermentor-produced biomass of the biocontrol fungi *Gliocladium virens* and *T. hamatum* protected eggplant, pepper, and zinnia seedlings from damping-off caused by the pathogen *R. solani* in a soilless mix. Also, they found the importance of spatial distribution of the biocontrol product in disease control.

Lewis, Fravel, and Papavizas (1995) worked on the effect of bran preparations of the potential biocontrol fungus *Cladorrhinum foecundissimum* on the survival and saprophytic growth of the soilborne plant pathogen *R. solani* in soils and soilless potting mix and on its ability to reduce incidence of damping-off of sugar beet, eggplant, and pepper caused by this pathogen. Bran preparations of some of the isolates prevented damping-off of eggplant and pepper in soilless mix and, depending on the rate of inoculum used, resulted in stands comparable to those in pathogen-free soilless mix.

Lewis and Larkin (1998) used five isolates of *Cladorrhinum foecundissimum* to control damping-off of pepper caused by *Rhizoctonia solani* and *Pythium ultimum*. The bran prepartions they used reduced symptom development. They tested five isolates of *Cladorrhinum foecundissimum*, which were added to the soilless mix as ten-day-old fresh bran preparations (1.0 percent w/w). The mixtures significantly reduced damping-off of pepper caused by *R. solani* strain R-23. After four weeks of growth, plant stands in the biocontrol-amended, pathogen-infested treatments were comparable to those in the noninfested controls. Alginate prill also reduced damping-off of pepper caused by *R. solani*, but only the stands of pepper were similar to that in the noninfested control. Alginate prill formulations also reduced populations of the pathogen and damping-off of pepper caused by *P. ultimum* (PuZ3). However, although the plant stands in the treatments were not as high as those in the noninfested controls, they were higher than those in the pathogen-infested controls. The treatments also reduced populations of *P. ultimum* in the soilless mix.

Lee et al. (1999), and Lee and Lee (2000) evaluated the effects of charcoal and charcoal wood extracts, alone and in combination with plant growth-promoting bacteria, on pepper plant growth, fruit and root development, and the inhibition of various pepper diseases. In their study, effects of charcoal and charcoal wood extracts on the bacterial and fungal population changes were also investigated.

Shin et al. (1999) selected *Pseudomonas* spp. and *Bacillus* spp., which are effective for the delay of occurrence of various fungal diseases during storage of pepper. In a related study, Lee and Jeong (2000) tested the effects of bacterial antagonists for the control of storage diseases and the extension of storage periods for hot pepper with preinoculation of antagonists and inoculation of storage disease pathogens at different intervals. As a result, several antagonists were selected for use in storage of pepper.

Cucumber

Fusarium

Cucumber (*Cucumis sativus* L.) is a crop of importance in Korea. However, *Fusarium* wilt of cucumber caused by *Fusarium oxysporum* f. sp. *cucumerinum* is a serious vascular disease in places where cucumber is cultivated. Biological control of *Fusarium* wilt of several crops has been accomplished by introduction of nonpathogenic *Fusarium* spp. in soil or in infection courts (Lee, 1994). There are many different mechanisms involved (Jee and Kim, 1987). Fluorescent pseudomonads have also been applied to seeds and rhizosphere soils or other infection courts for biological control (Weller, 1988; Alabouvette et al., 1996).

Jee and Kim (1987) effectively isolated bacteria and fungi antagonistic to *Fusarium oxysporum* f. sp. *cucumerinum* Owen. with modified Triple Layer Agar (TLA) technique from the rhizosphere soil where cucumber had been grown healthily in plastic filmhouse. Three predominant bacterial isolates selected were identified as *Pseudomonas fluorescens,* and *P. putida, Serratia* sp., and three fungal isolates were *Gliocladium* sp. *Trichoderma harzianum,* and *T. viride.* Antagonistic bacteria inhibited 26-45 percent of germination and 41-56 percent of germ tube elongation of microconidia of *F. oxysporum* f. sp. *cucumerinum* on Water Agar (WA). *P. fluorescens* was the strongest inhibitor. Several mycoparasitisms were observed on dual culture of WA between antagonistic fungi and *F. oxysporum* f. sp. *cucumerinum* such as coiling, penetration, overgrowing, and lysis. Mycelial lysis of the pathogen was the most severe at pH 4.6, followed by 3.6, 5.6, and 6.6 of the medium in a decreasing order. At pH 6.6, mycelia of the pathogen were not conspicuously damaged. However, the antagonistic fungi formed abundant chlamydospores. *Gliocladium* sp., and *T. harzianum* especially revealed the most excellent antagonism in vitro.

Park et al. (1988) selected strains of fluorescent pseudomonas to control *Fusarium* wilt of cucumber. Strains of fluorescent pseudomonas applied to the seed of cucumbers demonstrated different efficiencies in the biological control of *Fusarium* wilt. The reason for higher efficiency was not apparent in tests for in vitro antagonism, production of siderophores in iron-deficient medium, or by observation of inhibition of chlamydospore germination by the strains in rhizospheres of cucumbers. A more efficient biocontrol agent, *Pseudomonas putida* strain NIR, had higher rhizosphere competence and produced siderophores at higher iron levels than the inferior strains (A12) of the same species.

When Kim and Jee (1988) introduced antagonistic microorganisms to *Fusarium oxysporum* f. sp. *cucumerinum* into natural field soil artificially infested with the pathogen, the initial symptoms appeared late, and the progress of cucumber wilt was relatively slow. Disease severity and incidence were significantly reduced. In addition, cucumber growth in antagonist-treated plots was superior to that in untreated plot. *Pseudomonas fluorescens* and *Trichoderma harzianum* presented most predominant disease control effect over 50 percent compared with nontreated plots. There was no suppressive effect on the disease by antagonists in steam-sterilized soil, where all the plots were diseased and cucumber growth was remarkably retarded. Possible mechanism to explain such an observation are evidenced and discussed. Cucumber seedling growth was significantly promoted by antagonists, being about 20-50 percent greater in natural field soil and steam-sterilized soil. Increased cucumber growth was greater in field soil plots than in steam-sterilized plots. Antagonists that revealed outstanding inhibition to *F. oxysporum* f. sp. *cucumerinum* also promoted cucumber growth remarkably. Antagonists *Serratia* sp. and *P. putida* were successfully colonizing the cucumber rhizosphere over 1.1×10^4 cfu/g soil until 21 days after inoculation, while the population size abruptly declined in nonrhizosphere soil seven days after inoculation. Population densities of the pathogen, when introduced into natural field soil, increased tenfold in seven days in all plots, regardless of presence or absence of antagonists. However, pathogen propagules were greatly decreased in antagonist-treated soils thereafter. On the other hand, the population density already built up for first seven days in untreated plots persisted.

Cho et al. (1989) tested antagonistic activities of 56 *Gliocladium* isolates and nine *Trichoderma* isolates obtained from cucumber and strawberry fields, respectively, against *Fusarium oxysporum* f. sp. *cucumerinum* in vitro. A *Gliocladium* GC27 isolate that inhibited mycelial growth of the pathogen remarkably by nonvolatile antifungal antibiotics was identified as *Gliocladium virens. Trichoderma harzianum* T42 had the highest mycoparasitic effect and a low antibiosis to the pathogen. The antibiosis of *G. virens* GC27 to the pathogen was not affected by the type of nitrogen sources in the medium, but that of *T. harzianum* T42 with low antibiotic activity was increased by ammonium tartrate or NH_4NO_3. Addition of chitin, cell-wall preparation of the pathogen, cellulose, and organic sources, such as ground wheat bran, malt, rice straw, and corn, into the synthetic medium had no effect on the antibiotic activity of *T. harzianum* T42. Mycelial growth and conidial germination of the pathogens by *G. virens* GC27 were not affected by fungal cell-wall components, whereas they were inhibited greatly by wheat bran or malt addition to the mineral medium. Addition of chitin or wheat bran into the synthetic medium enhanced mycoparasitism of *T. harzianum* T42. The incorporation of *G. virens* GC27 or *T. harzianum* T42 cultured on wheat bran media into sterilized soil infested with the pathogen was much more effective in reducing the incidence of cucumber wilt caused by *F. oxsporum* f. sp. *cucumerinum* than the application of conidial suspension without organic food base in pot experiments. The wheat bran culture of *G. virens* GC27, added with or without inorganic nutrients, decreased disease incidence by 54.2 to 59.1 percent compared to that of control without the antagonist. Application of wheat bran culture of *T. harzianum* T42 mixed with inorganic nutrients also decreased the incidence of cucumber wilt by 52.3 to 59.7 percent, whereas without inorganic nutrients mixture, disease incidence was reduced by only 18.2 percent. The treatment of seeds with two antagonists and chitin addition to soil incorporated with antagonists did not have significant effects on the control of cucumber wilt.

Cho et al. (1992) showed the effect of organic sources on the growth of *F. oxysporum* f. sp. *cucumerinum,* and selected antagonistic bacterium *Pseudomonas gladioli* in vitro and in greenhouse tests. Various organic sources were examined for the ability to enhance the growth of the antagonist, and to inhibit growth of the pathogen as well. Among the twelve organic sources tested, desirable organic sources were not found. In in-vitro tests, barley bran, arrowroot leaf, and wheat bran powder stimulated growth of both *P. gladioli* PB-2 and the pathogen greatly, but bush clover leaf, paulowinia, and pine sawdust inhibited them. In greenhouse test with sterilized soil, *P. gladioli* PB-2 or B-83, alone or in combination with 1.0 percent addition of each of barley bran, arrowroot leaf, and paulowinia sawdust, failed to suppress the incidence of cucumber wilt. However, 0.5 percent addition of the mixture of paulowinia saw dust and barley bran at the ratio of 3:7 to the soil, where *P. gladioli* PB-2 or B-83 was applied, suppressed the disease incidence. In this case, the suppressive effect was slightly higher for isolate PB-2 than for B-83.

Chung and Ryou (1996) conducted studies from 1993 to 1995 to find out the effect on the inorganic and organic compounds as a soil amendment for the control of *Fusarium* wilt of cucumber caused by *F. oxysporum* f. sp. *cucumerinum*. In their study, 14 inorganic chemicals (1 percent, w/v), including $Al_2(SO_4)_3$, were added individually in vitro. The suppression by these chemicals was confirmed especially by Alum, CaO, and $Al_2(SO_4)_3$ that suppressed not only 20.9 to 25.0 percent on mycelial growth of the fungus, but also inhibited 72.8 to 97 percent on conidial germination. $Ca(NO_3)_2$ suppressed mycelial growth only, while KCl, K_2SO_4, NH_4NO_3, and Urea suppressed conidial germination. The seven chemicals were finally selected. Composed pine bark (CPB) suppressed definitely more than 90 percent on conidial germination in a different extract concentration (2.5 and 20 percent), although mycelial growth on the extract medium of CPB and milled alfalfa leaves (MAL) was not remarkable. The antagonist *Trichoderma* sp. (Tr-3), mixed with an amended soil (1 percent, w/w) containing composted pine bark, showed a good mycelial growth to compete with the causal fungus. And the antagonist *Pseudomonas* sp. (7-1-3) was also confirmed with its antagonistic ability with culture filtrate. It is known that a CPB soil amendment mixed with the two antagonists (1 percent, w/w) controlled *Fusarium* wilt of cucumber almost completely in greenhouse pots and a field experiment. It is therefore expected that biocontrol on *Fusarium* wilt of cucumber by a soil amendment can be applied to farmers' fields.

Yang and Kim (1994) studied on the cross-protection of *Fusarium* wilt of cucumber with the selection of nonpathogenic isolates and greenhouse tests. One hundred fifty-four out of two hundred sixty-two isolates of *F. oxysporum* obtained from healthy plant tissues of various crops and their rhizosphere soil were found to be nonpathogenic to cucumber plants. The nonpathogenic isolates were frequently found from sesame plant tissues and the rhizosphere soil, but less from healthy plant tissues of cucumber and watermelon. When the 154 nonpathogenic isolates were preinoculated into cucumber seedlings, and then challenge-inoculated with *Fusarium* wilt pathogen, 21 isolates protected effectively cucumber plants from *Fusarium* wilt infections. A year later, 9 out of 21 isolates fully sustained their protective effect. Among the nine isolates showing good protective effects, seven were isolates from cucumber plants. The nine isolates, except one, were not pathogenic to watermelon, chinese melon, tomato, and sesame.

Singh et al. (2003) used quantitative models to describe the efficacy of inundative biological control of *Fusarium* wilt of cucumber. *Fusarium* wilt of cucumber caused by *F. oxysporum* f. sp. *cucumerinum* is a serious vascular disease worldwide. Biological control of *Fusarium* wilt in several crops has been accomplished by introducing nonpathogenic *Fusarium* spp. and other biocontrol agents in soil or in infection courts. In this study, quantitative models were used to determine the biocontrol efficacy of inundatively applied antagonist formulations and the length of their effectiveness in controlling *Fusarium* wilt of cucumber.

Powdery Mildew

Powdery mildew of cucumber (Figure 11.2) has been the major hindrance for the production of cucumber in Korea, especially in the greenhouse system. Even though various chemical fungicides have been used quite successfully for the control of powdery mildews, there was need for selection of antagonists or selection of bioactive substances extracted from various beneficial plant species. Paik et al. (1994) worked with beneficial plant species for the selection of bioactive substances to control powdery mildews. Paik et al. (1996) further tested the effect of a bioactive substance extracted from *Rheum undulatum* for the control of cucumber powdery mildew. Effects of wettable powder formulations (a.i., 30 percent), made of the extract of *Rheum undulatum* (RK), and a related standard chemical, 1,8-dihydroxy anthraquinone (AK), on the control of the powdery mildew of cucumber caused by *Sphaerotheca fuliginea,* and their phytotoxicity on cucumber plants and toxicity on fish were investigated. In a polyethylene filmhouse test, the formulated RK and AK fungicides showed 100 percent control efficacy on the disease at 500 and 1,000 dilution concentrations; and in pot tests, their control efficacies were 75.0 to 100 percent, 67.0 to 75.0 percent, and 52.0 to 75.5 percent for AK, and 100 percent, 75.0 to 100 percent, and 52.0 to 81.6 percent for RK at 2,000, 3,000, and 5,000 dilution concentrations, respectively. In a field trial, 4,000 and 5,000 dilution concentrations of AK and RK suppressed disease incidence by 72.3 to 90.0 per-

FIGURE 11.2. Powdery mildew of cucumber *(Cucumis sativus)* caused by *Sphaerotheca fuliginea (left),* and *Ampelomyces* (Trade name: Topseed) treated cucumber *(right).*

cent. RK was not phytotoxic to cucumber plants up to the 250 dilution concentration, but AK showed phytotoxicity at the same dilution concentration. In the fish toxicity test, the doses for median tolerance limit (TLm) (50 percent survival for 48 hours) of these two formulations were exhibited above 2 ppm.

Other researchers, for example, Lee and Kim (2001) characterized the parasite *Ampelomyces quisqualis* 94013 against the powdery mildew fungus of cucumber (Figure 11.2). *Ampelomyces quisqualis* 94013(AQ94013) was selected for the use of biological control of cucumber powdery mildew caused by *Sphaerotheca fuliginea*. They also examined the parasitism processes with scanning electron microscopy and light microscopy of conidia of AQ94013 germinated on conidia, conidiophores, and hyphae of *Sphaerotheca fuliginea* four hours after inoculation. Appressorium-like structures were developed and attached to the hyphae of *S. fuliginea* 17 hours after inoculation. The hyphae of AQ94013 penetrated into the hyphae of *S. fuliginea* 24 hours after inoculation. Pycnidia of AQ94013 were produced in the hyphae and the basal part of conidiophores of *S. fuliginea* forty four hours after inoculation. The pycnidia of AQ94013 matured 48 hours after inoculation, and the conidia were discharged from the ostioles of the pycnidia 52 hours after inoculation. At the same time, the hyphae and conidiophores of *S. fuliginea* were distorted and died. Also, the concentrated culture filtrate and culture filtrate of AQ94013 had not suppressed the cucumber powdery mildew fungus as water treatment. Therefore, the mode of action of AQ94013 was assumed to be parasitism on powdery mildew fungi.

Anthracnose

Systemic resistance is the phenomenon in which a plant exhibits an increased level of resistance to pathogen infection after appropriate stimulation. Systemic resistance comprises systemic acquired resistance and induced systemic resistance. Induced systemic resistance-mediated plant protection is induced by colonization of the rhizosphere with plant growth-promoting rhizobacteria. Seed-coating or soil-drenching with PGPR protected plants against various pathogens, including fungi, bacteria, and viruses (Alstrom, 1991; Hoffland et al., 1996; Wei et al., 1996; Van Wee et al., 1997) in greenhouses and field conditions. Park et al. (2001) studied on the systemic resistance and expression of the pathogenesis-related genes mediated by the plant growth-promoting rhizobacterium *Bacillus amyloliquefaciens* EXTN-1 against anthracnose disease in cucumber. More than 800 strains of rhizobacteria were screened in the greenhouse. Among these strains, *Bacillus amyloliquefaciens* isolate EXTN-1 showed significant disease control efficacy on the plants. Induction of pathogenesis-related

(PR-1a) gene expression by EXTN-1 was assessed, using tobacco plants transformed with PR-1a::β-*glucuronidase* (GUS) construct. GUS activities of tobacco treated with EXTN-1 and salicylic acid-treated transgenic tobacco were significantly higher than those of tobacco plants with other treatments. Gene expression analyses indicated that EXTN-1 induced the accumulation of defense-related genes of tobacco. The results showed that some defense is expressed by treatment with EXTN-1, suggesting the similar resistance mechanism by salicylic acid.

Jeun et al. (2001) showed different mechanisms of induced systemic resistance and systemic acquired resistance against *Colletotrichum orbiculare* on the leaves of cucumber plants. Defense mechanisms against anthracnose disease caused by *Colletotrichum orbiculare* on the leaf surface of cucumber plants after pretreatment with PGPR, amino salicylic acid (ASA) or *C. orbiculare* were compared, using a fluorescence microscope. Induced systemic resistance was mediated by preinoculation in the root system with PGPR strain *Bacillus amylolquefaciens* EXTN-1 that showed direct antifungal activity to *C. gloeosporioides* and *C. orbiculare*. Also, systemic acquired resistance was triggered by pretreatments on the bottom leaves with amino salicylic acid or conidial suspension of *C. orbiculare*. The protection values on the leaves expressing SAR were higher compared with those expressing ISR. After preinoculation with PGPR strains, no change of the plants was found in the phenotype. On the other hand, necrosis or hypersensitive reaction (HR) was observed on the leaves of plants pretreated with ASA or the pathogen. After challenge inoculation, inhibition of fungal growth was observed on the leaves expressing both ISR and SAR. HR was frequently observed at the penetration sites of both the resistance-expressing leaves. Appressorium-formation was dramatically reduced on the leaves of plants pretreated with ASA, whereas EXTN-1 did not suppress the appressorium-formation. ASA also inhibited the conidial germination more strongly compared with EXTN-1. Conversely, EXTN-1 significantly increased the frequency of callus formation at the penetration sites, but ASA did not. The defense mechanisms induced by *C. orbiculare* were similar to those by done by ASA. Based on these results, it is suggested that resistance mechanisms on the leaf surface were different in the cucumber leaves expressing ISR and SAR, resulting in different protection values.

Damping-Off

Among the various diseases that affect cucumber, damping-off diseases caused by *Pythium* spp. and *Rhizoctonia solani,* and *Fusarium* wilts are the major diseases. Damping-off, especially, occurs during and/or at the germination of seeds, and right after germination of seeds by *Pythium* spp. and

Rhizoctonia spp. in the seedbed. In some cases, dual infection of *Pythium* spp. and *Rhizoctonia* spp. makes the matter worse. Baker (1987) and Hornby (1983) worked on biological control of damping-off diseases. Recently in Korea, Lee et al. (2003) tested the effect of organic amendments on the efficacy of biological control of seedling damping-off of cucumber with several microbial products. Several microbial biocontrol products (Greenbiotech Co., Paju, Korea), Green-all T *(Trichoderma harzianum),* Green-all S *(Bacillus* sp.), and Green-all G *(Streptomyces* sp.) were supplemented with organic amendments, such as sawdusts and rice hulls, to study the efficacy of the biological control of seedling damping-off of cucumber caused by *Pythium ultimum*. Sawdusts amended into potato dextrose agar alone could inhibit in vitro mycelial growth of *P. ultimum*. All the three microbial products of Green-all T, Green-all G, and Green-all S significantly reduced seedling damping-off. However, several amendments such as sawdusts and rice hulls into Green-all T and Green-all S products did not increase the efficacy of biological control compared with the nonamended treatment. In contrast, supplements of aminodoctor containing several amino acids (Greenbiotech Co., Korea) into Green-all G product significantly increased the efficacy of biological control of seedling damping-off, resulting in from 42 percent to 2 percent disease incidence in relation to seedling emergence. Also, amendment of sawdusts into *Trichoderma* product significantly increased the efficacy of biological control as disease index of 5.0, compared with nonamended control of 56.0 in Green-all T product alone. This indicates that organic amendments could increase the efficacy of biological control of cucumber seedling damping-off.

Tomatoes and Eggplants

Damping-Off

Lewis, Fravel, Lumsden, and Shasha (1995) contrived a method to apply biocontrol agents. They found pregelatinized starch-flour granules containing fermentor-produced biomass of the biocontrol fungi *Gliocladium virens* and *T. hamatum* protected eggplant, pepper, and zinnia seedlings from damping-off caused by the pathogen *R. solani* in a soilless mix. Also, they found the importance of spatial distribution of the biocontrol product in disease control.

Lewis, Fravel, and Papavisas (1995) worked on the effect of bran preparations of the potential biocontrol fungus *Cladorrhinum foecundissimum* on the survival and saprophytic growth of the soil-borne plant pathogen *R. solani* in soils and soilless potting mix and on its ability to reduce incidence of damping-off of sugar beet, eggplant, and pepper caused by this pathogen. Bran preparations of some of the isolates prevented damping-off of eggplant and pepper in soilless mix and, depending on the rate of inoculum used, resulted in stands comparable to those in pathogen-free soilless mix.

Lewis and Larkin (1998) tested five isolates of *Cladorrhinum foecundissimum* which were added to soilless mix as ten-day-old fresh bran preparations (1.0 percent w/w). The mixtures significantly reduced damping-off of eggplant caused by *Rhizoctonia solani* strain R-23. After four weeks of growth, plant stands in the biocontrol-amended, pathogen-infested treatments were comparable to those in the noninfested controls. Since plant stands were similar at two and four weeks, most of the disease was preemergence damping-off. The bran preparations also reduced the saprophytic growth of the pathogen, and there was an inverse correlation ($r2 = -0.94$) between saprophytic growth and eggplant stand. Added to soilless mix at a rate of 2.0 percent (w/w), alginate prill containing 20 percent fermentor-produced biomass of six biocontrol isolates of *C. foecundissimum* reduced ($P = 0.05$) damping-off of eggplant caused by *R. solani,* but only the prill with biomass of isolates Cf-1 or Cf-2 yielded plant stands (>80 percent) comparable to that in noninfested control. As with bran preparations, there was also an inverse correlation ($r2 = -0.80$) between saprophytic growth of R-23 and eggplant stand with the alginate prills. Alginate prill with the biomass of Cf-1 or Cf-2 also reduced ($P = 0.05$) damping-off of eggplant and pepper caused by other isolates (195, NG-2, DPR-1) of *R. solani*, but only the stands (>80 percent) of pepper were similar to that in the noninfested control. Alginate prill formulations of *C. foecundissimum* (Cf-1, Cf-2, and Cf-3) also reduced ($P = 0.05$) populations of the pathogen and damping-off of eggplant and pepper caused by *Pythium ultimum* (PuZ3). However, although the plant stands in the treatments were not as high as those in noninfested controls, they were higher than those in pathogen-infested controls. The treatments also reduced populations of *P. ultimum* in the soilless mix, so that there were inverse correlations between the pathogen population and eggplant stand ($r2 = -0.81$) and pepper stand ($r2 = -0.78$). Extruded flour/clay granules containing 5.0 percent biomass of Cf-1 and Cf-2, added to *R. solani*-infested soilless mix (2.0 percent), reduced ($P = 0.05$) damping-off of eggplant and pepper. However, only the Cf-2 treatments resulted in stands (>80 percent) equal to those in the noninfested controls for the crops after four weeks of growth. The influence of bran and alginate prill of Cf-1 or Cf-2 on the spatial spread of *R. solani* and its ability to incite damping-off of eggplant showed that prill with Cf-1 or Cf-2 and bran with Cf-2 were equally effective in reducing the spread of the pathogen from the point source of the inoculum to the center of the flats.

Wilt Diseases

Larena et al. (2003) used conidia of *Penicillium oxalicum* produced in a solid-state fermentation system at densities of 6×10^6 spores/g seedbed substrate to tomato seedbeds in water suspensions (T1: 5 days before sowing, or T2: 7 days before transplanting; 15 days after sowing), or in mixture with the production substrate (T3: 7 days before transplanting; 15 days after sowing). Treatments T2 and T3 significantly reduced *Fusarium* wilt of tomato in both greenhouse (artificial inoculation) [33 and 28 percent, respectively] and field conditions (naturally infested soils) [51 and 72 percent, respectively], while treatment T1 was efficient only in greenhouse (52 percent). *Verticillium* wilt disease reduction was obtained with T3 in two field experiments (56 and 46 percent, respectively), while T1 and T2 reduced disease only in one field experiment (52 percent for both T1 and T2). Treatment with conidia of *P. oxalicum* plus fermentation substrate (T3) resulted in better establishment of a stable and effective population of *P. oxalicum* in the seedbed soil and rhizosphere, providing populations of approx. 10^7 CFU/g soil before transplanting. Results indicate that it will be necessary to apply *P. oxalicum* at a rate of approx. 10^6 to 10^7 CFU/g in the seedbed substrate and rhizosphere before transplanting for effective control of *Fusarium* and *Verticillium* wilt of tomato, and that formulation of *P. oxalicum* has a substantive influence on its efficacy.

Jetiyanon and Kloepper (2002) selected mixtures of compatible PGPR strains with the capacity to elicit induced systemic resistance in bacterial wilt of tomato *(Lycopersicon esculentum)* caused by *Ralstonia solanacearum*. To examine compatibility, seven selected PGPR strains were individually tested for in vitro antibiosis against all other PGPR strains and against the tested pathogen, *R. solanacearum*. No in vitro antibiosis was observed among PGPR strains or against the pathogen. Twenty-one combinations of PGPR and seven individual PGPR were tested in greenhouse for induced resistance activity. Results indicated that four mixtures of PGPR and one individual strain treatment significantly reduced the severity of the disease compared with nonbacterized control: 16 mixtures reduced bacterial wilt of tomato. Most mixtures of PGPR provided a greater disease suppression than individual PGPR strains. These results suggest that mixtures of PGPR can elicit induced systemic resistance. Other researchers in Korea studied the effect of antagonists against *R. solancearum* (personal communication; Figure 11.3).

FIGURE 11.3. Bacterial wilt of tomato caused by *Ralstonia solanacearum (top)*, and antagonist-treated tomato plants *(bottom)*.

Cabbages, Radishes, Turnips, and Sugar Beets

Fusarium *Wilts*

Fusarium wilt diseases are responsible for yield losses on numerous crops. Some soils are known for their natural suppressiveness to *Fusarium* wilts. The soil microflora is responsible for the natural suppressiveness of these soils (Scher and Baker, 1980; Alabouvette, 1986). When biological control organisms were introduced to the disease conducive soil, *Fusarium* wilt was significantly suppressed (Kolepper et al., 1980; Alabouvette, 1986; Paulitz et al., 1987; Leeman, 1995). However, the inoculation of disease-suppressing microorganisms to the disease-conducive soil never reached the level of suppression observed in the natural suppressive soils, and the effects are often inconsistent (Weller, 1988). A multitude of factors could account for the inconsistent performance of the biocontrol agents due to the complex interactions among the antagonists, the pathogen, the host, and the environment (Weller, 1988; Schippers, 1992). In the Netherlands, *Fusarium* wilt of radish was a problem in continuous cropping of radish in green houses. In studies by Kloepper and Schroth (1978), the mechanisms responsible for the beneficial effects of separate or mixed treatment of three different bacterial strains on disease suppression to *Fusarium* wilt of radish were evaluated on different levels of disease incidence, and the relation between root colonization and suppression of *Fusarium* wilt by fluorescent pseudomonads in the rhizosphere was investigated as well.

Lee (1997) studied root colonization by beneficial *Pseudomonas* spp., and tested the suppression of *Fusarium* wilt of radish. Growth-promoting beneficial organisms, such as *Pseudomonas fluorescens* WCS374 (strain WCS374), *P. putida* RE10 (strain RE10), and *Pseudomonas* sp. EN415 (strain EN415), were used for microorganism-mediated induction of systemic resistance in radish against *Fusarium* wilt. In his bioassay, the pathogens and bacteria were treated into soil separately or concurrently, and the bacteria mixed with the different level of combination. Significant suppression of the disease by bacterial treatments was generally observed in pot bioassay. The disease incidence of the control recorded was 46.5 percent in the internal observation and 21.1 percent in the external observation, respectively. The disease incidence of *P. putida* RE10 recorded was 12.2 percent in the internal observation and 7.8 percent in the external observation, respectively. However, the disease incidence of *P. fluorescens* WCS374, which proved to be highly suppressive to *Fusraium* wilt, stood at 45.6 percent in the internal observation and 27.8 percent in the external observation, respectively. The disease incidence of *P. putida* RE10 mixed with *P. fluorescens* WCS374 or *Pseudumonas* spp. EN 415 was in the range of 10.0 to 21.1 percent. On the other hand, the disease incidence of *P. putida* RE10 mixed with *Pseudumonas* sp. EN415 was in the range of 7.8 to 20.2 percent. Colonization by *F. oxysporum* f. sp. *radicis* (FOR) was observed in the range of 2.4 to 5.1×10^3/g on the root surface and 0.7 to 1.3×10^3/g in the soil, but the numbers were not statistically different. As compared with 3.8×10^3/g root of the control, the colonization of infested FOR indicated 2.9×10^3/g root in separate treatment of *P. putida* RE10, and less than 3.8×10^3/g root of the control. Also, the colonization of FOR recorded 5.1×10^3/g root in mixed treatments of three bacterial strains such as *P. putida* RE10,

P. fluorescens WCS374, and Pseudomonas sp. EN415. The colonization of FOR in soil was less than that of FOR in root part. Based on soil or root part, the colonization of FOR didn't indicate a significant difference. The colonization of introduced three fluorescent pseudomonas was observed in the range of 2.3 to 4.0×10^7/g in the root surface and 0.9 to 1.8×10^7/g in soil, but the bacterial densities were significantly different. When growth-promoting organisms were introduced into the soil, the population of Pseudomonas sp. in the root part, treated with P. putida RE10, was similar in number to the control, and recorded the low numerical value as compared with any other treatment. The population density of Pseudomonas sp. in the treatment of P. putida RE10 indicated significant differences in the root part, but didn't show significant differences in soil. The population densities of the infested FOR and the introduced bacteria on the root were high in contrast to those of soil. P. putida RE10 and Pseudomonas sp. EN415, used in this experiment, appeared to induce the resistance of the host against Fusarium wilt.

Lee and Ogoshi (1991) studied the effect of seed bacterization of sugar beets on growth and suppression of damping-off in greenhouse and field. The effects of sugar beet seed bacterization on germination of seeds and suppression of damping-off during the seedling stage in greenhouse, and growth and disease incidence in two different fields located in Hokkaido National Agricultural Experiment Station (HNA) in Sapporo and Nippon Beet Sugar Mfg. Co. (NBS) in Obihiro, Japan were investigated. In greenhouse, at 17 days after the sowing, germination rates were considerably promoted by nine bacterial strains (mainly fluorescent pseudomonads) on HNA and by three strains in NBS. The damping-off rate of sugar beet examined by seed bacterization was 3.8 to 11.1 percent in HNA showing remarkable reduction over 21.6 percent compared with that of control. The effect of bacterization was not clear in NBS. In addition, the pathogen of the damping-off was examined and the isolation frequency of Pythium from damping-off was 6-16 plants of bacterized seedlings compared with 25 plants in control at HNA, suggesting that bacterization is very effective to Pythium damping-off. Upon transplanting, no clear effects were observed on the growth of sugar beet and on the disease development at harvest time in the two fields. During cultivation, the bacterial numbers on seeds (5×10^7 cfu/seed) decreased with time, and a significant difference was observed among the bacterial strains.

Root Rot

Sugar beet damping-off diseases are caused mainly by Aphanomyces cochlioides, Pythium sp., and Rhizoctonia solani in seedbeds (Lee and

Kobayashi, 1988). Bacterization of sugarbeet seed was found to be effective for production increase (Suslow and Schroth, 1982). In Lee and Kobayashi's (1988) studies on seed bacterization of sugar beets, a strain of *Pseudomonas* sp. (B-1218), which was antibiotic to *R. solani* (AG-1, AG-4) but not to *Pythium* spp., was isolated from the root surface of sugar beets. That strain was identified as *Pseudomonas cepacia* Palleroni and Holmes by bacterial physiological characteristics. The hyphal lysis and the branching of sugar beet pathogens, *R. solani* and *Aphanomyces cochlioides,* were observed when confronted with antagonistic bacteria (B-1218). An antibiotic product extracted from *P. cepacia* culture medium (King's B broth) with ethyl acetate strongly inhibited the growth of *R. solani.* Its antibiotic was identified as pyrrolnitrin by TLC, HPLC, and GC-MS. By the result, *P. cepacia* isolated from sugar beet root surface was thought to be a strain which produced the antibiotic substances to *R. solani.*

Damping-Off

Lewis, Fravel, and Papavisas (1995) worked on the effect of bran preparations of the potential biocontrol fungus *Cladorrhinum foecundissimum* on the survival and saprophytic growth of the soilborne plant pathogen *R. solani* in soils and soilless potting mix, and on its ability to reduce incidence of damping-off of sugar beet, eggplant, and pepper caused by this pathogen. In their studies, the bran inoculum of the antagonist, incubated for as long as 17 days, did not reduce the survival of the pathogen. However, bran inoculum older than six days prevented growth of the pathogen from the beet seed into soil.

Brittle Root Rot

Brittle root rot of Chinese cabbages, caused by *Aphanomyces raphani,* was the major soilborne disease in the alpine areas of Korea, especially in Kangwon Province, from 1994 to 1997. Lee (1996), Lee and Kim (1996), Lee et al. (1996), and Lee et al. (1997) studied biological control agents against the brittle root rot pathogen *(Aphanomyces raphani)* of Chinese cabbage in the alpine areas of Kangwon Province (Figure 11.4). Lee and others selected various kinds of antagonists, including bacteria. Among the antagonists they tested, *P. fluorescence* and other *Pseudomonas* sp. were quite inhibitory against *Aphanomyces raphani.* Other possible antagonists such as actinomycetes and *Trichoderma* spp. were found not to be effective.

FIGURE 11.4. Symptoms of brittle root rot disease of Chinese cabbage *(Brassica campestris* ssp. *pekinensis)* caused by *Aphanomyces raphani.*

Strawberries

Fusarium *Wilt*

Fusarium wilt of strawberry in Korea, especially in southern regions, is also a major problem in strawberry production, as in other parts of the world (Moon et al., 1988; Moon et al., 1990). As in other crops, the control of *Fusarium* wilt diseases of strawberry mainly depended on the use of fungicides. In recent years, however, there has been a great interest in the biological control of *Fusarium* wilt diseases in Korea.

Moon et al. (1988) obtained 105 *Trichoderma* isolates from strawberry fields by using selective media and tested in vitro for their antagonistic effects on *Fusarium oxysporum* f. sp. *fragariae.* Four isolates showed highly antagonistic effects on the pathogen. *Trichoderma* isolates T42 and T17, and T34 and T74 were identified as *Trichoderma harzianum* and *T. viride,* respectively. In the dual cultures of both *T. harzianum* and *F. oxysporum* f. sp. *fragariae* on agar media, *T. harzianum* parasitized the pathogen, leading to inhibition of the mycelial growth of the pathogen. The modes of mycoparasitism of *T. harzianum* appeared to be coiling around and its attachment on the host hyphae, microconidia, and macroconidia, and/or penetration into the hyphae or breaking the septa of both the hyphae and the conidia. It seems that *T. viride* produced nonvolatile antibiotics inhibiting mycelial growth of *F. oxysporum* f. sp. *fragariae* in the culture media where its antagonistic effect was relatively low.

In a subsequent study, Moon et al. (1988) conducted an experiment to examine the factors affecting mycoparasitism of *Trichoderma harzianum* isolate T42 or T17 and antibiosis of *T. viride* isolate T34 or T74 on *Fusarium oxysporum* f. sp. *fragariae* in relation to nutrition and physical environment in vitro. Maltose and $Ca(NO_3)_2$ were the best for the mycoparasitism of *T. harzianum* as a carbon and nitrogen source, respectively, and KH_2PO_4 was also an effective element. Mycoparasitism was enhanced in high-carbon and low-nitrogen medium. Addition of organic materials, such as corn., wheat bran, and rice straw powder increased mycoparasitism, whereas clover leaf powder amendment annuled the efficiency. Addition of cell walls of the pathogens, cellulose, or chitin in the cultures also enhanced the lysis of both the hyphae and conidia of the pathogen by *T. harzianum*. Optimal culture conditions for mycoparasitism appeared in the pH range of 3.5 to 5.5 at 25°C. *T. viride* produced nonvolatile antibiotics in the culture media, which inhibited mycelial growth of *F. oxysporum* f. sp. *fragariae,* but its antagonistic effect was relatively low. Carbon source was required for its inhibitory effect, while the other nutritional factors and environmental factors did not have effect. Moon et al. (1990) tested antagonistic activities of 184 bacterial isolates, obtained from several strawberry fields in Kim-hae area, against *F. oxysporum* f. sp. *fragariae* in vitro. An isolate, B-83 identified as *Pseudomonas gladioli* showed the most effective inhibition against mycelial growth and conidial germination of the pathogen. B-83 exhibited the best growth in wheat bran or wheat bran-vermiculite(1:1) media among five different organic sources tested at 28°C.

In a pot experiment, the incorporation of *P. gladioli* B-83 into field soil or sterile soil infested with the pathogen decreased incidence of strawberry wilt significantly by 60 percent compared with that of control without the antagonist. The addition of 0.5 percent(w/w) of wheat bran to sterile soil incorporated with *P. gladioli* B-83 reduced disease incidence as much as by 70 percent. Chitinase activity was not detected in *P. gladioli* B-83 cultured on colloidal chitin agar, but this antagonist lysed mycelia of the pathogen by submerged dual culture.

Bud Rot

Bud rot of strawberry caused by *R. solani* is also a major hindrance to the production of strawberry, especially under low-temperature conditions in greenhouse (Shin et al., 1994). The biological control of damping-off and root rot diseases of strawberry was reported to be possible by using of *Trichoderma* spp., nonpathogenic *Rhizoctonia* sp., and binucleate *Rhizoctonia* sp. (Cook and Baker, 1983; Chet, 1987; Murkerji and Garg, 1988; Hornby, 1990). However, bud rot diseases of strawberry occur under different environmental conditions. The causal agent of bud rot of strawberry, *R. solani* AG2-1 type, usually cause disease under low-temperature conditions in the greenhouse system.

In 1994, Shin et al. (1994) selected 40 out of 167 microbial isolates from the soil of controlled cultivation areas. The selected isolates inhibited the mycelial growth of *Rhizoctonia solani* AG2-1, the causal agent of strawberry bud rot, in vitro. Among the isolates, Kr013 and Kr020 showed suppressive effect against *R. solani* AG2-1 on the seedling of Chinese cabbage treated by root immersion, charcoal carrier granule, and drenching on 1.0 percent infested soil in pot. Furthermore, the corresponding effect was also revealed when the charcoal carrier granules of the isolates were treated on the seedling of strawberry that were planted on the planting hole in pot. To examine the effects of biological control in greenhouse, it was tested for the infection rates by using two different treatments. First, the strawberry runners were planted on the nursery soil mixed with 20 percent charcoal carrier granules of Kr013 and Kr020 isolates respectively, and grown for 20 days before transplanting. Then the young plants from the mother plant were separated and transplanted on the 1.0 percent infested soil. Another method was that charcoal carrier of Kr013 and Kr020 isolates applied to planting hole of 1.0 percent infested soil just before transplanting. Then the young plants were grown for 20 days on the sterilized nursery soil before transplanting. From the results, the effects of biological control were significantly higher on the former treatment (the infection rates were 7.3 and 5.7 percent, respectively) than on the latter treatment (the corresponding values were 16.7 and 15.7 percent, respectively). The antagonistic isolates of Kr013 and Kr020 were identified as *Pseudomonas cepacia* with the similarity of 55.0 percent and 60.0 percent, respectively, by using the Biolog GN Microplate system.

Gray Mold

Hang et al. (2005) selected *Bacillus subtilis* S1-0210 as a biocontrol agent against *B. cinerea* on strawberry. The isolate inhibited the mycelial growth of gray mold in in-vitro tests. A wettable powder formulation reduced the infection rate significantly with lower than 5 percent compared with higher than 70 percent of infection rates in untreated control. The formulation showed 85 to 89 percent control efficacies of gray mold incidence on fruits of strawberry in pots. They found that pretreatment of the agent was more effective in controlling gray mold on fruits and leaves than posttreatment at the early stage of disease development. In a field trial, 70 percent control efficacy was achieved with the same formulation.

Melons and Watermelons

Gummy Stem Rot

Watermelon [*Citrullus lanatus* (Thumb.) Matsum. & Nakai] is one of the most extensively planted fruit crops in greenhouses and fields of Korea. The gummy stem rot caused by the soilborne fungus *Didymella bryoniae* is one of the most serious diseases of watermelon, and is a major factor that limits production in subsequent years. In response to environmental and health concerns about extended use of chemicals, there is a considerable interest in finding alternative control methods for use in integrated disease management strategies. One of the alternative control methods is the use of beneficial microorganisms for biological control. Among the beneficial microorganisms, fluorescent pseudomonads, which are characterized by the production of yellow-green pigments that fluoresce under UV light and function as siderophores, are one of the most widely investigated bacteria. There have been many reports in recent years that plant growth-promoting *Pseudomonas* spp. induce systemic resistance to bacterial (Alstrom, 1991, 1995), fungal (Van Peer et al., 1991; Van Peer and Schippers, 1992; Zhou et al., 1992; Zhou and Paulitz, 1994), and viral pathogens (Maurhofer et al., 1994; Raupach et al., 1996).

Recently, the potential of PGPR-mediated ISR as a tool in integrated disease management is becoming more widely recognized, especially as the information regarding the occurrence of the phenomenon for an increasing number of plant-pathogen interactions and the knowledge of the underlying mechanisms of the resistance responses become more abundant, and the necessity of these kinds of research increases.

Lee et al. (2000b) selected five PGPR strains, WR8-3 *(Pseudomonas fluorescens)*, WR8-6 *(P. putida)*, WR9-9 *(P. fluorescens)*, WR9-11 *(Pseudomonas sp.)*, and WR9-16 *(P. putida)*. The selected isolates were tested on their growth promotion and control effect against gummy stem rot of watermelon. Strains WR8-3 and WR9-16 significantly increased the stem length of watermelon, and there was a little increase in the leaf area, fresh weight, and the root length when strains, WR8-3, WR9-9, and WR9-16 were treated. Generally, seed treatment was better for plant growth promotion than soil drench, but there was no significant difference. Seed treatment and soil drench of each bacterial strain also significantly reduced the mean lesion area (MLA) by gummy stem rot, but there was no significant difference between the two treatments. At initial inoculum densities of each strain ranging from 10^6 to 10^{15} cfu/g seed, approximately the same level of disease resistance was induced. But resistance induction was not induced at the initial inoculum density of 10^3 cfu/g seed. Resistance was induced by treating the strains, WR9-9, WR9-11, and WR9-16, on all of four watermelon varieties tested, and there was no significant difference in the decrease of gummy stem rot among the tested varieties. Populations of the strains treated initially at log 9 to 10 cfu/g seed decreased rapidly from planting day to one week after planting, but the population density was maintained above log 5.0 cfu/g soil for four weeks after planting. Generally no or very weak in vitro antagonism was observed with the strains treated, except WR9-11. Rifampicin-resistant bacteria which had been inoculated were not detected in the stems or leaves, suggesting that the bacterium and the pathogens remained spatially separated during the experiment. This is the first report of resistance induction in watermelon to gummy stem rot by PGPR stains.

In a different study, Lee et al. (2000a) studied the factors relating to induced systemic resistance in watermelon by plant growth-promoting *Pseudomonas* spp. The plant growth-promoting *Pseudomonas* strains, WR8-3 *(Pseudomonas fluorescens),* WR9-11 *(Pseudomonas* sp.), and WR9-16 *(P. putida),* which induced resistance systemically in watermelon to gummy stem rot, were investigated on their ISR-related characteristics. The pyoverdine production was repressed in the standard succinate medium by increasing the concentration of $FeCl_3$. But the iron-binding ability on chrome azurol S agar media (CAS) was observed only in the strains WR8-3 and WR9-16. When the two strains were mutated, the resulting iron-binding siderophore-negative mutants, WR8-3m and WR9-16m, failed to promote the growth of watermelon, and to induce resistance. Strains WR8-3 and WR9-16 slightly inhibited the growth of *Didymella bryoniae* at a low concentration of $FeCl_3$ on King's medium B, but not to exert control effect. Strain WR9-11 showed antagonism in the concentration of $FeCl_3$ from 0 to 1,000 μM. When the crude lipopolysaccharide of each strain was treated in the rhizosphere of watermelon, the mean lesion area was similar to that of the untreated control. Strains WR9-11 and WR9-16 produced the same level of hydrogen cyanide (HCN). Salicylic acid production was not detected in any of the strains.

Asparagus

Fusarium *Rot*

Lee (1994), Lee (1996), and Lee and Manning (1991a,b, 1992) selected *Fusarium solani* and avirulent *Fusarium oxysporum* for the biological management of root and crown rot caused by virulent *Fusarium* species on asparagus (Figure 11.5). In protection tests on plantlets with prior inoculations of *Fusarium solani* and avirulent *Fusarium oxysporum,* most protec-

FIGURE 11.5. Infections of various *Fusarium* species on asparagus *(Asparagus officinalis)*. Control *(Ctr1)*, avirulent *F. oxysporum (AvFo1)*, and *F. solani (FS)* did not cause any disease. However, *F. monilifome (Fm37)* showed severe disease symptoms and *F. oxysporum (Fo49)* showed mild disease symptoms.

tion of asparagus against virulent fusarial infections occurred when challenge isolates were inoculated five or seven days after the inoculation of the protective fusarial species. Avirulent *F. oxysporum* was more effective for protection against infection of *F. moniliforme* than it was against *F. oxysporum*. *F. solani* was more effective against infection of *F. oxysporum* than it was against *F. moniliforme*. In their studies, interactions of virulent and avirulent *Fusarium* species on clonal asparagus plantlet and mechanism involved in the protection of asparagus and avirulent *Fusarium* species against stem and crown rots were investigated.

Other Vegetables

Basal and Neck Rots of Onions

Lee et al. (2001) studied postharvest decay of onion bulbs. The onions were examined by inspecting the commercial packages in the market or in storage. Bulb rot incidence was unexpectedly high, and onion bulbs with 1st quality grade were rotten most severely by 51 percent, followed by 32 percent for 2nd and 21 percent for 3rd grades. This indicates that larger bulbs had higher incidence of bulb rots. Major pathogens associated with basal and neck rots were *Fusarium oxysporum* and *Aspergillus* sp. or *Botrytis allii*, respectively, of which basal rot was most prevalent and damaging during storage. Among the epiphytic microorganisms from onion plants, several *Bacillus* and *Paenibacillus* spp., and previously selected *Pseudomonas putida* and *Trichoderma harzianum* had inhibitory efficacy against bulb rot pathogens. Among these antagonists, *B. amyloliquefaciens* BL-3, *Paenibacillus polymyxa* BL-4, and *P. putida* Cha 94 were highly inhibitory to conidial germination of *F. oxysporum*. When *B. allii, P. putida* Cha 94, *B. amyloliquefaciens* BL-3, *P. polymyxa* BL-4, and *T. harzianum* TM were applied in the rhizoplane of onion at transplanting, populations of the antagonist decreased rapidly during the first month. However, among these antagonists, rhizoplane population densities of BL-3, Cha 94, and TM were consistently high thereafter, maintaining about 10^4 to 10^5 cells or spores per gram of onion root up to harvest time. The other bacterial antagonist BL-4 survived only for two months. TM was the most effective biocontrol agent against basal rot, with the number of rotten bulbs recorded at 4 percent, while that of the control was 16 percent. Cha 94 was effective for the first 20 days, but basal rot increased thereafter, and had about the same control efficacy as that of BL-3 and BL-4. When the antagonists were applied to the topping areas of onion bulbs at harvest, TM was the most effective in protecting the stored onion bulbs from neck rotting. The second effective antagonist was BL-3. TM and BL-3 completely suppressed the neck rot in another test, suggesting that the biocontrol of postharvest decay of onion using these microorganisms either at the time of transplanting or at harvesting may be promising.

CONCLUSIONS AND FUTURE PERSPECTIVES

The use of microorganisms for the control of other living organisms requires a thorough knowledge of their ecology. Therefore, except in the case of induced resistance, a biological control agent must occupy an ecological niche similar to that of the plant pathogen, and its mode of action (parasitism, antibiosis, induction of SAR, competition, etc.) must interfere both spatially and temporally with exact steps in the development of the pathogen. Several reviews have been published on the ecological principles in relation to the successful application of biological control of plant pathogens (Funck-Jensen and Lumsden, 1999; Paulitz and Belanger, 2001).

Since various biological control agents develop optimally within a defined spectrum of environmental conditions, field applications of these biocontrol agents can give disappointing results if various environmental conditions are not favorable to the antagonists. However, ecological considerations may be incompatible with market considerations. Therefore, the best marketing strategy is to promote a product against the diseases it best controls consistently for use under the conditions that will guarantee its efficacy (Paulitz and Belanger, 2001).

Given the very hopeful future of the use of microbes for the biological control of various plant pathogens, we can transfer the technology from the lab to the commercial grower and farmers by achieving the following:

1. More scientific efficacy tests with proper replication and statistical analysis are needed under applicable commercial conditions. Easy access to reliable and practical data for farmers and growers will be more persuasive than reliance on private company advertisements. These data should be accessible in the public domain.

2. Further studies are needed on the ecology and epidemiology of pathogens in the fields and in greenhouse. We have to know how the pathogens and the antagonists are introduced and how they are spread and interact. Also, we have to know the relationship between population densities and damages, the methods for the manipulations of environmental conditions to achieve successful biological control. Cost-effective methods for rapid detection of pathogens and for the production of antagonists should be achieved.

3. The challenge of production and formulation of biological control agents should be overcome (Bok et al., 1996; Fravel, 1999). Cost-effective production and formulation protocols of biocontrol agents usually require substantial investment to develop economic production and a formulation with adequate shelf life, stability, and titer. Even if when all these conditions are met, the formulated product may be incompatible with farming practice.

4. Scientists, plant pathologists and related companies involved in the development of microbial pesticides should view the future of biological control of plant diseases with optimism. A few products have already been registered in Korea and several more would be commercialized within the next few years. Success stories against a number of diseases will be important both to validate biological control of plant diseases and more important, to gain acceptance by farmers and people in general. Positive results from various parts involved in the development of microbial pesticides would, in turn, stimulate further research and investments in more biological control products, so that the reduction of chemical fungicides can be a quantifiable reality.

Recent progress in the area of biological control and sequential development procedures for biological control agents allow numerous other areas of research to be followed, opening ways for the development of the most effective and cost-effective biocontrol agents. The concepts of using combinations of microflora and antagonists to enhance activity by dissemination of the biological control agents, and applying combinations of biological control agents to broaden the spectrum of activity against one or several diseases require further studies (Whipps, 1997).

In recent years, commercialization of various microorganisms for the control of many crop diseases has been achieved in Korea. Since the laws for the registration of microbial pesticides were enacted in 2000, five different microbial fungicides were registered as of May 2004 (Table 11.1). Registered microbial fungicides for the control of diseases of various crop plants, such as tomato, strawberry, pumpkin, lawn grass, rice, and cucumber, were made with *Bacillus* sp., *Streptomyces* spp., *Ampelomyces* sp., or *Paenibacillus* sp. All the microbial fungicides were registered in 2003 and 2004. There have been several other candidate microbial fungicides for registration since May 2004.

The concept of combining systemic acquired resistance (SAR) and treatment with a microbial agent appears fairly new (Chen et al., 1996) and is also worth further research. The area of interactions between antagonists, saprotrophs, and soil flora in the rhizosphere and soil has been neglected. However, perhaps the most exciting area for further work involves the application of modern cellular and molecular methods to the biological control of diseases in all its aspects. Also, a considerable effort is currently underway to improve the biological control efficacy of microorganisms through the use of cellular and molecular biology. The following areas of cellular and molecular biology are used for biological control of various plant pathogens:

TABLE 11.1. List of microbial fungicides and their trade names registered in Korea as of May 2004.

Host	Pathogens	Antagonists	Year registered	Trade name
Tomato	*Botrytis cinerea*	*Bacillus subtilis* GB	2004	Green-all G[®]
Strawberry	*Botrytis cinerea*	*Streptomyces colombiensis* WYE20	2003	Mycocide[®]
Pumpkin	*Spharotheca fuliginea*	*Streptomyces colombiensis* WYE20	2004	Mycocide[®]
Turfgrass	*Rhizoctonia solani*	*Streptomyces colombiensis* WYE20	2004	Mycocide[®]
Turfgrass	*Rhizoctonia solani*	*Streptomyces goshikiensis* WYE324	2003	Safegrow[®]
Rice	*Rhizoctonia solani*		2004	Safegrow[®]
Cucumber	*Spharotheca fuliginea*	*Ampelomyces quisqualis* AQ94013	2004	Qpect[®]
Cucumber	*Spharotheca fuliginea*	*Paenibacillus polymixa* AC-1	2003	Topseed[®]

1. *Protoplast fusion.* Protoplast fusion has been used successfully to enhance the efficacy of biological control strains of *Trichoderma* (Harman et al., 1989; Sivan and Harman, 1991). In studies of Harman et al. (1989) and Sivan and Harman (1991), strain 1295-22, which resulted from a fusion between *T. harzianum* strains T12 and T45, grew more rapidly than either parental strain, was a more efficient seed protectant on a range of crops including bean, cotton, and sweetcorn, and was strongly rhizosphere-competent. This strain subsequently went on to be registered in the USA as F-stop, for the control of damping-off disease. However, effective biocontrol strains resulting from protoplast fusion are very rare (Migheli et al., 1995).

2. *Use of molecular biology for the identification of mode of action.* Molecular biology has helped identify the modes of action of many biocontrol agents. This has enabled targeted screening methods to be developed, thus leading to the search for further antagonists acting in the same or better way. Importantly, in the case of *Agrobacterium radiobacter,* used commercially for the control of crown gall of woody plants caused by *Agrobacterium tumefaciens,* the knowledge of the mode of action obtained through molecular biology has enabled the antagonist to be modified to maintain its use when resistance to the original antagonist strain K84 of *A. radiobacter* was observed (Jones et al., 1988; Ryder and Jones, 1990; Stockwell et al., 1996). These studies demonstrated that genetically modified microorganisms could be acceptable for commercial use if the mechanism of action is known and the genetic modifications carried out are clearly understood. If the extra level of legislation associated with the use of genetically modified microorganisms does not prove too much of a problem for registration in general, this work would promote future development of genetically modified disease biological control agents (Whipps, 1997).

3. *Introduction of useful genes into beneficial microbes.* A considerable effort is currently underway to improve the biocontrol efficacy of microorganisms through the use of molecular biology. For example, based on mode of action, a chitinase gene from *Serratis marcescens* has been introduced into several bacteria including *Escherichia coli, Pseudomonas fluorescens, P. putida,* and *Rhizobium meliloti,* as well as into the fungus *Trichoderma harzianum* (Whipps, 1997). Another chitinase gene from *T. harzianum* has also been introduced into *E. coli* (Whipps, 1997). These represent the first steps towards obtaining increased chitinase production in biocontrol strains. Further studies to introduce extra copies of the same or heterologous genes into biocontrol agents, or to obtain their constitutive expression to achieve enhanced biocontrol activity, will undoubtedly increase. Similarly, since gene sequences coding for the production or regulation of antibiotics and siderophores effective against several soilborne pathogens are known in

bacteria (Whipps, 1997), biocontrol strains improved by increased production of antibiotics would also be developed soon.

4. *Use of reporter or marker genes in studies of biocontrol.* The use of reporter or marker genes, such as *lux, lacZ, luc,* or ice nucleation, have also been valuable in studies of biocontrol. For example, when fused to promoters of antibiotic synthesis genes, transcription of these marker genes has enabled antibiotic production to be monitored in situ on the seed coat at levels undetectable by conventional means (Whipps, 1997). If developed further, this may enable biocontrol genes to be expressed by a predetermined environmental trigger such as soil nutrient level, water potential or temperature, or perhaps in response to a specific pathogen in the rhizosphere (Whipps, 1997). In addition, marker genes have been used in the environment to monitor gene transfer between microbes and to track bacteria and fungi (Whipps, 1997), providing ecological data which may assist both targeting and application methods of biocontrol agents and which are valuable in risk-assessment packages for the registration process.

5. *Monitoring of gene transfer.* In another case, marker genes have been used in the environment to monitor gene transfer between microbes and to track bacteria and fungi (Whipps, 1997), providing ecological data which may assist both targeting and application methods of biocontrol agents, and which are valuable in risk assessment packages for the registration process.

6. *Use of transgenic plants.* Finally, assuming the broad definition of biological control of Cook and Baker (1983), some brief reference to the use of transgenic plants for the control of soilborne diseases should be made. Through the use of molecular biology, many of the approaches and concepts parallel those adopted for antagonistic microorganisms. For example, chitinase and β-1,3-glucanase genes from a range of sources have been introduced into plants such as canola and tobacco to give enhanced resistance to pathogen attack (Whipps, 1997). Similarly, genes conferring resistance to toxins such as tabtoxin produced by pathogenic bacteria, lytic peptides such as B and other novel peptides inhibitory to a range of pathogenic bacteria, and oxalate-degrading enzymes providing resistance to *Sclerotinia sclerotiorum,* have all been cloned into plants, and this approach is continuing to develop (Whipps, 1997). If the expression of the genes is not debilitating to the plants, does not result in production of metabolites toxic to non-target organisms, and does not cause resistance development in the pathogen population excessively quickly, the approach has the advantage that no treatments are required once the seeds are produced. It would be interesting for both the scientist and the public to see the use of genetically modified microbial biocontrol agents and plants for disease resistance.

REFERENCES

Ahn, S. J. and Hwang, B. K. 1992. Isolation of antibiotic-producing actinomycetes antagonistic to *Phytopthora capsici* from pepper-growing soils. *Korean Journal of Mycology* 20(3): 259-268.

Alabouvette, C. 1986. *Fusarium* wilt suppressive soils from the Chateaurenard region: Review of a 10 year study. *Agronomie* 6: 273-284.

Alabouvette, C., Lemanceau, P., and Steinberg, C. 1996. Use of non-pathogenic *Fusarium oxysporum* and fluorescent pseudomonads to control *Fusarium* wilts, pp. 155-164. In *Proc. Int. Workshop Biol. Control Plant Dis.*, eds. Wenhua, T., Cook, R. J., and Rovira, A. Hokkaido University, Sapporo, Japan.

Alstrom, S. 1991. Induction of disease resistance in common bean susceptible to halo blight bacterial pathogen after seed bacterization with rhizosphere pseudomonads. *Journal General Applied Microbiology* 37: 495-501.

Alstrom, S. 1995. Induction of disease resistance in common bean susceptible to halo blight bacterial pathogen after seed bacterization with rhizosphere pseudomonads. *Journal General Applied Microbiology* 37: 495-501.

Baker, K. F. 1987. Evolving concepts of biological control of plant pathogens. *Annual Review Phytopathology* 25: 67-85.

Bok, S. H., Son, K. H., Lee, H. W., Choi, D., and Kim, S. U. 1996. Bioencapsulated biopesticides, pp. 303-309. In *Advances in Biological Control of Plant Diseases,* eds. Cook, R. J. and Rovira. A. Agricultural University Press. Beijing, China.

Chang, S. H., Lee, J. Y., Kim, K. D., and Hwang, B. K. 2000. Screening for in vitro antifungal activity of soil bacteria against plant pathogens. *Mycobiology* 28(4): 190-192.

Chang, Y. H., Chang, S. M., Lee, D. H., and Choi, J. 1996. Biological control of *Phytopthora* blight of red-pepper caused by *Phytopthora capsici:* I. Selection of a bacterial antagonist against *Phtyophthora capsici*. *Korean Journal of Environmental Agric*ulture 15(3): 289-295.

Chang, Y. H., Chang, S. M., Lee, D. H., Choi, J., and Lee, D. H. 1996a. Biological control of *Phytopthora* blight of red-pepper caused by *Phytophthora capsici:* II. Isolation and antifungal activity of the substances produced by *Pseudomonas* sp. A-183. *Korean Journal of Environmental Agriculture* 15(4): 399-405.

Chang, Y. H., Chang, S. M., Lee, D. H., Choi, J., and Lee, D. H. 1996b. Biological control of *Phytopthora* blight of red-pepper caused by *Phytophthora capsici:* III. Identification of the antifungal substance produced by *Pseudomonas* sp. A-183. *Korean Journal of Environmental Agriculture* 16(1): 1-6.

Chen, J., Jacobson, L. M., Handelsman, J., and Goodman, R. M. 1996. Compatibility of systemic acquired resistance and microbial biocontrol for suppression of plant disease in a laboratory assay. *Molecular Ecology* 5: 73-80.

Chet, I. 1987. *Innovative Approaches to Plant Disease Control.* John Wiley and Sons, New York.

Cho, C. T., Moon, B. J., and Ha, S. Y. 1989. Biological control of *Fusarium oxysporum* f. sp. *cucumerinum* causing cucumber wilt by *Gliocladium virens* and *Trichoderma harzianum*. *Korean Journal of Plant Pathology* 5(3): 239-249.

Cho, C. T., Son, S. L., and Moon, B. J. 1992. Suppression of cucumber wilt by antagonistic bacterium *Pseudomonas gladioli* and organic amendments. *Korean Journal of Plant Pathology* 8(1): 8-13.

Chung, B. K. and Ryou N. Y. 1996. Effect of a soil amendment for controlling *Fusarium* wilt of cucumber caused by *Fusarium oxysporum* f. sp. *cucumerinum*. *The Korean Journal of Mycology* 24(2): 93-103.

Cook, R. J. and Baker, K. F. 1983. *The Nature and Practice of Biological Control of Plant Pathogens*. APS Press. St. Paul, MN.

Fravel, D. R., Rhodes, D. J., and Larkin, R. P. 1999. Production and commercialization of biocontrol products, pp. 365-376. In *Integrated Pest and Disease Management in Greenhouse Crops*, eds. Albajes, R., Gullino, M. L., van Lenteren, J. C., and Elad, Y. Kluwer, Dordrecht, the Netherlands.

Funck-Jensen, D. and Lumsden, R. D. 1999. Biological control of soilborne pathogens, pp. 319-337. In *Integrated Pest and Disease Management in Greenhouse Crops*, eds. Albajes, R., Gullino, M. L., van Lenteren, J. C., and Elad, Y. Kluwer, Dordrecht, the Netherlands.

Hang, N.T.T., Oh, S.O., Kim, G. H., Hur, J. S., and Koh, Y. J. 2005. *Bacillus subtilis* S1-0210 as a biocontrol agent against *Botrytis cinerea* in strawberries. *Journal of Plant Pathology* 21(1): 59-63.

Harman, G. E., Taylor, A. G., and Stasz, T. E. 1989. Combining effective strains of *Trichoderma harzianum* and solid matrix priming to improve biological seed treatment. *Plant Disease* 73: 631-637.

Hoffland, E., Hakulinen, J., and van Pelt, J. A. 1996. Comparison of systemic resistance induced by avirulent and nonpathogenic *Pseudomonas* species. *Phytopathology* 86: 757-762.

Hornby, D. 1983. Suppressive soils. *Annual Review Phytopathology* 21: 65-85.

Hornby, D. 1990. *Biological Control of Soil-borne Plant Pathogens*. CAB International, Wallingford, Oxon. UK.

Hwang, B. K. and Kim, E. S. 1992. Protection of pepper plants against *Phytophthora* blight by an avirulent isolate of *Phytophthora capsici*. *Korean Journal of Plant Pathology* 8(1): 1-7.

Jee, H. J. and Kim, H. K. 1987. Isolation, identification and antagonistic of rhizospheric antagonists to cucumber wilt pathogen, *Fusarium oxysporum* f. sp. *cucumerinum* Owen. *Korean Journal of Plant Pathology* 3(3): 187-197.

Jee, H. J., Nam, C. G., and Kim, C. H. 1988. Studies on biological control of *Phytophthora* blight of red-pepper: I. Isolation of antagonists and evaluation of antagonists activity in vitro and in greenhouse. *Korean Journal of Plant Pathology* 4(4): 305-312.

Jetiyanon, K. and Kloepperb, J. W. 2002. Mixtures of plant growth-promoting rhizobacteria for induction of systemic resistance against multiple plant diseases. *Biological Control* 24(3): 285-291.

Jeun, Y. C., Park, K. S., and Kim C. H. 2001. Different mechanisms of induced systemic resistance and systemic acquired resistance against *Colletotrichum orbiculare* on the leaves of cucumber plants. *Mycobiology* 29(1): 19-26.

Jones, D. A., Ryder, M. H., Clare, B. G., Farrand, S. K., and Kerr, A. 1988. Construction of a Tra⁻ deletion mutant of pAg K84 to safeguard the biological control of crown gall. *Phytopathology* 73: 15-18.

Kang, S. C., Bark, Y. G., Lee, D. G., and Kim, Y. H. 1996. Antifungal activities of *Metarhizium anisopliae* against *Fusarium oxysporum, Botrytis cinerea,* and *Alternaria solani. The Korean Journal of Mycology* 24(1): 49-55.

Kim, C. H., Jee, H. J., Park, K. S., and Lee, E. J. 1990. Studies on biological control of *Phytophthora* blight of red-pepper: V. Performance of antagonist agents in fields. *Korean Journal of Plant Pathology* 6(2): 201-206.

Kim, C. H., Kim, K. D., and Jee, H. J. 1991. Enhanced suppression of red-pepper *Phytophthora* blight by combined applications of antagonists and fungicide. *Korean Journal of Plant Pathology* 7(4): 221-225.

Kim, H. K. and Jee, H. J. 1988. Influence of rhizosphere antagonists on suppression of cucumber wilt, increased cucumber growth and density fluctuation of *Fusarium oxysporum* f. sp. *cucumerinum* Owen. *Korean Journal of Plant Pathology* 4(1): 10-18.

Kim, K. C., Kim, D. Y., and Do, D. H. 1994. Biological control of *Pseudomonas* sp. for *Erwinia rhapontici* causing vegetables root rot. *J. Korean Soc. Food Nutr.* 23(1): 104-109.

Kim, K. C., Kim, H. S., Do, D. H., and Cho, C. M. 1992. Biological control of plant pathogen by *Pseudomonas* sp. *Korean Journal of Applied Microbial Biotechnology* 20(3): 263-270.

Kim, K. C., Yuk, C. S., and Do, D. H. 1990. Molecular cloning of bacteriocin gene and biological control of plant pathogen. *Korean Journal of Applied Microbial Biotechnology* 18(1): 98-102.

Kim, K. D., Nemec, S., and Musson, G. 2000. Effects of composts and soil amendments on physicochemical properties of soil in relation to *Phytophthora* root and crown rot of bell pepper. *Journal of Plant Pathology* 16(5): 283-285.

Kim, S. D., Yun, G. H., and Lee, E. T. 2001. Identification and antifungal antagonism of *Chryseomonas luteola* 5042 against *Phytophthora capsici. Korean Journal of Applied Microbial Biotechnology* 29(3): 178-186.

Kloepper, J. W., Leong, J., Meintz, T., and Schroth, M. 1980. *Pseudomonas* siderophores: A mechanism explaining disease suppressive soils. *Current Microbiology* 43: 317-320.

Kloepper, J. W., and Schroth, M. N. 1978. Plant growth promoting rhizobacteria on radishes, pp. 879-882. *Proceedings of the Fourth International Conferences on Plant Pathogenic Bacteria* (Vol. 2). Angers, France.

Larena, I., Sabuquillo, P., Melgarejo, P., and De Cal, A. 2003. Biocontrol of *Fusarium* and *Verticillium* wilt of tomato by *Penicillium oxalicum* under greenhouse and field conditions. *Journal of Phytopathology* 151(9): 507-512.

Lee, E. J., Jee, H. J., Park, K. S., and Kim, C. H. 1990. Studies on biological control of *Phytophthora* blight of red-pepper: V. Performance of antagonist agents in field under polyethylene filmhouse. *Korean Journal of Plant Pathology* 6(1): 58-64.

Lee, E. T. and Kim, S. D. 2000. Selection and antifungal activity of antagonistic bacterium *Pseudomonas* sp. 2112 against red-pepper rotting *Phytophthora capsici. Korean Journal Applied Microbial Biotechnology* 28(6): 334-340.

Lee, I.-K., Lim, C. -J., Kim, S. -D., and Yoo, I. -D. 1990. Metabolism and physiology: Antifungal antibiotic against fruit rot disease of red pepper from *Streptomyces parvullus*. *Korean Journal of Applied Microbial Biotechnology* 18(2): 142-148.

Lee, J. M., Do, E. S., Baik S. B., and Chun S. C. 2003. Effect of organic amendments on efficacy of biological control of seedling damping-off of cucumber with several microbial products.

Lee, J. T., Park S. H., Shin, C. K., Kwak Y. S., and Kim, H. K. 2001. Occurrence and biological control of postharvest decay in onion caused by fungi. *Journal of Plant Pathology* 17(3): 141-148.

Lee, M. W. 1997. Root colonization by beneficial *Pseudomonas* spp. and bioassay of suppression of *Fusarium* wilt of radish. *The Korean Journal of Mycology* 25(1): 10-19.

Lee, S. Y. and Kim, H. G. 2001. Parasitic characteristics of *Ampelomyces quisqualis* 94013 to powdery mildew fungus of cucumber. *The Korean Journal of Mycology* 29(2): 116-122.

Lee, W. H. and Kobayashi, K. 1988. Isolation and identification of antifungal *Pseudomonas* sp. from sugar beet roots and its antibiotic products. *Korean Journal of Plant Pathology* 4(4): 264-270.

Lee, W. H. and Ogoshi, A. 1991. The effect of seed bacterization of sugar beets on growth and suppression of damping-off in greenhouse and field. *Korean Journal of Plant Pathology* 7(2): 88-93.

Lee, Y. H., Lee, W. H., Shim H. K. and Lee, D. K. 2000a. Factors relating to induced systemic resistance in watermelon by plant growth-promoting *Pseudomonas* spp. *Journal of Plant Pathology* 17(3): 174-179.

Lee, Y. H., Lee, W. H., Shim H. K. and Lee, D. K. 2000b. Induction of systemic resistance in watermelon to gummy stem rot by plant growth-promoting rhizobacteria. *Journal of Plant Pathology* 16(6): 312-317.

Lee, Y. S. 1994. Biological management of virulent *Fusarium* species on asparagus with avirulent *Fusarium* species in vitro. *Korean Journal of Environmental Agriculture* 13(3): 288-300.

Lee, Y. S. 1996. Interactions of virulent and avirulent *Fusarium* species on clonal asparagus plantlet and mechanism involved in protection of asparagus and avirulent *Fusarium* species against stem and crown rots. *Korean Journal of Plant Pathology* 12(1): 47-57.

Lee, Y. S. 2000. Selection of plant growth promoting *Pseudomonas* species for several vegetable crops (Abstract). Fall Meeting of Korean Society of Plant Pathology, p. 46.

Lee, Y. S. and Jeong, C. S. 2000. Effects of bacterial antagonists for the control of fungal storage diseases and the extension of storage periods for hot pepper (*Capsicum annum* L.) (Abstract). Fall Meeting of Korean Society of Plant Pathology. p. 35.

Lee, Y. S., Kim, C. J., and Choi, J. K. 1996. Selection of antagonistic microbes against Chinese cabbage brittle root rot pathogen *Aphanomyces raphani* (Abstract). *Plant Disease and Agriculture* 2(1): 135.

Lee, Y. S., Kim, E. H., and Kim, M. J. 1977. Selection of antagonists of *A. raphani* and plant growth promoting *Pseudomonas* spp. on Chinese cabbage (Abstract). Proceedings for the 3rd Korean Allelopathy Research Association. p. 12.

Lee, Y. S. and Kim, J. J. 1996. Studies on the biological control agents against the brittle root rot pathogen *(Aphanomyces raphani)* of Chinese cabbage in the alpine areas. *Daesan Nonchong* 4: 93-99.

Lee, Y. S., Kim, K. S., Kim, H. J., Kim, J. W., Chung, S. K., and Lee, S. J. 1999. Effects of charcoal and charcoal wood extracts on the population change of bacteria and hot pepper plant growth, and root and fruit development. *Journal of Agricultural Science* 10: 78-81.

Lee, Y. S. and Lee, S. J. 2000. Effects of charcoal and charcoal wood extracts on the population change of plant growth promoting bacteria, and the hot pepper (*Capsicum annum* L.) plant growth, and root and fruit development (Abstract). Fall Meeting of Korean Society of Plant Pathology, p. 45.

Lee, Y. S. and Manning, W. J. 1991b. Reduction of root and crown rot of tissue-cultured asparagus plantlets, caused by *Fusarium moniliforme,* by prior inoculation with an avirulant isolate of *F. oxyaporum,* in vitro. (Abstract). *Phytopathology* 81(10): 1164.

Lee, Y. S. and Manning, W. J. 1991a. Susceptibility of tissue-cultured asparagus plantlets to Fusaria in vitro. (Abstract). *Phytopathology* 81(10): 1216.

Lee, Y. S. and Manning, W. J. 1992. Biological management of asparagus root and stem rot disease with avirulent isolates of *Fusarium.* (Abstract). Program and Abstracts of the Annual Meeting of the Korean Society of Environmental Agriculture, p. 12.

Leeman, M. 1995. Suppression of *Fusarium* wilt of radish by fluorescent spp., induction of disease resistance, co-inoculation with fungi and commercial application. PhD thesis. Utrecht University, Utrecht, the Netherlands.

Lewis, J. A., Fravel, D. R., Lumsden, R. D., and Shasha, B. S. 1995. Application of biocontrol fungi in granular formulations of pregelatinized starch-flour to control damping-off diseases caused by *Rhizoctonia solani. Biological Control* 5(3): 397-404.

Lewis, J. A., Fravel, D. R., and Papavizas, G. C. 1995. *Cladorrhinum foecundissimum*: a potential biological control agent for the reduction of *Rhizoctonia solani. Soil Biology and Biochemistry* 27(7): 863-869.

Lewis, J. A. and Larkin, R. P. 1998. Formulation of the biocontrol fungus *Cladorrhinum foecundissimum* to reduce damping-off diseases caused by *Rhizoctonia solani* and *Pythium ultimum. Biological Control* 12: 182-190.

Maurhofer, M., Hase, C., Meuwly, P., Metraux, J. P., and Defago, G. 1994. Induction of systemic resistance of tobacco to tobacco necrosis virus by the root-colonizing *Pseudomonas fluorescens* strain CHA0: Influence of the gacA gene and of pyoverdine production. *Phytopathology* 84: 139-146.

Migheli, Q., Whipps, J. M., Budge, S. P., and Lynch, J. M. 1995. Production of inter- and intra-strain hybrids of *Trichoderma* spp. by protoplast fusion and evaluation of their biocontrol activity against soil-borne and foliar pathogens. *Journal of Phytopathology* 143: 91-97.

Moon, B. J., Chung, H. S., and Cho, C. T. 1988. Studies on antagonism of *Tricho-derma* species to *Fusarium oxysporum* f. sp. *fragariae:* I. Isolation, identification and antagonistic properties of *Trichoderma* species. *Korean Journal of Plant Pathology* 4(2): 111-123.

Moon, B. J., Roh. S. H., and Cho, C. T. 1990. Biological control of *Fusarium* wilt of strawberry by antagonistic bacterium, *Pseudomonas gladioli,* in greenhouse. *Korean Journal of Plant Pathology* 6(4): 461-466.

Murkerji, K. G., and Garg, K. L. 1988. *Biocontrol of Plant Diseases* (Vol. 1, 2). CRC Press. Boca Raton, FL.

Paik, S. B. 1989. Screening for antagonistic plants for control of *Phytophthora* spp. in soil. *Korean Journal of Mycology* 17(1): 39-47.

Paik, S. B. Control of cucumber powdery mildew with anthraquinone derivatives. *Life Science Journal of Kunkuk University* 2: 63-65.

Paik, S. B. and Kim, E. -W. 1995. Screening for phyllospheral antagonistic microor-ganisms for control of red-pepper anthracnose *(Colletotrichum gloeosporioides)*. *Korean Journal of Mycology* 23(2): 190-195.

Paik, S. B., Kyung, S. H., Doh, E. S., Oh, Y. S., and Park, B. K. 1994. Screening and identification of fungicidal compounds derived from medicinal plants against cucumber powdery mildew. *Korean Journal of Environmental Agriculture* 13(3): 301-310.

Paik, S. B., Kyung. S. H., Kim, J. J. and Oh, Y. S. 1996. Effect of a bioactive sub-stance extracted from *Rheum undulatum* on control of cucumber powdery mil-dew. *Korean Journal of Plant Pathology* 12(1): 85-90.

Paik, S. B. and Oh, Y. S. 1990. Screening for antifungal medicinal plants controlling the soil borne pathogen, *Pythium ultimum. The Korean Journal of Mycology* 18(2): 102-108.

Park, C. S., Paulitz T. C., and Baker R. 1988. Attributes associated with increased biocontrol activity of fluorescent Pseudomonas. *Korean Journal of Plant Pathol-ogy* 4(3): 218-225.

Park, J. H. and Kim, H. K. 1989. Biological control of *Phytophthora* crown and root rot of greenhouse pepper with *Trichoderma hazianum* and *Enterobacter agglo-merans* by improved method of application. *Korean Journal of Plant Pathology* 5(1): 1-12.

Park, K. S., Ahn. I. P., and Kim, C. H. 2001. Systemic resistance and expression of the pathogenesis-related genes mediated by the plant growth-promoting rhizo-bacterium *Bacillus amyloliquefaciens* EXTN-1 against anthracnose disease in cucumber, *Colletotrichum orbiculare. Mycobiology* 29(1): 48-53.

Park, K. S., Jang, S. W., Kim, C. H., and Lee E. J. 1989. Studies on biological control of *Phytophthora* blight of red-pepper: III. Formulations of *Trichoderma harzia-num* and *Pseudomonas cepacia* antagonistic to *Phytopthora cepacia* and their preservation. *Korean Journal of Plant Pathology* 5(2): 131-138.

Paulitz, T. C. and Belanger, R. R. 2001. Biological control in greenhouse systems. *Annual Review of Phytopathology* 39. pp. 103-133.

Paulitz, T. C., Park, C. S., and Baker R. 1987. Biological control of *Fusarium* wilt of cucumber with non-pathogenic isolates of *Fusarium oxysporum. Canadian Journal of Microbiology* 33: 349-353.

Powell, K. A. and Jutsum, A. R. 1993. Technical commercial aspects of biocontrol products. *Pesticide Science* 37: 315-321.

Raupach, G. S., Liu, L., Murphy, J. F., Tuzun, S., and Kloepper, J. W. 1996. Induced systemic resistance in cucumber and tomato against cucumber mosaic cucumoviurs using plant growth-promoting rhizobacteria (PGPR). *Plant Disease* 80: 891-894.

Ryder, M. H. and Jones, D. A. 1990. Biological control of crown gall. In *Biological Control of Soil-borne Plant Pathogens.* pp. 45-63. ed. Hornby, D. CAB International, Wallingford, UK.

Scher, F. M. and Baker, R. 1980. Mechanism of biological control in a *Fusarium*-suppressive soil. *Phytopathology* 70: 412-417.

Schippers, B. 1992. Prospects for managements of natural suppressiveness to control soil-borne pathogens. pp. 21-34. In *Biological Control of Plant Diseases, Progress and Challenges for the Future.* eds. Tjamos, E. C., Papavizas, G. C., and Cook, R. J. NATO ASI series A: life sciences. Vol. 230. Plenum Press, New York and London.

Shen, S. -S., Choi, O. -H., Lee, S. -M., and Park, C. -S. 2002. In vitro and in vivo activities of a biocontrol agent, *Serratia plymuthica* A21-4, against *Phytophthora capsici. Journal of Plant Pathology* 18(4): 221-224.

Shen, S. -S., Choi, O. -H., Park, S. -H., Kim, C. -G., and Park C. -S. 2005. Root colonizing and biocontrol competency of *Serratia plymuthica* A21-4 against *Phytophthora* blight of pepper. *Journal of Plant Pathology* 21(1): 64-67.

Shen, S. -S., Park, S. -H., and Park, C. -S. 2005. Enhancement of biocontrol efficacy of *Serratia plymuthica* A21-4 against Phytophthora blight of pepper by improvement of inoculation buffer solution. *Journal of Plant Pathology* 21(1): 68-72.

Shin, D. B., Kobayashi, N. and Lee, J. T. 1994. Biological control of strawberry bud caused by *Rhizoctonia solani* AG2-1 with antagonistic microorganism. *Korean Journal of Plant Pathology* 10(2): 112-118.

Shin, H. S., Kim, K. S., Woo, S. J., Kim, H. J., Jeong, C. S., and Lee, Y. S. 1999. Selection of storage fungal pathogen of red pepper and their antagonistic bacteria. (Abstract). *Korean Journal of Horticulture Science and Technology* 17(2): 225.

Singh, P. P., Benbi, D. K., and Chung, Y. R. 2003. Use of quantitative models to describe the efficacy of inundative biological control of *Fusarium* wilt of cucumber. *Journal of Plant Pathology* 19(3): 129-132.

Sivan, A. and Harman, G. E. 1991. Improved rhizopshere competence in a protoplast fusion progeny of *Trichoderma harzianum. Journal of General Microbiology* 137: 23-29.

Stanghellini, M. E., K. D. H., Rasmussen, S. L., and Rorabaugh, P. A. 1996. Control of roor rot of peppers caused by *Phytophthora capsici* with a nonionic surfactant. *Plant Disease* 80: 1113-1116.

Stockwell, V. O., Kawalek, M. D., Moore, L. W., and Loper, J. E. 1996. Transfer of pAgK84 from the biocontrol agent *Agrobacterium radiobacter* K84 to *A. tumefaciens* under field conditions. *Phytopathology* 86: 31-37.

Suslow, T. W. and Schroth, M. N. 1982. *Rhizoctonia* of sugar beets: effects of seed germination and root colonization on yield. *Phytopathology* 72: 199-206.

Van Peer. R.. Niemann. G. J., and Schippers, B. 1991. Induced resistance and phytoalexin accumulation in biological control of *Fusarium* wilt of carnation by *Pseudomonas* sp. strain WCS417r. *Phytopathology* 81: 728-734.

Van Peer, R. and Schippers. B. 1992. Lipopolysaccharides of plant grwoth-promoting *Pseudomonas* sp. strain WCS417r induce resistance in carnation to *Fusarium* wilt. *Netherlands Journal of Plant Pathology* 98: 129-139.

Van Wee, S. C. M., Pieterse, C. M. J., Trijssenaar, A, vant Westende, Y. A. M., Hartig, F., and van Loon, L. C. 1997. Differential induction of systemic resistance in Arabidopsis by biocontrol bacteria. *Molecular Plant-Microbe Interactions.* 10: 716-724.

Wei. G., Kloepper, J. W., and Tuzun, S. 1996. Induced systemic resistance to cucumber diseases and increased plant growth by plant growth-promoting rhizobacteria under field conditions. *Phytopathology* 86: 221-224.

Weller, D. M. 1988. Biological control of soil borne plant pathogens in the rhizosphere with bacteria. *Annual Review of Phytopathology* 26: 379-407.

Whipps, J. M. 1997. Developments in the biological control of soil-borne plant pathogens. pp. 1-134. In *Advances in Botanical Research—Incorporting Advances in Plant Pathology.* Vol. 26. ed. Callow, J. A. Academic Press.

Yang, S. S. and Kim, C. H. 1994. Studies on cross protection of *Fusarium* wilt of cucumber: III. Selection of nonpathogenic isolates and their protective effect in the greenhouse. *Korean Journal of Plant Pathology* 10(1): 23-33.

Yoon. S-H., Lee, Y., Kim, Y-H.. Koo, B-S., Lim, J-H., Hwang, Y-S., and Choi, C. 1994. Purification of antifungal substance from *Pseudomonas maltophilia* sp. B14 inhibiting the growth of *Rhizoctonia solani.* RDA. *Journal of Agriculture Science* 36(1): 206-211.

Zhou, T. and Paulitz, T. C. 1994. Induced resistance in the biological control of *Pythium aphanidermatum* by *Pseudomonas* spp. on cucumber. *Journal of Phytopathology* 142: 51-63.

Zhou, T., Rankin, L.. and Paulitz, T. C. 1992. Induced resistance in the biological control of *Pythium aphanidermatum* by *Pseudomonas* spp. on European cucumber. *Phytopathology* 82: 1080.

Chapter 12

The Nature of Fungal Mycoparasitic Biocontrol Agents

C. Cortes-Penagos
V. Olmedo-Monfil
A. Herrera-Estrella

INTRODUCTION

Modern agriculture depends primarily on a small fraction of the many thousands of plant species grown globally. As plants have been redistributed from their centers of origin, pests have followed, resulting in devastating incidents of diseases such as potato late blight *(Phytophtora infestans)* (Fry et al., 1993). Plant diseases, caused by various pathogens such as viruses, bacteria, and fungi can lead to severe yield losses in agricultural crops. Recent worldwide estimates, however, show that regardless of the control methods available, a high percentage of the potential yield is still lost through diseases, pests, and weeds, especially in developing countries (Oerke, 1994). The risk of global spread of disease is increased by the reduced genetic diversity of modern crops, compared with that of the related wild species. Extreme examples are coffee (Monaco, 1977) and banana (Stover and Simmonds, 1987), of which single clones, propagated throughout the tropics, are susceptible to leaf rust caused by the fungus *Hemileia vastatrix* and black leaf streak caused by the fungus *Mycosphaerella fijiensis,* respectively. Long-distance dispersal (LDD) in the air of spores from phytopathogenic fungi has also contributed to the spread of several important diseases on a continental or global scale and allows the regular establishment of diseases in new areas (Brown and Hovmoller, 2002). Most of these involve rusts, which may be more likely than other diseases to be

Biological Control of Plant Diseases
© 2007 by The Haworth Press, Inc. All rights reserved.
doi:10.1300/5682_12

aerially dispersed across, and even between, continents because their spores are most able to defeat environmental challenges (Rotem et al., 1985).

Many plant pathogens have overcome pesticides and agricultural practices that once held them under control. At the same time, some effective chemicals, such as the fungicide methyl bromide, are being prohibited because of health and environmental concerns. During the past few decades, it has been established that disease control should not depend totally on chemical control, but that other resources available should be utilized more efficiently. Consistently, plant researchers are developing a series of new strategies of pest control, from high-tech genetic engineering to techniques that induce a plant to turn up its own defense mechanisms to relatively low-tech strategies designed to disrupt plant-pathogen interactions, including biological control.

Biological control has been used as a management tool for the control of crop and forest pests, and for the restoration of natural systems affected by emerging pests. The development of the concept of biological control occurred with the gradual accumulation of our biological and ecological knowledge of nature. Biological control based on fungal species can reduce the amount of inoculum or disease-producing activity of a plant pathogen, usually another fungus. This chapter compiles the knowledge accumulated in this area, presenting the most recent advances on the mechanisms involved in the establishment of parasitic interfungal associations, whether distance-mediated processes (antibiotic production) or contact-dependent mechanisms (mycoparasitism). The chapter also includes an overview of the developments achieved in their use as biocontrol agents and their future possibilities. The mycoparasitic associations are described from early events such as host recognition to late processes, including host death and the use of its cellular components as a source of nutrients. Most mycoparasites are considered in this review, with emphasis on those with greatest impact as biocontrol agents.

STRATEGIES OF PLANT PROTECTION

Protection of crops from diseases can ultimately improve agricultural production. The first line of defense against plant pathogens is natural resistance, which can often be transferred between species by crossbreeding. Resistant cultivars were almost the only strategy to avoid disease losses in the early days of plant protection. But breeding for resistance is time-consuming, and microbial pathogens quickly evolve to overcome the new challenge. In the past decade, the traditional methods of plant breeding were improved by the use of molecular markers and the incorporation of genetic engineering. The latter generated genetically modified (GM) crops, a new tool for plant protection that allows staying a step ahead of rapidly evolving pests. The most prominent development involved the transfer of a gene encoding an insect-killing protein (Bt) from the bacteria *Bacillus thuringiensis* into crop plants, including tomato, tobacco, and cotton (Fischhoff et al., 1987; Vaeck et al., 1987; Perlak et al., 1993). Today, Bt-protected corn, cotton, and potato have been commercialized in the United States, and one or more of these products are marketed in Argentina, Australia, Canada, France, Mexico, and South Africa (James, 1998, 1999), resulting in reduced application of chemical insecticides for some crops.

Despite the early success of GM crops, this technology still has some practical problems to solve. For example, one source of undesirable variability in transgenic plants is the random location of the transgene. Also, unspecified mutations generated during tissue culture, which is needed to grow transformed cells into whole plants, can induce unpredicted traits or loss of desirable qualities. Furthermore, the anti-GM movement, including ecologists and other scientists, has long expressed concerns about the potential impact of releasing GE plants into the environment. They have based their opposition mostly on the transplanting of genes from completely different organisms into crop plants. However, the complexity of ecological systems interferes with the prediction of risk, benefits, and inevitable uncertainties of GE plants. A comprehensive review of the potential impacts of releasing genetically engineered organisms into the environment is not given here, as several excellent and recent papers are available (Betz et al., 2000; Wolfenberger and Phifer, 2000). Currently, an alternative strategy to crop protection is the stimulation of plants' natural defense mechanisms, either through the use of chemicals or the insertion of genes from plant pathogens (Gozzo, 2003).

BIOLOGICAL CONTROL:
AN ECOLOGICAL APPROACH

The management of the environment to give plants a competitive advantage over pathogens represents one of the best strategies of plant protection, besides improving plants' own defenses. This may involve either the use of biological control organisms or their products that can eliminate potential pathogens, or the manipulation of some physical aspects of plant growth, such as light, temperature, or the seasons of planting and harvesting.

Biological control (or biocontrol) is the deliberate use of natural predators, antagonists, or competitor organisms to suppress a pest population, thereby making the pest less abundant and damaging than it would be in the absence of these organisms. Biocontrol of plant diseases can be achieved using a broad spectrum of organisms, from insects to fungi and bacteria. Two types of biological control are recognized: the classical and the inundative. Classical biological control can be defined as the importation of natural enemies for the control of exotic, invasive species. On the other hand, inundative control (or mass release) involves massive application of the biocontrol agent that can be marketed and employed in much the same way as a conventional chemical pesticide. This later strategy is used when natural enemies are expected to be insufficient to suppress pest population growth. The conceptual model underlying biocontrol was derived from the predator-prey theory (Smith and van den Bosch, 1967; van Driesche and Bellows, 1996) and is based on the notion that exotic species become invasive by evading the controlling influence of their natural enemies (Keane and Crawley, 2002). In this model, the introduction of any biocontrol agent results in a self-sustaining, balanced system in which the pest population is maintained below some economically or ecologically defined threshold, as its fitness or competitive edge is weakened (Evans and Ellison, 1990).

During the past two decades, there has been a tremendous increase in interest and research on biocontrol, driven by a search for more environmentally benign methods of plant protection. The use of biological control is not a new idea. Its conception was established between the late nineteenth and early twentieth centuries, as biological and ecological knowledge of the natural world was gradually accumulated. In successful biocontrol programs, the reduction of pest densities and recovery of adversely affected flora or fauna leads toward a system with better ecological balance and community structure. Following its establishment, successful natural enemies can provide enduring pest control; and they can replicate and disperse without continued human management. Cases of successful biological control programs against noxious insects, mites, weeds (aquatic and terrestrial), plant pathogens, and vertebrates are extensive, and well-documented examples exist (Mukerji and Garg, 1988; Symondson et al., 2002; Blossey and Hunt-Joshi, 2003; Werner and Pont, 2003).

BIOLOGICAL CONTROL BASED ON FUNGAL SPECIES

The ability of antagonistic fungi to inhibit the development of plant pathogens has been exploited in biocontrol of agricultural pests, including the most problematic groups: insects and fungi. Although there are other biocontrol agents, such as bacteria and viruses, there are several clear advantages in utilizing fungi: (1) most fungal species have the ability to directly infect the host by penetrating its outer surface, whereas virus and bacteria need to be ingested (in case of insect control) in order to be effective; and (2) the genetic diversity of fungi allows a number of biocontrol agents to be made from single species of fungi.

Some insect species, including many pests, are particularly susceptible to infection by naturally occurring, insect-pathogenic fungi. Many of the genera of the so-called entomopathogenic fungi that have been studied either belong to the Entomophthorales (Zygomycota) or to the Hyphomycetes (Deuteromycota). Recent reviews by Butt et al. (2001) and Shah and Pell (2003) provide further information about the principles and strategies of insect biological control mediated by fungi.

Filamentous fungal pathogens cause diseases on all agricultural crops around the world, resulting in millions of tons of crop losses and billions of dollars in lost revenue annually. Despite the many achievements of disease control, certain cultural practices have actually enhanced the destructive potential of fungal pathogens (Table 12.1). The majority of the fungal

TABLE 12.1. Emerging fungal plant diseases.

Disease	Pathogen	Host	Geographic distribution	Management
Blast	*Magnaporthe grisea*	Rice	Asia	Cultural, chemical
Downy mildew	*Peronosclerospora sorghi*	Sorgum, sweet corn	S.E. Asia, Africa, Australia	Natural resistance, chemical
Early blight	*Alternaria solani*	Potato, tomato	Worldwide	Chemical
Gray mold	*Botrytis cinerea*	Greenhouse crops	Worldwide	Cultural, chemical
Karnal bunt	*Tilletia indica*	Wheat	N. America, India	Cultural, chemical
Late blight	*Phytophthora infestans*	Potato, tomato	Worldwide	Natural resistance, chemical
Rust	*Puccinia melanocephala*	Sugarcane	Americas	Natural resistance, chemical
Rust	*Phakopsora pachyrhizi Phakopsora meibomiae*	Soybean	S.E. Asia, Russia, S. America	Chemical
Wilt	*Fusarium oxysporum*	Tomato	Worldwide	Cultural, chemical, natural resistance

species responsible of rusts, smuts, mildews, wilts, and other crop blights are susceptible to the antagonism of other fungi. The lifestyle of these later species is quite variable, ranging from species that are obligate parasites to those that attack only weakened hosts, to some that are apparently commensal to symbiotic fungi. Most of the fungal species that have been selected as biocontrol agents against phytopathogens establish parasitic relationships.

Biological control through fungal species is a potent means of reducing the damage caused by fungal plant pathogens. Commercialized systems for the biological control of plant diseases are few, although intensive activity is currently being geared toward the introduction of an increasing number of fungal biocontrol agents into the market (Table 12.2).

FUNGAL ANTAGONISM IN THE BIOLOGICAL CONTROL OF PHYTOPATHOGENIC FUNGI

The survival strategies of fungi in an ecological niche have allowed their use in the improvement of biocontrol schemes. It is difficult to determine the precise mechanisms working in an interfungal interaction in the field, due to the fact that most of studies concerning these antagonistic relationships have been carried out in laboratory or greenhouse, where the conditions are controlled in order to have a clear idea of how a particular interaction works. With the goal of analyzing and understanding the biocontrol systems, we, like many other authors, separate the elements affecting the growth and development of fungal plant pathogens that lead to effective disease control. Figure 12.1 summarizes the mechanisms displayed by biological control agents and relates them to the different stages of the fungal plant pathogens susceptible to being controlled during disease progress.

Indirect Mechanisms

Induced Resistance in Host Plants

Plants are not passive elements during attack by phytopatogens. They have their own defense mechanisms. These include the so-called hypersensitive response (HR), which causes cells to die in the immediate vicinity of infection sites, thereby preventing further pathogen spread, and cell wall accumulation of compounds (callose and phenolics) at infected sites, providing a physical barrier to pathogen entry. Moreover, the production of specific chemicals by the plant and increases in enzymatic activities (chitinases, peroxidases, commonly known as pathogenesis-related proteins) block pathogen growth and disease development. These defense mecha-

TABLE 12.2. Some commercial biocontrol products for use against soilborne crop diseases.

Biocontrol fungus	Trade name and manufacturer	Target pathogen (disease)	Crop
Ampelomyces quisqualis M-10	AQ10 Biofungicide, Ecogen Inc., Langhorne, Pennsylvania	Powdery mildew	Cucurbits, grapes, ornamentals, strawberries, tomatoes
Candida oleophila I-182	Aspire, Ecogen Inc., Langhorne, Pennsylvania	*Botrytis, Penicillium*	Citrus, pome fruit
Fusarium oxysporum (nonpathogenic)	Biofox C, S.I.A.P.A., Galliera, Bologna, Italy	*Fusarium oxysporum*	Basil, carnation, cyclamen, tomato
Trichoderma harzianum and T. polysporum	Binab T, Bio-innovation, Algaras, Sweden	Wilt and root rot pathogens, wood decay pathogens	Fruit, flowers, ornamentals, turf, vegetables
Coniothyrium minitans	Contans, Prophyta, Biologischer Pflanzenschutz, Malchow/Poel, Germany	*Sclerotinia sclerotiorum and S. minor*	Canola, sunflower, peanut, soybean, lettuce, bean, tomato
Coniothyrium minitans	Coniothyrin (Compañia rusa), Contans (Encore Technologies), KONI (Bioved)	*Sclerotinia sclerotiorum*	Cucumber, lettuce, capsicum, tomato and ornamental flowers in greenhouse
Fusarium oxysporum (nonpathogenic)	Fusaclean, Natural Plant Protection, Nogueres, France	*Fusarium oxysporum*	Basil, carnation, cyclamen, gerbera, tomato

(continued)

333

TABLE 12.2 *(continued)*

Biocontrol fungus	Trade name and manufacturer	Target pathogen (disease)	Crop
Streptomyces griseoviridis K61	Mycostop, Ag-Bio Development, Inc., Westminster, CO, and Kemira Agro., Finland	Root, stem and leaf pathogens *Pythium*, *Fusarium*	Greenhouse crops
Pythium oligandrum	Polygandron, Plant Protection Institute, Bratislavsk, S.R.	*Pythium ultimum*	Sugar beet
Trichoderma harzianum and *T. viride*	Promote, JH Biotech, Ventura, California	*Pythium, Rhizoctonia, Fusarium*	Greenhouse, nursery transplants, seedlings
Gliocadium catenulatum	Primastop, Kemira Agro, Finland	*Pythium, Rhizoctonia solani, Botrytis* and *Didymella*	Ornamental, vegetable and tree crops
Trichoderma harzianum	Rootshield, Bio-Trek T-22G, Planter Box Bioworks, Geneva, New York	*Pythium, Rhizoctonia, Fusarium, Sclerotinia homeocarpa*	Trees, shrubs, transplants, ornamentals, cabbage, tomato, cucumber, bean, corn, cotton, potato, soybean, turf
Talaromyces flavus V117b	Protus WG (Prophyta)	*Verticillium dahliae, V. alboatrum, Rhizoctonia*	Tomato, cucumber, strawberry, rape oilseed
Phlebia gigantea	Rotstop, Kemira Agro Oy, Helsinki, Finland	*Heterobasidium annosum*	Trees

Gliocladium virens GL-21	SoilGard (formerly GlioGard), Thermo Triology, Columbia, US	Damping-off and root pathogens, *Pythium*, *Rhizoctonia*	Ornamentals and food crops grown in greenhouses, nurseries, homes, interiorscapes
Trichoderma harzianum	Trichodex Makhteshim, Chemical Works, Beer Sheva, Israel	*Botrytis cinerea*, *Colletotrichum*, *Monilinia laxa*, *Plasmopara viticola*, *Rhizopus stolonifer*, *Sclerotinia sclerotiorum*	Cucumber, grape, nectarine, soybean, strawberry, sunflower, tomato
Trichoderma harzianum and *T. viride*	Trichopel*, Trichoject* Trieco *Agrimm Technologies, Christchurch, New Zealand	*Armillaria*, *Botryosphaeria*, *Fusarium*, *Nectria*, *Phytophthora*, *Pythium*, *Rhizoctonia*	Mustar, masoor, oilseeds, soybean, cotton, chilies, vegetables, tobacco, cardamon, coffe, tomato, sugar cane, citrus, and others
T. harziarum KRL-AG2 strain	T-22G, T-22HB (Bioworks)	*Phytium*, *Rhizoctonia*, *Fusarium*, *Sclerotinia*	Trees, shrubs, ornamentals, cabbage, tomato, cucumber

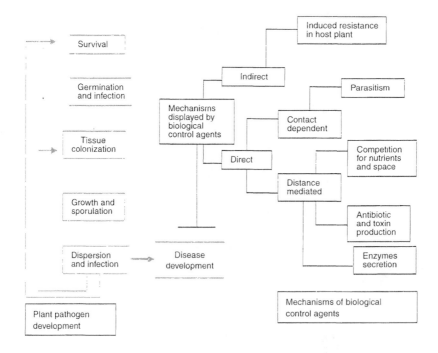

FIGURE 12.1. Main events controlling plant pathogen development leading to plant disease and mechanisms employed by biocontrol agents to antagonize the pathogen. *Source:* Adapted from Punja and Utkhede, 2003.

nisms are activated in local infection sites and in the whole plant, and all together comprise the plant systemic-acquired response (SAR), which is an unspecific defense system. SAR is triggered through elicitor molecules or inducing agents, liberated during infection (Dangl and Jones, 2001). The presence of fungi, such as *Trichoderma* spp., *Pythium oligandrum*, and nonpathogenic *Fusarium oxysposum*, turn on and keep in alert the plant defense system (Fuchs et al., 1997; Yedidia et al., 1999, 2000; Benhamou et al., 2002). The liberation of small molecules during normal fungal growth could be a key element in plant defense activation. Recently, Hanson and Howell (2004) analyzed *Trichoderma virens* culture filtrates and found the presence of protein molecules, between 18 and 30 kDa, with the capacity to act as elicitors of plant defense responses in cotton roots. Moreover, a biocontrol-ineffective *T. virens* strain that does not produce those molecules was not able to activate plant responses. On the other hand, it is possible that when nonpathogenic fungi internally colonize root tissues without severe damage, they activate SAR and help prevent plant disease.

Direct Mechanisms

Mycoparasitism

In nature, some fungi establish antagonistic interactions with other fungal species. This kind of relationship is referred to as mycoparasitism and represents a useful tool in biological control due to the fact that many of the hosts are phytopathogens of field crops (Papavizas, 1985; Chet et al., 1997). The study of parasitic interfungal associations has been conducted according to its classification in two main classes: biotrophic and necrotrophic, based on parasites' aggressiveness toward their hosts. The first class groups antagonistic species that utilize nutrients and compounds produced by its host with no evident damage, at least during the early stages. The establishment of biotrophic associations requires high specificity in recognition between the host and the mycoparasite (Jeffries and Young, 1994). For example, ultrastructural analysis of the interaction between *Tetragoniomyces uliginosus* and its host *Rhizoctonia solani* revealed the formation of a micropore that llows a direct cytoplasmatic connection (Bauer and Oberwinkler, 1990). A particularity of biotrophic mycoparasites is that they are obligate parasites, because they are incapable of surviving without their hosts, and they have a limited host range. *Piptocephalis unispora* is a biotrophic mycoparsite that forms haustoria as attack structures, and its host range is restricted to the subgenus *Micromucor* (Cuthbert and Jeffries, 1984). *Sporodesmiun sclerotivorum* and *Terastoperma oligocladum* produce haustoriumlike infection structures to attack sclerotia of *Sclerotium cepivorum, Botrytis cinerea,* and some species belonging to the genus *Sclerotinia* (Ayers and Adams, 1981; Bullock et al., 1986). The species grouped in this class are not suitable for biocontrol, because their host dependence makes them difficult to manipulate.

The most successful species in biological control are those that establish necrotrophic interfungal associations. Fungi grouped in this class are facultative parasites that can grow by using organic compounds of dead material, through a saprotrophic lifestyle. But they are also able to parasitize living hosts in order to obtain the necessary nutrients. The majority of the commercial biocontrol products listed in Table 12.2 are formulated by using necrotrophic mycoparasites (*Ampelomyces quisqualis, Coniothyrium minitans, Pythium oligandrum, Talaromyces flavus, Trichoderma* spp., and *Gliocadium* spp). This fact reflects the relevance of this particular parasitic process as a successful tool in the development of strategies to control fungal diseases. Necrotrophic mycoparasites have wide host ranges and display more general attack mechanisms, which involve host death and the utilization of the host biomass as a nutrient source. Some *Arthrobotrys* species kill *R. solani* and other fungi by hyphal interference without forming specialized structures, but contact between hyphae of both fungi is a necessary condition (Tzean and Estey, 1978; Persson et al., 1985). In other interfungal associations, the mycoparasite penetrates the host hyphae and reproductive structures through mechanical pressure carried out by specialized structures in conjunction with biochemical degradation of physical barriers through the production of lytic enzymes. Mycoparsites with this characteristic are classified as invasive necrotrophes (Jeffries and Young, 1994). *Nectria inventa* is an invasive species that attacks spores, vegetative, and reproductive structures of several fungi by secretion of adhesive substances and apressorium formation (Tsuneda and Skoropad, 1980). *Phytium oligandrum,* a very aggressive parasite, penetrates and grows inside the host hyphae, such as *Botrytis cinerea* and *Sclerotinia sclerotiorum* (Deacon, 1976). Mycoparasitism by *Trichoderma* involves a positive chemotropism of hyphae toward its host, coiling around host hyphae, penetration of the host by appressorium-like structures, and the production of cell wall–degrading enzymes (Herrera-Estrella and Chet, 2003). Directed *Trichodema* growth toward *R. solani* hyphae has been observed (Chet et al., 1981); however, there is no clear information about the nature of the compounds involved in this behavior. Even when *Trichoderma* attacks a wide host range, a specific recognition event controlling coil formation seems to be necessary. Barak et al. (1985) proposed that lectins present on the host surface participated in the molecular mechanisms leading to the formation of specialized structures during a *Trichoderma*-host interaction. Ten years later, Inbar and Chet (1995) demonstrated that *Trichoderma* was able to grow with a similar pattern to that shown in an interfungal interaction, by using a mimetic system. Nylon fibers simulated host hyphae and lectins purified from *Sclerotium rolfsii* were the recognition molecules responsible for the induction of coiling and specialized structure formation. By using the same biomimetic system, Omero et al. (1999) found that chitin oligomers were also capable of activating those morphological changes in *Trichoderma.* In an attempt to better understand the mechanisms controlling coiling and structure formation as an important aspect in mycoparasitism, several studies have been carried out at the biochemical and genetic levels. Omero et al. (1999) indicated that a signal transduction pathway involving cAMP and hetrotrimeric G-proteins participate in the regulation of coiling. Moreover, transgenic *T. atroviride* strains overexpressing the *tga1* gene, encoding for the α subunit of a hetrotrimeric G-protein, showed a higher coiling frequency and appeared to be much better in the biocontrol of

R. solani, compared with the wild type, under laboratory conditions (Rocha-Ramírez et al., 2002). Secretion of hydrolytic enzymes is one of the main characteristics of species belonging to the genus *Trichoderma.* However, there is no strong evidence directly linking a specific enzyme with host penetration. Inbar and Chet (1995) detected an increase in the activity of a 102 kDa intracellular glucosaminidase in mycelia growing under the bio-mimetic system. Nevertheless, the role of this chitinase during parasitism, especially in penetration, is still unknown. The border line between parasit-ism and saprophytic behavior is unclear, because many of the enzymes are produced in a contact-independent manner, and a strong link exists between nutritional limitation and enzyme production. It is difficult to separate the enzymatic activities from the parasitic process; it is possible that many lytic enzymes affect the hosts by simply making them more susceptible to pene-tration. Production of lytic enzymes and its influence on biocontrol efficacy is discussed later in this chapter (hydrolytic enzymes).

Competition for Space and Nutrients

The soil and its microbial communities can form a very complex and dy-namic structure. Nutrients and space availability are two of the main limita-tions in the colonization of a particular ecological niche. In this sense, those microorganisms with a high growth rate and optimal nutrient utilization must have clear advantages for survival. Many of the fungi exploited as biocontrol systems fulfill the characteristics mentioned above. If biological control agents are well established in plants, it will be difficult for phyto-pathogenic fungi to have access to infection sites in the plant. In this sense, the mycorrhizal associations can also provide biocontrol by preventing pathogen entry through roots (Hooker et al., 1994). Nitrogen, carbon, and iron competition have been associated with *Fusarium* suppression by *Tri-choderma* and nonpathogenic *Fusarium* species (Sivan and Chet, 1989; Mandeel and Baker, 1991; Couteadier, 1992). There is no information with regard to mechanisms involved in competition for nitrogen and carbon sources. Probably, the establishment of this competition is derived from a better capacity of the biological control agents in substrate colonization (Baker, 1991). Iron competition is the best-understood mechanism. Iron is commonly present in an insoluble form. Some bacterial and fungal species have developed a system for iron uptake. This consists in the production of siderophores, peptides working as iron-binding ligand, and the transport for both molecules inside the cell. *Pseudomonas* species constitute the best prokaryotic system studied in relation to siderophore synthesis. The rele-vance of siderophore production on plant disease suppression has been demonstrated by using *Pseudomonas* mutant lacking the ability to produce siderophores (Duijff et al., 1994; Buysens et al., 1996). In fungi, there is little information about siderophore production and biocontrol. In *T. virens,* three classes of hydroxamate siderophores have been identified (Jalal et al., 1986). In addition, the *Psy1* gene, encoding a peptide-synthetase probably involved in siderophore production, has been cloned from *T. virens.* Nevertheless, elimination of *Psy1* did not affect the biocontrol capacity of the mutants toward *R. solani* and *Pythium ultimun* (Wilhite et al., 2001). Thus, even when the evidence suggests the participation of fungal siderophores in iron competition as a biocontrol mechanism, there is no strong evidence relating the two events. An important aspect to keep in mind is that no siderophore-producing organisms can use siderophores previously liberated in soil, unless they have the right transport system (Raajimakers et al., 1995). This situation could be useful to the BCAs applied but could be extended to plant pathogens, making iron competition an unstable tool for disease control.

Antibiotics and Toxins

Most fungi produce antimicrobial secondary metabolites, either as part of their normal growth and developmental program or in response to a biotic stress (parasitism). The production of toxins and antibiotics by fungi is very simply demonstrated in vitro. Numerous agar plate-based assays have been developed to detect volatile and nonvolatile antibiotic production by putative biocontrol agents and to quantify their effects on plant pathogens. However, the relevance of antibiotic production during in vivo biological control remains unclear. Species of *Trichoderma* are well-known biological control agents that produce a range of antibiotics that are active against pathogens in vitro and, consequently, antibiotic production has commonly been suggested as an advantageous trait for these fungi (Ghisalberti and Sivasithamparam, 1991).

The peptaibols are a family of antibiotic peptides isolated from members of the genus *Trichoderma* and related genera, such as *Emericellopsis* and *Gliocladium,* which exhibit antibacterial and antifungal properties. They range between 5 and 20 residues in length. Peptaibols are amphipathic in nature, and this property allows many of them to form voltage-dependent ion channels in lipid bilayer membranes (Sansom, 1991; Rebuffat et al., 1992). These channels are responsible for causing cytoplasmatic leakage, resulting in the loss of soluble proteins from the hyphae, leading to cell death. Over 200 peptailbols sequences have been reported so far, most of which are compiled in the Peptaibol Database at www.cryst.bbk.ac.uk/peptaibol. A useful review about sequence features and modeling structures of different peptaibols was recently published (Chugh and Wallace, 2001).

Peptaibols exhibit a range of antibiotic functions against different target organisms, although there does not appear to be a clear correlation between members of the same subfamily (based on sequence analysis) and antibiotic activity or target. It is pertinent to notice that peptaibols produced by *Trichoderma* species have shown to act synergistically with the production of cell wall–degrading enzymes to inhibit the growth of fungal pathogens (Lorito et al., 1996; Woo et al., 2002). Peptaibols, like many other peptide-toxins in bacteria and fungi, are produced nonribosomally by large multifunctional peptide synthases. Recently, Wiest et al. (2002) cloned the gene responsible for the production of paptaibols in *T. virens (tex1)*. Consistently, the disruption of the gene abolished the production of all peptaibol isoforms by this species. Although biocontrol capacity was not tested on *tex1⁻* strains, unraveling the role of peptaibols in *T. virens* fungal antagonism seems to be closer.

A large number of *Trichoderma* species are capable of producing an array of metabolites other than peptaibols, which show antifungal activity. Production of gliotoxin, a modified cyclic phenylalanine-serine dipeptide, is associated with *T. virens* suppression of damping-off caused by *P. ultimum* (Lewis et al., 1991; Lumsden et al., 1992). *Trichoderma harzianum* and *T. koningii* both produce 6-n-pentyl-2H-pyran-2-one (6PP) or related analogues that showed potent inhibitory properties against sclerotia-forming pathogens, and considerably reduced the rate of damping-off in lettuce seedlings by *R. solani* (Dickinson et al., 1989). The previously mentioned 6PP has also been associated with the volatile compounds produced by *T. virens* (Moss et al., 1975). These antifungal compounds together with some other metabolites (e.g., antraquinones and sesquiterpenes) may have a significant role, either in parasitism or in the competitive saprophytic ability of *Trichoderma*, which is critical for its survival in soil.

Necrotrophic associations between fungi may include hyphal interference reactions, defined as the inhibitory effect of the mycelium of a fungus that grows in close proximity of another species, reducing its rate of growth and causing cytoplasmic disruption (Ikediugwu and Webster, 1970). Low-molecular-weight metabolites have been associated with this phenomenon, which is widespread among mycoparasites used in biocontrol. Some detailed descriptions of the cytological changes occurring during hyphal interference have been described for several fungal interactions. However, the nature of the diffusive factor has not been determined in most of the cases. Butt-rot of pine trees, caused by the bracket fungus *Heterobasidion annosum*, is prevented by *Peniophora gigantea* by a combination of hyphal interference and direct competition. The interference reaction involves a reduction in the number of mitochondria in the cytoplasm of *H. annosum* (Ikediugwu, 1976). Hyphal interference was also suggested to be the mechanism of antagonism of *Erysiphe graminis* and *Sphaerotheca fuliginea* by *Tilletiopsis* spp. (Hoch and Provvidenti, 1979; Klecan et al., 1990). Both *Erysiphe* and *Sphaerotheca* are obligate parasites of higher plants and cause powdery mildew diseases. The phenomenon of hyphal interference is probably underlying the antagonistic interactions of several soil fungi and may constitute the basis for the successful biocontrol of a plant-pathogenic fungus.

Hyphal interference is mediated by a diffusive metabolite that seems to be induced by the close proximity of the host hyphae. In contrast, several necrotrophic interactions involve freely diffusive material, which can affect hyphae some distance away. For example, *Dycima pulvinata*, a destructive mycoparasite of *Cladosporium fulvum*, produces deoxyphomenone (sesquiterpene) which induces cytoplasmic disruption (Tirilly et al., 1991). Similarly, the diketopiperazine antibiotic gliovirin, produced by *Gliocadium virens*, was shown to cause disintegration and coagulation of the cytoplasm of *P. ultimum* (Howell and Stipanovic, 1983).

Considerable evidence has been accumulated for the role of antifungal metabolites in the antagonistic interactions of fungi. However the emphasis on regulatory mechanisms has been done primarily on degradative enzymes (discussed in the next section).

Hydrolytic Enzymes

Production and secretion of lytic enzymes is referred to as one of the most important characteristics of mycoparasites used in biocontrol. The major enzymatic activities reported for different fungal species have as targets the structural components of the cell wall; among them are chitinases, glucanases, glucosaminidases, and proteases. Studies concerning hydrolytic enzyme production have analyzed the process at the biochemical level, evaluating the enzymatic profile shown by mycoparsites in the presence of their hosts or structural components of them, such as cell walls, chitin, and glucan. In the *T. atroviride* complex enzymatic system, composed by six intracellular chitinases (Haran et al., 1995), seven extracellular β-1,3-glucanases (Vazquez-Garciduenas et al., 1998) and a basic protease (Geremia et al., 1993) have been reported. Studies at the molecular level are focused on cloning, sequencing, analyzing, and eliminating or overexpressing lytic enzyme-encoding genes from mycoparasites in order to define their relevance in biological control. *T. virens* is the species with the larger number of cloned genes encoding lytic enzymes; among them are five endochitinase-encoding genes, *ech1, ech2, ech3, cht1,* and *cht2;* three endoglucanase-encoding genes. *bgn1, bgn2,* and *bgn3*; two N-acetyl-β-D-glucosaminidase-encoding genes, *nag1* and *nag2;* and one protease-encoding gene. *tvsp1* (Kim et al., 2002; Pozo et al., 2004). Overproduction of a single enzyme leads to improvement of strains in biocontrol. Transgenic *Trichoderma* lines overproducing a 33kDa chitinase were more effective in the control of *R. solani* in vitro (Limón et al., 1995). *T. atroviride* lines carrying extra copies of the *prb1* gene, encoding a basic protease, provided better protection against *R. solani* in cotton plants (Flores et al., 1997). The overexpression in *T. logibrachiatum* of the gene *egl1*, encoding a β-1,4 endoglucanase, allowed a more effective disease suppression in cucumber plants infected with *P. ultimun* (Migheli et al., 1998). In many cases, the resulting transgenic lines did not show changes when they were host challenged. *T. atroviride* mutants in the *ech42* gene, encoding a 42 kDa endochitinase, were not affected in their ability to control *P. ultimun, R. solani,* and *S. rolfsii.* The same *ech42* mutant was less efficient when tested against *B. cinerea* (Carsolio et al., 1999; Woo et al., 1999), indicating that the successful establishment of the antagonistic interaction is host dependent. Moreover, there are many factors affecting the process, such as physiological and environmental conditions under which the interactions are tested, that affect host susceptibility and pathogen aggressiveness. In fact, analysis based on submerged cultures indicated that the expression of several lytic enzymes encoding genes was blocked if the culture media contained glucose as carbon source (De la Cruz et al., 1993; Carsolio et al., 1994; Garcia et al., 1994; Peterbauer et al., 1996) or primary nitrogen sources (de las Mercedes et al., 2001; Olmedo-Monfil et al., 2002). The requirement of this nutritional limitation for the production of enzymes could indicate that those enzymes are simply part of the saprophytic behavior. Nevertheless, an increase in the expression level of genes over that observed under nutritional-restrictive conditions has also been reported. This specific induction was triggered by the presence of host cell walls, indicating the occurrence of a complex regulatory mechanism for the control of lytic enzymes production during the mycoparasitic process (Olmedo-Monfil et al., 2002). It has been proposed that the molecule responsible for the induction of gene expression must be some sort of degradation product(s) from the fungal cell wall (Cortes et al., 1998; Zeilinger et al., 1999). Contact of the mycoparasite with its host is not required for liberation of inducers. Cortés et al. (1998) analyzed the expression of the genes *prb1* and *ech42* in a *T. atroviride-R. solani* confrontation in petri dishes by using a double cellophane membrane as a physical barrier between both fungi. Under this no-contact condition, it was possible to detect a similar expression pattern to that observed in a direct confrontation for the two tested genes. Kullning et al. (2000) determined that the size of the diffusive molecule was higher than 12 kDa. N-acetyl glucosamine has been also tested as inducer of enzyme production. Omero et al. (1999) observed an increase in chitinase activities due to the addition of this compound to *Trichoderma* mycelia. Zeilinger et al.

344 BIOLOGICAL CONTROL OF PLANT DISEASES

(1999) reported differences in the timing of gene expression in *T. atroviride:* the *ech42* gene was expressed before contact with *R. solani,* while *nag1* expression was observed only after host contact. A *T. atroviride* strain mutated in the *nag1* gene was unable to induce the expression of *ech42,* suggesting that the activity of this enzyme is involved in the release of the inducer (Brunner et al., 2003). These two later results seem to be contradictory; the authors argued that a basal undetectable expression of *nag1* could allow *ech42* expression before contact with the host. Even though this important issue has been addressed, there is no solid indication yet of the nature of the inducers of lytic enzyme production. It is likely that different types of inducers exist, activating particular groups of hydrolytic enzymes. Actually, Mach et al. (1999) reported that the expression of two important chitinases in *T. atroviride* is controlled through different signals.

Mycoparasites, like all living organisms, have intracellular regulatory mechanisms that allow them to respond properly to environmental variations. Signals indicating the presence of a potential host must be transduced into the cell. The mitogen-activated protein kinase pathway (MAPK) is a conserved regulatory pathway among eukaryotes and comprises one of the most extensively studied signal transducing pathways (Kultz, 1998). The participation of a MAPK pathway involving a MAP kinase homologous to Kss1 from the yeast *Saccharomyces cerevisiae* has an essential role on pathogenicity of important phyto-pathogenic fungus, such as *Magnaporthe grisea* (Xu and Hamer, 1996), *F. oxysporum* (Di Pietro et al., 2001), and *B. cinerea* (Zheng et al., 2000). Mutation of the genes encoding MAP kinase homologous to Kss1 caused reduction or even complete loss of pathogenicity in these fungal species. Moreover, in the case of *F. oxysporum* and *B. cinerea* mutant strains, a decrease in the production of lytic enzymes involved in plant penetration was also observed. With regard to mycoparasites, it has been reported for *T. atroviride* that the expression of the *prb1* gene, in response to nitrogen deprivation, was blocked by the presence of a specific inhibitor of the MAPK pathway (Olmedo-Monfil et al., 2002). This biochemical approach allowed linking lytic enzymes production with a MAPK signaling pathway. Recently, two groups working independently, cloned from different isolates of *T. virens* the gene encoding a MAP kinase homologous to Kss1, which was named *tmkA* or *tmk1;* the nucleotide sequence of the gene showed greater similarity to the *pmk1* gene from *B. cinerea* (Mendoza-Mendoza et al., 2003; Mukherjee et al., 2003). Mukherjee et al. (2003) analyzed the *tmkA* loss-of-function mutants in confrontation assays. Against *R. solani,* the *tmkA* mutants displayed a similar pathogenic behavior to the wild type, concerning coiling and host lysis. Only a reduction in sclerotia colonization was detected in the *tmkA* mutants, while the mutants, in an in vitro confrontation with *S. rolfsii,* showed a reduction in their antagonistic properties and were unable to parasitize sclerotia. Mendoza-Mendoza et al. (2003) analyzed the expression pattern of the lytic enzyme-encoding genes in *tvk1* null mutants. By using two induction systems, confrontation in plate against *R. solani* or submerged cultures containing *R. solani* cell walls, they observed an increase in gene expression levels for all the tested genes, which correlated with higher enzymatic activities. The effects of the *tvk1* mutation in gene expression were more rapid and involved stronger responses, maintaining the expression pattern of the genes in response to nutrient limitation and presence of the host. The impact of the overexpression of these enzymes in biocontrol efficacy was also tested under greenhouse conditions, in cotton plants infected with *R. solani* or *P. ultimum*. The *tvk1* mutant strains did not show significant differences in mortality when plants were infected with *R. solani*, as compared with the parental strain. However, those plants protected with the *tvk1* mutants showed less severe disease symptoms in their root systems. Protection provided by *tvk1* mutant strains to *Pythuim*-infected plants was much better than that obtained with the parental strain and even significantly higher than the protection obtained by treatment of the seeds with the commercial fungicide Apron. Differences in the antagonistic capacity reported in the same kind of mutant *tvk1/tvkA* against *R. solani* could be due to the conditions used. Greenhouse experiments provide conditions completely different to those present in a plate confrontation. An alternative explanation is that different strains of the same species use different mechanisms to exert biocontrol. The results derived from studying the *tvk* mutants bring to light the importance of lytic enzymes production in the biological control process by mycoparasites.

CONCLUSIONS

Pest management—whether based on genetic, cultural, biological, or chemical control practices, or their combination into integrative pest management (IPM) strategies—has successfully increased crop production and improved yield stability over the past 30 to 40 years. However, the increasing awareness of the benefits of biological control, as well as of the economic and environmental costs of chemical pesticides, is a powerful driving force reorienting plant protection policies worldwide.

Biological control based on microorganisms has been a component of IPM strategies in developing countries for more than a decade, although successful approaches for microbial control have evolved worldwide (Kogan, 1998; Abate et al., 2000). In most developing countries, despite considerable popularity and optimistic predictions for the commercialization of biological control agents, the anticipated demand for these products has not materialized. The market for microbial pesticides is growing but still represents only under 1 percent of the total crop protection market, and products based on *B. thuringensis* (Bt) accounted for most of it (Lisansky, 1997). The various reasons for the lack of product development and market penetration of biological control include expectations that the multinational agrochemical companies would take a lead in product development; traditional low-input farming systems in developing countries; and practice of an inappropriate model for biocontrol strategies based on small research teams lacking the multidisciplinary knowledge required. Future success of biological control will depend not only on continued investment on research projects but also in their integration with governmental crop improvement programs.

A variety of organisms cause and prevent plant diseases. Fungi are the most significant and widespread. It is estimated that there are over 5,000 pathogenic species of fungi, many of which are economically important. The knowledge of the interaction between phytopathogenic fungi and natural fungal enemies has opened the possibility to exploit the potential use of such predators in pest control. Examples of successful fungal species are now available (Table 12.2). However, more information is needed about the basic ecology of interfungal relationships in nature to overcome some of the tough problems associated with biological control programs. It seems clear that the development of strong, host-specific fungal biological control agents should be the paradigm for future biocontrol strategies in crop protection.

Genomics, or the ability to analyze whole genomes, offers unprecedented opportunities to obtain new information that will allow employment of effective disease management programs with enhanced safeguards to the environment. A complete understanding of disease management will require not only the knowledge of the interaction between plant hosts and pathogens but also of the biocontrol agent genotypes. Sequence data from the genes for such microorganisms (including fungi) are essential to advance our knowledge of antagonism and the interaction of predator and host-gene products, and will provide information on how these organisms evolved and spread, that is, the true nature of biological control.

REFERENCES

Abate, T., A. van Huis, and J.K. Ampofo. (2000). Pest management in traditional agriculture: An African perspective. *Annual Review of Entomology* 45: 631-659.

Ayers, W.A. and P.B. Adams. (1981). Mycoparasitism of sclerotial fungi by *Teratosperma oligocladum. Canadian Journal of Microbiology* 27: 886-892.

Baker, R. (1991). Diversity in biological control. *Crop Protection* 10: 85-94.

Barak, R., Y. Elad, D. Mirelman, and I. Chet. (1985). Lectins: A possible basis for specific recognition in the interaction of *Trichoderma* and *Sclerotium rolsii*. *Phytopathology* 75: 458-462.

Bauer, R. and F. Oberwinkler. (1990). Direct cytoplasm-cytoplasm connection: An unusual host-parasite interaction of the tremelloid mycoparsite *Tetragoniomyces ulginosus*. *Protoplasma* 154: 157-160.

Benhamou, N., C. Garand, and A. Goulet. (2002). Ability of nonpathogenic *Fusarium oxysporum* strain Fo47 to induce resistance against *Pythium ultimum* infection in cucumber. *Applied and Environmental Microbiology* 68: 4044-4060.

Betz, F.S., B.G. Hammond, and R.L. Fuchs. (2000). Safety and advantages of *Bacillus thuringensis*-protected plants to control insect pests. *Regulatory Toxicology and Pharmacology* 32: 156-173.

Blossey, B. and T.R. Hunt-Joshi. (2003). Belowground herbivory by insects: Influence on plants and aboveground herbivores. *Annual Review of Entomology* 48: 521-547.

Brown, J.K.M. and M.S. Hovmoller. (2002). Dispersal of pathogens on the global and continental scales and its impact on plant disease. *Science* 297: 537-541.

Brunner K., C. Peterbauer, R.L. Mach, M. Lorito, S. Zeilinger, and C. Kubicek. (2003). The Nag1 N-acetylglucosaminidase of *Trichoderma atroviride* is essential for chitinase induction by chitin and of major relevance in biocontrol. *Current Genetics* 43: 289-295.

Bullock, S., P.B. Adams, H.J. Willets, and W.A. Ayers. (1986). Production of haustoria by *Sporidesmium sclerotivorum* in sclerotia of *Sclerotinia minor*. *Phytopathology* 76: 101-103.

Butt, T.M., C. Jackson, and N. Magan. (eds). (2001). *Fungi As Biocontrol Agents: Progress, Problems and Potential*. CABI International, Wallingford.

Buysens, S., K. Heungens, J. Poppe, and M. Höfte. (1996). Involvement of pyochelin and pyoverdin in suppression of *Pythium*-induced damping-off of tomato by *Pseudomonas aeruginosa* 7NSK2. *Applied and Environmental Microbiology* 62: 865-871.

Carsolio, C., N. Benhamou, S. Haran, C. Cortés, A. Gutiérrez, I. Chet, and A. Herrera-Estrella. (1999). Role of the *Trichoderma harzianum* endochitinase gene, *ech42*, in the mycoparasitism. *Applied and Environmental Microbiology* 65: 929-935.

Carsolio, C., A. Gutiérrez, B. Jiménez, M. Van Montagu, and A. Herrera-Estrella. (1994). Characterization of *ech-42*, a *Trichoderma harzianum* endochitinase gene expressed during mycoparasitism. *Proceedings of the National Academy of Sciences USA* 91: 10903-10907.

Chet, I., G.E. Harman and R. Baker. (1981). *Trichoderma hamtum*: its hyphal interactions with *Rhizoctonia solani* and *Phytium* spp. *Microbial Ecology* 7: 29-38.

Chet, I., J. Inbar, and Y. Hadar. (1997). Fungal antagonists and mycoparasites. In *The Mycota IV. Enviromental and Microbial Relationships,* eds. Wicklow, Soderstrom. pp. 165-184.

Chugh, J.K. and B.A. Wallace. (2001). Peptaibols: Models for ino channels. *Biochemical Society Transactions* 29: 565-570.

Cooke, R.C. and J.M. Whipps. (1980). The evolution of nutrition in fungi parasitic on terrestrial plants. *Biological Reviews* 55: 341-362.

Cortés, C., A. Gutiérrez, V. Olmedo, J. Inbar, I. Chet, and A. Herrera-Estrella. (1998). The expression of genes involved in parasitism by *Trichoderma harzianum* is triggered by a diffusible factor. *Moleculer General Genetics* 260: 218-225.

Couteaudier, Y. (1992). Competition for carbon in soil and rhizosphere, a mechanism involved in biological control of Fusarium wilts. In *Biological Control of Plant Diseases,* eds. E.C. Tjamos, A.C. Papavizas, and R.J. Cook. Plenum Press, New York, pp. 99-104.

Cuthbert, A. and P. Jeffries. (1984). Mycoparasitism by *Piptocephalis unispora* within the Mortierellaceae. *Transactions British Mycological Society* 83: 700-702.

Dangl, J.L. and J.D.G. Jones. (2001). Plant pathogens and integrated defence responses to infection. *Nature* 411: 826-833.

Deacon, J.W. (1976). Studies of *Phytium oligandrum,* an agressive parasite of other fungi. *Transactions British Mycological Society* 66: 383-391.

De la Cruz, J., M. Rey, J.M. Lora, A. Hidalgo-Gallego, F. Dominguez, J.A. Pintor-Toro, A. Llobell, and T. Benitez. (1993). Carbon source control on β-glucanases, chitobiase and chitinase from *Trichoderma harzianum. Archives Microbiology* 159: 1-7.

De las Mercedes, D., M.C. Limón, R. Mejías, R.L. Mach, T. Benítez, J.A. Pintor-Toro, and C. Kubicek. (2001). Regulation of chitinase 33 *(chit33)* gene expression in *Trichoderma harzianum. Current Genetics* 38: 335-342.

Dickinson, J.M., J.R. Hanson, P.B. Hitchcock, and N. Claydon. (1989). Structure and biosynthesis of harzianopyridone, an antifungal metabolite of *Trichoderma harzianum. Journal of the Chemical Society Perkin Transactions* 1: 1885-1887.

Di Pietro A., F.I. García-MacEira, E. Meglecz, and M.I. Roncero. (2001). A MAP kinase of the vascular wilt fungus *Fusarium oxisposum* is essential for root penetration and pathogenesis. *Molecular Microbiology* 39: 1140-1152.

Duijff, B.J., P.A.H.M. Bakker, and B. Schlppers. (1994). Suppression of *Fusarium* wilt of carnation by *Pseudomonas putida* WCS358 at different levels of disease incidence and iron availability. *Biocontrol Science and Technology* 4: 279-288.

Elad, Y., I. Chet, and Y. Hadar. (1980). *Trichoderma harzianum:* A biocontrol agent affective against *Sclerotium rolsii* and *Rhizoctonia solani. Phytopathology* 70: 119-121.

Evans, H.C. and C.A. Ellison. (1990). Classical biological control of weeds with micro-organisms: Past, present, prospects. *Aspects of Applied Biology* 24: 39-49.

Fischhoff, D.A., K.S. Bowdish, F.J. Perlack, P.G. Marrone, S.M. McCormick, J.G. Nidermeyer, D.A. Dean, et al. (1987). Insect tolerant transgenic tomato plants. *Biotechnology* 5: 807-813.

Flores, A., I. Chet, and A. Herrera-Estrella. (1997). Improved biocontrol activity of *Trichoderma harzianum* by overexpression of the proteinase-encoding gene *prb1. Current Genetics* 31: 30-37.

Fry, W.E., S.B. Goodwin, A.T. Dyer, J.M. Matuszack, A. Drenth, P.W. Tooley, L.S. Sujkowski, et al. (1993). Historical and recent migrations of *Phytophtora infestans:* Chronology, pathways and implications. *Plant Disease* 77: 653-660.

Fuchs, J.G., Y. Moenne-Loccoz, and G. Defago. (1997). Nonpathogenic *Fusarium oxysporum* strain Fo47 induces resistance in the suppression of fusarium wilt in tomato. *Plant Disease* 81: 492-496.

García, I., J.M. Lora, J. de la Cruz, T. Benítez, A. Llobell, and J.A. Pintor-Toro. (1994). Cloning and characterization of a chitinase *(chit42)* cDNA from the mycoparasitic fungus *Trichoderma harzianum. Current Genetics* 27: 83-89.

Geremia, R.A., G.H. Goldman, D. Jacobs, W. Ardiles, S.B. Vila, M. Van Montagu, and A. Herrera-Estrella. (1993). Molecular characterization of the proteinase-encoding gene *prb1*, related to mycoparasitism by *Trichoderma harzianum. Molecular Microbiology* 8: 603-613.

Ghisalberti, E.L. and K. Sivasithamparam. (1991). Antifungal antibiotics produced by *Trichoderma spp. Soil Biology and Biochemistry* 23: 1011-1020.

Gozzo, F. (2003). Systemic acquired resistance in crop protection: From nature to a chemical approach. *Journal of Agricultural and Food Chemistry* 51: 4487-4503.

Hanson, L.E. and C.R. Howell. (2004). Elicitors of plant defense response from bicontrol strains of *Trichoderma virens. Phytopathology* 94: 171-176.

Haran, S., H. Schickler, A. Oppenheim, and I. Chet. (1995). New components of the chitinolytic system of *Trichoderma harzianum. Mycological Research* 99: 441-446.

Herrera-Estrella, A. and I. Chet. (2003). The biological control agent *Trichoderma:* From fundamentals to applications. In *Handbook of Fungal Biotechnology,* ed. D. Arora. Dekker, New York.

Hoch, H.C. and R. Provvidenti. (1979). Mycoparasitic relationships: Cytology of the *Sphaerotheca fuliginea-Tilletiopsis* sp. interaction. *Phytopathology* 69: 359-362.

Hooker, J.E., M. Jaizme-Vega and D. Atkinson. (1994). Biocontrol of plant pathogens using arbuscular mycorrhizal fungi. In *Impact of Mycorrhizas on Sustainable Agriculture and Natural Ecosystems,* eds. S. Gianinazzi and H. Schuepp. Birkhauser Veerlag Basel, Switzerland, pp. 191-200.

Howell, C.R. and R.D. Stipanovic. (1983). Gliovirin, a new antibiotic from *Gliocladium virens,* and its role in the biological control of *Pythium ultimum. Canadian Journal of Microbiology* 29: 321-324.

Ikediugwu, F.E.O. (1976). The interface in hyphal interference by *Peniophora* against *Heterobasidion annosum. Transactions of the British Mycology Society* 66: 291-296.

Ikediugwu, F.E.O. and J. Webster. (1970). Hyphal interference in a range of coprophilous fungi. *Transactions of the British Mycology Society* 54: 205-210.

Ikediugwu, F.E.O. and J. Webster. (1994). Biomimics of fungal cell recognition by use of lectin coated nylon fibers. *Journal of Bacteriology* 174: 1055-1059.

Inbar, J. and I. Chet. (1995). The role of recognition in the induction of specific chitinases during mycoparasitism by *Trichoderma harzianum. Microbiology* 141: 2823-2829.

Jalal, M.A., S.K. Love, and D. van der Helm. (1986). Siderophore mediated iron (III) uptake in *Gliocladium virens:* 1. Properties of cis-fusarinine, trans-fusarinine, dimerum acid, and their ferric complexes. *Journal of inorganic Biochemistry* 28: 417-430.

James, C. (1998). *Global Review of Commercialized Transgenic Crops: 1998,* ISAAA Briefs No. 8. ISAAA, Ithaca, NY.

James, C. (1999). *Preview—Global Review of Commercialized Transgenic Crops: 1999,* ISAAA Briefs No. 12. ISAAA, Ithaca, NY.

Jeffries, P. and T.W.K. Young. (1994). *Interfungal Parasitic Relationships.* International Mycological Institute and CAB International, Oxon.

Keane, R.M. and M.J. Crawley. (2002). Exotic plant invasion and the enemy release hypothesis. *Trends in Ecology and Evolution* 4: 164-170.

Kim, D.J., J.M. Baek, P. Uribe, C.M. Kenerley, and D.R. Cook. (2002). Cloning and characterization of multiple glycosil hydrolase genes from *Trichoderma virens.* *Current Genetics* 40: 374-384.

Klecan, A.L., S. Hippe, and S.C. Somerville. (1990). Reduced growth of *Erysiphe graminis* f.sp. *hordei* induced by *Tilletiopsis pallescens. Phytopathology* 80: 325-331.

Kogan, M. (1998). Integrated pest management: Historical perspective and contemporary developments. *Annual Review of Entomology* 43: 243-270.

Kullning, C., R. Mach, M. Lorito, and C. Kubicek. (2000). Enzyme difussion from *Trichoderma atroviride* (=*T. harzianum* P1) to *Rhizoctonia solani* is a prerequisite for triggering of Trichoderma *ech42* gene expression before mycoparasitic contact. *Applied and Environmental Microbiology* 66: 2232-2234.

Kultz, D. (1998). Phylogenetic and functional classification of mitogen- and stress-activated protein kinases. *Journal of Molecular Evolution* 46: 571-588.

Lau, G. and J.E. Hammer. (1996). Regulatory genes controlling MPG1 expression and pathogenicity in the rice blast fungus *Magnaporhe grisea. Plant Cell* 8: 771-781.

Lewis, J.A., D.P. Roberts, and M.D. Hollenbeck. (1991). Induction of cytoplasmic leakage from *Rhizoctonia solani* hyphae by *Gliocladium virens* and partial characterization of a leakage factor. *Biocontrol Science and Technology* 1: 21-29.

Limon, M., J.A. Pintor-Toro, and T. Benítez. (1995). Increased antifungal activity of *Trichoderma harzianum* transformants that overexpress a 33KDa chitinase *Phytopathology* 89: 254-261.

Lisansky, S. (1997). Microbial biopesticides. In *Microbial Insecticides: Novelty or Necessity?* British Crop Protection Council Proceedings/Monograph Series, No 68, pp. 3-10.

Lorito, M., V. Farkas, S. Rebuffat, B. Bodo, and C.P. Kubicek. (1996). Cell wall synthesis is a major target of mycoparasitic antagonism by *Trichoderma harzianum. Journal of Bacteriology* 178: 6382-6385.

Lumsden, R.D., J.C. Locke, S.T. Adkins, J.F. Walter, and C.J. Ridout. (1992). Isolation and localization of the antibiotic toxin gliotoxin produced by *Gliocladium virens* from alginate pills in soil and soilless media. *Phytopathology* 82: 230-235.

Mach, R.L., C.K. Peterbauer, K. Payer, S. Jaksits, S. Woo, S. Zeilinger, C.M. Kullnig, M. Lorito, and C. Kubicek. (1999). Expression of two major chitinase genes of *Trichoderma atroviride* (*T. harzianum* P1) is triggered by different regulatory signals. *Applied and Environmental Microbiology* 65: 1858-1863.

Mandeel, Q. and R. Baker. (1991). Mechanisms involved in biological control of *Fusarium* wilt on cucumber with strains of no-pathogenic *Fusarium oxysporum*. *Phytopathology* 81: 462-469.

Mendoza-Mendoza, A., M.J. Pozo, D. Grzegorski, P. Martínez, J.M. García, V. Olmedo-Monfil, C. Cortés, Ch. Kenerley, and A. Herrera-Estrella. (2003). Enhanced biocontrol activity of *Trichoderma* through inactivation of a mitogen-activated protein kinase. *Proceedings of the National Academy of Sciences USA* 100: 15965-15970.

Migheli, Q., L. Gonzalez-Candelas, L. Dealessi, A. Camponogara, and D. Ramon-Vidal. (1998). Transformants of *Trichoderma longibrachiatum* overexpressing the β-1,4, endoglucanase, gen *egl1* show enhanced biocontrol of *Pythium ultimum* on cucumber. *Phytopathology* 88: 673-677.

Monaco, L.C. (1977). Consequences of the introduction of coffee and leaf rust into Brazil. *Annals of the New York Academy of Science* 287: 57-71.

Moss, M.O., R.M. Jackson, and D. Rogers. (1975). The characterisation of 6-(Pent-1-enyl)-α-pyrone from *Trichoderma viride*. *Phytochemistry* 14: 2706-2708.

Mukerji, K.G. and K.L. Garg (eds). (1988). *Biocontrol of Plant Diseases,* Vol. 1 CRC Press Inc., Boca Raton, FL.

Mukherjee, PK., J. Latha, R. Hadar, and B.A. Horwitz. (2003). TmkA, a mitogen-activated protein kinase of *Trichoderma virens*, is involved in biocontrol properties and repression of conidiation in the dark. *Eukaryotic Cell* 2: 446-455.

Oerke, E.C., H.W. Dehne, F. Schonbeck, and A. Weber. (1994). Conclusion and perspectives. In *Crop Production and Crop Protection: Estimated Losses in Major Food and Cash Crops,* eds. E.C Oerke, H.W. Dehne, F. Schonbeck, and A. Weber. Elsevier, Amsterdam. pp. 742-770.

Olmedo-Monfil, V., A. Mendoza-Mendoza, I. Gomez, C. Cortes, and A. Herrera-Estrella. (2002). Multiple environmental signals determining the transcriptional activation of the mycoparasitism related gene *prb1* in *Trichoderma atroviride*. *Molecular and General Genomics* 267: 703-712.

Omero, C., I. Inbar, V. Rocha-Ramírez, A. Herrera-Estrella, and B. Horwitz. (1999). G protein activators and cAMP promote mycoparasitic behaviour in *Trichoderma harzianum*. *Mycological Research* 103: 1637-1642.

Papavizas, G.C. (1985). *Trichoderma* and *Gliocadium*: Biology, ecology and potential for biocontrol. *Annual Review of Phytopathology* 23: 23-54.

Perlak, F.J., T.B. Stone, Y.M. Muskopf, L.J. Petersen, G.B. Parker, S.A. McPherson, J. Wyman, et al. (1993). Genetically improved potatoes: Protection from damage by Colorado potato beetles. *Plant Molecular Biology* 22: 313-321.

Persson, Y., M. Veenhuis, and B. Nordbring-Hertz. (1985). Morphogenesis and significance of hyphal coiling by nematode-trapping fungi in mycoparasitic relationships. *FEMS Microbiology Ecology* 31: 283-291.

Peterbauer, C., M. Lorito, C.K. Hayes, G.E. Harman, and C.P. Kubicek. (1996). Molecular cloning and expression of nag1 (N-acetyl-β-glucosaminidase-encoding) gene from *Trichoderma harzianum*. *Plant Current Genetics* 29: 812-820.

Pozo, M.J., J.M. Baek, J.M. Garcia, and C.M. Kenerley. (2004). Functional analysis of *tvsp1*, a serine protease-encoding gene in biocontrol agent *Trichoderma virens*. *Fungal Genetics and Biology* 41: 336-348.

Punja, Z.K. and R.S. Utkhede. (2003). Using fungi and yeast to manage vegetable crop diseases. *TRENDS in Biotechnology* 21: 400-407.

Raajimakers, J.M., I. van der Sluis, M. Koster, P.A.H.M. Bakker, P.J. Weisbeek, and B. Schippers. (1995). Utilization of heterologous siderophores and rhizospere competence of fluorescnet *Pseudomonas* spp. *Canadian Journal of Microbiology* 41: 126-135.

Rebuffat, S., H. Duclohier, C. Auvin-Guette, G. Molle, G. Spach, and B. Bodo. (1992). Membrane-modifying properties of the pore-forming peptaibols saturnisporin SA IV and harzianin HA V. *FEMS Microbiology Immunology* 105: 151-160.

Rocha-Ramírez, V., C. Omero, I. Chet, B. Horwitz, and A. Herrera-Estrella. (2002). A *Trichoderma atroviride* G protein a subunit gene, *tga1*, involved in myco-parasitic coiling and conidiation. *Eukaryotic Cell* 1: 594-605.

Rotem, J., B. Wooding, and D.E. Aylor. (1985). The role of solar radiation, especially in the mortality of fungal spores. *Phytopathology* 75: 510-514.

Sansom, M.S.P. (1991). The biophysics of peptide models of ion channels. *Progress in Biophysics and Molecular Biology* 55: 139-235.

Shah, P.A. and J.K. Pell. (2003). Entomopathogenic fungi as biological control agents. *Applied Microbiology and Biotechnology* 61: 413-423.

Sivan, A. and I. Chet. (1989). The possible role of competition between *Trichoderma harzianum* and *Fusarium oxysporum* on rhizosphere colonization. *Phytopathology* 79: 198-203.

Smith, R.F. and R. van den Bosch. (1967). Integrated control. In *Pest Control: Biological Physical, and Selected Chemical Methods,* eds. W.W. Kilgore and R.L. Doutt. Academic Press, New York, pp. 295-342.

Stover R.H. and N.W. Simmonds. (1987). *Bananas.* 3rd ed. Longman, New York.

Symondson, W.O.C., K.D. Sunderland, and M.H. Greenstone. (2002). Can generalist predators be effective biocontrol agents. *Annual Review of Entomology* 47: 561-594.

Tirilly, Y., F. Lambert, and D. Thouvenot. (1991). Bioproduction of [^{14}C] deoxy-phomenone, a fungistatic metabolite of the hyperparasite *Dycima pulvinata.* *Phytochemistry* 30: 3963-3965.

Tsuneda, A. and W.P. Skoropad. (1980). Interactions between *Nectria inventa,* a destructive mycoparasite and fourteen fungi associated with rapeseed. *Transactions of the British Mycological Society* 74: 501-507.

Tzean, S. and R. Estey. (1978). Nematode-trapping fungi as mycopathogens. *Phytopathology* 68: 1266-1270.

Vaeck, M., A. Reybnaerts, J. Hofte, S. Jansens, M. DeBeuckeleer, C. Dean, M. Zabeau, M. Van Montagu, and J. Leemans. (1987). Transgenic plants protected from insect attack. *Nature* 328: 33-37.

van Driesche, R.G. and T.S. Bellows Jr. (1996). *Biological Control.* Chapman and Hall, New York.

Vázquez-Garcidueñas, S., C. Leal-Morales, and A. Herrera-Estrella. (1998). Analysis of the β-1,3glucanolytic system of the biocontrol agent *Trichoderma harzianum.* *Applied and Environmental Microbiology* 64: 1442-1446.

Werner, D. and A.C. Pont. (2003). Dipteran predators of Simuliid blackflies: a worldwide review. *Medical and Veterinary Entomology* 17: 115-132.

Wiest, A., D. Grzegorski, X. Bi-Wen, C. Goulard, S. Rebuffat, D.J. Ebbole, B. Bodo, and C. Kenerley. (2002). Identification of Peptaibols from *Trichoderma virens* and cloning of a peptaibol synthetase. *Journal of Biological Chemistry* 277: 20862-20869.

Wilhite, S.E., R.D. Lumsden, and D.C. Straney. (2001). Peptide synthetase gene in *Trichoderma virens*. *Applied and Environmental Microbiology* 67: 5055-5062.

Wolfenberger, L.L and R.R. Phifer. (2000). The ecological risks and benefits of genetically engineered plants. *Science* 290: 2088-2093.

Woo, S., B. Donzelli, F. Scala, R. Mach, G. Harman, C. Kubicek, G. Del Sorbo, and M. Lorito. (1999). Disruption of *ech42* (endochitinase-encoding) gene affects biocontrol activity in *Trichoderma harzianum* strain. *Molecular Plant-Microbe Interactions* 12: 419-429.

Woo, S., V. Fogliano, F. Scala. and M. Lorito. (2002). Synergism between fungal enzymes and bacterial antibiotics may enhance biocontrol. *Antonie Van Leeuwenhoek* 81: 353-356.

Xu, J.R. and J.E. Hamer. (1996). MAP kinase and cAMP signaling regulate infection structure formation and pathogenic growth in the rice blast fungus *Magnapothe grisea*. *Genes and Development* 10: 2696-2707.

Yedidia, I., N. Benhamou, and I. Chet. (1999). Induction of defense responses in cucumber plants *(Cucumis sativus L.)* by the biocontrol agent *Trichoderma harzianum*. *Applied and Environmental Microbiology* 65: 1061-1070.

Yedidia, I., A.K. Srivastava, Y. Kapulnik, and I. Chet. (2000). Induction and accumulation of PR protein activity during early stages of root colonization by the mycoparasite *Trichoderma harzianum* strain T-203. *Plant Physiology and Biochemistry* 38: 863-873.

Zeilinger, S., C. Galhaup, K. Payer, S.L. Woo, R.L. Mach, C. Feteke, M. Lorito, and C.P. Kubicek. (1999). Chitinase gene expression during mycoparasic interaction of *Trichoderma harzianum* with its host. *Fungal Genetics and Biololgy* 26: 131-140.

Zheng, L., M. Campbell, J. Murray, S. Lam, and J.R. Xu. (2000). The BMP1 gene is essential for pathogenicity in the gray mold fungus *Botrytis cinerea*. *Molecular Plant-Microbe Interactions* 13: 724-732.

Chapter 13

Applications of Plant Tissue Culture for Studies of Fruit Tree Defense Mechanisms

Eva Wilhelm

INTRODUCTION

The first work on the culture of isolated plant cells and tissues in artificial nutrient solutions was attempted by Gottland Haberlandt (1902) in Graz, Austria, more than 100 years ago, and his experiments have become enormously important for fundamental and practical applications (Laimer and Rücker, 2003). Tissue culture techniques are contributing to our steadily increasing knowledge of plant cell physiology and molecular biology. Tissue culture methods offer possibilities for precisely controlling physical and chemical environments, thus providing a good system for studying plant-microbe interactions. Large numbers of host cells may be exposed to microbes without excessive tissue injury; cell numbers and inoculum density may be precisely controlled; and precursors and products of interaction may be added to, or removed from, the culture system with ease. Furthermore, in vitro cultures of plants have an obvious advantage for studies with fungi which cannot be grown on artificial media. This fact was originally recognized by Morel, who in the early 1940s infected callus from vine *(Vitis vinifera)* with zoospores of the downy mildew fungus *Plasmopara viticola* and tried to select in vitro for resistant callus cultures. Since then, plant tissue culture methods have been increasingly applied in studies of plant-microbe interactions. As a consequence, in 1980, the Federation of British Plant Pathologists published a book on *Tissue Culture Methods for Plant Pathologists*, describing the merits of these techniques (Ingram and Helge-

Biological Control of Plant Diseases
© 2007 by The Haworth Press, Inc. All rights reserved.
doi:10.1300/5682_13

son, 1980). Plant tissue cultures provide ideal systems for symbionts, such as mycorrhizal fungi (especially, vesicular arbuscular mycorrhiza), which cannot be grown on synthetic medium without a host plant. Cell cultures have been widely used in studies of phytoalexin production and elicitation (Dixon, 1983). An important and practical application of tissue cultures in plant pathology is the elimination of viruses, mycoplamas, and other microbial pathogens by meristem culture.

Although experiments with tissue cultures of woody species have been going on for decades, it took more than 30 years to regenerate a complete plant from a tree species. This was achieved by Winton in 1968 in leaf explants of *Populus trichocarpa* Torr. et Gray. Even nowadays, tissue culture of certain fruit tree species or cultivars remains challenging, particularly due to effects of ontogenetic aging. Several authors have reviewed the applications of plant tissue culture in studying tree defense mechanisms (Diner and Karnosky, 1987; Ostry and Skilling, 1992). Generally, investigations on host-parasite interactions within trees are greatly hampered by the multifactorial basis of many tree diseases and their long generation times. When the primary disease-causing agent is the pathogen of interest, these problems can be partially resolved by utilizing a tissue culture system.

One of the most important requirements of any study dealing with the basic events of plant-microbe interactions is a simple model system able to be manipulated. Some systems, in particular callus, cell suspension cultures as well as shoot cultures, have proven to be very suitable for plant-microbe interaction studies. Direct organogenesis via axillary shoot formation from preformed axillary buds has been shown to be highly useful for several trees–pathosystems. Once a good multiplication system for certain cultivars has been established, the production of large numbers of genotypically identical shoots and whole plants within a short period of time can be achieved with relative ease. A major prerequisite of a useful tissue culture system is that the expression of resistance or susceptibility among undifferentiated cells is similar to that of the whole plant. This is more likely to be the case when tissue integrity remains high. Furthermore, in vitro shoots can serve as suitable tools in the context of quick and reliable screening methods for either testing the disease resistance of cultivars or for verifying the virulence level of pathogens. Specifically, chestnut is a good example of a fruit tree, about whose defense reactions most current knowledge has been gained by using in vitro shoots.

In cell suspension and callus cultures the tissue organization level is reduced, yet this has been shown to be advantageous in performing elicitation studies. It is necessary for most of the cells to respond to a certain treatment in a similar manner in order to make subtle changes in metabolism visible. Elicitation of phytoalexin responses is thus amplified, compared with local infection events, in organs and tissue. Furthermore, phytoalexins and other metabolites can be isolated rather easily, while enzymes of specific pathways are coordinately induced, and mRNAs coordinately transcribed.

Several research studies with apple cell suspension cultures have contributed toward elucidating the complex interaction between pathogens and host responses.

IN VITRO SCREENING AND SELECTION
FOR DISEASE RESISTANCE IN FRUIT TREES

One important research area of tissue culture today aims at the selection of plants resistant to disease, insects, or stress. The search for plants resistant to biotic or abiotic stress can be accelerated considerably by using cell culture assays, since in vitro assays are generally of much shorter duration, thus making higher throughputs possible. Most early work used cell cultures as a system for either artificially inducing of mutagenesis or for stimulating somaclonal variation to the end of screening for resistance against certain pathogens. Crop plants have been of major interest, and several approaches have been successful (Daub, 1986), but also screening of fruit trees for resistance to pathogens using tissue culture methods has yielded some practical results.

Apple scab, a serious disease, is caused by the fungus *Venturia inaequalis,* which can attack both foliage and fruits. Scab resistance has been achieved by means of traditional breeding programs, with the result that a number of resistant apple cultivars are available today. Most of these cultivars carry the *Vf* resistance derived from *Malus floribunda.* Yepes and Aldwinckle (1993) reported selection for resistance to *V. inaequalis,* using detached leaves from apple shoots grown in vitro. Light microscopy studies have indicated that the reactions of resistant and susceptible cultivars in vitro mimicked reactions in the greenhouse and field. Leaves from resistant cultivars showed no symptoms, whereas susceptible cultivars displayed chlorosis and became necrotic after five weeks. Fireblight, another serious disease of the pome fruits (apple and pear), is caused by the bacterium *Erwinia amylovora.* Fireblight resistance has been evaluated by using *E. amylovora* bacteria labeled with the *Gfp* (green fluorescent protein) gene, using leaves taken from apples grown in vitro (Hanke and Geider, 2002). Using this approach, it has become possible to differentiate the various levels of resistance by means of the colonization of detached leaves. Various transgenic genotypes have been able to be assessed with respect to their level of resistance. Another in vitro system useful for studying the interaction between *E. amylovora* and pear (*Pyrus communis* L.) has been developed recently (Abdollahi et al., 2004). The method uses bromocresol green as a pH indicator. The pattern of acidification in the medium (rate and intensity) differed significantly, depending on cultivar susceptibility, inoculation method of the shoots (apical or basal), and the presence of pathogen. Inoculations of apple shoot cultures (*Malus* sp.) with cedar-apple rust fungus resulted in responses similar to those known for resistant and susceptible cultivars ex vitro (Joung et al., 1987). Papaya shoot cultures *(Carica papaya)* challenged with sporangial solutions of *Phytophthora palmivora* (Sharma and Skidmore, 1988) and in vitro responses were similar to whole-plant responses in the field. Excised shoot tips of peach *(Prunus persica)* were used to detect a phytotoxin of *Leucostoma* by exposing shoot cultures to different fractions of cell-free filtrates of the fungus (Svircev et al., 1989). Peach shoots treated with a specific fraction of the filtrate developed canker-like symptoms. Downy mildew *(Plasmopara viticola)* resistance in grapevine has been characterized recently, using in vitro plants. A strong correlation was found between peroxidase activitiy in leaves of in vitro plants and the resistance of grapevine plants to *P. viticola* (Kortekamp and Zyprian, 2003). First indications that phenolic compounds are involved in the resistance to *P. viticola* were made with grapevine callus (Dai et al., 1995).

Mutation induction and tissue culture for the purpose of improving fruits has been reviewed recently (Predieri, 2001). Selection of variants with increased resistance to pests, diseases, and herbicides (Brar and Jain, 1998) has also been achieved via somaclonal variation. Several studies have been conducted on fruit crops, and successful selections with a wide range of improved traits have been reported. Somaclones exhibiting resistance to *E. amylovora* have been obtained from a certain pear cultivar (Viseur, 1990) and from apple (Donovan et al., 1994; Chevreau et al., 1998). Somaclones resistant to *Phytophtora cactorum* have been selected from apple rootstocks M26 and MM106 (*Malus pumila* Mill.) (Rosati et al., 1990). Somaclones showing in vitro resistance to *Colletotrichum gleosporiensis* have been found in mango (Litz et al., 1991). *Citrus* spp. somaclones resistant to *Phoma tracheiphila* (Deng et al., 1995) have also been discovered. Grapevine somaclones showing an increased resistance to *Botrytis cinerea* and *P. viticola* in the field have been found as well (Kuksova et al., 1997). Nevertheless, while field validation has often either not been reported or has yet failed to show effective improvement of genotypes, some research has already led to valuable results. For example, Hammerschlag et al. (1995) established a regeneration protocol for peach (*Prunus persica* L. Batsch), and then selected for resistance to bacterial spot (*Xanthomonas campestris* pv. *prusi*) by using a culture filtrate of the pathogen containing a toxic metabolite known to be involved in disease development (Hammerschlag, 1988a,b,

1990). At the end of three years of field observations, only some of the selected plants had retained high levels of disease resistance (Hammerschlag et al., 1995). One peach somaclone proved to be a true mutant, as its progeny also demonstrated high levels of bacterial spot resistance. This somaclone also exhibited resistance to bacterial canker (*Pseudomonas syringae* pv. *syringae*) (Hammerschlag, 2000).

GENERAL PLANT DEFENSE MECHANISMS

Generally, there are three main reasons why a plant may be an inappropriate host for pathogens. Plants do not support the lifestyle of an invading pathogen and thus does not serve as a substrate for microbial growth. Plants are a so-called nonhost. Plants possess preformed structural barriers, or are equipped with antimicrobial compounds that prevent pathogen ingression and spread. These barriers are termed constitutive structural and biochemical defenses. The plant may recognize the pathogen and initiate its endogenous, multicomponent defense system. This form of plant defense comprises sophisticated systems of perception and signal transduction, which finally lead to the activation of defense mechanisms, the so-called induced defenses. Inducible plant defense mechanisms share common features which may be categorized according to three stages: recognition of the pathogen by the plant—elicitation, intracellular signal transduction leading to changes of gene expression, and initiation of defense responses (Nürnberger, 1999). Once a pathogen elicitor has been recognized by its plant receptor, defense responses are initiated. Molecules derived from the pathogen cell walls interact with the host at the plasma membrane level and alter membrane potentials along with calcium and protein kinase-mediated pathways. Virulent pathogens generally fail to elicit, or suppress the establishment of a resistance reaction. The pathogen usually colonizes the host and causes severe damage. During the interaction between an incompatible pathogen and the host, defense mechanisms are established. An additional redox signal, nitric oxide (NO), is discussed as signal compound in pathogen defense, functioning analogously to H_2O_2 and, in particular, as a stimulator of genes of the phenylpropanoid pathway and cell death.

Pathogen recognition may lead to a systemic reaction in the plant. This is mediated by diffusive signal factors transported via the vessels of the plant or even by gaseous compounds; the latter may mediate communication not only between distant plant parts but also from plant to plant. One of the best-characterized signaling molecules triggering defense responses is salicylic acid (SA) (Malamy et al., 1990); it is known to play a key role in systemic-acquired resistance (SAR) induction (Métraux et al., 1990). Moreover, SA-independent signaling pathways have been described, in which jasmonate, ethylene, and methyljasmonat (MA) act as signal factors. Finally, signaling compounds interact with regulators of transcription factors that specifically alter gene transcription. The defense responses can include a wide number of events: hypersensitive response (HR); synthesis of antimicrobial compounds; barrier formation; tylosis formation; cell wall lignification and suberization; synthesis of cell wall strengthening, and pathogen-related (PR) proteins and anti-fungal enzymes (Agrios, 1997; Dorey et al., 1997).

Considerable success has been achieved in cloning avirulence and resistance genes (R genes). R-gene products interact with avirulence gene products to initiate the defense response. The first-plant R genes were cloned at the beginning of the 1990s by positional cloning. In the meantime, eight different classes comprising many different R genes have been isolated from barley, rice, tomato, potato, tobacco, pepper, lettuce, flax, *Arabidopsis*, and maize (Hammond-Kosack and Parker, 2003). So far, no R genes have been identified for woody plants, except those detected in mapping programs for disease resistance, but for which no specific function is known, for example, chestnut–chestnutblight (Kubisiak et al., 1997) or apple–scab (Patocchi, Vinatzer, 1999; Patocchi, Gianfranceschi, and Gessler, 1999).

Wound and Pathogen Defense Is Interlinked in Plants

Many important and serious diseases are caused by pathogens that initiate infections in wounds. Trees are especially susceptible to pathogens attacking wounds caused by insects, humans, fire, lightning, wind, hail, animals, or nutritional and physiological disorders. There is a striking amount of overlap, particularly between defenses against wounding and against pathogens, in spite of many differences. Studies have shown that in *Arabidoposis,* responses to wounding and pathogen infection are often similar (Durrant et al., 2000; Reymond et al., 2000). The discovery of mutants has allowed further dissection of local and systemic-signaling networks and begun to highlight the complex interplay between defense molecules such as SA, NO, reactive oxygen intermediates (ROI), JA, and ethlyene (Hammond-Kosack and Parker, 2003; Shah, 2003). Protein kinases, the central elements of signal transduction in cells, are another link, as are calcium-mediated signal pathways due to their participation in wound and pathogen defense reactions. It has become evident that pathways, pathogen and wound responses, are not completely independent of each other, and that the induction of one influences the other (Maleck and Dietrich, 1999).

While trees have been investigated with respect to wounding, in most cases, leaf or needle tissue has been analyzed (Gleadow and Woodrow, 2000; Richard et al., 2000). Only relatively few works report on wound reactions of stems, and these focus mainly on histological, chemical, and ecological effects (Tomlin et al., 1998, 2000; Christiansen et al., 1999). Eklund and Little (1998) showed that ethylene represents a signalling compound in the bark of trees which triggers defense reactions, and Lev-Yadun and Aloni (1990) hypothesized that auxin and ethylene are the major factors controlling first periderm formation in woody stems in the wound-healing process. Furthermore, the presence of a low amount of SA in chestnut shoots and an increased concentration of this compound in response to pathogen inoculation has been demonstrated for in vitro shoot cultures (Schafleitner et al., 1999). The presence of at least some signaling compounds in the bark of trees playing crucial roles in herbaceous plants indicates some similarity between stress response pathways in both tissues. However, until now, no reports on either the presence or action of JA or systemin in woody tissue have been presented. Hence, it is not possible to draw any conclusions with regard to similarities or dissimilarities of signaling pathways among woody and leafy plant tissues.

In a manner analogous to herbaceous plants, the importance of elicitors and the expression of lytic enzymes in trees have also been documented. In experiments with peach bark, Biggs and Peterson (1990) applied 14 different chemical treatments to wounds, including acid extracts from peach-leaf cell walls and fungal cell-wall extracts (from *Leucostoma persoonii*), alone and in combination, as well as ethylene, abscisic acid, chitosan, calcium ion, and cellobiose. Fungal cell wall extract and cellobiose stimulated bark lignin formation by a factor of ten compared with the control. These results show that elicitors of pathogens also play a crucial role in triggering defense responses in bark tissue.

Cytological results obtained from infected in vitro grapevine plants confirmed that enhanced *Plasmopara* resistance is associated with an expression of distinct reactions in a chronological order. The production of reactive oxygen species was followed by a hypersensitive response. In addition, an increased activity of peroxidase in cells flanking the infected area and within the vascular tissue was observed, which was accompanied by accumulation of phenolic compounds. Gomez et al. (2003) reported on a nonspecific lipid-transfer protein genes expression in grape (*Vitis* sp.) cells in response to fungal elicitor treatment. Nonspecific lipid transfer proteins (nsLTPs) are small, basic cystein-rich proteins believed to be involved in plant defense mechanisms. The expression of three nsLTP genes was investigated in 41B-rootstock grape cell suspension, in response to various defense-related signal molecules. Ergosterol (a fungi-specific sterol), JA, cholesterol, and sitosterol promoted nsLTPs mRNA accumulation, whereas SA had no effect. One isoform also reduced *Botrytis* mycelium growth in vitro.

Phenolic metabolism has often been linked with host resistance to disease. However, inconclusive results have been achieved in this regard, and no significant, quantitative, or qualitative differences among the cultivars or any difference in composition among new and old leaves has been demonstrated. Cell suspension cultures of apple cultivars susceptible and resistant to scab have been used as a model system by Hrazdina et al. (1997) and Borejsza-Wysocki et al. (1999) for studying phytoalexin production in response to elicitation. A new compound, "Malusfuran," 2,4-methoxy-3-hydroxy-9-glucosyloxydibenzofuran, which has not been described before, was identified via HPLC-analysis. Biphenyl and dibenzofuran compounds were identified as the major metabolites of elicitor-treated cell suspension cultures of the scab-resistant cultivar in their studies performed in 1999. However, quite recently Hrazdina (2003) reported that these metabolites could not be detected in the intact plants. These results indicate a differential response of plant organs and cell suspension cultures to elicitor treatment or pathogen invasion. In order to mimic infection by the fungus *V. inaequalis*, cell suspension cultures of apple *(M. domestica)* were treated with cellulases, pectinases, and other possible elicitors derived from the fungal cell wall. Marked changes in the phenol contents were observed, indicating the participation of these enzymes in triggering the elicitor responses (Lux-Endrich et al., 2002).

Inoculation of apple cell cultures with *E. amylovora* strains differing in virulence (*hrp* mutants) reproduced only partly the protective effects found in the whole plant, and it was concluded that cell cultures are not a suitable model for investigating mechanisms leading to protection (Faize et al., 2002). The activity of peroxidases in apple shoot cultures challenged with *E. amylovora* was studied, but no direct relationship to susceptibility among cultivars was observed (Keck et al., 2002). The involvement of three pathogenicity factors of *E. amylovora* has been verified with microshoots of a susceptible pear cultivar (Venisse et al., 2003). *E. amylovora* induces in its susceptible host plants an oxidative burst, as does an incompatible pathogen. The elicitation of this phenomenon is the result of the combined action of two Hrp effectors of the bacteria, HrpN and DspA. In addition, it could confirm that desferrioxamine, the siderophore of *E. amylovora*, is necessary for the bacteria to be able to tolerate high levels of hydrogen peroxide.

CHESTNUT (CASTANEA SPP.)
AND THE MAJOR PATHOGENS

The genus chestnut comprises the species *Castanea sativa* (European chestnut), *C. dentata* (American chestnut), *C. molissima* (Chinese chestnut), and *C. crenata* (Japanese chestnut). Further chestnut species without economic importance are the chinkapins, native to China and the United States, comprising *C. henry, C. ozakensis, C. seguini,* and *C. pumila.* In Europe, chestnut cultivation has a long tradition in many countries, mostly around the Mediterranean. Most of the present chestnut areas are the result of a secondary anthropogenic introduction of the species by the Romans. Chestnut blight and ink disease are the major threats for the survival of chestnut today.

The chestnutblight fungus *Cryphonectria parasitica* (Murr.) Barr., an ascomycete, is endemic to China and Japan on *C. mollissima* and *C. crenata.* These chestnut species exhibit a high degree of resistance against the blight. In the 1880s, *C. parasitica* was accidentally imported to the United States, and it virtually extinguished the highly susceptible American chestnut (Anagnostakis, 1987). About 60 years ago, this fungus was introduced to Europe, where *C. sativa* was dying in groves from this disease (Heiniger and Rigling, 1994). The fungus is transmitted via spores. After germination, the hyphae invade the tree by wounds in the bark. Chestnut blight destroys the bark and the cambium and causes wilting and death of distal tree parts by interrupting the transport of water and nutrients. Once the integrity of the bark tissue has been severely affected and destroyed, the tree does not dispose of any further means of defense and dies. The pathogenicity mechanisms of *C. parasitica* are still a matter of controversial discussion. Oxalate production and cell wall components-degrading enzymes, such as cutinase, esterase, cellulase, polygalacturonase, and protease, have been discussed as putative virulence factors (Bazzigher, 1955; McCarrol and Thor, 1985; Roane et al., 1986; Vareley et al., 1992). The possible production of toxins, such as skyrin, regulosin, diaporthin, and cryparin, has not really been able to be confirmed as being involved in pathogenesis (Bazzigher, 1953; Carpenter et al., 1992). The methods for controlling chestnut blight that have been applied so far are general phytosanitary measures, such as cutting off infested branches and felling trees, and the application of hypovirulence (Seemann, 2001). Hypovirulent strains of this fungus harbor a double-stranded RNA (dsRNA) virus, which is transmitted via hyphal anastomosis, thus converting virulent *C. parasitica* phenotype strains into hypovirulent ones (Heiniger et al., 2001).

Another serious phytopathological threat, known as "ink disease," is still causing the death of a large number of chestnut trees all around Europe, and in the past decade, its incidence both in natural forests and orchards has increased (Vannini and Vettraino, 2001). Within sweet chestnut, at least two species of *Phytophthora, P. cambivora,* and *P. cinnamomi,* are responsible for this destructive disease (Cacciola et al., 2001). The pathogens cause brownish-black lesions on the root and flame-shaped necroses on the stem. The trees die when the stem is girdled or when most of the roots have died.

Recently, a strong recrudescence of the disease has been recorded, especially in Italy and on the Iberian Peninsula, and large surfaces of natural and planted stands are today endangered in other countries as well (Vettraino et al., 2001). Currently, there are no control strategies available for the ink disease beyond phytosanitary measures.

Much work has been done on the pathogens, their mode of pathogenicity and their genetic diversity, but much less research has been performed toward investigating the tree's defense mechanisms

Defense Responses of Chestnut

Research studies on the defense responses of chestnut have mainly focused on the chestnut seed proteins and on the chestnut blight pathogen *C. parasitica.* Several chestnut seed proteins, such as castamollin, cystatin (Pernas et al., 2000). chitinases (Gomez et al., 2002), and small heat-shock proteins (Collada et al., 1993), have been described to be involved in biotic and abiotic stress tolerance. Antifungal proteins such as cystatin (Pernas et al., 1999) and chitinases of chestnut seeds (Collada et al., 1993; Allona et al., 1996) were isolated and characterized. All of these polypeptides belong to structurally diverse families associated with plant defensive responses and accumulate at high levels in chestnut seeds. All of them have stress-inducibe homologues in chestnut stems, roots, and/or leaves (Connors et al., 2002; Gomez et al., 2002).

Several studies tried to elucidate the physiological and histopathological basis of resistance in susceptible versus resistant chestnut species (Hebard and Kaufman, 1978). Bark extracts or plant components have been tested for their inhibition of fungal growth. Fungal polygalacturonase was inhibited 15 times more by bark extracts of resistant chestnut *(C. molissima)* than by extracts of susceptible *C. dentata* (Gao and Shain, 1995). The bark extract of resistant chestnuts inhibited growth of *C. parasitica* more than the extract harvested from the susceptible chestnuts (Shain and Spalding, 1995). A negative correlation between polyphenol-oxidase acitivity of the host and the lesion size caused by the fungus was observed by Vannini et al., 1993. Tannins of resistant or susceptible chestnut do not differentially stimulate the growth of *C. parasitica,* indicating that the tannin composition is not involved in resistance, contrary to former reports (Anagnostakis, 1992).

Most of our current knowledge on defense reactions of chestnut to biotic stress has been obtained via application of tissue culture methods based either on axillary shoots or on callus cultures (Hebard and Kaufman, 1978; Piagnani et al., 1997, 2002). However, early attempts to use chestnut callus cultures for distinguishing between resistant and susceptible lines failed (Hebard and Kaufman, 1978). They encountered several problems, such as the lack of invasive fungal colonization and the friable nature of callus tissue, as well as difficulties with the stabilization of the growth of the cultures.

The effect of crude extracts of virulent and hypovirulent *C. parasitica* strains on growth and physiological activities of chestnut calli was investigated on cell cultures obtained from either a susceptible *C. sativa* or a resistant *C. mollissima* chestnut species. The decrease of callus growth and physiological activities (O_2 uptake and H^+-ATPase activity) caused by the virulent strain was attributed to toxic compounds, whereas a stimulatory effect of the hypovirulent strain on the overall callus metabolism was observed. No differences in susceptibility to culture filtrate of the virulent fungus was detected between *C. sativa* and *C. mollissima*, implying that resistance components are probably more likely to be acting on fungal colonization rather than to involve a host-cell tolerance to the toxins produced by *C. parasitica* virulent strains. However, the two chestnut species showed a different response with regard to O_2 uptake and H^+-ATPase.

In vitro shoot cultures of chestnut have offered an ideal system for monitoring physiological host responses at biochemical and molecular levels (Schafleitner and Wilhelm, 1997, 1999, 2000, 2001, 2002a,b; Wilhelm et al., 1998; Vannini et al., 1999; Schafleitner et al., 1999). To elucidate the modes of interaction between the plant, pathogen, and bacterium, dual and triple cultures were established. The level of SA as well as gene expression patterns upon wounding, the induction pattern of extracellular PR protein-levels (chitinases and β-1,3 glucanases), and also protein patterns were monitored during the different disease manifestations expressed locally in the shoots and systemically in the leaves.

PR Proteins and Gene Expression

PR proteins are known to be important for plant defense mechanisms (Bowles, 1990). Chitinases and β-1,3 glucanases inhibit several fungal pathogens (Schlumbaum et al., 1986). PR protein activity (chitinase and β-1,3 glucanase) was shown to be more strongly induced in the bark of resistant chestnuts *(C. mollissima)* than in the bark of the susceptible American chestnut after being challenged with ethylene (Gao and Shain, 1995).

Vannini et al. (1999) purified four proteins (C1, C2, C3, and C4) with a chitin-binding domain from in vitro chestnut plantlets. Three of the proteins, C2, C3, and C4, displayed chitinase activity. C3 from the N-terminal sequence is a Class I basic chitinase displaying high-sequence homology with other Class I chitinases from chestnut cotyledons and other plants. Antifungal tests showed that C1, C3, and C4 inhibit hyphal growth of the chestnut blight fungus in culture. The proteins were much more effective in inhibiting the hypovirulent form of the fungus than the virulent form, suggesting that the hypovirus induced effect upon virulence may be due to changes in the susceptibility of *C. prasitica* to host defense responses.

In vitro shoots of chestnut *(C. sativa)* were inoculated with three different kinds of biotic stress: virulent *C. parasitica*, hypovirulent *C. parasitica*, and a nonpathogenic endophytic *Bacillus subtilis*. These gram-positive, spore-forming, facultative anaerobic bacteria can be commonly found in nature and as a tissue-culture contaminant. *B. subtilis* is known to act as an antagonist against several phytopathogenic fungi (*Vitis-Eutypa*, Ferreira et al., 1991). A strain of *B. subtilis* (FZB 24) has been registered as a "plant strengthening product" since 1998 and is commercially available in Germany. Field experiments with *B. subtilis* have demonstrated biological control of fire blight, caused by *E. amylovora*, on *M. domestica* (Laux et al., 1999). Chestnut shoots cultured in vitro were inoculated with *B. subtilis*, and the tissue was then challenged with fungal pathogens. Development of disease symptoms was scored in vitro, and the levels of extracellular PR-proteins (chitinase, β-1,3-glucanase) were monitored. The multiplication rate of bacterized chestnut shoots in comparison to noninoculated shoots was assessed. To verify the results of this bioassay, in vivo tests with chestnut stems were also performed. Preinoculation of in vitro chestnut shoots with *B. subtilis* resulted in significant protection against *C. parasitica* (Wilhelm et al., 1998). These effects were attributed to bacterial elicitation, leading to a temporary increase in the level of PR proteins, an occurrence which may be related to the phenomenon of SAR (systemic acquired resistance). Induced resistance responses to microbial inoculants are well documented for plants grown in vivo (Kloepper et al., 1997), while this phenomenon is also known to be caused by specific strains of nonpathogenic rhizobacteria-colonizing plant roots (Van Loon et al., 1998).

Experiments with virulent and hypovirulent *C. parasitica* showed that the defense mechanisms of chestnut shoots (PR proteins) are more strongly induced by the hypovirulent form of the fungus as compared with the virulent fungus (Schafleitner and Wilhelm, 1997). It has been hypothesized, therefore, that the lower virulence level of the fungus and the stronger defense mechanisms of the tree together lead to the high survival rate after infection with hypovirulent *C. parasitica*. Similarly, it was found that chestnut in vitro stems contain chitinase and β-1,3-glucanase enzymes, and their activity increases under different biotic stress conditions (Schafleitner and Wilhelm, 1997). All forms of biotic stress elicited within the host tissue a general increase in PR protein levels. Wounding alone did not cause a significant increase beyond the normal constitutive expression of the two PR proteins. Studies with hybrid poplar and *Citrus sinensis* have demonstrated that mechanical wounding induces chitinases in the leaves (Clarke et al., 1994;

Derckel et al., 1996). Plant hormones mediate effects in plants over long distances, and ethylene promotes gene activation of defense-related processes (Ward et al., 1991; Ryals et al., 1994). Ethylene production in plant tissue increases temporarily following wounding, a phenomenon occurring during subculture manipulation, when shoots are cut at their base to be multiplied, rooted, or inoculated. Depending on the type of culture vessel and closure, plant tissues may be exposed to high ethylene levels throughout the entire culture period. Plants possess at least two separate systems of resistance induction: one involving pathogens through SA and another involving wounding, mediated by ethylene and methyl jasmonate. This explains why PR proteins can be induced by both pathogens and wounding (Derckel et al., 1996). These findings are also valid for the chestnut-*Cryphonectria* pathosystem (Schafleitner and Wilhelm, 1999; Schafleitner et al., 1999). Very little attention has been given to active defense mechanisms of trees, while little is known about the role of plant growth regulators (PGRs) in the defense mechanisms of trees (Johnson, 1987). Exogenous application of PGRs has been shown to influence the expression of genes encoding PR proteins (Memelink et al., 1987; Xu et al., 1994; Leubner-Metzger and Meins, 1999). In addition to the wounding effect, the shoot multiplication media were supplemented with a low concentration of BAP (0.9 μM), a cytokinin commonly used. The resulting, high-background level of PR protein may be explained as a general stress response due to the tissue culture system itself. In vitro chestnut shoots produced PR protein levels approximately 100 times higher than greenhouse plants of the same genotype (unpublished results).

Cryphonectria parasitica is a wound parasite, and the primary host recognition events it is involved in occur outside of the cells. In many host-pathogen interactions, the gene products induced during the defense response are secreted to the extracellular space (Bowles, 1990). Therefore, the apoplast is of key importance. In these studies, the site of inoculation was the stem, and intercellular fluid was recovered from the shoots in vitro, while leaves were analyzed separately. This was the first report on intercellular fluids (IFs) recovered from a plant tissue culture system for the purpose of observing tree defense responses. Previous reports on defense-related gene products derived from the IF of annual plants had identified chitinases, β-1,3 glucanases (Hammond-Kosack, 1992), hydroxyprolin-rich-glycoproteins, and numerous additional PR proteins of unknown function. The results obtained are in accordance with previous findings demonstrating that extracellular chitinases are mostly acidic. Time-course studies of chestnut-microbe interaction have revealed that PR protein patterns are differentially induced, depending on the kind of fungal infection. Infection with a hypovirulent *C. parasitica* strain resulted in earlier and stronger induction of PR proteins, locally and systemically, than infection with the virulent strain (Schafleitner and Wilhelm, 1997). Hypovirulent *C. parasitica* induces in the host a more rapid recognition process that may lead to earlier activation of defense mechanisms. This may contribute to the different fates of infections. Several other PR proteins as well as SA (Schafleitner et al., 1999; Schafleitner and Wilhelm, 2001) are known to be involved in chestnut defense responses, yet it is unclear as to whether these components really play a key role in resistance against chestnut blight.

Wound-Responsive Genes

In *C. sativa* shoots grown in tissue culture, 26 wound-responsive genes were able to be isolated by means of mRNA display; their expression upon wounding and infection with the chestnut blight fungus was found to differ (Schafleiter and Wilhelm, 2002b). The functions of the isolated genes were attributed to signaling, stress- and pathogen-response, cell wall modification, protein- and sterol-metabolism, and intracellular transport. For five of the isolated cDNAs, no function of the corresponding genes could be deduced. Temporal expression assessment upon wounding and inoculation with the chestnut blight fungus by reverse Northern dot blot hybridization revealed that the expression profiles of most of the wound-responsive genes were altered upon *C. parasitica* inoculation. Pathogen-elicited gene induction as well as suppression of wound-responsive gene expression was observed. It was thus concluded that *C. parasitica* does not inhibit the gene expression of the host in general, but rather influences the transcript accumulation in a specific way. Expression analysis involving greenhouse-grown chestnut trees verified that several of these isolated genes are indeed involved in the wound response of bark tissue (Schafleitner and Wilhelm, 2002a). The putative wound-responsive signaling genes (serine-threonine protein kinase, two calmodulin genes), a novel wound-responsive putative chaperone gene, and a new family of proline-rich proteins revealed strong temporal induction of these genes upon wounding.

LIMITATIONS AND PROBLEMS

There are obvious advantages to tissue culture systems, but there are also limitations. It is important to remember that callus and cell suspension cultures are physiologically different from whole plants. In shoot cultures, protective tissue such as cuticle may be different or even absent, the tissues are often actively growing, and therefore, resemble only meristematic areas of the plant, and sometimes resistance is not expressed in culture. Plant growth regulators may alter the biochemical host response, while events associated with resistance may be different in callus and whole plants. When growth regulators, either from the medium or produced from the pathogen, interact with plant tissues, confounding of results may ensue. Using an in vitro selection system, Vardi et al. (1986) attempted to identify resistant *Citrus* plants by applying culture filtrates of *Phytophthora citrophthora* (root rot of *Citrus* spp.). The results obtained were contrary to known resistance of *Citrus* lines, because an auxin-like substance had been produced in the fungal culture filtrate. Elicitation studies using apple and pear cell suspension cultures could not be confirmed with intact plants (Faize et al., 2002; Hrazdina, 2003). In addition, Lux-Endrich et al. (2002) reported that many different factors, including elicitor concentration, media composition, cell line, growth stage, and growth behavior of the cell cultures, have been shown to influence the responses of the apple cells.

CONCLUSIONS

The suitability of tissue culture systems for studies of host responses of trees on the biochemical and molecular levels has been able to be demonstrated. The future affords great possibilities in view of the prospects for combining forward and reverse genetics, biochemistry, and physiology. Although our knowledge in the field of developmental tree physiology is steadily growing, thus contributing to the improvement of cell culture methods, the systems are often not as trivial as previously thought. Results are sometimes difficult to reproduce, and unknown factors may be involved, among them significant influence on the part of the operator. Nevertheless, great opportunities along with additional applications for studying host-parasite interactions in woody species will be provided by the growing knowledge obtained from marker genes and targeted molecular tree breeding programs. New strategies for future disease control concepts will take into account more strongly the complex biochemical and molecular communication taking place during host-pathogen interaction. Often, it will be easier to gain such insights by the use of suitable tissue culture systems. However, for practical purposes, it is of the utmost importance to additionally verify any information generated in the laboratory in vivo, that is, under natural field conditions.

REFERENCES

Abdollahi, H., M. Ruzzi, E. Rugini, and R. Muleo. (2004). In vitro system for study-ing the interaction between *Erwinia amylovora* and genotypes of pear. *Plant Cell Tissue and Organ Culture* 79: 203-212.

Agrios, A. (ed). (1997). *Plant Pathology*, 4th Edition. Academic Press, San Diego, CA.

Allona, I., C. Collada, R. Casado, J. Paz-Ares, and C. Aragoncillo. (1996). Bacterial expression of an active class 1b chitinase from *Castanea sativa* cotyledons. *Plant Molecular Biology* 32: 1171-1176.

Anagnostakis, S.L. (1987). Chestnut blight: The classical problem of an introduced pathogen. *Mycologia* 29: 23-37.

Anagnostakis, S.L. (1992). Chestnut bark tannin assays and growth of chestnut blight fungus on extracted tannin. *Journal of Chemical Ecology* 18: 1365-1373.

Bazzigher, G. (1953). Beitrag zur Kenntnis der Endothia parasitica (Murr.) And., dem Erreger des Kastaniensterbens. *Phytopathologische Zeitschrift* 21: 105-132.

Bazzigher, G. (1955). Über Tannin-und Pehnolspaltende Fermente dreier para-sitischer Pilze. *Phytopathologische Zeitschrift* 29: 299-304.

Biggs, A.R. and C.A. Peterson. (1990). Effects of chemicals applied to peach bark wounds on accumulation of lignin and suberin and susceptibility to *Leucostoma persoonii*. *Phytopathology* 80: 861-865.

Borejsza-Wysocki, W., C. Lester, A.B. Attygalle, and G. Hrazdina. (1999). Elicited cell suspension cultures of apple *(Malus × domestica)* cv. Liberty produce biphenyl phytoalexins. *Phytochemistry* 50: 231-235.

Bowles, D.J. (1990). Defense related proteins in higher plants. *Annual Reviews of Biochemistry* 59: 873-907.

Brar, D.S. and S.M. Jain. (1998). Somaclonal variation and in vitro selection to plant improvement. In *Somaclonal Variation and Induced Mutations in Crop Improve-ment,* eds. S.M. Jain, D.S. Brar, and B.S. Ahloowalia. Kluwer Academic Publish-ers, Dordrecht, the Netherlands, pp. 15-37.

Cacciola, S.O., N.A. Williams, D.E.L. Cooke, and J.M. Duncan. (2001). Molecular identification and detection of *Phytophthora* species on some important Medi-terranean plants including sweet chestnut. *Forest Snow and Landscape Research* 76: 351-356.

Carpenter, C.E., R.J. Mueller, P. Kazmierczak, L. Zhang, D.K. Villalon, and N.K. Van Alfen. (1992). Effect of a virus on accumulation of a tissue-specific cell-surface protein of the fungus *Cryphonectria (Endothia) parasitica*. *Molecular Plant-Microbe Interactions* 4: 55-61.

Chevrau, E., M.N. Brisset, J.P. Paulin, and D.J. James. (1998). Fireblight resistance and genetic trueness-to-type of four somaclonal variats from the apple cultivar Greensleeves. *Euphytica* 104: 199-205.

Christiansen, E., P. Krokene, A.A. Berryman, V.R. Franceschi, T. Krekling, F. Lieu-tier, A. Lönneborg, and H. Solheim. (1999). Mechanical injury and fungal infec-tion induce acquired resistance in Norway spruce. *Tree Physiology* 19: 399-403.

Clarke, H.R.G., J.M. David, S.M. Wilbert, H.D. Bradshaw, and M.P. Gordon. (1994). Wound induced and developmental activation of a poplar tree chitinase gene promotor in transgenic tobacco. *Plant Molecular Biology* 25: 799-815.

Collada, C., R. Casado, and C. Aragoncillo. (1993). Endochitinases from *Castanea crenata* cotyledons. *Journal of Agricultural Food Chemistry* 41: 1716-1718.

Connors, B.J., N.P. Laun, C.A. Maynard, and W.A. Powell. (2002). Molecular characterization of a gene encoding a cystatin expressed in the stems of American chestnut *(Castanea dentata)*. *Planta* 215: 510-514.

Dai, G.H., C. Andary, L. Monodolot-Cosson, and D. Boubals. (1995). Involvement of phenolic compounds in the resistance of grapevine callus to downy mildew *(Plasmopara viticola)*. *European Journal of Plant Pathology* 101: 541-547.

Daub, M.E. (1986). Tissue culture and the selection of resistance to pathogens. *Annual Review of Phytopathology* 24: 159-186.

Deng, Z.N., A. Gentile, F. Domina, E. Nicolosi, and E. Tribulato. (1995). Selecting lemon protoplasts for insensitivity to *Phoma tracheiphila* toxin and regenerating tolerant plants. *Journal of the American Society for Horticultural Science* 120: 902-905.

Derckel, J.P., J.C. Audran, B. Haye, B. Lambert, and L. Legendre. (1996). Characterization, induction by wounding and salicylic acid, and activity against *Botrytis cinerea* of chitinases and beta-1,3-glucanases of ripening grape berries. *Physiologia Plantarum* 104: 56-64.

Diner, A.M. and D.F. Karnosky. (1987). Tissue culture application to forest pathology and pest control. In *Cell and Tissue Culture in Forestry*, Vol 2, eds. J.M. Bonga and D.J. Durzan. Martinus Nijhoff Publishers, Dordrecht, the Netherlands, pp. 351-373.

Dixon, A. (1983). Plant tissue culture methods in the study of phytoalexin induction. In *Tissue Culture Methods for Plant Pathologists*, eds. D.S. Ingram and J.P. Helgeson (eds.) Balckwell Scientific Publication, Oxford, UK, pp.185-196.

Donovan, A.M., R. Morgan, C. Valobra-Piagnani, M.S. Ridout, D.J. Janes, and C.M.D. Garrett. (1994). Assessment of somaclonal variation in apple: I. Resistance to the fire blight pathogen, *Erwinia amylovora. Journal of Horticultural Science* 69: 105-113.

Dorey, S., F. Baillieul, M.A. Pierrel, P. Saindrenan, B. Fritig, and S. Kauffmann. (1997). Spatial and temporal induction of cell death, defense genes and accumulation of salicylic acid in tobacco leaves reacting hypersensitively to a fungal glycoprotein elicitor. *Molecular Plant-Microbe Interactions* 10: 646-655.

Durrant, W.E., O. Rowland, P. Piedras, K.E. Hammond-Kosack, and J.D.G. Jones. (2000). cDNA-AFLP reveals a striking overlap in race-specific resistance and wound response gene expression profiles. *Plant Cell* 12: 963-977.

Eklund, L. and C.A. Little. (1998). Ethylene evolution, radial growth and carbohydrate concentrations in *Abies balsamea* shoots ringed with ethrel. *Tree Physiology* 18: 383-391.

Faize, M., M.N. Brisset, J.S. Venisse, J.P. Paulin, and M. Tharaud. (2002). Protective effects against *Erwinia amylovora* induced by *hrp* mutants in the whole plant are only partially mimicked in cultivated cells. *European Journal of Plant Pathology* 108: 547-553.

Ferreira, J.H.S., F.N. Matthee, and A.C. Thomas. (1991). Biological control of *Eutypa lata* on grapevine by an antagonistic strain of *Bacillus subtilis. Phytopathology* 81: 283-287.

Gao, S. and L. Shain. (1995). Activity of polygalacturonase produced by *Cryphonec-tria parasitica* in chestnut bark and its inhibition by extracts from American and Chinese chestnut. *Physiological and Molecular Plant Pathology* 46: 199-213.

Gleadow, R.M. and I.E. Woodrow. (2000). Temporal and spatial variation in cyano-genic glycosides in *Eucalyptus cladocalyx. Tree Physiology* 20: 591-598.

Gomez, E., E. Sagot, C. Gaillard, L. Laquitaine, B. Poinssot, Y.H. Sanejouand, S. Delrot, and P. Coutos-Thevenot. (2003). Nonspecific lipid-transfer protein genes expression in grape (*Vitis* sp.) cells in response to fungal elicitors. *Molecular Plant-Microbe Interactions* 16: 456-464.

Gomez, L., I. Allona, R. Casado, and C. Aragoncillo. (2002). Seed chitinases. *Seed Science Research* 12: 217-230.

Haberlandt, G. (1902). Culturversuche mit isolierten Pflanzenzellen. Sitzungsber. Akademie der Wissenschaften Wien. *Mathematisch Naturwissenschaftliche Klasse* 111: 69-91.

Hammerschlag, F.A. (1988a). Screeening peaches in vitro for resistance to *Xanthomonas campestris* pv. pruni. *Journal of the American Society for Horticultural Science* 113: 164-166.

Hammerschlag, F.A. (1988b). Selection of peach cells for insensitivity to culture filtrates of *Xanthomonas campestris* pv. pruni and regeneration of resistant plants. *Theoretical and Applied Genetics* 76: 865-869.

Hammerschlag, F.A. (1990). Resistant responses of plants regenerated from peach callus to *Xanthomonas campestris* pv. pruni. *Journal of the American Society for Horticultural Science* 115: 1034-1037.

Hammerschlag, F.A. (2000). Resistant responses of peach somaclone 122-1 to *Xanthomonas campestris* pv. pruni and to *Pseudomonas syringae* pv. *syringae. Horticultural Science* 35: 141-143.

Hammerschlag, F.A., D. Ritchie, D. Werner, G. Hashimi, L. Krusberg, R. Meyer, and R. Huettel. (1995). In vitro selection of disease resistance in fruit trees. *Acta Horticulture* 392: 19-26.

Hammond-Kosack, K.E. (1992). Preparation and analysis of intercellular fluid. In *Molecular Plant Pathology—A Practical Approach,* Vol. 2, eds. S.J.Gurr, M.J. McPherson, and D.J. Bowles. IRL, Oxford University Press, Oxford, UK, pp. 15-21.

Hammond-Kosack, K.E. and J.E. Parker. (2003). Deciphering plant-pathogen communication: Fresh perspectives for molecular resistance breeding. *Current Opinion in Biotechnology* 14: 177-193.

Hanke, V. and K. Geider. (2002). A new approach to evaluate fireblight resistance in vitro. *Acta Horticulture* 590: 397-399.

Hebard, F.V., F.J. Griffin, and J.R. Elkins. (1984). Developmental histopathology of cankers incited by hypovirulent and virulent isolates of *Endothia parasitica* on susceptible and resistant chestnut trees. *Phytopathology* 74: 140-149.

Hebard, F.V. and P.B. Kaufman. (1978). Chestnut callus cultures: Tannin content and colonization by *Endothia parasitica.* In *Proceedings American Chestnut Symposium.* West Virginia. University Books, Morgantown, pp. 63-79.

Heiniger, U. and D. Rigling. (1994). Biological control of chestnut blight in Europe. *Annual Review of Phytopathology* 32: 581-599.

Heiniger, U., S. Schmid, and D. Rigling. (2001). Hypovirus and mitochondria transfer between *Cryphonectria parasitica* strains. *Forest Snow and Landscape Research* 76: 397-401.

Helgeson, J.P. and G.T. Haberlach. (1983). Disease resistance studies with tissue cultures. In *Tissue Culture Methods for Plant Pathologists*, eds. D.S. Ingram and J.P. Helgeson. Blackwell Scientific Publication, Oxford, UK, pp.179-184.

Herrmann, K.M. (1995). The shikimate pathway as an entry to aromatic secondary metabolism. *Plant Physiology* 107: 7-12.

Hrazdina, G. (2003). Response of scab-susceptibel (McIntosh) and scab-resistant (Liberty) apple tissues to treatment with yeast extract and *Venturia inaequalis*. *Phytochemistry* 64: 485-492.

Hrazdina, G., W. Borejsza-Wysocki, and C. Lester. (1997). Phytoalexin production in an apple cultivar resistant to *Venturia inaqualis*. *Phytopathology* 87: 868-876.

Ingram, D.S. and J.P. Helgeson. (eds.) (1980). *Tissue Culture Methods for Plant Pathologists*. Blackwell Scientific Publications, Oxford.

Johnson, J.D. (1987). Stress physiology of forest trees: The role of plant growth regulators. In *Hormonal Control of Tree Growth*, eds. S.V. Kossutz and S.D. Ross. Martinus Nijhoff Publishers, Dordrecht. *Plant Growth Regulation* 6: 193-215.

Joung, H., S.S. Korban, and R.M. Skirvin. (1987). Screening shoot cultures of *Malus* for cedar-apple rust infection by in vitro inoculation. *Plant Disease* 71: 1119-1122.

Keck, M., S. Richter, B. Suarez, E. Kopper, and E. Jungwirth. (2002). Activity of peroxidases in plant material infected with *Erwinia amylovora*. *Acta Horticulture* 590: 343-347.

Kloepper, J. W., S. Tuzun, and G.W Zehnder. (1997). Multiple disease protection by rhizobacteria that induce systemic resistance—historical perspective. *Phytopathology* 87: 136-137.

Kortekamp, A. and E. Zyprian. (2003). Characterization of *Plasmopara*-resistance in grapevine using in vitro plants. *Journal of Plant Physiology* 160: 1393-1400.

Kubisiak, T.L., F.V. Hebard, C.D. Nelson, J. Zhang, R. Bernatzky, H. Huang, S.L. Anagnostakis, and R.L. Doudrick. (1997). Molecular mapping of resistance to blight in an interspecific cross in the genus *Castanea*. *Phytopathology* 87: 751-759.

Kuksova, V.B., N.M. Piven, and Y.Y. Gleba. (1997). Somaclonal variation and in vtiro induced mutagenesis in grapevine. *Plant Cell Tissue and Organ Culture* 49: 17-27.

Laimer, M. and R. Rücker. (eds.) (2003). *Plant Tissue Culture 100 Years Since Gottlieb Haberlandt*. Springer Wien, New York.

Laux, P., K. Hofer, and W. Zeller. (1999). Untersuchungen im Freiland zur Bekämpfung des Feuerbrandes mit bakteriellen Antagonisten. In *Beitragsband, 2nd Symposium on Phytomedizin und Pflanzenschutz im Gartenbau*, ed. G. Bedlan, Wien, pp. 53-54.

Leubner-Metzger, G. and F. Meins Jr. (1999). Functions and regulations of plant β-1,3-glucanases (PR-2). In *Pathogenesis-Related Proteins in Plants*, eds. S.K. Datta and S. Muthukrishnan. CRC Press LLC, Boca Raton, FL, pp. 49-76.

Lev-Yadun, S. and R. Aloni. (1990). Polar patterns of periderm ontogeny, their relationship to leaves and buds, and the control of cork formation. *IOWA Bulletin* 11: 289-300.

Litz, R.E., W.H. Mathews, R.C. Hendrix, and C. Yurgalevitch. (1991). Mango somatic cell genetics. *Acta Horticulture* 291: 133-140.

Lux-Endrich, A., D. Treutter, and W. Feucht. (2002). Response of apple *(Malus domestica)* cell suspension cultures cv. Alkmene on elicitation with biotic elicitors. *Journal of Applied Botany* 76: 121-126.

Malamy, J., J.P. Carr. D.F. Klessig, and I. Raskin. (1990). Salicylic acid: A likely endogenous signal in the resistance response of tobacco to viral infection. *Science* 250: 1002-1004.

Maleck, K. and R.A. Dietrich. (1999). Defense on multiple fronts: How do plant cope with diverse enemies? *Trends in Plant Science* 4: 215-219.

McCarrol, D.R. and E. Thor. (1985). Pectolytic, cellulolytic and proteolytic activities expressed by cultures of *Endothia parasitica,* and inhibition of these activities by components extracted from Chinese and American chestnut inner bark. *Physiological Plant Pathology* 26: 367-378.

McComb, J.A., J.M. Hinch, and A.E. Clarke. (1987). Expression of field resistance in callus tissue inoculated with *Phytophthora cinnamomi. Phytopathology* 77: 347-351.

Memelink, J.J., H.C. Hodge, and R.A. Schilperoot. (1987). Cytokinin stress changed the developmental regulation of several defence related genes in tobacco. *The EMBO Journal* 6: 3579-3583.

Metraux, J.P., H. Signer, J. Ryals, E. Ward, M. Wyss-Benz, J. Gaudin, K. Raschdorf, E. Schmid, W. Blum, and B. Inverardi. (1990). Increase in salicylic acid at the onset of systemic acquired resistance in cucumber. *Science* 250: 1004-1006.

Morel, G. (1944). Le développement du mildiou sur des tissus de vigne cultures in vitro. *Comptes rendus hebdomadaire des séances de l'Academie des sciences* Paris 218: 50-52.

Morel, G. (1948). Recherches sur la culture associée de parasites obligatoires et de tissus végétaux. *Annales des Éphiphyties Série Pathologie Végétale* 14: 1-112.

Nürnberger, T. (1999). Signal perception in plant pathogen defense. *Cellular Molecular Life Sciences* 55: 167-182.

Ostry, M.E. and D.D. Skilling. (1992). Applications of tissue culture for studying tree defense mechanisms. In *Defense Mechanisms of Woody Plants Against Fungi,* eds. R.A. Blanchette and A.R. Biggs. Springer Verlag, Berlin, pp. 405-423.

Patocchi, A., L. Gianfranceschi, and C. Gessler. (1999). Towards the map-based cloning of *Vf*: Fine and physical mapping of the *Vf* region. *Theoretical and Applied Genetics* 99: 1012-1017.

Patocchi, A., B.A. Vinatzer, L. Gianfranceschi, S. Tartarini, H.B. Zhang, SA. Sansavini, and C. Gessler. (1999). Construction of a 550 kb BAC contig spanning the genomic region containing the apple scab resistance gene *Vf. Molecular General Genetics* 262: 884-891.

Pernas, M., E. Lopez-Solanilla, R. Sanchez-Mong, G. Salcedo, and P. Rodriguez-Palenzuela. (1999). Antifungal activity of a plant cystatin. *Molecular Plant-Microbe Interactions* 12: 624-627.

Pernas, M., R. Sanchez-Mong, and G. Salcedo. (2000). Biotic and abiotic stress can induce dystatin exporession in chestnut. *FEBS Letters* 467: 206-210.

Piagnani, C., G. Assante, P. Scalisi, Zocchi, and A. Vercesi. (2002). Growth and physiological responses of chestnut calli to crude extracts of virulent and hypovirulent strains of *Cryphonectria parasitica*. *Forest Pathology* 32: 43-53.

Piagnani, C., F. Faoro, S. Sant, and A. Vercesi. (1997). Growth and ultrastructural modifications to chestnut calli induced by culture filtrates of virulent and hypovirulent *Cryphonectria parasitica* strains. *European Journal of Forest Pathology* 27: 23-32.

Predieri, S. (2001). Mutation induction and tissue culture in improving fruits. *Plant Cell, Tissue and Organ Culture* 64: 195-210.

Reymond, P., H. Weber, M. Damond, and E.E. Farmer. (2000). Differential gene expression in response to mechanical wounding and insect feeding in *Arabidopsis*. *Plant Cell* 12: 707-720.

Richard, S., G. Lapointe, R.G. Rutledge, and A. Seguin. (2000). Induction of chalcone synthase expression in white spruce by wounding and jasmonate. *Plant Cell Physiology* 41: 982-987.

Roane, M.K., G.J. Griffin, and J.R. Elkins. (1986). *Chestnut Blight, Other Endothia Diseases and the Genus Endothia*. APS Press, St Paul, MN.

Rosati, P., B. Mezzetti, M. Ancherani, S. Foscolo, S. Predieri, and F. Fasolo. (1990). In vitro selection of apple rootstock somaclones with *Phytophthora cactorum* resistance with culture filtrates of the fungus. *Acta Horticulture* 265: 123-128.

Ryals, J., S. Uknes, and E. Ward. (1994). Systemic acquired resistance. *Plant Physiology* 104: 1109-1112.

Ryan C.A. (2000). The systemin signaling pathway: Differential activation of plant defensive genes. *Biochimia Biophysica Acta* 1477: 112-21.

Schafleitner, R., A. Buchala, and E. Wilhelm. (1999). Class III chitinase expression and salicylic acid accumulation in chestnut (*Castanea sativa* L) after challenge with hypovirulent and virulent *Cryphonectria parasitica* (Murr.) Barr. *Phyton* 39: 191-196.

Schafleitner, R. and E. Wilhelm. (1997). Effect of virulent and hypovirulent *Cryphonectria parasitica* (Murr.) Barr on the intercellular pathogen related proteins and on total protein pattern of chestnut (*Castanea sativa* mill.). *Physiological and Molecular Plant Pathology* 51: 323-332.

Schafleitner, R. and E. Wilhelm. (1999). Chestnut blight: Monitoring the host response with heterologous cDNA probes. *Acta Horticulture* 494: 481-486.

Schafleitner, R. and E. Wilhelm. (2001). Assessment of stress gene expression in chestnut (*Castanea sativa* Mill.) upon pathogen infection (*C. parasitica* (Murr.) Barr.) and wounding. *Forest Snow and Landscape Research* 76: 409-414.

Schafleitner, R. and E. Wilhelm. (2002a). Isolation of wound inducible genes from *Castanea sativa* stems and expression analysis in the bark tissue. *Plant Physiology and Biochemistry* 40: 235-245.

Schafleitner, R. and E. Wilhelm. (2002b). Isolation of wound-responsive genes from chestnut (*Castanea sativa* Mill.) microstems by mRNA display and their differential expression upon wounding and infection with the chestnut blight fungus (*Chryphonectria parasitica* (Murr.) Barr.). *Physiological and Molecular Plant Pathology* 61: 339-348.

Schlumbaum, A.L., F. Mauch, U. Vögeli, and T. Boller. (1986). Plant chitinases are potent inhibitors of fungal growth. *Nature* 324: 365-367.

Seemann, D. (2001). Plant health and quarantine regulations of the European Union for *Cryphonectria parasitica*. *Forest Snow and Landscape Research* 76: 402-404.

Shah, J. (2003). The salicylic acid loop in plant defense. *Current Opinion in Plant Biology* 6: 365-371.

Shain, L. and R.J. Spalding. (1995). Quantitation of chitinase and b-1.3 glucanase in bark of American and Chinese chestnut. Abstr. 1995 APS Annual Meeting, Pittsburgh PA., *Phytopathology* 85: 1142.

Sharma, N.K. and D.I. Skidmore. (1988). In vitro expression of partial resistance to *Phytophthora palmivora* by shoot cultures of papaya. *Plant Cell Tissue and Organ Culture* 14: 187-196.

Svircev, A.M., A.R. Biggs, N. Miles, and C. Chong. (1989). Isolation, purification and characterization of a phytotoxin from liquid cultures of *Leucostoma persoonii* and *L. cincta* (Abstr). *Phytopathology* 79: 1154.

Tomlin, E.S., R.I. Alfaro, J.H. Borden, and H.E. Fangliang. (1998). Histological response of resistant and susceptible white spruce to simulated white pine weevil damage. *Tree Physiology* 18: 21-28.

Tomlin, E.S., E. Antonejevic, R.I. Alfaro, and J.H. Borden. (2000). Changes in volatile terpene and diterpene resin acid composition of resistant and susceptible white spruce leaders exposed to simulated white pine weevil damage. *Tree Physiology* 20: 1087-1095.

Van Loon, L.C., P.A.H.M. Bakker, and C.M.J. Pieterse. (1998). Systemic resistance induced by rhizosphere bacteria. *Annual Review of Phytopathology* 36: 453-483.

Vannini, A., C. Caruso, L. Leonardi, E. Rugini, E. Chiarot, C. Caporale, and V. Buonocore. (1999). Antifungal properties of chitinases from *Castanea sativa* against hypovirulent and virulent strains of the chestnut blight fungus *Cryphonectria parasitica*. *Physiological and Molecular Plant Pathology* 55: 29-35.

Vannini, A., E. Rugini, and P. Magro. (1993). Polyphenoloxidase activity of *Castanea sativa* shoots growing in vitro infected with virulent and hypovirulent strains of *Cryphonectria parasitica*. In *Plant Signals in Interaction with Other Organisms*, eds. J. Schulz and I. Raskin. American Society of Plant Physiologists, Rockville, MD, pp. 258-261.

Vannini, A. and A.M. Vettraino. (2001). Ink disease in chestnut: Impact on the European chestnut. *Forest Snow and Landscape Research* 76: 345-350.

Vardi, A., E. Epstein, and A. Breiman. (1986). Is the *Phytophthora citrophthora* culture filtrate a reliable tool for the in vitro selection of resistant citrus variants? *Theoretical and Applied Genetics* 72: 569-574.

Vareley, D.A, G.K. Podila, and S.T. Hiremath. (1992). Cutinase in *Cryphonectria parasitica*, the chestnut blight fungus: Supression of cutinase gene expression in isogenic hypovirulent strains containing double-stranded RNAs. *Molecular Cell Biology* 21: 4539-4544.

Venisse, J.S., M.A. Barny, J.P. Paulin, and M.N. Brisset. (2003). Involvement of three pathogenicity factors of *Erwinia amylovora* in the oxidative stress associated with compatible interaction in pear. *FEBS Letters* 537: 198-202.

Vettraino, A.M., G. Natali, N. Anselmi, and A. Vannini. (2001). Recovery and pathogenicity of *Phytophthora* species associated with a resurgence of ink disease in *Castanea sativa* in Italy. *Plant Pathology* 50: 90-96.

Viseur, J. (1990). Evaluation of fire blight resistance of somaclonal variants obtained from the pear cultivar "Durondeau." *Acta Horticulture* 273: 275-284.

Ward, E.R., S.J. Uknes, S.C. Williams, S.S. Dincher, D.L. Wiederhold, D.C. Alexander, P. Ahlgoy, J.P. Metraux, and J.A. Ryals. (1991). Coordinate gene activitiy in response to agents that induce systemic acquired resistance. *Plant Cell* 3: 1085-1094.

Wilhelm, E., W. Arthofer, R. Schafleitner, and B. Krebs. (1998). *Bacillus subtilis* an endophyte of chestnut *(Castanea sativa)* as antagonist against chestnut blight *(Cryphonectria parasitica)*. *Plant Cell, Tissue and Organ Culture* 52: 105-108.

Winton, L.L. (1970). Shoot and tree production from aspen tissue cultures. *American Journal of Botany* 57: 904-909.

Xu, Y., P.L. Chang, D. Liu, M.L. Narasimhan, K.G. Raghothama, P.M. Hasegawa, and R.A. Bressan. (1994). Plant defense genes are synergistically induced by ethylene and methyl-jasmonate. *Plant Cell* 6: 1077-1085.

Yepes, L.M. and H.S. Aldwinckle. (1993). Selection of resistance to *Venturia inaequalis* using detached leaves from in vitro grown apple shoots. *Plant Science* 93: 211-216.

Chapter 14

Microbial Chitinases:
Effective Biocontrol Agents

Sandhya Chandran
Binod Parmeswaran
Ashok Pandey

INTRODUCTION

Synthetic chemical fungicides, insecticides, and pesticides have been extensively used over the past 50 years due to their effectiveness and ease of use. The widespread use of chemical pesticides causes serious health problems to humans and other life forms and can also induce pathogen resistance among them. It also results in hazardous environmental pollution. The public concern about the safety of these chemicals has led to more rigorous regulatory requirements. As a result, many of the most effective chemical pesticides have failed to pass registration. The removal of these chemical pesticides from the market coupled with the emergence of herbicides and insecticide-resistant pests has heightened interest in the development of biologically based pest control strategies (Jackson, 1997).

Biological control, in the classical sense, is the purposeful introduction of parasites, predators, and/or pathogenic microorganisms by humans to reduce or suppress populations of plant or animal pests. This concept of classical biological control is not new, having been practiced in many forms around the world since the earliest days of recorded history. Biological control has been and is currently used as a viable management strategy for insect pests, unwanted plants, and the control of nuisance reptiles and mammals. Plant protection is a major challenge to the agricultural field worldwide, with fungi being one of the main causes of significant yield loss. The extensive use of chemical fungicide for controlling fungal diseases has led to growing concern about the environment. In addition, the high cost of chemicals has encouraged farmers and researchers to look for substitutes, such as the use of biocontrol agents and fungus-resistant crop cultivars. In the past few years, numerous microorganisms with antifungal activities and their active factors have been identified and the mechanisms by which antifungal factors inhibit growth of potentially pathogenic fungi have been demonstrated (Wang et al., 1999). Although intensive activity toward the development of means of biological control is currently taking place, commercial products are few. The genetic approach of breeding to produce crops which are resistant to fungal diseases has proven successful; however, this time-consuming process is expensive and makes it difficult to react adequately to the evolution of new, virulent fungal races (Schickler and Chet, 1997). The augmentative method of biological control involves the massive application of a pest-specific, indigenous pathogen to weed or insect infested crops (Wilson, 1969; Templeton, 1982; Charudattan, 1991). In many ways, the inundative application of fungal biopesticides resembles the use of chemical pesticides since the agent is applied as needed and must contact the pest (Auld and Morin, 1995). The use of a stable, aggressive pathogen, which is capable of consistently killing the pest host under field conditions, is a requirement. By using an indigenous isolate, the biological control registration costs are reduced due to less stringent regulatory requirements. A limited number of commercial products using fungi in biological control are presently in use in the United States and around the world (Charudattan, 1991; Moore and Prior, 1993; Jackson, 1997).

Plants can be systemically protected against disease caused by some pathogens by limited infection or physiological/biochemical stress. This phenomenon has been termed *induced resistance* (Ross, 1961; Kuc, 1995). Several mechanisms have been suggested to contribute to plant induced resistance. They include the hypersensitive reaction (HR), the production of phytoalexins and pathogenesis-related (PR) proteins, and the deposition of lignin and hydroxyproline-rich glycoprotein (Ji and Kuc, 1996). One mechanism contributing to induced resistance is the reduced penetration by the fungus into plant tissue. The possible reason for this reduction of penetration may be the result of multiple mechanisms (Kuc, 1995). β-1,3 glucanase and chitinase activities are correlated with induced resistance (Metraux and Boller, 1986; Metraux et al., 1989; Irving and Kuc, 1990; Ji and Kuc, 1996). Since numerous PR-proteins have been reported to have antifungal activity (Niderman et al., 1995), it is possible that some chitinase isoforms may be antifungal when interacting with PR proteins other than β-1,3-glucanase and that they may also have an important function in releasing elicitors for activation of other defense mechanisms (Ji and Kuc, 1996).

As pathogenic fungi and insects contain chitin in their protective covers, induction of chitinases in plants is the main defense response. Chitinases are hydrolytic enzymes that catalyze chitin degradation and have been detected in bacteria, fungi, and plants and in the digestive systems of coelenterates, nematodes, polychaets, molusks, and arthropods.

In this chapter, we discuss the importance of microbial chitinases as biocontrol agents, strategies for improving their production, as well as mechanisms for biocontrol activity.

CHITINASE AS A BIOCONTROL AGENT

One of the strategies to increase plant tolerance to fungal pathogens is the constitutive overexpression of proteins involved in plant defense mechanisms. Most transgenic plants exhibit increased tolerance to fungal diseases relative to their nontransgenic counterparts. Chitinases are considered to be a defense-related protein in higher plants, which protects the plant against fungal pathogens by degrading chitin, a major component of the cell walls of many fungi. Upon interaction with a pathogen, plants initiate a complex network of defense mechanisms, among which is a dramatic increase in chitinase activity (Schickler and Chet, 1997). The combined expression of chitinases with other plant defense proteins such as glucanases and ribosome-inactivating proteins further enhances the plant's resistance to fungal attack (Schickler and Chet, 1997). The chitinase-producing organisms could be used directly in biological control of fungi or indirectly by using purified protein or through gene manipulation (Mahadevan and Crawford, 1997). Nowadays a great amount of effort is devoted to the cloning of several chitinase genes in order to incorporate them, by using gene transfer technology, into disease-resistant transgenic plants. But the problem of gene silencing is not completely solved. Another way to use enzymes for plant defense is to cultivate soil microorganisms expressing chitinolytic activity as plant protectors against soilborne pathogens. The target of chitinolytic enzymes is widespread. The degradation of chitin leads to growth inhibition and death of microorganisms (fungi, insects, worms etc.); moreover, the chitin oligomers are positively charged and possess antibacterial and antimutagenic properties.

MICROBIAL SOURCES OF CHITINASES

Due to the potential applications of chitinase as biocontrol agent, it is very important to study the organisms that can produce the enzyme. Chitinolytic microbes occur widely in nature and are the preferred source of chitinase because their production cost is low and enzyme contents of microbes are more predictable and controllable. Another factor is easy availability of raw materials with constant composition for their cultivation. The ability of a microbial community to degrade chitin is also important for the recycling of nitrogen in the soil.

The major chitinase-producing bacteria are *Vibrio alginolyticus, Streptomyces kurssanovii, Serratia marcescens, Serratia plymuthica, Bacillus circulans, Aeromonas cavie, Streptomyces lydicus, Stenotrophomonas maltophilia, Paenibacillus illinoisensis,* and *Paenibacillus illinoisensis,* of which *S. marcescens, S. plymuthica, A. cavie, S. lydicus, Stenotrophomonas maltophilia,* and *Paenibacillus illinoisensis* proved to be potent biocontrol agents.

Major chitinase-producing fungi used as biocontrol agents are *Trichoderma harzianum, Myrothecium verrucaria,* and *Trichoderma reesei.*

CLASSIFICATION OF CHITINASES

According to the recommendation by the International Union of Biochemistry and Molecular Biology (IUBMB) in 1992, complete hydrolysis of chitin is performed by an enzymatic system consisting of three categories of chitinases based on the mode of action (Patil et al., 2000).

1. Endochitinases—Random hydrolysis of the chain (EC 3.2.1.14)
2. Chitobiase—Hydrolysis of terminal nonreducing sugar (EC 3.2.1.29)
3. β-N-acetylglucosaminidase—Successive removal of sugar unit from the nonreducing end (EC 3.2.1.52). The classification of chitinolytic enzymes is not well defined.

Henrissat (1991) grouped chitinases and N-acetyl-hexoaminidases into three families, 18, 19, and 20, based on amino acid sequences and three dimensional structures. The family 18 chitinases are ubiquitous, but the family 19 chitinases have been identified mostly in plants. Plant family 19 chitinases are thought to constitute part of the defense mechanism against fungal pathogens and to have antifungal properties in vitro. N-acetyl hexosaminidase (EC 3.2.1.30) from *Vibrio harveyi* and N-acetyl hexosaminidases (EC 3.2.1.52) from human and *Dictyostelium discoideum* are comprised in family 20.

Chitinases fall into three broad classes, as proposed by Shinshi et al. (1990). Class I chitinases are usually localized in the vacuole and are potent growth inhibitors in vitro of many fungi. They are basic and contain cysteine-rich N-terminal domains with putative chitin-binding properties.

Class II chitinases consists of a monomeric catalytic domain with strong homology to the catalytic domain of the class I chitinases, but lack the cysteine-rich domain. They are generally acidic and extracellular and can be detected in the apoplastic fluid or culture medium of protoplasts (Schickler and Chet, 1997). Class III chitinases are extracellular hydrolases whose conserved catalytic domain amino acid sequence differs from the conserved sequence of class I or class II chitinases. Most of the class III chitinases are classified as such on the basis of homology to previously described lysozymes with chitinase activity. Class IV chitinases lack the C-terminal extension, and are therefore assumed to be accumulated extracellularly. Thus, these chitinases may fulfill an antifungal role similar to that of class I within the apoplast.

PRODUCTION METHODS FOR CHITINASES

Recent biochemical research on plant disease control focused on two prime objectives. They were (1) selection and identification of microorganisms with antibacterial or antifungal activities and production, isolation, and characterization of the specific antifungal factors within these microorganisms; and (2) determination of the operative mechanisms of these antimicrobial agents. Analyses of the various production methods for fungal biopesticides have been the subject of several in-depth reviews (Churchill, 1982; Latge and Moletta, 1988; Goettel and Roberts, 1992; Jackson et al., 1996). In general, state two methods exist for producing chitinases through fermentation: solid, fermentation (SSF) and liquid culture fermentation or submerged fermentation (SmF).

Solid-State Fermentation

The production of fungal spores using solid-state fermentation is the widely accepted method because, in nature, most fungi form conidia on aerial hyphae. It is a low-cost fermentation process, particularly suitable to the needs of developing countries. One of the major advantages of SSF is that usually it is carried out using naturally occurring agricultural by-products such as straw, bran, etc. (Pandey, 1992; Pandey et al., 2001; Lonsane et al., 1992). Agroindustrial residues offer potential advantages as substrates, since they supply nutrients to the microbial culture growing in it as well as anchorage for the cells (Pandey and Soccol, 2000; Pandey, Soccol, and Mitchell, 2000; Pandey, Soccol, Nigam, Brand, et al., 2000; Pandey, Soccol, Nigam, and Soccol, 2000; Pandey, Soccol, Nigam, Soccol, et al., 2000). This technique can easily be carried out in the laboratory, and often the propagules produced in an aerial environment, conidia, tend to be more tolerant to desiccation and more stable as a dry preparation compared to spores produced in submerged culture (Bartlett and Jaronski, 1988; Silman et al., 1993). However, this system is susceptible to water content, pH, and oxygen gradients. Also, the scale-up of SSF processes to a commercial level is difficult due to problems associated with substrate sterilization, gas exchange, temperature control, maintenance of pure culture, and product recovery from the substratum (Mitchell, Krieger, et al., 2000; Mitchell, Pandey, et al., 2000).

Optimization of Process Parameters for Improving the Enzyme Production

During medium optimization for the production of biological control agents propagule stability (desiccation tolerance, shelf life) and propagule efficacy as a biocontrol agent must be considered. In SSF, nutritional components for the microorganism provide low-cost, complex substrates. Use of this directed optimization strategy not only aids in the development of production media for specific biocontrol agents but also provides nutritional information which will be useful in developing production media for other microbial biocontrol agents (Jackson, 1997).

The process of SSF is influenced by both biological and physico-chemical factors. All these factors are intimately related and cannot be considered as independent from one another.

Biological Factors

These factors are related with the biology, metabolic process, and reproduction of a living species or organism. These determine the behaviour of the particular species in a specific way (Pandey et al., 2001).

Physico-Chemical Factors

These factors include physical and chemical factors such as temperature, pH, moisture and water activity, aeration and agitation, particle size, and carbon and nitrogen sources.

Temperature

The incubation temperature is largely characteristic of the organism, irrespective of the type of solid support involved in SSF (Nagendra and Chandrasekaran, 1996). *Trichoderma harzianum* exhibited a better growth and chitinase production at 30°C in a wheat bran medium (Nampoothiri et al., 2003). Studies by Wang and Hwang (2001) on *Bacillus* chitinase using shrimp and crab shell powder as solid medium also show 30°C as the optimal temperature. Low enzyme activity at the highest temperature might be due to increased cell turnover of proteins and nucleic acids leading to less energy being available for growth-associated functions (Forage et al., 1985).

pH

pH is another important physical factor in any fermentation process. Each microorganism possesses a pH range for its growth and activity. Hence, controlling pH of the medium in SSF is important. Agroindustrial residues generally possess an excellent buffering capacity (Pandey et al., 2001). This was supported by the results obtained by Nampoothiri et al. (2003). In SSF initial pH of the medium can be controlled by adjusting the pH of wetting salt solution. In *Trichoderma longibrachiatum* the initial pH value of the wetting salt solution of 2-5 gave best chitinolytic activity (Kovacs et al., 2003).

Moisture and Water Activity (a$_u$)

Moisture is an important factor, intimately related with the characteristics of the biological material. A great majority of viable cells are characterized by moisture content of 70 to 80 percent. In the case of fungi the range could be as wide as 20 to 70 percent. The respective pressure between the mixture of substrate and the gaseous phase in equilibrium is termed as water activity. The optimal moisture of 66.7 percent was found to be best in the case of *T. longibrachiatum* (Kovacs et al., 2003). Maximum chitinase activity by *Trichoderma harzianum* was observed at initial moisture of 65.7 percent (Nampoothiri et al., 2003). The influence of moisture on chitinase yield by *Verticillium lecanii* ATCC 26854 showed 75 percent initial moisture content as the optimum level (Matsumoto et al., 2001). The metabolic activities of the culture and consequently product synthesis were variously affected by low and high moisture contents. In the case of very less moisture catabolic heat evolution causes water evaporation, which leads to rapid decrease in moisture content and affects the growth and enzyme production.

Aeration and Agitation

These two factors have a determinant influence in SSF. During SSF there exists heat gradients owing to the heterogeneous nature of the system. This substrate heterogeneity may cause air canalization. When air canalization occurs, there will be nonaerated zones that could produce nondesired cell metabolism. Forced aeration and agitation are possible solutions to solve substrate heterogeneity.

Particle Size

An optimum particle size is needed for growth and chitinase production in SSF. In a wheat-bran-based medium, average particle size of <425 μm produced maximum chitinase activity (Suresh and Chandrasekharan, 1999). Reduction in particle size provides larger surface area for microbial growth, but interparticle porosity is less. In the case of larger particles, porosity is greater while surface area is less.

Carbon Source and Carbon/Nitrogen Relationship

The main energy source that will be available for the growth of the microorganism is the carbon source. It was found that though high constitutive levels of enzyme production were observed when *Trichoderma longibrachiatum* IMI 92027 was grown on different carbon sources other than chitin, addition of chitin to SSF medium induced the production of chitinase considerably (Kovacs et al., 2003). In *Trichoderma harzianum* chitinase production was enhanced by the addition of colloidal chitin instead of chitin flakes when rice husk is used as the substrate (Binod et al., 2003).

Nitrogen is another important factor that determines the growth of microorganisms. The ratio between the mass of carbon and nitrogen (C/N) is most crucial for the production of specific enzymes. Besides, when dealing with fungi, it is necessary to take into account how the C/N ratio could induce or delay sporulation. No significant effect on chitinase yield was observed by different nitrogen additives in wheat bran chitin medium (Kovacs et al., 2003). However, Nampoothiri et al. (2003) reported that supplementation of 2.0 percent (w/w) yeast extract to wheat-bran-based medium showed enhancement in chitinase yield by *Trichoderma harzianum*.

Submerged Fermentation

At present, liquid culture fermentation is the most economical method for producing most microbial biocontrol agents. The medium in the submerged fermentation is liquid which remains in contact with the microorganism. A supply of oxygen is essential in submerged fermentation. There are four main ways of growing microorganisms in submerged fermentation: (1) batch culture; (2) fed batch culture; (3) perfusion batch culture; and

(4) continuous culture. The production of antibiotics, amino acids, ethanol, and organic acids by submerged culture fermentation has provided an extensive knowledge base for optimization processes and for designing fermentation vessels for the liquid culture production of biopesticides. Production methods for bakers' yeast, distillers' yeast, and bacterial starter cultures for the diary industry have demonstrated that living biomass derived from liquid culture fermentations can be economically produced and can be stabilized as dry preparations. These commercial successes using liquid culture fermentation have strengthened industry's acceptance of this method. Three of the four fungal biopesticides registered for commercial use in North America are produced using liquid culture fermentation (Jackson, 1997). In submerged fermentation, a homogenous nutritional environment can be maintained and monitored. The homogeneity of a liquid medium simplifies production and processing methods. It also aids in the development of optimized nutritional conditions for production. In addition, environmental factors such as temperature, aeration, and pH can be easily controlled compared to solid substrate fermentation. Controlled nutritional and environmental conditions, process scale-up capabilities, quality assurance issues, ease of product recovery, and lower production costs makes this production method a widely accepted one.

Factors Influencing Chitinase Production
Under Submerged Fermentation

The optimization of nutritional conditions for the production of fungal biocontrol agents in submerged culture involves development of a medium which maximizes not only propagule yield but also propagule fitness as a biocontrol agent. After defining a medium that supports adequate growth, nutrients are varied in a directed way and their impact on spore yield and spore fitness can be assessed. A production medium can be formulated from the optimized defined medium.

Carbon source. In most fermentation processes it was found that addition of carbon sources other than chitin reduced chitinase production but supported growth. Generally chitin concentration in the range of 1 to 1.5 percent was found to be suitable for chitinase production (Felse and Panda, 2000). Monreal and Reese (1969) suggested that the most probable inducers of chitinase in *Serratia marcescens* are soluble oligomers derived from chitin, but not the monomer. But St Leger et al. (1986) demonstrated that the most effective inducer of chitinase in *Metarhizium anisopliae* was N-acetylglucosamine, when supplied at a rate of about 20 μg $\cdot$ ml^{-1} per h. Among the various pentoses and hexoses studied with *Streptomyces* species, arabinose doubled enzyme production while glucose repressed enzyme synthesis (Gupta et al., 1995). No chitinase production was observed when *Stachybotrys elegans* was grown on glucose, sucrose, or N-acetyl glucosamine (Tweddel et al., 1994). Frandberg and Schnurer (1994) reported that production of chitinase by *B. pubuli* K1 grown on chitin was repressed by addition of glucose, starch, β-glucan, and glycerol. High chitinase activity was found in *Trichoderma harzianum* cultures supplied with chitin but not with other polymers such as cellulose and chitosan. Neither chitobiose nor N-acetyl glucosamine promoted enzyme production in this particular strain (Ulhoa and Peberdy, 1991).

Nitrogen source. In industrial microbiology, the nitrogen source affects the synthesis of enzymes involved in both primary and secondary metabolism (Shapiro, 1989; Merrick and Edwards, 1995). The best nitrogen source for the production of chitinase by *Pseudomonas stulzeri* YPL-1 was found to be peptone (3 percent) (Ho-Seong and Sang-Dal, 1994). Similarly, supplementation of nitrogen sources (0.42 percent w/v) such as peptone and tryptone in the fermentation medium showed a marked increase in chitinase production by *Trichoderma harzianum* (Sandhya et al., 2004). Exclusion of urea from the production medium showed increased chitinase production (Kapat et al., 1996).

pH. Unlike SSF, pH of the fermentation media in submerged fermentation is easily controllable due to its homogeneous nature. Reports showed that most of the fungal chitinases are produced under slightly acidic pH. In the case of *Trichoderma harzianum* 39.1, optimal pH of 6.0 was found to be best for chitinase production (Ulhoa and Peberdy, 1991).

Aeration and agitation. Aeration and agitation will not only help in homogenization of the media but also lead to increase in substrate consumption, growth, and subsequent enzyme production. It enhances the phenomenon of mass transfer and will also help in the uniform distribution of organisms but have negative effects such as rupture of cells, change in morphological state, decrease in productivity, etc. The best combination for extracellular chitinase by *Trichoderma harzianum* in batch mode was found to be an aeration rate of 1.5 ll (−1) per minute and agitation rate of 224 rpm (Felse and Panda, 1999). Frandberg and Schnurer (1994) reported that chitinase production by *Bacillus pabuli* was highest at 10 percent oxygen. A possible explanation could be that oxygen concentrations above 10 percent inactivated chitinase while activating protease. Mild agitation and aeration conditions are suitable for large-scale process development and was supported by the observations of Khoury et al. (1997).

Temperature. The metabolic activity of the microorganisms is proportional to the heat liberated during fermentation. A temperature range of 28°C to 30°C was found to best for chitinase production. The highest production of chitinase by *Bacillus pabuli* K1 was observed at 30°C (Frandberg and Schnurer, 1994).

Improving Production of Chitinase Using Gene Manipulation

Genetic improvement plays an important role in the production of chitinases. The conventional procedures for the strain improvement are mutation and protoplast fusion. UV irradiation and nitrosoguanidine treatment yielded a nonpigmented chitinase and chitobiase overproducing mutant of *Serratia marcescens* (Joshi et al., 1989). Kole and Altosarr (1985), using a nonpigmented, stable mutant of *Serratia marcescens* designated BL 40, showed about 167 percent increase in chitinase activity over the wild-type strain under similar conditions.

Genetic engineering programs aim at developing disease resistance plants by introducing genes encoding chitinase. Extensive literature is available on molecular cloning for chitinases either to increase biocontrol efficiency, to prepare highly active chitinase or even transgenic plants to increase pathogen resistance (Wiwat et al., 1996; Tantimavanich et al., 1997; Watanalai et al., 1997). These genes have originated from both bacteria (Jones, 1986) and plants (Shinishi et al., 1987; Broglie et al., 1991; Nishizawa and Hibi, 1991). The bacterial chitinase gene from *Serratia marcescens* (*chi-A*) was transferred into tobacco plants by Jones (1988) and Suslow et al. (1988), and the transformants showed significantly higher chitinase activity and more resistance to *Alternaria longipes* than that of the control plant. Table 14.1 provides a list of sources and host for chitinase genes involved in improving biocontrol efficiency through molecular cloning.

Genes from mycoparasitic fungi were found to be a rich source for controlling disease in plants (Lorito et al., 1998). A fungal chitinase gene from *Rhizopus oligosporus* confers antifungal activity of transgenic tobacco (Terakawa et al., 1997). *Trichoderma* parasitic fungus-fungus interactions involve a number of chitinases, both enzymes and genes, and chitinase-mediated molecular mechanisms that can provide interesting opportunities to increase biocontrol of pests (Kubicek and Harman, 1998). In fact, the *Trichoderma* system is revealing several biocontrol genes and gene products useful for increasing the antifungal activity of *Trichoderma* itself, other microbes, or plants. The co-transformation of *Trichoderma reesei* protoplasts with *Aphanocladium album* chitinase was reported. The 6.5-fold higher activity expressed in transformant was found to be useful in bioremediation and biocontrol (Deane et al., 1999).

The discovery of the compounds that activate the mycoparasitic gene expression cascade may create new tools for boosting production of useful

TABLE 14.1. List of sources and host for chitinase genes involved in improving biocontrol efficiency through molecular cloning.

Source	Host	Biocontrol efficacy	References
Bacillus licheniformis	Bacillus thuringiensis ssp. aizawai	Tested against Spodoptera exigua larvae	Tantimavanich et al. (1997)
Bacillus licheniformis	B. thuringiensis	Can be effectively used against Aedes aegypti larvae	Watanalai et al. (1997)
Enterobacter agglomerans	E. coli	Control of root infecting fungi, Fusarium oxysporum, and Rhizoctonia solani	Chernin et al. (1997)
Kurthia zopfii	E. coli	Control of powdery mildew of barley (Erysiphe graminis)	Ikeda et al. (1996)
Rhizopus oligosporus	Tobacco	The infection with Sclerotinia sclerotiorum and Botrytis cinerea was suppressed	Terakawa et al. (1997)
Trichoderma atroviride	Tobacco and potato	Biocontrol capabilities against Alternaria alternata, A. solani, and Botrytis cinerea	Lorito et al. (1998)

Source: Adapted and modified from Patil et al. (2000).

enzymes and stimulating/modulating biocontrol in vivo, therefore reducing the need to apply genetically modified agents (Woo et al., 2001). Woo et al. (1998) found a strong reduction of biocontrol against Botrytis cinerea, but, interestingly, a significant increase in the biocontrol efficacy of chit42-disrupted strain in soils heavily infested with R. solani. These studies show that the manipulation of a chitinase producing gene can alter the biocontrol mechanism both in a positive as well as in a negative way.

Large-Scale Production of Chitinases

Bioreactors are employed for large-scale production of chitinases. When compared to shake flask cultures, it required limited labor, was easy to control, ensured high oxygen transfer and better temperature distribution, superior productivity, and high efficiency. Some disadvantages include high power requirement, foam build up, high capital investment, difficult sterilisation system and occasional serious contamination problems. However, optimization and control of a bioreactor can best be accomplished with a model–predictive control technique which helps to calculate the response as a function of initial conditions, input, and/or set point changes. In order to study combined effect of factors involved in the optimization process, statistical approaches such as factorial design and response surface analysis can be carried out (Binod et al., 2004).

Mechanisms of Biocontrol by Chitinases

The interaction between a pathogen and its host plant initiates a complex network of defense mechanisms, including the synthesis of polymers forming physical barriers, such as lignin and cellulose; the synthesis of antimicrobial metabolites (phytoalexins), and the synthesis of pathogenesis-related (PR) proteins, chitinases being one among them (Benhamou, 1995; Hammond and Jones, 1996). Different mechanisms have been responsible for biocontrol activity, which include competition for space and nutrients, secretion of chitinolytic enzymes, mycoparasitism, and production of inhibitory compounds (Haram et al., 1996; Zimand et al., 1996). It is therefore complicated to elucidate the specific roles of chitinases in plant defense, despite the fact that the inducibility of chitinases and chitinase genes as a result of pathogen attack is very well documented (Punja and Zhang, 1993). Studies done by Wubben et al. (1996) showed that the interaction between *Cladosporium fulvum* and tomato, resistance against fungus correlates with early transcription induction of genes encoding apoplastic chitinase and 1,3 β-glucanase and the accumulation of these proteins in inoculated tomato leaves. The time for chitinase induction is also dependent on the specific host-pathogen interaction, and varies from minutes to 15 to 20 h (Punja and Zhang, 1993). After induction, the time course of chitinase activity is in the range of several days (Graham and Sticklen, 1994). This time frame suggests that the role of chitinases in plant defense is mainly to reduce pathogen growth and sporulation at later disease stages, rather than to be involved in the early events of the host-pathogen interaction. The dramatic increase in chitinase activity as a result of PR induction, together with evidence of chitinase antifungal activity in vitro, strongly support the correlative observation that chitinases are key enzymes in antifungal plant defense (Schickler, 1997).

The activity of chitinase is highly specific. In in vitro assays, a class I chitinase from *Arabdopsis* was effective against *Trichoderma reesie* but not against commercially important pathogens such as *Fusarium oxysporum, Alternaria solani, Sclerotium rolfsii,* or *Phytophthora megasperma* (Verburg and Huynh, 1991; Graham and Sticklen, 1994). In many cases, chitinase is only an effective fungicide in vitro, when applied in combination with 1,3 β-glucanase (Schlumbaum et al., 1986; Mauch et al., 1988).

To have antifungal activity, first, a chitinase must bind to fungal cell walls, and second, it must degrade the chitin in it. However, no report of a binding manner of plant chitinase to the fungal cell walls has been found. A plant class I chitinase has at least two binding parts, the N-terminal chitin-binding domain and the catalytic domain, while class II has only one in the catalytic domain. It is not clear whether the CB domain takes part in or reinforces the antifungal activity of class I chitinase (Taira et al., 2002). Chitin and β-glucan fibers are synthesized during apical growth in filamentous fungi, simultaneously in the tip of the growing hypha. In the fungal mature cell walls, at a distant part from the hyphal tip, the polysaccharides are cross-linked to form mixed chitin-glucan fibers and may be overlaid by other polysaccharides and protein layers (Bartnicki-Garcia, 1973; Benitez et al., 1975; Schoffelmeer et al., 1999). At the hyphal tip, the exposed nascent chitin chains are only accessible to hydrolysis by chitinase, whereas the chitin layer in the mature cell walls is inaccessible to degradation by the enzyme (Collinge et al., 1993). Studies by Taira et al. (2002) on antifungal activity of rye seed chitinase shows that basic class I chitinase bound to hyphal tips and lateral walls and septa, consisting of mature cell walls, by mainly ionic interaction of the Cat domain and by hydrophobic interaction of CB domain and degrade mature chitin fibers as well as nascent chitin by its hydrolytic action. On the other hand, basic class II chitinase bound only to the hyphal tip by mainly ionic interaction by itself, followed by degradation of only nascent chitin. As a result, the basic class I chitinase more effectively inhibits fungal growth than basic class II chitinase.

COMMERCIALIZATION OF CHITINASES AS BIOCONTROL AGENTS

For effective industrial production of biological control agents, research efforts must be shifted from the discovery of potential biocontrol agents to solving the production, storage, and efficacy problems. Development of low-cost, stable products which give consistent control under field conditions is also important. A collaborative research between plant and insect pathologists, microbiologists, fermentation specialists, biochemists, and formulation scientists is necessary for solving these problems. Studies by Jackson (1997) demonstrated that a multidisciplinary research approach is required if significant progress is to be made in overcoming the constraints which impede the commercialization of these agents. Figure 14.1 describes the important steps involved in the commercialization of biocontrol agents.

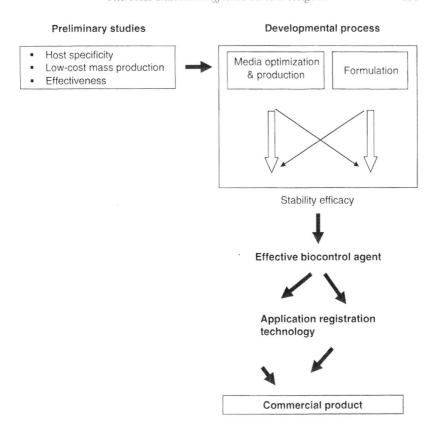

FIGURE 14.1. Important steps involved in the commercialization of biocontrol agents. *Source:* Adapted and modified from Jackson (1997).

The successful use of solid substrate methods for the producing fungal biopesticides relies on either a market for the product, which can tolerate a high input, or a market where low-cost manpower is available for production, such as the case of production in third world countries. Mycotrol, a product of Mycotech Corporation, consists of conidia of the fungus *Beauveria bassiana*. This commercial bioinsecticide is produced by solid substrate fermentation, and the successful commercialization of this organism is based on the extremely high concentration of spores produced on solid substrate and on the high value crops (vegetables, cotton) on which this product is used (Bartlett and Jaronski, 1988; Jackson, 1997).

The success in employing chitinases for application as a biocontrol agent depends on supply of highly active preparations at reasonable cost. The first such application became commercially available in the mid-1990s. Most of the suppliers use either natural microbes or overproducing mutants for production of efficient preparations. Usukizyme, a commercial fungal cell wall lytic enzyme preparation from *Trichoderma viride,* was developed at Usuki Bio Center, Japan. Microbial chitinases from different sources *(Streptomyces* spp., *Serratia marcescens)* are commercially available from Sigma. The approximate cost of one unit is $30.5 (1,200 units/g solid).

CONCLUSION

From this discussion, it is clear that microorganisms are the preferred source of chitinase. Several reports are available on the application of microbial chitinase in improving insect and disease resistance of crops, which proves its versatility as a biocontrol agent. However, the high cost of commercial chitinase restricts its large-scale application in agricultural fields. Hence, there is a necessity to develop hyperproducer microbial strains and use inexpensive raw materials for their production through fermentation technology. Extensive studies on the mechanism of these hydrolytic enzymes have revealed that chitinases are directly involved in inhibition of spore germination as well as in degradation of fungal hyphal tips.

REFERENCES

Auld, B.A. and L. Morin. (1995). Constraints in the development of bioherbicides. *Weed Technology* 9: 638-652.

Bartlett, M.C. and S.T. Jaronski. (1988). Mass production of entomogenous fungi for biological control of insects. In *Fungi in Biological Control Systems,* ed. M.N. Burge. Manchester University Press, Manchester, UK, pp. 61-85.

Bartnicki-Garcia, S. (1973). Fungal cell wall composition. In *Handbook of Microbiology,* Vol. 2, ed. S. Bartnicki-Garcia. Chemical Rubber Co., Cleveland, OH, pp. 201-214.

Benhamou, N. (1995). Ultrastructural and cytochemical aspects of the response of egg plant parenchyma cells in direct contact with *Verticillium*-infected xylem vessels. *Physiological and Molecular Plant Pathology* 46: 321-338.

Benitez, T., T.G. Villa, and A.I. Garcia. (1975). Chemical and structural differences in mycelial and regeneration walls of *Trichoderma viride. Archieves for Microbiology* 105: 277-282.

Binod, P., C. Sandhya, and A. Pandey. (2004). Chitinolytic enzymes. In *Enzyme Technology,* eds. A. Pandey, C. Webb, C.R. Soccol, and C. Larroche. Asiatech Publishers, Inc., New Delhi.

Binod, P., G. Szakacs, and A. Pandey. (2003). Chitinase—A potentially important lytic enzyme from *Trichoderma harzianum.* In *International Conference on*

Emerging Frontiers at the Interface of Chemistry and Biology, ed. A. Pandey. Trivandrum, India, pp. 63-64.

Broglie, K., I. Chet, M. Holliday, R. Cressman, P. Biddle, S. Knowlton, C.J. Mauvains, and R. Broglie. (1991). Transgenic plants with enhanced resistance to the fungal pathogen *Rhizoctonia solani. Science* 254: 1194-1197.

Charudattan, R. (1991). The mycoherbicide approach with plant pathogens. In *Microbial Control of Weeds,* ed. D.O. TeBeest. Chapman and Hall, New York, pp. 24-57.

Chernin, L.S., L. de la Fuente, and V. Sobolev. (1997). Molecular cloning, structural analysis, and expression in *Escherichia coli* of a chitinase gene from *Enterobacter agglomerans. Applied Environmental Microbiology* 63: 834-839.

Churchill, B.W. (1982). Mass production of microorganisms for biological control. In *Biological Control of Weeds with Plant Pathogens,* eds. R. Charudattan and H.L. Walker. John Wiley and Sons, New York, pp. 139-156.

Collinge, D.B., K.M. Kragh, J.D. Mikkelsen, K.K. Nielsen, U. Rasmussen, and K. Vad. (1993). Plant chitinases. *Plant Journal* 3: 31-40.

Deane, E.E., J.M. Whipps, J.M. Lynch, and J.F. Peberdy. (1999). Transformation of *Trichoderma reesei* with a constitutively expressed heterologous fungal chitinase gene. *Enzyme Microbial Technology* 24: 419-424.

Felse, P.A. and T. Panda. (1999). Self directing optimization of parameters for extra cellular chitinase production by *T. harzianum* in batch mode. *Process Biochemistry* 34: 563-566.

Felse, P.A. and T. Panda. (2000). Production of microbial chitinases—A revisit, *Bioprocess Engineering* 23: 127-134.

Forage, RG., D.E.F. Harrison, and D.E. Pitt. (1985). Effect of environment on microbial activity. *Comprehensive Biotechnology* 1: 251-280.

Frandberg, E. and J. Schnurer. (1994). Chitinolytic properties of *Bacillus pabuli* K1. *Journal of Applied Bacteriology* 76: 361-367.

Goettel, M.S. and D.W. Roberts. (1992). Mass production, formulation and field application of entomopathogenic fungi. In *Biological Control of Locusts and Grasshoppers,* eds. C.J. Lomer and C. Prior. CAB International, Wallingford, UK, pp. 230-238.

Graham, L.S. and M.B. Sticklen. (1994). Plant chitinases. *Canadian Journal of Botany* 72: 1057-1083.

Gupta, R., R.K. Saxena, P. Chaturvedi, and A.S. Viridi. (1995). Chitinase production by *streptomyces viridificans* its potential in fungal cell wall lysis. *Journal of Applied Bacteriology* 78: 378-383.

Hammond, K.K.E. and J.D.G. Jones. (1996). Resistance gene-dependent plant defense responses. *Plant Cell* 8: 1773-1791.

Haram, S., H. Schickler, A. Oppenheim, and I. Chet. (1996). Differential expression of *Trichoderma harzianum* chitinases during mycoparasitism. *Phytopathology* 86: 980-985.

Henrissat, B. (1991). A classification of glycosyl hydrolases based on amino acid sequence similarities. *Biochemical Journal* 280: 309-316.

Ho-Seong, L. and K. Sang-Dal. (1994). The production and enzymatic properties of chitinase from *Pseudomonas stulzeri* YPL-1 as a biocontrol agent. *Journal of Microbiology and Biotechnology* 4: 134-140.

Ikeda, S., H. Toyoda, Y. Matsuda. M. Kurokawa, T. Tamai, K. Yoshida. C. Kami, T. Ikemoto, and M. Enomoto. (1996). Cloning of chitinase gene chiSH1 cloned from gram-positive bacterium *Kurthia zopfii* and control of powdery mildew of barley. *Nippon Shokubutsu Byori Gakkaiho* 62: 11-16.

Irving, H.R. and J. Kuc. (1990). Local and systemic induction of peroxidase, chitinase and resistance in cucumber plants by K_2HPO_4. *Physiological and Molecular Plant Pathology* 37: 355-366.

Jackson, M.A. (1997). Optimizing nutritional conditions for the liquid culture production of effective fungal biological control agents. *Journal of Industrial Microbiology* 19: 180-187.

Jackson. M.A., D.A. Schisler, P.J. Slininger, C.D. Boyette. R.W. Silman, and R.J. Bothast. (1996). Fermentation strategies for improving the fitness of a bioherbicide. *Weed Technology* 10: 645-650.

Ji, C. and J. Kuc. (1996). Antifungal activity of cucumber β-1,3-glucanase and chitinase. *Physiological and Molecular Plant Pathology* 49: 257-265.

Jones, J.D.G. (1986). Isolation and characterization of genes encoding two chitinase enzymes from *Serratia marcescens. EMBO Journal* 5: 467-473.

Jones, J.D.G. (1988). Expression of bacterial chitinase protein in tobacco leaves using two photosynthetic gene promoters. *Molecular and General Genetics* 212: 536-542.

Joshi, S., M. Kozlowski, S. Richens, and D.M. Comberbach. (1989). Chitinase and chitobiase production during fermentation of genetically improved *Serratia marcescens. Enzyme and Microbial Technology* 11: 289-296.

Kapat, A., S.K. Rakshit. and T. Panda. (1996). Optimisation of carbon and nitrogen sources in the medium and environmental factors for enhanced production of chitinase by *Trichoderma harzianum. Bioprocess Engineering* 15: 13-20.

Khoury, C., M. Minier, N. Van Huynch, and F. Le Goffic. (1997). Optimal dissolved oxygen concentration for the production of chitinases by *Serratia marcescens. Biotechnology Letters* 19: 1143-1146.

Kole, M.M. and I. Altosarr. (1985). Increased chitinase production by nonpigmented mutant of *Serratia marcescens. FEMS Microbiology Letters* 26: 265-269.

Kovacs, K., G. Szakacs, and A. Pandey. (2003). Screening of *Trichoderma* strains capable of producing chitinase in solid-state fermentation and optimization of fermentation parameters. In *International Conference on Emerging Frontiers at the Interface of Chemistry and Biology,* ed. A. Pandey. Trivandrum, India.

Kubicek, C.P. and G.E. Harman. (1998). *Trichoderma* and *Gliocladium.* Taylor and Francis Ltd., London.

Kuc, J. (1995). Induced systemic resistance—An overview. In *Induced Resistance to Disease in Plants,* eds. R. Hammerschmidt and J. Kuc. Kluwer Academic Publishers, Dordrecht, the Netherlands, pp. 169-175.

Latge, J.P. and R. Moletta. (1988). Biotechnology. In *Atlas of Entomopathogenic Fungi*, eds. R.A. Sampson, H.C. Evans, and J.P. Latge. Springer-Verlag, Berlin, pp. 187-192.

Lonsane, B.K., G.S. Castenada, M. Raimbault, S. Roussos, G.V. Gonzalez, N.P. Ghildyal, M. Ramakrishna, and M.M. Krishnaiah. (1992). Scale-up strategies for solid-state fermentations. *Process Biochemistry* 27: 259-273.

Lorito, M., S.L. Woo, I.G. Fernandez, G. Colucci, G.E. Harman, J.A. Pintor-Toro, E. Filippone, et al. (1998). Genes from mycoparasitic fungi as a source for improving plant resistance to fungal pathogens. *Proceedings of the National Academy of Science USA* 95: 7860-7865.

Mahadevan, B. and D.L. Crawford. (1997). Properties of the chitinase of the antifungal biocontrol agent *Streptomyces lydicus* WYEC108. *Enzyme and Microbial Technology* 20: 489-493.

Matsumoto, Y., S. Revah, G. Saucedo, G.M. Hall, and K. Shirai. (2001). Chitinase production in solid state fermentation from shrimp waste silage. In *Chitin Enzymology*, ed. RAA Muzzarelli. Atec, Italy, pp. 381-389.

Mauch, F., B. Mauch-Mani, and T. Boller. (1988). Antifungal hydrolases in pea tissue: II. Inhibition of fungal growth by combination of chitinase and β-1,3-glucanase. *Plant Physiology* 88: 936-942.

Merrick, M.J. and R.A. Edwards. (1995). Nitrogen control in bacteria. *Microbiology Review* 59: 604-622.

Metraux, J.P. and T. Boller. (1986). Local and systemic induction of chitinase in cucumber plants in response to viral, bacterial and fungal infections. *Physiological and Molecular Plant Pathology* 28: 161-169.

Metraux, J.P., W. Burkhart, M. Moyer, S. Dincher, W. Middlesteadt, S. Williams, G. Payne, M. Carnes, and J. Ryals. (1989). Isolation of a complementary DNA encoding a chitinase with structural homology to a bifunctional lysozyme/chitinase. *Proceedings of the National Academy of Sciences USA* 86: 896-900.

Mitchell, D. A., N. Krieger, D.M. Stuart, and A. Pandey. (2000). New developments in solid-state fermentation: II. Rational approaches for bioreactor design and operation. *Process Biochemistry* 35: 1211-1225.

Mitchell, D.A., A. Pandey, P. Sangsurasak, and N. Krieger. (2000). Scale-up strategies for packed-bed bioreactors for solid-state fermentation. *Process Biochemistry* 35: 167-178.

Monreal, J. and E.T. Reese. (1969). The chitinase of *Serratia marcescens*. *Canadian Journal of Microbiology* 15: 689-696.

Moore, D. and C. Prior. (1993). The potential of mycoinsecticides. *Biocontrol News Information* 14: 31N-40N.

Nagendra, P.G. and M. Chandrasekaran. (1996). L-glutaminase production by marine *Vibrio costicola* under solid state fermentation using different substrates. *Journal of Marine Biotechnology* 4: 176-179.

Nampoothiri, K.M., T.V. Baiju, C. Sandhya, A. Sabu, G. Szakacs, and A. Pandey. (2003). Process optimization for antifungal chitinase production by *Trichoderma harzianum*. *Process Biochemistry* 39: 1583-1590.

Niderman,T., I. Genetet, T. Bruyere, R. Gees, A. Stintzi, M. Legrand, B. Fritig, and E. Mosinger. (1995). Pathogenesis-related PR-1 Proteins are antifungal. *Plant Physiology* 108: 17-27.

Nishizawa, Y. and T. Hibi. (1991). Rice chitinase gene: cDNA cloning and stress-induced expression. *Plant Science* 76: 211-218.

Pandey, A. (1992). Recent process developments in solid state fermentation. *Process Biochemistry* 27: 109-117.

Pandey, A., P. Selvakumar, C.R. Soccol, and P. Nigam. (1999). Solid-state fermentation for the production of industrial enzymes. *Current Science* 77: 149-162.

Pandey, A. and C.R. Soccol. (2000). Economic utilization of crop residues for value addition—A futuristic approach. *Journal of Scientific and Industrial Research* 59: 12-22.

Pandey, A., C.R. Soccol, and D.A. Mitchell. (2000). New developments in solid-state fermentation: I. Bioprocesses and products. *Process Biochemistry* 35: 1153-1169.

Pandey, A., C.R. Soccol, P. Nigam, D. Brand, R. Mohan, and S. Roussos. (2000). Biotechnological potential of coffee pulp and coffee husk for bioprocesses. *Biochemical Engineering Journal* 6: 153-162.

Pandey, A., C.R. Soccol, P. Nigam, and V.T. Soccol. (2000). Biotechnological potential of agro-industrial residues: I. Sugarcane bagasse. *Bioresource Technology* 74: 69-80.

Pandey, A., C.R. Soccol, P. Nigam, V.T. Soccol, P.S.V. Luciana, and R. Mohan. (2000). Biotechnological potential of agro-industrial residues: II. Cassava bagasse. *Bioresource Technology* 74: 81-87.

Pandey, A., C.R. Soccol, J.A. Rodriguez-leon, and N. Poonam. (2001). *Solid State Fermentation in Biotechnology Fundamentals and Application.* Asiatech Publisher, New Delhi.

Patil, R.S., V. Ghormade, and M.V. Deshpande. (2000). Chitinolytic enzymes: An exploration. *Enzyme and Microbial Technology* 26: 473-483.

Punja, Z.K. and Y.Y. Zhang. (1993). Plant chitinases and their role in resistance to fungal diseases. *Journal of Nematology* 25: 526-540.

Ross, A.F. (1961). Systemic acquired resistance induced by localized virus infections in plants. *Virology* 14: 340-358.

Sandhya, C., L.K. Adapa, K.M. Nampoothiri, P. Binod, G. Szakacs, and A. Pandey. (2004). Extracellular chitinase production by *Trichoderma harzianum* in submerged fermentation. *Journal of Basic Microbiology* 44: 49-58.

Schickler, H. and I. Chet. (1997). Heterologous chitinase gene expression to improve plant defense against phytopathogenic fungi. *Journal of Industrial Microbiology and Biotechnology* 19: 196-201.

Schlumbaum, A., F. Mauch, U. Vogeli, and T. Boller. (1986). Plant chitinases are potent inhibitors of fungal growth. *Nature* 324: 365-367.

Schoffelmeer, E.A.M., F.K. Klis, J.H. Sietsma, and B.J.C. Cornelissen. (1999). The cell wall of *Fusarium oxysporum. Fungal Genetic and Biotechnology* 27: 275-282.

Shapiro, S. (1989). Nitrogen assimilation in actinomycetes and the influence of nitrogen nutrition on actinomycete secondary metabolism. In *Regulation of*

Secondary Metabolism in Actinomycetes, ed. S. Shapiro. CRC Press, Boca Raton, FL, pp. 135-211.

Shinishi, H., D. Mohnen, and F. Meins. (1987). Regulation of a plant pathogenesis-related enzyme: Inhibition of chitinase and chitinase mRNA accumulation in cultured tobacco tissue by auxin and cytokinin. *Process National Academy of Sciences USA* 84: 89-93.

Shinshi, H., J.M. Neuhaus, J. Ryals, and F. Meins. (1990). Structure of tobacco endochitinase gene: Evidence that different chitinase genes can arise by transposition of sequences encoding a cystein-rich domain. *Plant Molecular Biology* 14: 405-412.

Silman, R.W., R.J. Bothast, and D.A. Schisler. (1993). Production of *Colletotrichum truncatum* for use as a mycoherbicide: Effects of culture, drying and storage on recovery and efficacy. *Biotechnology Advances* 11: 561-575.

St Leger, R.J., R.M. Cooper, and A.K. Charnley. (1986). Cuticle degrading enzymes of entomopathogenic fungi: Regulation of production of chitinolytic enzymes. *Journal of General Microbiology* 132: 1509-1517.

Suresh, P.V. and M. Chandrasekharan. (1999). Utilization of prawn waste for chitinase production by the marine fungus *Beauveria bassiana* by solid state fermentation. *World Journal of Biotechnology* 14: 655-660.

Suslow, T.V., D. Matsubara, J. Jones, R. Lee, and P. Dunsmuir. (1988). Effect of expression of bacterial chitinase on tobacco susceptibility to leaf brown spot. *Phytopathology* 78: 1556.

Taira, T., T. Ohnuma, T. Yamagami, Y. Aso, M. Ishiguro, and M. Ishihara. (2002). Antifungal activity of rye *(Secale cereale)* seed chitinases: The different binding manner of class I and class II chitinases to the fungal cell walls. *Bioscience Biotechnology and Biochemistry* 66: 970-977.

Tantimavanich, S., S. Pantuwatana, S. Bhumiratana, and W. Panbangred. (1997). Cloning of a chitinase gene in to *Bacillus thuringiensis* spp. *aizawai* for enhanced insecticidal activity. *Journal of General and Applied Microbiology* 43: 341-347.

Templeton, G.E. (1982). Status of weed control with plant pathogens. In *Biological Control of Weeds with Plant Pathogens,* eds R. Charudattan and H.L. Walker. John Wiley and Sons, New York, pp. 29-44.

Terakawa, T., N. Takaya, H. Horiuchi, M. Koike, and M. Takagi. (1997). A fungal chitinase gene from *Rhizopus oligosporus* confers antifungal activity to transgenic tobacco. *Plant Cell Reports* 16: 439-443.

Tweddel, R.J., S.H. Jabag-Hare, and P.M. Charest. (1994). Production of chitinase and β-1,3 glucanase by *Stachybotrys elegans* a mycoparacite of *Rhizoctonia solani. Applied Environmental Microbiology* 60: 489-495.

Ulhoa, C.J. and J.F. Peberdy. (1991). Regulation of chitinase synthesis in *Trichoderma harzianum. Journal of General Microbiology* 137: 2163-2169.

Verburg, J.G. and Q.K. Huynh. (1991). Purification and characterization of an antifungal chitinase from *Arabidopsis thaliana. Plant Physiology* 95: 450-455.

Wang San-Lang and Jau-Ren Hwang. (2001). Microbial reclamation of shellfish wastes for the production of chitinases. *Enzyme and Microbial Technology* 28: 376-382.

Wang San-Lang, Tsu-Chau Yieh, and Ing-Lung Shih. (1999). Purification and characterization of a new antifungal compound produced by *Pseudomonas aeruginosa* K-187 in a shrimp and crab shell powder medium. *Enzyme and Microbial Technology* 25: 439-446.

Watanalai, P., T. Srisurang, S. Syjaree, P. Nirat, P. Somsak, and B. Amaret. (1997). Molecular cloning of chitinase gene and its application for biological control. *Biotechnology Sustainable Util Biol Resource Trop* 11: 252-258.

Wilson. C.L. (1969). Use of plant pathogens in weed control. *Annual Review Phytopathology* 7: 411-434.

Wiwat, C., M. Lertcanawanichakul, P. Siwayapram, S. Pantuwatana, and A. Bhumiratana. (1996). Expression of chitinase-encoding genes from *Aeromonas hydrophilia* and *Pseudomonas maltophilia* in *Bacillus thuringiensis* spp. *isrealiensis, Gene,* 179: 119-126.

Woo, S., R. Ciliento, D. Piacenti, R. Hermosa, F. Scala, A. Zoina and M. Lorito. (2001). *Trichoderma* chitinases and the encoding genes as molecular tools for enhancing biocontrol. In *Chitin Enzymology 2001,* ed. R.A.A. Muzzarelli. Atec, Italy, pp. 47-55.

Woo, S.L., B. Donzelli, F. Scala. R.L. Mach, G.E. Harman, C.P. Kubicek, G. Del Sorbo, and M. Lorito. (1998). Disruption of *ech42* (endochitinase-encoding) gene affects biocontrol activity in *Trichoderma harzianum* strain. *Molecular Plant-Microbe Interactions* 12: 419-429.

Wubben, J.P., C.B. Lawrence, and P.J.G.M. De Wit. (1996). Differential induction of chitinase and 1,3-beta-glucanase gene expression in tomato by *Cladosporium fluvum* and its race-specific elicitors. *Physiology and Molecular Plant Pathology* 48: 105-116.

Zimand, G.. Y. Elad, and I. Chet. (1996). Effect of *Trichoderma harzianum* on *Botrytis cinerea* pathogenicity. *Phytopathology* 86: 1255-1260.

Chapter 15

Fungal Phytopathogen Suppression Using Siderophoregenic Bioinoculants

S.B. Chincholkar
B.L. Chaudhari
M.R. Rane
P.D. Sarode

INTRODUCTION

Phytopathogens can be suppressed by any of the three methods, physical, chemical, or biological. While each method has some advantages along with some disadvantages, the biological method is the only faithful method for generations to come, considering sustainability would demand further standardization in the context of appropriate molecules, microorganisms, and methods involved. In the long-lasting efforts to control phytopathogens, research is now moving toward standardization of biological control methods. Through some recent communications, the ecofriendly nature of bio- inoculants has been discussed critically (Boland and Brimner, 2004; Kiss, 2004). This is an indication of the maturity of "research on biocontrol agents." At this juncture, this is an attempt to review application of siderophore-producing microorganisms to control fungal phytopathogens.

Siderophores have been defined as "low-molecular-weight bimolecules secreted by micro-organisms under iron deficiency conditions" (Lankford, 1973). In addition, the following features of these molecules can be listed:

1. Mineralize and sequester specifically iron from insoluble forms (Lankford, 1973)
2. Sequester some other ions, for example, Pu (III), Pu(IV), Th(IV), Pb(II), Eu(III), Al(III), Zn(II), Ga(III), and Cr(III) (Boukhalfa et al., 2003)

Biological Control of Plant Diseases
© 2007 by The Haworth Press, Inc. All rights reserved.
doi:10.1300/5682_15

3. Ferrated forms are taken up by producer organisms (Gomez and Sansom, 2003), plants (Duijff et al., 1994), and occasionally by nonproducer organisms (Poole and McKay, 2003)
4. Induce systemic resistance in plants (De Meyer et al., 1999)

Owing to the specificity of these features, siderophores are being exploited in agriculture, medicine, environment, and industry (Chincholkar et al., 2005). Tables 15.1 and 15.2 give a concise collection of variety of siderophores produced by bacteria and fungi, respectively.

In earlier literature, designations like siderochromes, sideramines, sideromycins, and ionophores have been used to describe siderophores. However, now these terms have been unanimously replaced by term "siderophore" which means in Greek: *sideros* = iron and *phores* = bearer. Although the term siderophore is held in reserve for the iron-free ligands only (Neilands, 1981), in some cases, the name had been given to iron-complexed siderophores (coprogen, ferrichrome, ferrioxamine). The corresponding iron-free form, therefore, needs the prefix desferri or deferri, for example, desferricoprogen. Whereas, the siderophores which represent the iron-free form, for example, enterobactin, staphyloferrin, and rhizoferrin, need the prefix ferri or ferric when iron is bound to it (Neilands, 1981).

SIDEROPHORE-MEDIATED BIOCONTROL MECHANISMS

Various mechanisms involved in biocontrol by virtue of bimolecules involve antibiosis, induced systemic resistance, biosynthesis of lytic extracellular enzymes, and competition for acquisition of minerals and other nutrients. In the case of siderophores, their significance in competition for iron is well established. However, in a few instances, induction of systemic resistance has also been attributed to these molecules (Cook et al., 1995; Handelsman and Stab, 1996; Walsh et al., 2001; Whipps, 2001). Table 15.3 provides examples of siderophoregenic microorganisms reported to have biocontrol activity.

COMPETITION FOR IRON

Iron in the form of ferric hydroxide is an abundant element in earth's crust; however, this is an insoluble form. The concentration of the available form of iron is less than 10^{-18} M. Thus, iron nutrition is a challenge for microorganisms, and siderophore production is the most suitable aid for assimilation of iron under aerobic conditions (Chincholkar et al., 2000).

TABLE 15.1. Profile of bacterial siderophores.

Name of siderophore	Producer organism	Reference
Acinetobactin	*Acinetobacter baumannii*	Yamamoto et al., 1994
Acinetoferrin	*Acinetobacter haemolyticus*	Okujo et al., 1994
Aerobactin	*Aerobacter aerogenes*	Gibson and Magrath, 1969
Agrobactin	*Agrobacterium* sp.	Sonoda et al., 2002
Alcaligin	*Bordetella bronchiseptica*	Hou et al., 1998, Brickman and Armstrong, 2002
Alcaligin E	*Alcaligenes eutrophus*	
Alterobactin	*Alteromonas luteoviolaces*	Reid et al., 1993
Aminochelin	*Azotobacter vinelandii*	Khodr et al., 2002
Amonbactins	*Aeromonas hydrophila*	Stintzi and Raymond, 2000
Anguibactin	*Vibrio anguillarum*	Actis et al., 1986
Aquachelin	*Halomonas aquamarina* strain DS40M3	Xu et al., 2002
Asterobactin	*Nocardia asteroids*	Nemoto et al., 2002
Azotobactin	*Azotobacter vinelandii*	Tindale et al., 2000
Azoverdin	*Azomonas macrocytogenes* ATCC 12334	Wasielewski et al., 2002
Bisucaberin	*Vibrio salmonicida*	Hou et al., 1998
Cepabactin	*Pseudomonas cepacia*	Meyer et al., 1989
Chrysobactin	*Erwinia chrysanthemi*	Neema et al., 1993, Rauscher et al., 2002
Coelichelin	*Streptomyces coelicolor*	Challis and Ravel, 2000
Corynebactin	*Corynebacterium glutamicum*	Budzikiewicz et al., 1997
Desferrithiocin	*Streptomyces antibioticus*	
Exochelin MS	*Mycobacterium smegmatis*	Sharman et al., 1995
Ferribactin	*Pseudomonas fluorescens*	Maurer et al., 1968
Ferrioxamine	*Streptomyces pilosus* *Enterobacter aglomerans*	Berner and Winkelmann, 1990
Ferrioxamine B and E	*Streptomyces viridosporus*	Imbert et al., 1995
Fluvibactin	*Vibrio fluvialis*	Yamamoto et al., 1993
Heterobactin	*Rhodococcus erythropolis* IGTS8	Carran et al., 2001
IC202A	*Streptoalloteichus* sp. 1454-19	Iijima et al., 1999
Legiobactin	*Legionella pneumophila*	Liles et al., 2000
Marinobactin	*Marinobacter* sp. strains DS40M6 and DS40M8	Xu et al., 2002, Martinez et al., 2000
Mycobactin	*Mycobacterium tuberculosis*	Snow, 1970

TABLE 15.1 *(continued)*

Name of siderophore	Producer organism	Reference
Mycobactin P	*Mycobacterium phlei*	Snow and White, 1969
Myxochelin A	*Angiococcus disciformis*	Kunze et al., 1989
Myxochelin B	*Stigmatella aurantiaca*	Silakowski et al., 2000
Ornibactin	*Burkholderia cepacia*	Sokol et al., 1999
Petrobactin	*Marinobacter hydrocarbonoclasticus*	Barbeau et al., 2002
Protochelin	*Azotobacter vinelandii*	Cornish and Page, 2000
Pseudobactin	*Pseudomonas putida* B 10	Teintz et al., 1981
Pyoverdin	*Pseudomonas* sp.	Vossen and Taraz, 1999
Quinolobactin	*Pseudomonas fluorescens* ATCC 17400	Mossialos et al., 2000
Rhizobactin	*Rhizobium meliloti*	Persmark et al., 1993
Schizokinen	*Bacillus megaterium*	Mullis et al., 1971
Siderochelin A	*Nocardia* sp.	Liu et al., 1981
Staphyloferrin A	*Staphylococci*	Konetschny-Rapp et al., 1990
Vibriobactin	*Vibrio cholerae*	Griffiths et al., 1984
Vibrioferrin	*Vibrio parahaemolyticus*	Funahashi et al., 2002
Vicibactin	*Rhizobium leguminosarum*	Yeoman et al., 1999
Vulnibactin	*Vibrio vulnificus*	Webster and Litwin, 2000
Yersiniabactin	*Yersinia enterocolitica, Yersinia pestis*	Brem et al., 2001, Perry et al., 1999
Yersiniophore	*Yersinia enterocolitica*	Chambers et al., 1996

Almost all aerobic organisms are known to produce siderophores, few have the ability to employ siderophores of other organisms (Poole and McKay, 2003). Although in several instances, siderophore-borne iron competition has been claimed as an effective biocontrol mechanism, it is speculated that siderophores can act as a biocontrol molecule only under soluble iron-scarce environments and when phytopathogen(s) fail to utilize those ferreted siderophores. The reviewers' group has successfully employed siderophoregenic *Pseudomonas aeruginosa* 4365 in the control of groundnut root pathogens, *viz. Aspergillus niger, Aspergillus flavus* (Figure 15.1), *Aspergillus oryzae, Fusarium oxysporum, Sclerotium rolfsii,* and *Alternaria alternata.* Siderophore-rich supernatants exhibited antifungal activity and using both bore well as well as a disc method; however, similar studies with XAD-purified pyoverdine showed reduced biocontrol activity (Manwar

TABLE 15.2. Profile of fungal siderophores.

Name of siderophore	Producer organism	Reference
Coprogen	*Curvularia lunata* NCIM 716	Chaudhari, 1998
Coprogen B	*Histoplasma capsulatum*	Howard et al., 2000
Dimerum acid	*Histoplasma capsulatum* *Stemphylium botyrosum*	Timmerman and Woods, 2001 Manulis et al., 1987
Ferrichrome	*Penicillium parvum*	Winkelmann and Braun, 1981
Ferrichrome A	*Ustilago sphaerogena*	Emery, 1971
Ferrichrome C	*Neurospora crassa*	Horowitz et al., 1976
Ferrichrysin	*Cuunighamella blaskesleeana*	Patil et al., 1999
Ferricrocin	*Aspergillus nidulans*	Haas et al., 2003
Fusarinine	*Histoplasma capsulatum*	Howard et al., 2000
Fusarinine A and B	*Fusarium roseum*	Sayer and Emery, 1968
Fusigen	*Aspergillus nidulans*	Haas et al., 2003
Malionichrome	*Fusarium roseum*	Emery, 1980
Methyl coprogen B	*Histoplasma capsulatum*	Howard et al., 2000
Neocoprogen I and II	*Curvularia lunata*	Hossain et al., 1987
Rhizoferrin	*Rhizopus microsporus Rhizopus arrhizus*	Drechsel et al., 1991 Yehuda et al., 1996 Carrano et al., 1996
Rhodotorulic acid	*Rhodotorula mucilaginosa*	Andersen et al., 2003
Triacetylfusarinine C	*Aspergillus nidulans*	Haas et al., 2003

et al., 2004). Recent studies showed that the blue band on the XAD column reported by Manwar (2001) contained pyocyanin and phenazine-1-carboxylic (Rane et al., unpublished data). Naphade (2002) described in vitro antagonistic activity of siderophores produced by *Enterobacter cloacae* against *A. niger, A. flavus. Alternaria* spp., and *F. oxysporum,* using bore well method on Czapeck Dox agar. A 20 µl Chrom Azurol S (CAS) positive cell-free extract of culture broth of *E. cloacae* was used in this experiment, and the zone of inhibition observed had a diameter of 30, 20, 33, and 35 mm, respectively. Sayyed (2002) reported in vitro antagonistic activity of siderophore-positive culture supernatants and XAD-purified siderophore fraction of *Alcaligenes faecalis* against *A. niger, A. flavus, F. oxysporum,* and *Alternaria alternata.* These studies (involving two organisms) strongly indicate that siderophores have the ability to antagonize fungal phytopathogens, and competition for iron is the principal mechanism involved in antagonism. Observations of Rane et al. (unpublished data) indicate that antagonism showed by pyoverdin-rich supernatants may be the result of syn-

TABLE 15.3. Recent examples of siderophoregenic microorganisms reported to have biocontrol activity.

Siderophore producer strain	Organism controlled	Plant disease	Reference
Rhizobium meliloti	Macrophomina phaseolina	Charcoal rot in groundnut	Arora et al., 2001
Enterobacter cloacae	Aspergillus niger, Aspergillus flavus, Fusarium oxysporum, and Alternaria spp.	. . .	Naphade, 2002
Pseudomonas fluorescens	Rhizoctonia solani	Rice sheath blight	Nagarajkumar et al., 2004
Pseudomonas spp. GRP3A, PRS9	Colletotrichum dematium, Rhizoctonia solani, and Sclerotium rolfsii	Maize	Sharma and Johri, 2003
Pseudomonas sp. EM85	Macrophomina phaseolina, Fusarium moniliforme, and Fusarium graminearum	Maize root diseases	Pal et al., 2001
Proteus sp.	Fusarium oxysporum	Mungo beans	Barthakur, 2000
Rhodotorulla strains	Botrytis cinerea	Gray mold on a wide variety of host plants including numerous commercial crops	Calvente et al., 2001
Pseudomonas aeruginosa (GRC1)	Macrophomina phaseolina and Fusarioum oxysporum	. . .	Gupta et al., 1999
Pseudomonas aeruginosa	Aspergillus niger, Aspergillus flavus, Aspergillus oryzae, Fusarium oxysporum, Sclerotium rolfsii, and Alternaria alternata	. . .	Manwar, 2001

ergy between pyoverdins and phenazines, which are coproduced in minute amounts.

Lemanceau et al. (1992, 1993) showed that pseudobactin produced by *Pseudomonas putida* WCS 358 significantly improves biological control of Fusarium wilt caused by nonpathogenic *F. oxysporum* Fo47b10. In these studies involving purified pseudobactin and nonpathogenic and pathogenic *F. oxysporum* strains, it was observed that nonferrated pseudobactin reduced the growth of both the *F. oxysporum* strains whereas ferric pseudobactin did not, indicating clearly that antagonism by pseudobactin was related to iron competition. It was also observed that in the presence of pseudobactin, competition between two *F. oxysporum* strains for carbon led to discouragement of pathogenic strains. Such observation confirms that siderophores may strongly regulate community structure in the plant rhizosphere.

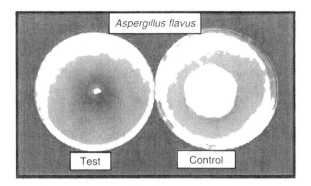

FIGURE 15.1. In vitro antagonistic activity exerted by Pseudomonas ID 4365 against *Aspergillus flavus.*

Experiments involving siderophore-negative mutation are another way to confirm whether siderophores really have antagonistic potential in the three-component system. Krishnaraj et al. (2001) have evaluated the role of fluorescent siderophores in biological control against *Ralstonia solana-cearum* causing bacterial wilt in tomato, using Tn^5 mutants of fluorescent *Pseudomonas* sp. They observed that mutants (Sid⁻) of two siderophore-producing pseudomonas strains, which either fail to produce siderophores or produce them in less quantity, have reduced activity to control bacterial wilt disease of tomato, speculating the activity of siderophore in biocontrol. Buysens et al. (1996), in a hydroponic system, compared the wild-type strain (*P. aeruginosa* 7NSK2) with mutants deficient in the production of pyochelin and/or pyoverdin to characterize the role of these siderophores in the protection mechanism. All mutant strains retained at least to some extent the ability to protect germinating tomato seeds against *Pythium*-induced seed and radicle rot (preemergence damping-off), possibly by competing with the pathogen for seed exudates. However, the production of either pyochelin or pyoverdin proved to be necessary to achieve high levels of protection against *Pythium*-induced postemergence damping-off. Moreover, the role of pyoverdin and pyochelin were mutually exchangeable.

Thomashow (1996) has reviewed the role of siderophores in plant disease suppression and opined that competition for iron is not the only mechanism involved when pyoverdine synthesis is correlated with biocontrol activity, and experiments involving siderophore negative mutants may lead to wrong conclusion. In experiments performed by Keel et al. (1989), reduced control of black root rot of tobacco by Pvd⁻ mutants of *P. fluorescence*

CHA0 was attributed more to the loss of HCN (synthesis of which requires iron) than to the competition.

In almost all studies on biological control involving fluorescent pseudomonads, iron competition has been reported as a possible mechanism. Arora et al. (2001) have described application of siderophore-producing strains of *Rhizobium meliloti* for biocontrol of *Macrophomina phaseolina,* the causal agent of charcoal rot of groundnut. During these studies, twelve root-nodulating *Rhizobium meliloti* strains were isolated from the medicinal plant *Mucuna pruriens*. Out of these twelve strains, two siderophore-producing strains, RMP_3 and RMP_5, showed ability to inhibit *Macrophomina phaseolina,* and hence were selected for further studies. A marked enhancement in seed germination, seedling biomass, nodule number, and nodule fresh weight of *M. phaseolina*-infected groundnut plants inoculated with the strains RMP_3 and RMP_5 compared with noninoculated and noninfected controls has been reported. This study is particularly important as the organism showing biocontrol activity has nitrogen-fixing ability.

INDUCED SYSTEMIC RESISTANCE

All plants have natural defense mechanisms against pathogen attacks. A virulent pathogen avoids triggering these mechanisms, suppresses resistance reactions, or evades the effects of active defenses. Disease can be reduced if these defense mechanisms are triggered by leaf-necrosis pathogens or the rhizobacteria prior to infection, and this phenomenon is commonly known as systemic acquired resistance (SAR) and induced systemic resistance (ISR), respectively (Bakker et al., 2003). Various nonpathogenic rhizobacteria have been shown to induce systemic resistance in plants, thereby providing protection against a broad spectrum of phytopathogenic fungi, bacteria, and viruses. Several bacterial determinants are claimed to produce systemic resistance and include siderophores, salicylic acid, and the O-antigenic side chain of the bacterial outer membrane protein lipopolysaccharide (LPS). Colonization of tobacco roots by *P. fluorescens* CHA0 reduces leaf necrosis caused by tobacco necrosis virus (TNV) and induces physiological changes in the plant (e.g., increase in salicylic acid and PR proteins in the leaves). A pyoverdine-negative mutant of CHA0 could only partially induce resistance against TNV (Uknes et al., 1993; Notz, 2002). The concept that biocontrol agents might induce resistance in the host was first suggested based on experiments showing bacterial treatments protected potato tubers from subsequent infections by *P. solanacearum* (Kempe and Sequeira, 1983). Maurhofer et al. (1994) showed that the biocontrol agent *P. fluorescens* strain CHA0 induces SAR-asociated proteins, confers systemic resistance to a viral pathogen, and induces accumulation of salicylic acid, which plays a role in signal transduction in SAR. Pyoverdine mutants of this strain failed to induce systemic resistance. De Meyer (1999) reported this kind of systemic resistance induced by the rhizobacterium *Pseudomonas aeruginosa* 7NSK2 based on the production of salicylic acid in bean roots. In another study, Audenaert et al. (2002) concluded that pyochelin produced by *P. aeruginosa* 7NSK2 are the determinants for induced resistance to *Botrytis cinerea* in wild-type tomato.

Press et al. (2001) studied the role of iron in rhizobacteria-mediated ISR of cucumber. It was observed that the level of ISR by *Serratia marcescens* strain 90-166 varied with the iron concentration of the potting mix in which cucumber was grown. ISR mediated by this strain was improved significantly when external iron availability to the host plant was reduced through the addition of iron chelator EDDHA. The extent of reduced ISR with increased iron availability indicated the role of siderophores in ISR. Further confirmation was obtained by producing siderophore-negative mutants of *S. marcescens* strain 90-166, which showed no ISR activity.

CONCLUSION

Considering the fact that physico-chemical methods for phytopathogen suppression have numerous limitations (some methods are polluting, others are costly, and others have limitations owing to climatic conditions of the region), chemical methods have become unacceptable to consumers, and the rate of development of new kinds of synthetic chemical pesticides is reduced. Therefore, biological control, less explored so far, remains the only hopeful method for phytopathogen suppression.

Many siderophores have shown antagonistic activity in pure form. Moreover, cell-free siderophoregenic-fermented broth of most fluorescent *Pseudomonas* strains has antagonistic activity through iron competition. Few other strains also have the ability to induce systemic resistance in plants. Unfortunately, due to the complexity of biotic and abiotic factors that play a role in biological control, practical results with biocontrol of soilborne pathogens have been varied. Future research to ensure successful introduction of antagonistic microorganisms against pathogens and varying environmental factors contributing to sufficient establishment of rhizobacterial strains and efficient expression of their antagonistic activity will lead to a success story. This also holds out hope for siderophoregenic bioinoculants.

Siderophores in some cases turn out to have double potential. First, the pathogen is weakened in the rhizosphere by depriving it for iron, and then they help generate enhanced defensive capacity, that is, ISR, in the host plant.

In the light of recent data overviewed, it becomes clear that instead of having a pessimistic view regarding the biocontrol activity of siderophores, one can now be optimistic about realization of their potential. The structural diversity of microbial siderophores has not yet been correlated with biocontrol aspects. Hence, it is necessary to exploit the wide array of siderophores for a noble cause.

REFERENCES

Actis L.A., W. Fish, J.H. Crosa, K. Kellerman, S.R. Ellenberger, F.M. Hauser, and J. Sanders-Loehr. (1986). Characterization of anguibactin, a novel siderophore from *Vibrio anguillarum. Journal of Bacteriology* 167: 57-65.

Andersen D., J.C. Renshaw, and M.G. Wiebe. (2003). Rhodotorulic acid production by *Rhodotorula mucilaginosa. Mycological Research* 107: 949-956.

Arora N.K., S.C. Kang, and D.K. Maheshwari. (2001). Isolation of siderophore-producing strains of *Rhizobium meliloti* and their biocontrol potential against *Macrophomina phaseolina* that causes charcoal rot of groundnut. *Current Science* 81: 673-677.

Audenaert K., T. Pattery, P. Cornelis, and M. Hofte. (2002). Induction of systemic resistance to *Botrytis cinerea* in tomato by *Pseudomonas aeruginosa* 7NSK2: Role of salicylic acid, pyochelin, and pyocyanin. *Molecular Plant-Microbe Interactions* 15: 1147-1156.

Bakker P.A.H.M., L.X. Ran, C.M.J. Pieterse, and L.C. van Loon. (2003). Understanding the involvement of rhizobacteria mediated induced resistance in biocontrol of plant diseases. *Canadian Journal of Plant Pathology* 25: 5-9.

Barbeau K., G. Zhang, D.H. Live, and A. Butler. (2002). Petrobactin, a photoreactive siderophore produced by the oil-degrading marine bacterium *Marinobacter hydrocarbonoclasticus. Journal of American Chemical Society* 124: 378-379.

Barthakur M. (2000). Plant growth promotion and fungicidal activity in a sideroproducing strain of *Proteus* sp. *Folia Microbiologica* 45: 539-543.

Berner I. and G. Winkelmann. (1990). Ferrioxamine transport mutants and the identification of the ferrioxamine receptor protein (FoxA) in *Erwinia herbicola (Enterobacter agglomerans). Bio Metals* 2: 197-202.

Boland G.J. and T. Brimner. (2004). Nontarget effects of biological control agents. *New Phytologist* 163: 455-457.

Boukhalfa H., J. Lack, S.D. Reilly, L. Hersman, and M.P. Neu. (2003). Siderophore production and facilitated uptake of iron and plutonium in *P. putida. AIP Conference Proceedings* 673: 343-344.

Brem D., C. Pelludat, A. Rakin, C.A. Jacobi, and J. Heesemann. (2001). Functional analysis of yersiniabactin transport genes of *Yersinia enterocolitica. Microbiology* 147: 1115-1127.

Brickman T.J. and S.K. Armstrong. (2002). Alcaligin siderophore production by *Bordetella bronchiseptica* strain RB50 is not repressed by the BvgAS virulence control system. *Journal of Bacteriology* 184: 7055-7057.

Budzikiewicz H., A. Boessenkamp, K. Taraz, A. Pandey, and J.M. Meyer. (1997). Corynebactin, a cyclic catecholate siderophore from *Corynebacterium glutamicum* ATCC 14067 (*Brevibacterium* sp. DSM 20411) *Zeitschrift Fhr Naturfordchug* 52: 551-554.

Buysens S., K. Heugens, J. Poppe, and M. Hofte. (1996). Involvment of pyochelin and pyoverding in suppression of *Pythium*-induced damping-off of tomato by *Pseudomonas aeruginosa* 7nsk2. *Applied Environmental Microbiology* 62: 865-871.

Calvente V., M.E. de Orellano, G. Sansone, D. Benuzzi, and M.I. Sanz de Tosetti. (2001). Effect of nitrogen source and pH on siderophore production by *Rhodotorula* strains and their application to biocontrol of phytopathogenic moulds. *Journal of Industrial Microbiology and Biotechnology* 26: 226-229.

Carran C.J., M. Jordan, H. Drechsel, D.G. Schmid, and G. Winkelmann. (2001). Heterobactins: A new class of siderophores from *Rhodococcus erythropolis* IGTS8 containing both hydroxamate and catecholate donor groups. *BioMetals* 14: 119-125.

Carrano C.J., H. Drechsel, D. Kaiser, G. Jung, B. Matzanke, G. Winkelmann, N. Rochel, and A.M. Albrecht-Gary. (1996). Coordination chemistry of the carboxylate type siderophore rhizoferrin: The iron(III) complex and its metal analogs. *Inorganic Chemistry* 35: 6429-6436.

Challis G.L. and J. Ravel. (2000). Coelichelin, a new peptide siderophore encoded by the *Streptomyces coelicolor* genome: Structure prediction from the sequence of its non-ribosomal peptide synthetase. *FEMS Microbiol Lett.* 187:111-114.

Chambers C.E., D.D. McIntyre, M. Mouck, and P.A. Sokol. (1996). Physical and structural characterization of yersiniophore, a siderophore produced by clinical isolates of *Yersinia enterocolitica. BioMetals* 9: 157-167.

Chaudhari B.L. (1998). Studies on siderophores of *Curvularia lunata* NCIM 716, PhD thesis, North Maharashtra University, Jalgaon.

Chincholkar S.B., B.L. Chaudhari, S.K. Talegaonkar, and R.M. Kothari. (2000). Microbial iron chelators: A sustainable tool for the biocontrol of plant disease. In *Biocontrol Potential and Its Exploitation in Sustainable Agriculture,* eds. R.K. Upadhyay, K.G. Mukerji, and P.C. Chamola. Kluwer Academic Publishers, pp. 49-70.

Chincholkar S.B., M.R. Rane, and B.L. Chaudhari. (2005) Siderophores: Their biotechnological applications. In *Basic Research and Biotechnological Applications: Microbes,* eds. G.K. Podila and A. Varma. I.K. International, New Delhi.

Cohen A.R., R. Galanello, D. J. Pennell, M. J. Cunningham, and E. Vichinsky. (2004). Thalassemia. *American Society of Hematology,* 1-34.

Cook R.J., L.S. Thomashow, D.M. Weller, D. Fujimoto, M. Mazzola, G. Bangera, and D. Kim. (1995). Molecular mechanisms of defense by rhizobacteria against root disease. *Proceeding National Academy of Science USA* 92: 4197-4201.

Cornish A.S. and W.J. Page. (2000). Role of molybdate and other transition metals in the accumulation of protochelin by *Azotobacter vinelandii*. *Applied Environmental Microbiology* 66:1580-1586.

De Meyer G., K. Capieau, K. Audenaert, A. Buchala, J.P. Metraux, and M. Hofte. (1999). Nanogram amounts of salicylic acid produced by the rhizobacterium *Pseudomonas aeruginosa* 7NSK2 activate the systemic acquired resistance pathway in bean. *Molecular Plant-Microbe Interactions* 12: 450-458.

Drechsel H., J. Metzger, S. Freund, G. Jung, J.R. Boeleart, and G. Winkelmann. (1991). Rhizoferrin—A novel siderophore from the fungus *Rhizopus microsporus* var. *rhizopodiformis*. *BioMetals* 4: 238-243.

Duijff B.J., W.J. de Kogel, P.A.H.M. Bakker, and B. Schippers. (1994). Influence of pseudobactin 358 on the iron nutrition of barley. *Soil Biology Biochemitry* 26: 1681-1688.

Emery T. (1971). Role of ferrichrome as a ferric ionophore in *Ustilago sphaerogena*. *Biochemistry* 10: 1483-1488.

Emery T. (1980). Malionichrome, a new iron chelate from *Fusarium roseum*. *Biochem Biophysics Acta* 629: 382-390.

Funahashi T., K. Moriya, S. Uemura, S. Miyoshi, S. Shinoda, S. Narimatsu, and S. Yamamoto. (2002). Identification and characterization of pvuA, a gene encoding the ferric vibrioferrin receptor protein in *Vibrio parahaemolyticus*. *Journal of Bacteriology* 184: 936-946.

Gibson F. and D.I. Magrath. (1969). The isolation and characterization of a hydroxamic acid (aerobactin) formed by *Aerobacter aerogenes* 62-I. *Biochem Biophysics Acta* 92: 175-184.

Gilis A., M.A. Khan, P. Cornelis, J.M. Meyer, M. Mergeay, and D. van der Lelie. (1996). Siderophore mediated iron uptake in *Alcaligenes entrophus* CH 34 and identification of ale β encoding the feric iron—alcaligin E receptor. *Journal of Bacteriology* 178: 5499-5507.

Gomez J.D.F. and M.S.P. Sansom. (2003). Acquisition of siderophores in Gramnegative bacteria. *Nature Reviews. Molecular Cell Biology* 4: 105-116.

Griffiths G.L., S.P. Sigel, S.M. Payne, and J.B. Neilands. (1984). Vibriobactin, a siderophore from *Vibrio cholerae*. *Journal of Biology Chemistry* 259: 383-385.

Gupta C.P., A. Sharma, R.C. Dubey, and D.K. Maheshwari. (1999). *Pseudomonas aeruginosa (GRC1)* as a strong antagonist of *Macrophomina phaseolina* and *Fusarium oxysporum*. *Cytobios* 99: 183-189.

Haas H., M. Schoeser, E. Lesuisse, J.F. Ernst, W. Parson, B. Abt, G. Winkelmann, and H. Oberegger. (2003). Characterization of the *Aspergillus nidulans* transporters for the siderophores enterobactin and triacetylfusarinine C. *Biochemistry Journal* 371: 505-513.

Handelsman J. and E.V. Stabb. (1996). Biocontrol of soilborne plant pathogens. *The Plant Cell* 8: 1855-1869.

Horowitz N.H., G. Charlang, G. Horn, and N.P. Williams. (1976). Isolation and identification of the conidial germination factor of *Neurospora crassa*. *Journal of Bacteriology* 127: 135-140.

Hossain M.B., M.A.F. Jalal, B.A. Benson, C.L. Barnes, and D. van der Helm. (1987). Structure and confirmation of two coprogen type siderophores: Neocoprogen I and Neocoprogen II. *Journal of American Chemistry Society* 109: 4984-4954.

Hou Z., K.N. Raymond, B. O'Sullivan, T.W. Esker, and T. Nishio. (1998). A pre-organized siderophore: Thermodynamic and structural characterization of alcaligin and bisucaberin, microbial macrocyclic dihydroxamate chelating agents. *Inorganic Chemistry* 37: 6630-6637.

Howard D.H., R. Rafie, A. Tiwari, and K.F. Faull. (2000). Hydroxamate siderophores of *Histoplasma capsulatum*. *Infect Immunology* 68: 2338-2343.

Iijima M., T. Someno, M. Amemiya, R. Sawa, H. Naganawa, M. Ishizuka, and T. Takeuchi. (1999). IC202A, a new siderophore with immunosuppressive activity produced by *Streptoalloteichus* sp. 1454-19: II. Physico-chemical properties and structure elucidation. *Journal of Antibiotics* (Tokyo) 52: 25-28.

Imbert M., M. Bechet, and R. Blondeau. (1995). Comparison of the marine siderophores produced by some species of *Stremptomyces*. *Current Microbiology* 31: 129-133.

Keel C., C. Voisard, C.H. Berling, G. Khar, and G. Defago. (1989). Iron sufficiency, a prerequisite for suppression of tobacco black root rot by *P. fluorescens* strain CHA0 under gnotobiotic conditions. *Phytopathology* 79: 584-589.

Kempe J., and L. Sequeria. (1983). Biological control of bacterial wilt of potatoes: Attempts to induce resistance by treating tubers with bacteria. *Plant Disease* 67: 499-503.

Khandelwal S.R., A.V. Manwar, B.L. Chaudhari, and S.B. Chincholkar. (2002). Siderophoregenic bradyrhizobia boost yield of soybean. *Applied Biochemistry and Biotechnology* 102-103: 155-168.

Khodr H.H., R.C. Hider, and A.K. Duhme-Klair. (2002). The iron-binding properties of aminochelin, the mono (catecholamide) siderophore of *Azotobacter vinelandii*. *Journal of Biology and Inorganic Chemistry* 7: 891-896.

Kiss L. (2004). How dangerous is the use of fungal biocontrol agents to nontarget organisms? *New Phytologist* 163: 453-455.

Konetschny-Rapp S., G. Jung, J. Meiwes, and H. Zahner. (1990). Staphyloferrin A: A structurally new siderophore from staphylococci. *European Journal of Biochemistry* 20:191: 65-74.

Krishnaraj P.U., J.H. Kulkarni, and K.S. Jagadeesh. (2001). Evaluation of the role of fluorescent siderophore in the biological control of bacterial wilt in tomato using Tn[5] mutants of fluorescent *Pseudomonas* sp. *Current Science* 81: 25-26.

Kunze B., N. Bedorf, W. Kohl, G. Hofle, and H. Reichenbach. (1989). Myxochelin A, a new iron-chelating compound from *Angiococcus disciformis* (Myxobacterales): Production, isolation, physico-chemical and biological properties. *Journal of Antibiotics* (Tokyo) 42: 14-17.

Lankford C.E. (1973). Bacterial assimilation of iron. *Critical Rev Microbiology* 2: 273.

Leeman M., F.M. Den Ouden, J.A. van Pelt, A. Cornelissen, Matamala-Garros, P.A.H.M. Bakker, and B. Schippers. (1996). Suppression of Fusarium wilt of radish by co-inoculation of fluorescent *Pseudomonas* spp. and root-colonizing fungi. *European Journal of Plant Pathology* 102: 21-31.

Lemanceau P., P.A.H.M. Bakker, W.J. de Kogel, and C. Alabouvette. (1992). Effect of pseudobactin 358 production by *Pseudomonas putida* WCS358 on suppression of Fusarium wilt of carnations by nonpathogenic *Fusarium oxysporum* Fo47. *Applied Environmental Microbiology* 58: 2978-2982.

Lemanceau P., P.A.H.M. Bakker, W.J. de Kogel, C. Alabouvette, and B. Schippers. (1993). Antagonistic effect on nonpathogenic *Fusarium oxysporum* strain Fo47 and pseudobactin 358 upon pathogenic *Fusarium oxysporum* f. sp. *dianthi*. *Applied Environmental Microbiology* 59: 74-82.

Liles M.R., T.A. Scheel, and N.P. Cianciotto. (2000). Discovery of a nonclassical siderophore, legiobactin, produced by strains of *Legionella pneumophila*. *Journal of Bacteriology* 182: 749-757.

Liu W.C., S.M. Fisher, J.S. Wells Jr., C.S. Ricca, P.A. Principe, W.H. Trejo, D.P. Bonner, et al. (1981). Siderochelin, a new ferrous-ion chelating agent produced by *Nocardia*. *Journal of Antibiotics* (Tokyo) 34: 791-799.

Manulis S., Y. Kashman, and I. Barash. (1987). Identification of siderophores and siderophoremediated uptake of iron in *Stemphylium botryosum*. *Phytochemistry* 26: 1317-1320.

Manwar, A.V. (2001). Application of microbial iron chelators (siderophores) for improving yield of groundnut. PhD thesis, North Maharashtra University, Jalgaon.

Manwar A.V., S.R. Khandelwal, B.L. Chaudhari, J.M. Meyer, and S.B. Chincholkar. (2004). Siderophore production by a marine *Pseudomonas aeruginosa*, and its antagonistic action against phytopathogenic fungi. *Applied Biochemistry and Biotechnology* 118: 243-251.

Martinez J.S., G.P. Zhang, P.D. Holt, H.T. Jung, C.J. Carrano, M.G. Haygood, and A. Butler. (2000). Self-assembling amphiphilic siderophores from marine bacteria. *Science* 287: 1245-1247.

Maurer B., A. Muller, W. Keller-Schierlein, and H. Zahner. (1968). Metabolic products of microorganisms: Ferribactin, a siderochrome from *Pseudomonas fluorescens*. Migula. *Archieves Mikrobiology* 60: 326-339.

Maurhofer M., C. Hase, P. Meuwly, J.P. Metraux, and G. Défago. (1994). Induction of systemic resistance of tobacco to tobacco necrosis virus by the root-colonizing *Pseudomonas fluoresecens* strain CHA: Influence of the gac A gene and pyoverdine production. *Phytopathology* 84: 139-146.

Meyer J.M., D. Hohnadel, and F. Halle. (1989). Cepabactin from *Pseudomonas cepacia*, a new type of siderophore. *Journal of General Microbiology* 135: 1479-1487.

Mossialos D., J.M. Meyer, H. Budzikiewicz, U. Wolff, N. Koedam, C. Baysse, V. Anjaiah, and P. Cornelis. (2000). Quinolobactin, a new siderophore of *Pseudomonas fluorescens* ATCC 17400, the production of which is repressed by the cognate pyoverdine. *Applied Environmental Microbiology* 66: 487-492.

Mullis K.B., J.R. Pollack, and J.B. Neilands. (1971). Structure of schizokinen, an iron-transport compound from *Bacillus megaterium*. *Biochemistry* 10: 4894-4898.

Nagarajkumar M., R. Bhaskaran, and R. Velazhahan. (2004). Involvement of secondary metabolites and extracellular lytic enzymes produced by *Pseudomonas fluorescens* in inhibition of *Rhizoctonia solani*, the rice sheath blight pathogen. *Microbiology Research* 159: 73-81.

Naphade B.S. (2002). Studies on the application of siderophoregenic *Enterobacter cloacae* as biocontrol agent. PhD thesis, North Maharashtra University, Jalgaon.

Neema C., J.P. Laulhere, and D. Expert. (1993). Iron deficiency induced by chrysobactin in Saintpaulia leaves inoculated with *Erwinia chrysanthemi*. *Plant Physiology* 102: 967-973.

Neilands J.B. (1981). Microbial iron compounds. *Annual Review Biochemistry* 50: 715-731.

Nemoto A., Y. Hoshino, K. Yazawa, A. Ando, Y. Mikami, H. Komaki, Y. Tanaka, and U. Grafe. (2002). Asterobactin, a new siderophore group antibiotic from Nocardia asteroides. *Journal of Antibiotics* (Tokyo) 55: 593-597.

Notz R.E. (2002). Biotic factors affecting 2,4-diacetylphloroglucinol biosynthesis. In *The Model Biocontrol Strain Pseudomonas fluorescens*. Thesis submitted to Swiss Federal Institute of Technology, Zürich.

Okujo N., Y. Sakakibara, T. Yoshida, and S. Yamamoto. (1994). A structure of acinetoferrin, a new citrate-based dihydroxamate siderophore from *Acinetobacter haemolyticus*. *Biometals* 2: 170-176.

Pal K.K., K.V. Tilak, A.K. Saxena, R. Dey, and C.S. Singh. (2001). Suppression of maize root diseases caused by *Macrophomina phaseolina, Fusarium moniliforme* and *Fusarium graminearum* by plant growth promoting rhizobacteria. *Microbiology Research* 156: 209-223.

Patil B.B., R.D. Wakharkar, and S.B. Chincholkar. (1999). Siderophores of *Cunninghamella blakesleeana* NCIM 687. *World Journal of Microbial Biotechnology* 15: 233-235.

Perry, R.D., P.B. Balbo, H.A. Jones, J.D. Fetherston, and E. DeMoll. (1999). Yersiniabactin from *Yersinia pestis*: Biochemical characterization of the siderophore and its role in iron transport and regulation. *Microbiology* 145: 1181-1190.

Persmark M., P. Pittman, J.S. Buyer, B. Schwyn, P.R. Gill, and J.B. Neilands. (1993). Isolation and structure of rhizobactin 1021, a siderophore from the alfalfa symbiont *Rhizobium meliloti* 1021. *Journal American Chemical Society* 115: 3950-3956.

Poole K. and G.A. McKay. (2003). Iron acquisition and its control in *Pseudomonas aeruginosa*: Many roads lead to Rome. *Frontiers in Bioscience* 8: 661-686.

Press C.M., J.E. Loper, and J.W. Kloepper. (2001). Role of iron in rhizobacteria-mediated induced systemic resistance of cucumber. *Phytopathology* 91: 593-598.

Rane M.R., B.S. Naphade, R.Z. Sayyed, and S.B. Chincholkar. (2004). Methods for microbial iron chelator (siderophore) analysis. In *Basic Research and Applications of Mycorrhizae, Microbiology*, series eds. G.K. Podila and A. Varma. I. K. International, New Delhi, pp. 475-492.

Rauscher L., D. Expert, B.F. Matzanke, and A.X. Trautwein. (2002). Chrysobactin-dependent iron acquisition in *Erwinia chrysanthemi*: Functional study of a homolog of the *Escherichia coli* ferric enterobactin esterase. *Journal of Biology Chemistry* 277: 2385-2395.

Reid R.T., D.H. Live, D.J. Faulkner, and A. Butler. (1993). A siderophore from a marine bacterium with an exceptional ferric ion affinity constant. *Nature* 366: 455-458.

Sayeed R.Z. (2002). Studies on the secondary metabolites of *Alcaligenes faecalis* with special reference to its aplicability in plat nutrition and disease management. PhD thesis, North Maharashtra University, Jalgaon.

Sayer J.M. and T. Emery. (1968). Structures of naturally occurring hydroxamic acids, Fusarinine A and B. *Biochemistry Journal* 7: 184-190.

Sharma A. and B.N. Johri. (2003). Growth promoting influence of siderophore-producing *Pseudomonas* strains GRP3A and PRS9 in maize (*Zea mays* L.) under iron limiting conditions. *Microbiology Research* 2003 158: 243-248.

Sharman G.J., D.H. Williams, D.F. Ewing, and C. Ratledge. (1995). Isolation, purification and structure of exochelin MS, the extracellular siderophore from *Mycobacterium smegmatis*. *Biochemistry Journal* 305: 187-196.

Silakowski B., B. Kunze, G. Nordsiek, Blöcker, et al. (2000). The myxochelin iron transport regulon of the myxobacterium *Stigmatella aurantiaca* Sg a15. *European Journal of Biochemistry* 267: 6476-6485.

Snow G.A. (1970). Mycobactins: Iron-chelating growth factors from mycobacteria. *Bacteriology Review* 34: 99-125.

Snow G.A. and A.J. White. (1969). Chemical and biological properties of mycobactins isolated from various mycobacteria. *Biochemistry Journal* 115: 1031-1043.

Sokol P.A., P. Darling, D.E. Woods, E. Mahenthiralingam, and C. Kooi. (1999). Role of ornibactin biosynthesis in the virulence of *Burkholderia cepacia:* Characterization of pvdA, the gene encoding L-ornithine N(5)-oxygenase. *Infect Immunology* 67: 4443-4455.

Sonoda H., K. Suzuki, and K. Yoshida. (2002). Gene cluster for ferric iron uptake in *Agrobacterium tumefaciens* MAFF301001. *Genes Genetics and Systematics* 77: 137-146.

Stintzi A. and K.N. Raymond. (2000). Amonabactin-mediated iron acquisition from transferrin and lactoferrin by *Aeromonas hydrophila:* Direct measurement of individual microscopic rate constants. *Journal of Biology Inorganic Chemistry* 5: 57-66.

Teintz M., M.B. Hossain, C.L. Barnes, J. Leong, and D. van der Helm. (1981). Structure of ferric pseudobactin, a siderophore from a plant growth promoting *Pseudomonas*. *Biochemistry* 20: 6446-6457.

Thomashow L.S. (1996). Biological control of plant root pathogens. *Current Opinion in Biotechnology* 1996, 7: 343-347.

Timmerman M.M. and J.P. Woods. (2001). Potential role for extracellular gluta-thione-dependent ferric reductase in utilization of environmental and host ferric compounds by *Histoplasma capsulatum*. *Infect Immunology* 69: 7671-7678.

Tindale A.E., M. Mehrotra, D. Ottem, and W.J. Page. (2000). Dual regulation of catecholate siderophore biosynthesis in *Azotobacter vinelandii* by iron and oxidative stress. *Microbiology* 146: 1617-1626.

Uknes, S., A. M. Winter, T. Delaney, B. Vernooij, A. Morse, L. Friedrich, G. Nye, et al. (1993). Biological induction of systemic acquired resistance in *Arabidopsis*. *Molecular Plant-Microbe Interactions* 6: 692-698.

Vossen W and K. Taraz. (1999). Structure of the pyoverdin PVD 2908—a new pyoverdin from *Pseudomonas* sp. 2908. *Bio Metals*. 12: 323-329.

Walsh U.F., J.P. Morrisey, and F. O'Gara. (2001). *Pseudomonas* for biocontrol of phytopathogens: From functional genomics to commercial exploitation. *Current Opinion in Biotechnology* 12: 289-295.

Wasielewski E., R.A. Atkinson, M.A. Abdallah, and B. Kieffer. (2002). The three-dimensional structure of the gallium complex of azoverdin, a siderophore of *Azomonas macrocytogenes* ATCC 12334, determined by NMR using residual dipolar coupling constants. *Biochemistry* 41: 12488-12497.

Webster A.C. and C.M. Litwin. (2000). Cloning and characterization of vuuA, a gene encoding the *Vibrio vulnificus* ferric vulnibactin receptor. *Infect Immunology* 68: 526-534.

Whipps J.M. (2001). Microbial interactions and biocontrol in the rhizosphere. *Journal of Experimental Botany* 52: 487-511.

Winkelmann G. and V. Braun. (1981). Stereoselective recognition of ferrichrome by fungi and bacteria. *FEMS Microbiology Letters* 11: 237-241.

Xu G., J.S. Martinez, J.T. Groves, and A. Butler. (2002). Membrane affinity of the amphiphilic marinobactin siderophores. *Journal of American Chemical Society* 124: 13408-13415.

Yamamoto S., N. Okujo, Y. Fujita, M. Saito, T. Yoshida, and S. Shinoda. (1993). Structures of two polyamine-containing catecholate siderophores from *Vibrio fluvialis*. *Journal of Biochemistry* 113: 538-544.

Yamamoto S., N. Okujo, and Y. Sakakibara. (1994). Isolation and structure elucidation of acinetobactin, a novel siderophore from *Acinetobacter baumannii*. *Arch Microbiology* 162: 249-254.

Yehuda Z., M. Shenker, V. Romheld, H. Marschner, Y. Hadar, and Y. Chen. (1996). The role of ligand exchange in the uptake of iron from microbial siderophores by gramineous plants. *Plant Physiology* 112: 1273-1280.

Yeoman K.H., A.G. May, N.G. deLuca, D.B. Stuckey, and A.W. Johnston. (1999). A putative ECF sigma factor gene, rpoI, regulates siderophore production in *Rhizobium leguminosarum*. *Molecular Plant-Microbe Interactions* 12: 994-999.

Index

A. brassicicola, 199
A. euteiches, 25, 50, 52, 62, 198, 201
A. fistulosum, 286
A. longipes, 389
A. niger, 193, 208, 404
A. oryzae, 404
A. raphani, 307
A. rhizogenes, 156
A. sativum, 197
A. solani, 172, 287, 331, 390, 391
A. tumefaciens, 156, 157, 158, 159,
    160, 162, 164, 165, 173, 317
Abelmoschus esculentus, 200
Acaulospora laevis, 54
Achromobacter sp., 195
Acinetobacter sp., 4
Acquired physiological immunity, 85
Acremonium brevie, 265
Aedes aegypti, 390
Aeromonas cavie, 382
Agrobacterium radiobacter, 156, 165,
    317
Agrobacterium sp., 156, 158, 159, 160,
    165, 178
Alcaligenes faecalis, 405
Allium cepa, 197, 286
Allium sp., 123, 124
Alternaria alternata, 197, 198, 390,
    404, 405, 268
Alternaria spp., 10, 405
Ampelomyces quisqualis, 247, 248,
    253, 255, 299, 333, 337
Ampelomyces sp., 316
Aphanocladium album, 389
Aphanomyces cochlioides, 284, 306,
    307
Aphanomyces sp., 24, 28, 197, 198,
    202, 207
Apis mellifera, 257

Appressoria, 171
Appressorium, 20, 299, 300, 338
Arabidopsis thaliana, 86, 87, 88
Arabidopsis, sp. 86, 87, 117, 360, 391
Arbuscule, 21, 27, 28, 29, 30, 31, 32,
    33, 57, 58, 60
Armillaria mellea, 168
Armillaria sp., 335
Arthrobacter sp., 83, 195
Arthrobotrys sp., 338
Ascochyta sp., 232
Aspergillus flavus, 404, 405
Aspergillus sp., 132, 232, 313
Athelia bombacina, 247, 251
Aureobasidium pullulans, 249, 251,
    252, 254
Azadirachta indica, 197, 198

B. brongniartii, 236
B. campestris, 286
B. cereus, 194
B. cinerea, 155, 168, 170, 171, 172,
    173, 192, 196, 198, 199, 201,
    203, 202, 204, 206, 233, 234,
    235, 240, 242, 246, 247, 248,
    249, 250, 251, 252, 253, 255,
    256, 257, 258, 259, 265, 267,
    287, 310, 331, 335, 337, 338,
    343, 344, 358, 390, 409
B. circulans, 382
B. juncea, 138
B. licheniformis, 130, 390
B. megaterium, 4
B. mycoides, 246, 255
B. napus, 138, 202
B. oleracea var. capitata, 286
B. pubuli, 388, 389

Index

423

Malus domestica, 362, 366
Malus sp., 358
Medicago sativa, 198
Meloidogyne domestica, 362, 366
Meloidogyne sp., 53
Metarhizium anisopliae, 287, 387
Metchnikowia pulcherrima, 246
Micromonopora carbonacea, 199
Micromucor sp., 337
Micropropagation, 158, 160
Microshaeropsis ochracea, 247, 248
Microsphaeropsis sp., 251
Monilinia fructicola, 155, 240, 206, 240, 242, 247, 252, 256, 267
Monilinia sp., 268
Mucor pyriformis, 249, 254
Mucuna pruriens, 408
Muscodor albus, 247, 256
Mycolysis, 82
Mycoparasite, 126, 127, 248, 328, 337, 338, 341, 342, 343, 345
Mycoparasitism, 140, 167, 169, 170, 294, 296, 308, 309, 328, 337, 338, 391
Mycorrhiza-helper bacteria, 10, 36
Mycorrhizosphere, 18, 19, 26, 35, 37, 55, 56
Mycosphaerella fijiensis, 327
Mycotoxins, 82, 155, 166, 253
Myrothecium verrucaria, 382

Necrotorphic, 167, 337, 338, 341, 342
Nectria inventa, 338
Nectria sp., 335

O. cernua, 155
O. cumana, 152
O. ramosa, 152, 155
Ocimum sanctum, 197, 198
Orobanche aegyptiaca, 152, 155
Orobanche spp., 152, 153

P. aurantiaca, 78
P. aureofaciens, 78, 80
P. capsici, 199, 283, 284, 287, 290, 291

P. cepacia, 289, 307, 310
P. chlororaphis, 8
P. cinnamomi, 25, 35, 50, 200, 363
P. citrophthora, 369
P. coloraium, 200
P. corrugata, 246
P. expansum, 128, 155, 203, 206, 240, 247, 248, 250, 256, 264, 265, 267, 268
P. fluorescens, 8, 11, 78, 79, 80, 86, 87, 89, 104, 105, 106, 110, 113, 114, 116, 117, 199, 200, 201, 202, 203, 204, 205, 246, 259, 291, 294, 295, 305, 306, 307, 311, 312, 317, 407, 408
P. frequentan, 193, 204, 251
P. fumosoroseus, 236
P. funiculosum, 193, 205
P. gladioli, 246, 296, 309
P. guilermondii, 193, 249, 255
P. infestans, 327, 331
P. italicum, 206
P. maltophilia, 286
P. megasperma, 391
P. nicotianae var. *parasitica*, 28, 49, 50
P. nigricans, 127, 130
P. oligandrum, 10, 193, 201, 334, 336, 337, 338
P. oxalicum, 193, 200, 303
P. macerans, 8
P. melanocephala, 331
P. membranefaciens, 246
P. palmivora, 358
P. parasitica, 25, 61, 50, 205, 289
P. penetrans, 54
P. polymyxa, 206, 314
P. putida, 4, 117, 194, 195, 196, 202, 246, 294, 295, 305, 306, 311, 312, 314, 317, 406
P. solanacearum, 56, 408
P. stulzeri, 388
P. syringe, 4, 116, 117, 155, 173, 208, 242, 246, 248, 250, 264, 265, 266, 267
P. syringae pv. *lachrymans*, 202
P. syringae pv. *syringae*, 359
P. terrestris, 29, 50
P. ultimum, 25, 50, 52, 53, 81, 83, 117, 201, 286, 293, 294, 301, 302, 334, 340, 341

#0071 - 230317 - C0 - 212/152/24 [26] - CB - 9781560223276